Analysis of Multifactorial Disease

The HUMAN MOLECULAR GENETICS series

Series Advisors

D.N. Cooper, *Institute of Medical Genetics, University of Wales College of Medicine, Cardiff, UK*

S.E. Humphries, *Division of Cardiovascular Genetics, University College London Medical School, London, UK*

T. Strachan, *Department of Human Genetics, University of Newcastle upon Tyne, Newcastle upon Tyne, UK*

Human Gene Mutation
From Genotype to Phenotype
Functional Analysis of the Human Genome
Molecular Genetics of Cancer
Environmental Mutagenesis
HLA and MHC: Genes, Molecules and Function
Human Genome Evolution
Gene Therapy
Molecular Endocrinology
Venous Thrombosis: from Genes to Clinical Medicine
Protein Dysfunction in Human Genetic Disease
Molecular Genetics of Early Human Development
Neurofibromatosis Type 1: from Genotype to Phenotype
Analysis of Triplet Repeat Disorders
Molecular Genetics of Hypertension
Human Gene Evolution
Analysis of Multifactorial Disease

Forthcoming title
Transcription Factors

Analysis of Multifactorial Disease

T. Bishop
ICRF Genetic Epidemiology Laboratory, St James's Hospital, Leeds, UK

P. Sham
Institute of Psychiatry, King's College London, Denmark Hill, London, UK

BIOS

First published 2000

A CIP catalogue record for this book is available from the British Library.

ISBN 1 85996 113 4

BIOS Scientific Publishers Ltd
9 Newtec Place, Magdalen Road, Oxford OX4 1RE, UK
Tel. +44 (0)1865 726286. Fax +44 (0)1865 246823
World Wide Web home page: http: //www.bios.co.uk/

Production Editor: Fran Kingston
Typeset by Saxon Graphics Ltd, Derby, UK
Printed by Biddles Ltd, Guildford, UK, www.biddles.co.uk

Contents

Contributors

Asherson, P. Social Genetic and Developmental Psychiatry Research Centre, Institute of Psychiatry, 1 Windsor Walk, Denmark Hill, London SE5 8AF, UK

Badzioch, M. Unit of Genetic Epidemiology, International Agency for Research on Cancer, 150 cours Albery Thomas, F-69372 Lyon, Cedex 08, France

Barrett, J. ICRF Genetic Epidemiology Laboratory, Cancer Genetics Building, St James's University Hospital, Beckett Street, Leeds LS9 7TF, UK (formerly at ARC Epidemiology Research Unit, University of Manchester, UK)

Bishop, T. ICRF Genetic Epidemiology Laboratory, St James's Hospital, Beckett Street, Leeds LS9 7TF, UK

Collier, D.A. Section of Genetics, Institute of Psychiatry, De Crespigny Park, Denmark Hill, London SE5 8AF, UK

Cookson, W.O.C.M. Asthma Genetics Group, University of Oxford, Wellcome Trust Centre for Human Genetics, Roosevelt Drive, Oxford OX3 7BN, UK

Craddock, N. Division of Neuroscience, University of Birmingham, Queen Elizabeth Psychiatric Hospital, Birmingham B15 2QZ, UK

Field, L.L. British Columbia Research Institute for Children's and Women's Health, 950 W28 Avenue, Vancouver, BC, V5Z 4H4, Canada

Goldgar, D. Chief, Unit of Genetic Epidemiology, International Agency for Research on Cancer, 150 cours Albert Thomas, 69008 Lyon, France

Hanson, R.L. Diabetes and Arthritis Epidemiology Section, National Institute of Diabetes & Digestive and Kidney Diseases, 1550 East Indian School Road, Phoenix, Arizona 85014, USA

Humphries, S. Centre for Cardiovascular Genetics, University College London, The Rayne Institute, 5 University Street, London WC1E 6JJ, UK

John, S. Research and Development Genetics, AstraZeneca, Alderley Park, Cheshire SK10 4TG, UK (formerly at ARC Epidemiology Research Unit, University of Manchester, UK)

Jones, I. Division of Neuroscience, University of Birmingham, Queen Elizabeth Psychiatric Hospital, Birmingham B15 2QZ, UK

Knowler, W.C. Diabetes and Arthritis Epidemiology Section, National Institute of Diabetes & Digestive and Kidney Disease, 1550 East Indian School Road, Phoenix, Arizona 85014, USA

Lathrop, G.M. Directeur General, Centre National de Genotypage, 2 rue Gaston Crémieux, CP 5721, 91057 Evry, France

McGuffin, P. Social, Genetic and Developmental Psychiatry Research Centre, Institute of Psychiatry, De Crespigny Park, Denmark Hill, London SE5 8AF, UK

Makoff, A.J. Department of Psychological Medicine, Institute of Psychiatry, 1 Windsor Walk, Denmark Hill, London SE5 8AF

Malhi, G.S. Social, Genetic and Developmental Psychiatry Research Centre, Institute of Psychiatry, De Crespigny Park, Denmark Hill, London SE5 8AF, UK

Moore, J. Department of Child and Adolescent Psychiatry, Institute of Psychiatry, 5 Windsor Walk, Denmark Hill, London SE5 8BB, UK

Nashef, L. Neuroscience, King's College Hospital, Denmark Hill, London SE5 9RS, and Department of Neurology, Kent & Canterbury Hospital, Canterbury, Kent CT1 3NG, UK

Owen, M.J. University of Wales College of Medicine, Neuropsychiatric Genetics Unit, Heath Park, Cardiff, CF14 1FF, UK

Palmer, L.J. Channing Laboratory, Harvard Medical School, 181 Longwood Avenue, Boston, MA 02115-5804, USA

Rice, J.P. Department of Psychiatry, Washington Univeristy School of Medicine, 660 South Euclid Avenue, St Louis, MO 63110, USA

Roks, C. Department of Epidemiology and Biostatistics, Erasmus University Medical Centre, PO Box 1738, DR 3000, Rotterdam, The Netherlands

Saccone, N.L. Department of Psychiatry, Washington University School of Medicine, 660 South Euclid Avenue, St Louis, MO 63110, USA

Sham, P. Institute of Psychiatry, King's College London, De Crespigny Park, Denmark Hill, London SE5 8AF, UK

Suarez, B.K. Department of Psychiatry, Washington Univeristy School of Medicine, 660 South Euclid Avenue, St Louis, MO 63110, USA

Taylor, G. Regional DNA Laboratory, Ashley Wing, St James' Hospital, Leeds LS9 7TF, UK

Treasure, J. Eating Disorders Unit, Institute of Psychiatry, De Crespigny Park, Denmark Hill, London SE5 8AF, UK

Van Duijn, C. Department of Epidemiology and Biostatistics, Erasmus University Medical Centre, DR 3000, Rotterdam, The Netherlands

Weeks, D.E. University of Pittsburgh, Graduate School of Public Health, Department of Human Genetics, A300 Crabtree Hall, 130 DeSoto Street, Pittsburgh, PA 15261, USA

Abbreviations

A2M	alpha-2 macroglobuline
ACE	angiotensin-converting enzyme
AD	alzheimer's disease
ADNFLE	autosomal dominant frontal lobe elilepsy
AFBAC	affected-family-based control method
AGFAP	antigen genotype frequency among patients
AGRP	agouti-related protein
AN	anorexia nervosa
APM	affected-pedigree-member
apo	apolipoprotein
ApoE	apolipoprotein E
APP	amyloid precursor protein
AR	airway responsiveness
ARMS	amplification refractory mutation system
ASP	affected sibling pair
BAC	bacterial artificial chromosome
BECTS	benign elilepsy of childhood with centrotemporal spikes
BED	binge eating disorder
BFIC	benign familial infantile convulsions
BFNC	benign familial neonatal convulsions
BMI	body mass index
BN	bulimia nervosa
CAE	childhood absence epilepsy
CCM	chemical cleavage of mismatch
CD	Crohn's disease
CDK4	cyclin-dependent kinase 4
CGH	comparative genome hybridization
CI	confidence interval
CNS	central nervous system
COMT	catechol-*O*-methyltransferase
CRH	corticotropin-releasing hormone
CTLA-4	cytotoxic T lymphocyte associated-4
DGGE	denaturing gradient gel electrophoresis
DHPLC	denaturing high performance liguid chromatography
DNS	dysplastic nevus syndrome
DRD2	dopamine D2 receptor
DRS	dose–response slope
DSM-IV	diagnostic and Statistical Manual-IV
DZ	dizygotic
EA-2	episodic ataxia type 2

ECRAF	European Consortium of Rheumatoid Arthritis Families
ECTIM	Etude Cas-Tèmoin de l'Infarctus du Myocarde
EEG	electroencephalograph
EGMSA	grand mal seizures on awakening
ESR	estrogen receptor
FAMMM	familial atypical multiple mole melanoma
FAP	familial adenomatous polyps
FAT	fatty acid translocase
FATP	fatty acid transport protein
FCH	familial combined hyperlipidemai
FDB	familial defective Apo B100
FEV_1	forced expiratory volume in 1 s
FGF3	fibroblast growth factor 3
FH	familial hypercholesterolemia
FHM	familial hemiplegic migraine
FISH	fluorescent *in situ* hybridization
FVC	forced vital capacity
GABA	[gamma]-aminobutyrate
GAD	glutamic acid decarboxylase
GAD	generalized anxiety disorder
GCK	glucokinase gene
GEFS+	generalized epilepsy with febrile seizures plus
GGR	genotypic risk ratio
GI	gastrointestinal
GST	glutathione-*S*-transferase
GSW	generalized spike/wave
GTCS	generalized tonic clonic seizures
HDL	high density lipoprotein
HLA	human leukocyte antigen
Hlod	heterogeneity lod
HNPCC	hereditary non-polyposis coli
HPA	hypothalamic-pituitary-adrenal
HRR	haplotype relative risk
hSERT	human serotonin transporter
5-HT	serotonin
I/D	insertion–deletion polymorphism
IA-2	islet cell antigen 512
IAA	insulin antibodies
IBD	identical-by-descent
IBD	inflammatory bowel disease
ICA	islet cell antibodies
ICCA	infantile convulsions and choreoathetosis
ICD-10	International Classification of Diseases-10
IDDM	insulin-dependent diabetes mellitus
IgE	immunoglobulin E
IGE	idiopathic generalized epilepsy
IGF2	insulin-like growth factor 2

IGH	immunoglobulin heavy chain
IHD	ischemic heart disease
IL	interleukin
ILAE	International League Against Epilepsy
INS	insulin gene
IS	insulin sensitivity
ISIS-3	Third International Study of Infarct Survival
JME	juvenile myoclonic epilepsy
LADA	latent autoimmune diabetes in adults
LA-PCR	long accurate polymerase chain reaction
LBM	lean body mass
LD	linkage disequilibrium
LDB	linkage database
LDL	low density lipoprotein
LDLR	LDL receptor
LINE	long interspersed nuclear elements
lod	logarithm of the odds ratio
LOH	loss of heterozygosity
Lp(a)	lipoprotein a
LPL	lipoptotein lipase
MAE	myoclonic astatic epilepsy of early childhood
MD	major depression
MHC	major histocompatibility complex
MI	myocardial infarction
MIN	microsatellite instability
MLS	maximum LOD score
MODY	maturity-onset diabetes of the young
mRNA	messenger RNA
MTHFR	methylene tetrahydrofolate reductase
MZ	monozygotic
NCS	National Co-morbidity Study
NEAT	nonexercise activity thermogenesis
NIDDM	noninsulin dependent diabetes mellitus
NMDA	*N*-methyl-D-aspartate
NPHS	Northwick Park Heart Study
NPY	Neuropeptide Y
OR	odds ratio
PC_{20}	provocative concentration
PCR	polymerase chain reaction
PD	panic disorder
PD_{20}	provocative dose
PFGE	pulsed-field gel electrophoresis
PPAR	peroxisome proliferator activated receptor
PPR	photoparoxysmal response
PPRE	peroxisome proliferator responsive element
PSEN-1	presenilin-1
PSEN-2	presenilin-2

PV_H	paraventricular nucleus of the hypothalamus
QTL	quantitative trait locus
RA	rheumatoid arthritis
RAS	random affected sib pair
RDA	representational difference analysis
RE	restriction enzyme
RED	repeat expansion detection
RF	rheumatoid factor
RFLP	restriction fragment length polymorphism
RMR	resting metabolic rate
RQ	respiratory quotient
RR	relative risk
SAD	seasonal affective disorder
SAGE	serial analysis of gene expression
SCA6	spinocerebellar ataxia type 6
SE	shared epitope
SMEI	severe myoclonic astatic epilepsy
SML	single Major Locus Model
SNP	single nucleotide polymorphism
SPA	spontaneous physical activity
SSCP	single-strand conformational polymorphism
TBF	total body fat
TCR	T-cell receptor
TDT	transmission disequilibrium test
TEE	total energy expenditure
TH	tyrosine hydroxylase
TNF	tumor necrosis factor
UC	ulcerative colitis
UPDB	Utah population database
VAPSE	variation affecting protein sequence or expression
VLDL	very low density lipoprotein
VNTR	variable number tandem repeat
WGS	whole genome screen
WHR	waist-to-hip ratio
YAC	yeast artificial chromosome

Preface

The 1990s saw a major addition to the methods available to epidemiologists with an interest in disease etiology. The approach of identifying the responsible genes without prior knowledge of their function through family studies gave a new and successful angle to these investigations. While the approach was not intrinsically novel, the information available to facilitate such studies was; this information on the positions of genetic landmarks along chromosomes and, ultimately, the actual DNA sequence and its variation among individuals is the direct product of the Human Genome Project. This approach to gene cloning focused not on the etiology of the disease as a whole, but on the causes of the disease in particular multiple case families, those in which inheritance of disease could reasonably be interpreted as being due to a mutation in a single gene. While it was clear that such families were rare, indeed, for many diseases, extremely rare, the expectation was that the insight gained by understanding the particular gene or genes in these families would provide clues to the overall etiology.

Towards the end of the 1990s, the focus moved from the rare high-penetrance forms of the disease with clear inheritance to multifactorial, or 'complex' disease. These diseases showed family aggregation, some evidence for the role of genes and also some evidence for the role of non-genetic factors. Furthermore, the population prevalence of these diseases identified them as being major causes of human morbidity and/or mortality. Attempts at replicating the approaches to high-penetrance genes have met with limited success, but have also produced a large number of uninformative or inherently contradictory findings.

We proposed this book as editors with research interests in study design and statistical methodology, as well as a focus on particular diseases. Our intention was to review the approaches adopted in the study of multifactorial disease. We asked the authors to provide background to the epidemiological characteristics of each disease or trait (such as descriptive statistics on incidence and worldwide variation, as well as risk factors), the basis for the proposition that genes played a role, the extent of biological insight into mechanisms, the types of genetic approaches adopted, and their results. While we have tried to make the book as complete as possible, in terms of knowledge of the specific genes identified or the latest genome scans, we have erred towards offering an understanding of the rationale for the different approaches. We have included three introductory chapters (on molecular biology, study design and statistical methodology) to assist the reader in understanding some of the more technical aspects of the studies, as well as removing some of the potentially repetitive information from the disease-focus chapters.

Finally, we recognize that a number of diseases or traits have been omitted from the list of contents. Such a book as this could never be a complete encyclopedia of all research into multifactorial disease. We attempted to explore multifactorial disease from a number of different angles and include a broad spectrum of phenotypes. For instance, we chose to explore diseases with differing prevalences and degrees of familial aggregation. Similarly, the phenotypes chosen vary in terms of diagnostic precision and understanding of pathophysiological mechanisms. By providing such a variety we hope to present a balanced view of the present status and future prospects of the application of genetic approaches to multifactorial diseases.

Tim Bishop and Pak Sham

Foreword

As the third canonical millennium begins, genetic epidemiologists are wrestling with complex inheritance. This has turned out to be much more difficult than we had foreseen. After all, a reliable method for linkage mapping of major loci was developed between 1931 and 1955 by a handful of workers. Since then the number of mapped loci has increased by 45% per year and the cloning paradigm has been refined. Surely the effort devoted by hundreds if not thousands of researchers on complex inheritance since 1935 will be rewarded. Alas, despite considerable progress in methods and results, most studies continue to be inconclusive and frequently controversial. This book illustrates different methodologies as they have been applied to various diseases, from which we can try to identify the major obstacles and ways round them.

One approach to the plethora of alternative methods is to ask what essential features distinguish success with major loci from confusion about minor loci. Every reader of this book will form his own opinion. However, three principles are evident, which we may term reliability, power, and comprehensibility.

Reliability is the probability that a significant localization to a candidate region of w centimorgans be true. It obviously depends on the level of significance, α, the occupancy probability, ϕ, that a random region of size w contains a disease locus and the power, P, of the test. For major loci, this problem was solved by taking the region as an ideal chromosome for which the recombination fraction is uniformly distributed between 0 and ½, assuming that the occupancy probability is 0.05, and calculating power P for a sequential test with parameters A, B determining reasonable errors α and β, where $A = 0.001$ was considered because of the low value of ϕ, with β taken as 0.01 to assure a high exclusion rate. On these assumptions the reliability exceeds 0.95 unless the number of informative meioses is very small. In practice, the bound on α as $<1/A$ is conservative, and so the reliability for major loci has turned out to be nearly 1. Unfortunately, the parameters are much less well determined for the oligogenes that are important in complex inheritance; which grade imperceptibility into polygenes with effects so small that detection with realistic samples is not feasible. An analogy has been conjectured between linkage testing and Brownian motion, but it leaves many questions unanswered. Why should a multipoint test based on a vector of posterior probabilities be "just like" motion of a weightless particle? How can the theory fail to be explicit about heterozygosity, parental testing, the distribution of distances between markers, and other realities of the test? How can the number of independent regions exceed the number of centimorgans in the map, as the model for dense mapping predicts, even though markers within 1 cM are almost perfectly correlated? How can a stationary distribution be justified for a typically small sample? The Ornstein-Uhlenbeck theory for Brownian motion may be applicable to linkage mapping, but at present the analogy is pure speculation. Until reliability is approached as realistically as for major loci, it will not be clear whether the conventionally significant lod of 3 is too small, too large, or about right for oligogenic linkage. Consequently, the probative value of any particular lod is in doubt.

Power is the most restrictive parameter of reliability. Even with a parsimonious model and an efficient test, the expected sample size required to give high power for a locus accounting for 1/10 of the genetic effect is several thousand affected sib pairs. No single study or national consortium could hope to reach that number. This places a premium on methods of analysis that permit pooling of lods over samples, just as standard lod tables gave that facility to major loci. The classical case dealt with lods from pairs of loci at fixed values of sex-average recombination. In large samples the evidence can be presented more concisely, and this has been useful for high-density maps of markers that have been typed in different samples. However, multipoint tests of significance are more reliable and powerful, and so a standard lod table for complex inheritance should correspond to ordered marker loci, instead of recombination fractions. It is not necessary that the same markers be typed in each sample, nor phenotypes be defined in the same way, nor ascertainment be consistent. Only through such a standard lod table may samples be pooled with enough power to map oligogenes reliably. So far multipoint mapping has not been extended to allelic association, where multiple pairwise methods work well for major loci. Pairs of markers may be more informative.

Comprehensibility has been an outstanding feature of major locus analysis. What could be clearer than the ratio of two likelihoods, corresponding to a maximum likelihood location within multiple markers vs. the null hypothesis of infinite distance? Multiple generations, inbreeding, untested parents, and genetic heterogeneity demand sophisticated software, but the biological and statistical models are simple. Complex inheritance threatens to destroy this transparency. Many investigators do not realize that exclusion mapping in a negative region never fails to rule out a locus with high heritability but always fails to exclude a locus that accounts for as little as 10% of the heritability. The effect of a disease locus must be estimated at each point in the standard lod table, not merely examined for arbitrary perameters fixed over all markers. The object of analysis often seems to be a graph, which can never be pooled with other data. The appeal of methods that reflect negative lods to zero, like the 'possible triangle' of MLS, is dispelled by the realization that such biased statistics cannot be pooled. There are at least three different ways to estimate significance in GENEHUNTER, with unknown properties in small samples and often disagreeing by an order of magnitude. Two populations should not be assumed to have different disease loci merely because a nominally significant locus has been claimed in one and not the other. After all, effects of the polymorphisms we know about, like ABO and HLA, are similar among ethnic groups. Several measures of 'information' are restricted to markers without considering phenotypes, whereas the maximum likelihood method gives a metric that factors neatly into the two components and explains why extremely discordant pairs are more informative than extremely concordant pairs. Few reports on complex inheritance are explicit about how parameters were estimated, how many degrees of freedom were used, and how many phenotypes and conditional likelihoods were examined to give results that may be nominally significant. Such obscurity is fostered by journals that discourage detail on material and methods.

Over the years population structure has interested mathematicians and anthropologists as well as geneticists. Major genes with significant effects on fitness, even if recessive, usually trace to founders within a few millennia. Either such a gene is absent from an isolate, or it is unusually frequent. Therefore Finns, Ashkenazi Jews, and other populations with few founders are treasuries for genetic epidemiology. On the other hand, polymorphisms with little effect on fitness may trace to founders millions of years ago, perhaps with migration out of Africa as the last bottleneck. Oligogenes must be drawn preferentially from polymorphisms, since a gene that is both rare and of small effect is unlikely to be identified as disease-related. Isolates offer no special advantage for studying polymorphisms, and linkage information declines as homozygosity increases, while ethnic hybrids offer some advantage if family-based methods are used. The situation is reversed for allelic association, but on current evidence the advantage is small and too variable to be relied upon. Simulation studies are inconclusive, attributing allelic association to a founder effect or subsequent drift according to whether initial disequilibrium is assumed or not. Only empirical evidence can resolve whether isolates offer the same advantages for oligogenes as for major genes.

Complex inheritance is being studied partly because its presence is a challenge and partly because it is a logical extension of progress with major loci. Venture capitalists hope for short-term gain, contrary to experience with genes of greater effect. The strongest motive for studying complex inheritance may be that it will lead to better understanding of the different metabolic pathways underlying the heterogeneous diseases that today bear names like asthma, hypertension and coronary heart disease. Progress with major loci has been so rapid that we forget the optimism with which geneticists once pursued 'the gene' for cancer, dyslexia or Alzheimer's disease. In the long term, increased understanding must contribute to prevention and treatment.

This book will have served its purpose if, by exemplifying how controversial current approaches have often been, it furthers the insights of its readers and thereby contributes to the reliability, power and comprehensibility of research on complex inheritance.

Newton E. Morton

This book is dedicated to the memories of
David Fulker and Alistair D. Stewart
for their contribution to
our lives and to our
science

1

Principles of molecular genetics

G. Taylor

1. Introduction

This chapter reviews aspects of molecular genetics and DNA technology from their perspective as tools for the study of populations. For information about wider aspects of molecular genetics, the reader is referred to Lewin, 1999, and to Strachan and Read, 1999.

1.1 DNA and genome organization

Crick's Central Dogma of molecular biology (Crick, 1970) states that the genetic material is DNA which can self-replicate and can also be transcribed into RNA, which can in turn be translated into proteins. In a simplified version of this dogma, DNA could be viewed as a passive repository of genetic information, expressed as structural and functional proteins via RNA intermediates. Although this principle is a useful first approximation, several additional attributes such as gene splicing, reverse transcription and T and B cell gene rearrangements complicate the model. Nevertheless, for present purposes we can ignore these and consider the human diploid genome as simply a large collection of inherited markers that can be used to study individuals, pedigrees and populations.

1.2 Structure and properties of nucleic acids

The human genome, estimated to contain about 80 000 genes, consists of 24 linkage groups, corresponding to the 22 autosomes and the sex chromosomes. The DNA content of the haploid genome is approximately 3×10^9 base pairs, each diploid cell nucleus containing about 5×10^{-12} g DNA. In addition to the nuclear genome, the mitochondrial genome contains 16 569 base pairs of DNA and is maternally inherited. Whereas almost all of the mitochondrial genome codes for gene products, only about 5% of the nuclear genome is coding. DNA is a chemically monotonous backbone of deoxyribose linked by phosphate groups attached to the 5′ carbons and 3′ carbons of the deoxyribose. The strand is therefore

Analysis of Multifactorial Disease, edited by T. Bishop and P. Sham.

asymmetrical about the sugar-phosphate axis so it can be assigned a 5′ end and a 3′ end. The genetic code is contained in the sequence of purine (adenine, A and guanine, G) and pyrimidine (thymine, T and cytosine, C) bases attached. With the exception of single-stranded viral genomes, DNA normally exists as a double-stranded entity. A double-stranded helix is formed by two single-stranded fragments aligning with the hydrophilic sugar-phosphate on the exterior and the hydrophobic base in the center of the helix where they form hydrogen bonds with complementary bases such that A binds to T and G binds to C. As the two strands align with one strand 5′→3′ and the other 3′→5′ the structure is said to be antiparallel. It is this self-complementary structure that enables DNA to act as a template for its own replication and for the synthesis of RNA and subsequent gene expression.

DNA has several advantages as a source of genetic markers: it is chemically stable, and so can survive lengthy periods of storage. The main causes of DNA degradation are hydrolysis by enzymes (nucleases) and purine damage by ultraviolet (uv) radiation. Nucleases are abundant in all cells, so DNA is best stored once extracted from tissue, although fixed tissues, desiccated tissues and dried bloodspots can provide DNA suitable for analysis. Genomic DNA is present in the same form throughout the organism, so DNA extracted from peripheral lymphocytes should be equivalent to DNA from buccal cells in the same individual. There are many enzymes and proteins that interact with DNA and can be adapted for experimental use. For example type II restriction endonucleases cleave DNA at specific sequences; whereas DNA polymerases, which catalyze DNA replication or repair, have been harnessed for the *in vitro* synthesis and labeling of DNA. Finally, the self-complementary nature of DNA means that DNA probes can be used to detect and bind to target sequences which are complementary to their own. This property is fundamental to all sequence analysis techniques.

Genetic markers can be derived from either coding (intragenic) or noncoding regions of the genome. A useful marker is polymorphic (i.e. exists in more than one state) but may not necessarily have any effect on phenotype. Vogel and Motulsky (Vogel and Motulsky, 1986) define a polymorphism as a Mendelian trait that exists in the population with the frequency of the rarer of the two alleles being greater than 1–2%. Rarer variants are arbitrarily defined as monogenic traits. Mutation, on the other hand, could be taken to mean any sequence variation from a reference standard, although it has also been defined as a *de novo* change in genomic sequence (Schena, 1999) in contrast to polymorphism, defined by these authors as any inherited sequence change. In common usage polymorphism is often used to refer to a presumed nonpathogenic mutation. For present purposes I will use the term mutation to describe any DNA sequence variation from a reference sequence, and a polymorphism as a mutation which is present in a sufficiently large proportion of a population to provide a useful genetic marker. Genetic markers can be any sequence variant and can be determined by a variety of laboratory techniques, some of which are outlined below. For reasons of economy, generic methods with simple technical manipulation and the potential for automation are preferred. The most widely used genetic markers are microsatellites and single nucleotide polymorphisms (SNPs). Microsatellites are runs of repetitive DNA, usually di-, tri- or tetra-nucleotide repeats flanked by unique sequences. The human genome is estimated to have about 10^5 microsatellite repeats and they are thought to be distributed widely throughout the genome. The repeat length of a microsatellite is often variable (i.e. polymorphic) within populations, yet has a relatively low new mutation rate (10^{-3}–10^{-5}), providing a rich source of multi-allelic markers. SNPs are likely to be even more abundant, although will mostly be di-allelic and hence a single SNP would be less informative than a single microsatellite marker.

2. Genotyping

It is beyond the range of current technology to deliver a complete genomic sequence on a case-by-case, let alone population, basis. So the practical experimental issue becomes that of selecting the most acceptable approximation to a genome sequence scan for each particular study. This depends largely on the nature of the hypothesis under investigation: genome-wide linkage using large pedigrees, association studies, or candidate gene testing. Total genomic searches require a set of polymorphic markers spanning the entire genome, whereas testing one or a few genes requires a more detailed sequence analysis of a more limited region.

As sequence changes can abolish or create cleavage sites for the dozens of commercially available restriction endonucleases (REs), RE polymorphisms were the first tools used for genetic mapping and diagnosis, in combination with Southern blotting of genomic DNA. Although there are still some applications, for example mapping large deletions or rearrangements, for which Southern blotting is the best method, the polymerase chain reaction (PCR) is now the method of choice for routine genotyping.

2.1 PCR

The PCR is a patented modification of the primer extension reaction that can be used to generate large quantities of DNA of fragment sizes ranging from 50 to over 40 000 base pairs in length (*Figure 1*). The process is widely used and simple to perform. In a typical configuration a pair of single-stranded DNA primers (usually 18–30 bases long) are designed to flank the region to be amplified. These primers are each complementary to opposite strands of the DNA target and direct polymerase activity (which proceeds from a double-stranded region of DNA in the 3′ direction) towards the primer binding site of the complementary strand. Thus two primer extension reactions occur in opposite directions, overlapping between the primer binding sites. Primer binding is enabled by first heating the template DNA to 95°C, which causes the strands to separate (denaturation), followed by cooling to a temperature (usually between 50 and 65°C) at which both primers can bind specifically to their complementary targets. Primer concentration is in great excess so that primer–template hybrids form faster than template can reanneal. Inclusion of thermostable DNA polymerase, buffer and nucleotide triphosphates in the reaction then enables synthesis of new DNA starting from the primer, doubling the amount of DNA between the primers. If this process of denaturation, primer annealing and primer extension is repeated, again the

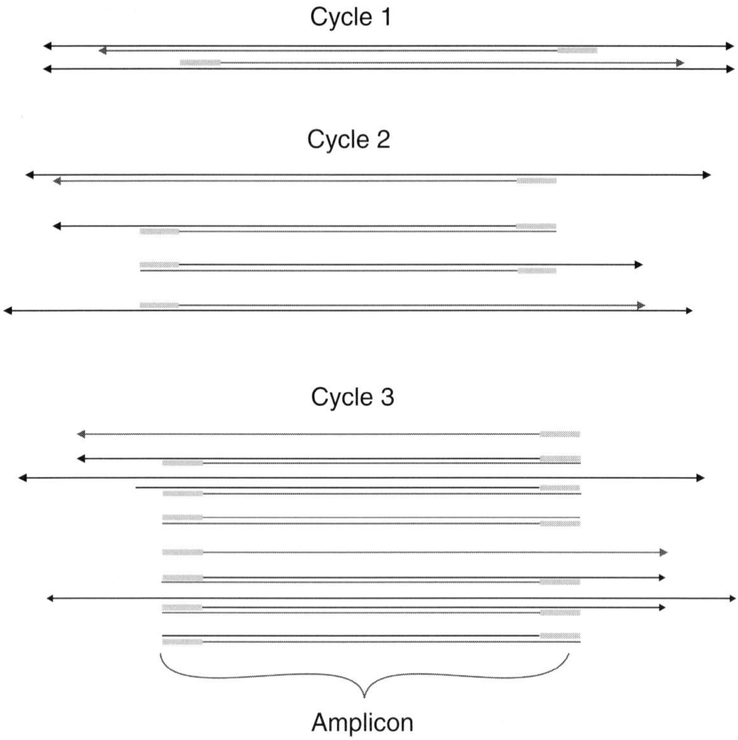

Figure 1. Expected products after the first, second and third cycles of heat denaturation and primer extension of a PCR. The region flanked by the primers (gray tint) is amplified exponentially at each cycle. Under typical conditions a 100 000 amplification of the amplicon can be achieved.

amount of DNA is doubled. This process is typically cycled 20–40 times to produce microgram amounts of specific fragments of DNA referred to as amplicons. In standard PCR the thermostable polymerase is usually from *Thermus aquaticus* (*Taq* polymerase) or a related thermophillic eubacteria. This enzyme can synthesize fragments up to 5000 bases and withstand the high temperatures required in each cycle to denature double-stranded DNA. The amplicon population (PCR product) has an average sequence that corresponds to that of the template, but because of the error rate of *Taq* polymerase (about 1 error in 10^4 bases) any individual product may have sequence variation artefacts. These would not be seen unless the product was cloned, or if very small amounts of starting material were used, in which case an error arising in an early cycle could be passed on to a substantial proportion of the final product. DNA polymerases from archaea such as *Pyrococcus furiosus* (*Pfu* polymerase) have a proofreading activity that gives rise to higher fidelity during the PCR but are not as widely used. Their extension rates tend to be slower and they can digest single-stranded primers, making them a little more difficult to work with.

Multiplex PCR refers to the situation in which several primer pairs for different amplicons are included in the same reaction. This is essential for high throughput applications, for example several high order multiplex PCRs have to be run to provide a single SNP DNA chip with material to analyze. Long accurate PCR (LA-PCR), first demonstrated by Barnes (Barnes, 1994) works by combining *Taq* polymerase and *Pfu* polymerase in the same reaction, making it possible to amplify very long (up to 40 000 base) templates. Although this is still a long way from the cloning capacity of some modern vectors, LA-PCR amplicons can be an order of magnitude greater than those routinely seen with standard PCR.

The specificity of hybridization is conferred by the sequence at the 3′ end of the primers. The 5′ end can be modified by adding additional sequences (which become copied during the PCR) or by adding chemical tags such as fluorescent dyes or biotin. The amplicons produced are then fluorescent or biotin-containing which can be used to simplify downstream handling. Amplification efficiency is not impaired by additional sequences and tags, although the cost of primer synthesis may be increased. Although PCR has had a tremendous impact on genetic testing, it does have two major weaknesses, the size limitation of amplicons ($\leq$ 30 kbp from genomic DNA) and the risk of allelic dropout, where a primer on one allele fails to bind, perhaps due to a polymorphism or deletion at a primer binding site on one of the chromosomes. Allelic dropout would then lead to false calling of genotypes as homozygous. Such false results might only be revealed by either apparent noninheritance seen by pedigree analysis or by a quantitative PCR assay. A further issue that must be considered is the risk of PCR contamination. If a large number of amplifications are conducted using the same primers, there is a risk that amplicons could find their way into reagents and equipment and be re-amplified in subsequent experiments, giving false genotype results. A negative (no template DNA) control is mandatory in any PCR experiment. Those individuals who believe that their PCR is so good that they can amplify product from distilled water need to start again with fresh reagents and better precautions against contamination. Nevertheless PCR is so powerful and simple in practice that it is the default option for genotyping. Genotyping or mutation detection methods typically use PCR amplicons as their starting point.

2.2 Genotype analysis using PCR

Restriction fragment length polymorphism (RFLP) is now primarily of historical interest, although it is a robust method. Amplicons are generated to flank a polymorphic RE site and subjected to digestion and the presence or absence of the site can be determined by agarose gel electrophoresis of the digested amplicon and visualization using ethidium bromide staining and UV illumination. Artificial restriction sites can be produced by incorporating modifications into one of the amplimers to increase the range of polymorphisms that can be examined. Gains in throughput can be achieved using high density gels such as the MADGE (microtitre array diagonal gel electrophoresis) format (Day *et al.*, 1995).

2.3 Microsatellite analysis

Linkage mapping was accelerated by the description in 1989 of microsatellites, a novel type of polymorphic marker (*Figure 2*) (Weber and May, 1989). The term satellite DNA was originally coined to describe repetitive DNA identified as a satellite of different density in a cesium chloride density gradient. Satellite DNA

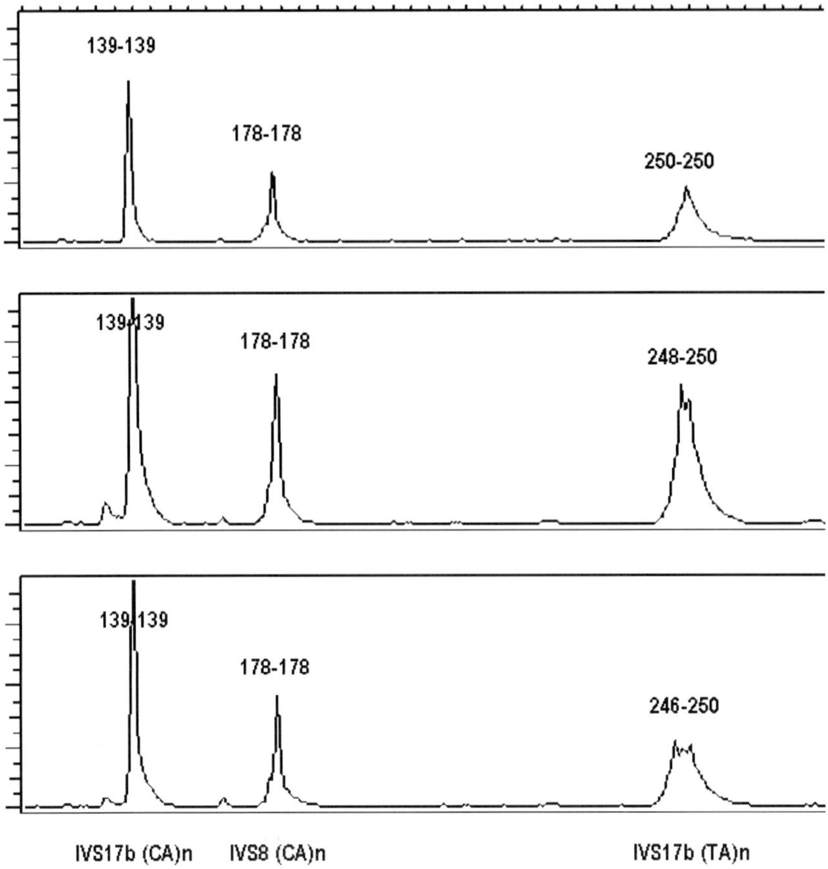

Figure 2. Microsatellite products sized using automated fragment analysis software (PE-Applied Biosystems). Three microsatellites from the ABCC7 gene were co-amplified. The peaks correspond to microsatellite amplicons. Sample 1 is homozygous at all three loci, whereas IVS17b (TA) is heterozygous in samples 2 and 3.

consists of repeated motifs of several tens to several hundreds of bases, such as Alu and LINE sequences. Microsatellite repeats are much shorter, usually dinucleotide repeats, but also mono-, tri- and tetranucleotide repeats. The useful feature of microsatellites is that the repeat length often varies between chromosomes. The origin of this length polymorphism is thought to be strand slippage during replication. One strand may form a short hairpin and produce a copy of different length in the daughter strand. The PCR was gaining widespread usage when microsatellites were first described and they were ideally suited for analysis by designing PCR primers (amplimers) which bind to unique sequences flanking the microsatellite repeat motif. Amplicons need only be sized to within the resolution of the repeat motif to provide a genotype (*Figure 3*). This requires sizing fragments of about 100–300 bases in length with an accuracy of ± 1 base. The most accurate way to do this is to use some form of automated fragment analysis equipment such as the Perkin-Elmer or Amersham-Pharmacia automated DNA sequencers.

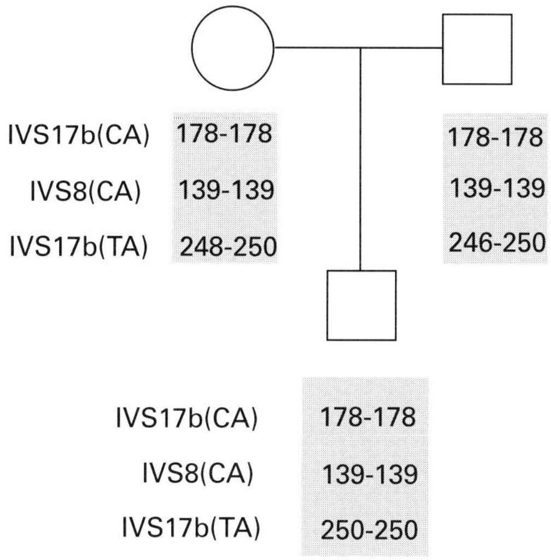

Figure 3. Microsatellite alleles used as markers in a small pedigree.

2.4 SNPs

Recently interest has returned to single nucleotide polymorphisms, of which RE polymorphisms are a subset. SNPs are di-allelic (although there is no reason in principle why a particular base could not be substituted by more than one alternative) and so are less informative individually than microsatellites, but far more abundant (Nickerson *et al.*, 1992). The human genome may contain millions of SNPs, however they are probably more abundant in noncoding regions of the genome. Several efforts are now underway to produce genomic SNP maps. The main appeal of SNPs is the prospect of high throughput automated analysis using array or chip technology (Fan *et al.*, 1997), however a variety of generic mutation detection techniques can be adapted for SNP detection.

2.5 Mutation detection

Mutation detection techniques can be divided into those which test for known mutations and those which scan for any mutation in a given target region. Broader aspects of mutation detection such as gene dosage, gross rearrangements and methylation can be excluded for present purposes. For the detection of known mutations (mutation testing or genotyping) there are several well-tried methods, including the amplification refractory mutation system (ARMS), solid-phase minisequencing, TaqMan and molecular beacons which are outlined below. Space does not permit a comprehensive review of all available methods, although there are several publications dealing with them (Cotton, 1997; Taylor, 1997). The prime considerations in any approach to mutation detection are sensitivity (the proportion of mutations that can be detected) and specificity (absence of false positives). Cost per genotype and throughput are also important factors.

ARMS. This is a modification of conventional PCR in which one of the primer pairs is designed to have the polymorphic base in the template at its 3′ position. *Taq* polymerase is unable to extend from a mismatched base, and so the generation of a PCR product only occurs if the 3′ base in the primer matches the template. The technique can be multiplexed to type up to 20 SNPs simultaneously.

Minisequencing. Minisequencing determines the base immediately 3′ to a primer by extending the primer by one base only (Sokolov, 1990). Readily adaptable to microtiter and array formats it has considerable potential as a high throughput system.

TaqMan and molecular beacons. TaqMan is a closed system assay that can be adapted for quantification as well as genotyping. Single nucleotide differences are detected in PCR products by the sequence specific hybridization of a probe which contains both a fluorochrome and a quencher. When hybridized, the quencher molecule is cleaved and the bound fluorochrome can be detected by a fluorescence assay. As it is possible to have different colored fluorochromes, the probes can be differentially labeled, allowing both alleles of a SNP to be typed in the same tube (Livak, 1999). Molecular beacons are hairpin shaped structures which contain a fluorochrome and a quencher on the 5′ and 3′ ends. When in free solution the fluorochrome is quenched but upon hybridization to a specific target the hairpin opens and the molecule becomes fluorescent. They can be used in a closed system for allelic discrimination of PCR products (Tyagi *et al.*, 1998) .

2.6 Mutation scanning

Mutation scanning aims to identify any mutation in a region, usually a PCR amplicon. A range of methods have been described, the most widely used being single-strand conformational polymorphism (SSCP), denaturing gradient gel electrophoresis (DGGE), denaturing high performance liquid chromatography (DHPLC) and mismatch cleavage techniques including chemical cleavage of mismatch (CCM) and various enzymatic methods (*Table 1*).

SSCP. SSCP (Orita *et al.*, 1989) relies on the formation of secondary structures by single-stranded DNA. These structures are determined by intrastrand interactions resulting in the formation of hairpins and looped regions. The interactions themselves are determined by the DNA sequence, so that a single base substitution may result in an alternative secondary structure. Under suitable electrophoresis conditions mobility changes caused by sequence changes can be resolved with sensitivities ranging from 60% more than 95% (Liu *et al.*, 1999; Sheffield *et al.*, 1993).

DGGE and DHPLC. These two methods each rely on the formation of heteroduplex DNA, one strand being a reference strand and the complementary strand being a test. If there is a mismatch between the reference and test strand, the base pairing is disrupted at that position. This disruption will reduce the melting temperature of the double-stranded molecule compared with that of a perfectly matched molecule. In DGGE (Myers *et al.*, 1987) the reduced melting temperature

Table 1 Mutation scanning methods

Method	Outline	Advantages	Disadvantages
SSCP	Mutations affect mobility of single-stranded DNA on gel electrophoresis	Simple, improved sensitivity using automated sequencers	Less than 100% efficient Does not localize or define nature of mutation
DGGE	Double-stranded DNA is progressively denatured by electrophoresis through a gradient	Simple. Close to 100% sensitivity	Special G + C clamped primers needed Does not locate mutation
DHPLC	Double-stranded DNA is progressively denatured by reverse-phase chromatography through a gradient	Simple. Autoloading for hands-free operation	Less than 100% efficient Does not localize or define nature of mutation
CCM	Heteroduplex mismatches are vulnerable to KMnO4 or hydroxylamine attack	Most sensitive method available Locates the mutation	Multi-step process, hazardous chemicals
EMC	Heteroduplex mismatches are vulnerable to enzymatic cleavage	Technically easier than CCM	Less than 100% sensitivity and specificity
Dideoxy sequencing	DNA sequenced by chain termination	Complete characterization of mutation	May not detect all heterozygotes
Arrayed primer extension	DNA is sequenced by primer extension	Rapid, chip based	Relatively untried
Affymetrix gene chip	DNA is sequenced by hybridization	Rapid, chip based	Less than 100% sensitivity

EMC, enzyme mismatch cleavage

of a heteroduplex is revealed by electrophoresis through a gradient of increasing temperature or chemical denaturant. Mutations cause the heteroduplex to denature before homoduplexes and their mobility is retarded. In DHPLC (Underhill *et al.*, 1997) as the concentration of denaturant is increased, mismatched heteroduplexes elute from the column before homoduplexes. The sensitivity of DHPLC has been reported to exceed 95% (Jones *et al.*, 1999) and with correctly designed primers DGGE can achieve 100% sensitivity (Wu *et al.*, 1998).

Enzymatic or chemical cleavage of mismatches. As well as altering the melting profile of double-stranded DNA, a mismatched base pair is more vulnerable to cleavage by enzymatic or chemical means. As the cleavage site will be at or close to the mismatched base, the size of the cleaved products indicates the location of the mismatch. The first cleavage method reported used ribonuclease (Myers *et al.*, 1985), and this method has been steadily refined to produce a nonradioactive method (Goldrick *et al.*, 1996) which uses several ribonucleases to achieve sensitivities of more than 80% with high specificity. T4 endonuclease VII has also been used to selectively cleave mismatches (Youil *et al.*, 1995). However the most sensitive and specific method of mismatch cleavage remains by chemical modification of mismatched bases (Cotton *et al.*, 1988), now modified for solid phase purification (Hansen *et al.*, 1996) and fluorescent detection (Freson *et al.*, 1998) .

2.7 DNA sequencing

As the cost of DNA sequencing falls and throughput increases, there is a need to continually evaluate other mutation detection methods against DNA sequencing, both in terms of throughput and accuracy. Whilst one might intuitively expect that DNA sequencing would be the most accurate way to detect mutations, this is not supported by experimental evidence (Ahrendt *et al.*, 1999), where both automated sequencing and sequencing by hybridization failed to detect all of the mutations seen by manual sequencing. Sequencing quality control has generally been with respect to the accuracy of base calling of clones, which would not be expected to carry heterozygous mutations. It may be that allelic discrimination on DNA chips requires enzymatic activity to increase the specificity needed, for example by the use of arrayed primer extension (Pastinen *et al.*, 1997) although it is too early to see if this will be a cost effective approach.

3. Appendix. Mapping and sequencing of the human genome

Genetic analysis of human populations is being transformed by the information generated from the human genome sequencing project. The main thrust of this program has been via a loosely coordinated consortium of academic centers operating through the Human Genome Organization and the Genome Database in the first instance. Several national genome sequencing programs have been established and there have also been minor contributions from commercial organizations. The order (but not the precise distances) physical and linkage maps will converge into 24 chromosomal linkage groups with the completion of the human genomic sequence. The ultimate physical map of the human genome is the complete DNA sequence, the determination of all base pairs on each chromosome. The completed map will enable understanding of the mechanisms of inherited diseases and of normal gene function. As the full sequence becomes known, emphasis is shifting to gene function.

3.1 The linkage map

The first human linkage map consisted primarily of RFLP markers (Donis-Keller *et al.*, 1987), but have now been replaced by high density microsatellite maps (Dib *et al.*, 1996) and the developing SNP maps (Lai *et al.*, 1998). A SNP map would have the advantage of ready adaptability to array based analysis, although increasingly high throughput analysis of microsatellites is also possible using capillary based sequencing devices.

3.2 Physical mapping

Physical mapping at higher resolution than cytogenetic banding (several megabases) has used cloning vehicles such as yeast or bacterial artificial chromosomes that can contain large quantities of inserted DNA sequenced tagged sites and irradiation hybrid mapping panels.

Genomic libraries. Libraries are collections of cloned DNA which may be made using a variety of cloning vectors. Technological improvements now make possible the cloning of large DNA pieces, using artificially constructed chromosome vectors that carry human DNA fragments as large as 1 Mb. These vectors are maintained in yeast cells as artificial chromosomes (YACs). YAC clone insert sizes range from a few hundred kilobases to more than 1 megabase. The entire human genome could, in principle, be contained in 3000 1 megabase YACS, although in practice libraries are constructed to include a 5–10-fold redundancy to increase the chance of a given region of the genome being included. A physical map of most of the human genome represented in overlapping YAC clones has been produced (Chumakov *et al.*, 1995). A more detailed map of a large YAC insert can be produced by subcloning, a process in which fragments of the original insert are cloned into smaller-insert vectors. Cosmid vectors can carry about 40 kb of genomic DNA and are the workhorses of physical mapping. Most genomic sequencing starts from cosmids that have been mapped and aligned to form a contiguous array along a chromosome region (contigs).

Because some YAC regions are unstable, large-capacity bacterial vectors human genomic libraries have also been developed (Kim *et al.*, 1996). Bacterial or phage artificial chromosomes (BACs or PACs) are derived from a bacterial episomal mating factor and can package larger DNA fragments of up to 250 kbp. Individual chromosomes can be separated by means of fluorescence-activated cell sorting (FACS) and used as starting points for chromosome-specific gene libraries.

Pulsed-field gel electrophoresis (PFGE). The development of PFGE methods improved the mapping and cloning of large DNA molecules (Schwartz and Cantor, 1984). Whereas conventional gel electrophoretic methods separate pieces less than 40 kb (1 kb = 1000 bases) in size, PFGE separates molecules up to 10 Mb, allowing the application of both conventional and new mapping methods to larger genomic regions.

Irradiation hybrids. The GeneBridge4 radiation hybrid panel (Gyapay *et al.*, 1996) panel of hybrids is distributed as DNA, which can be used as a target for

PCR amplification. The hybrid mapping panel consists of 86 clones of the whole human genome, and can be used for ordering markers within 5000 kb of each other, as well as establishing the distance between these markers. It is distributed as aliquots of 1 μg of DNA from each hybrid, with appropriate human and hamster control DNA. Data on markers previously mapped to the panel is available on-line from the European Bioinformatics Institute

Fluorescent in situ *hybridization (FISH).* Until recently, even the best chromosomal maps could be used to locate a DNA fragment only to a region of about 10 Mb, the size of a typical band seen on a chromosome. Improvements in FISH methods allow orientation of DNA sequences that lie as close as 2–5 Mb (Korenberg *et al.*, 1992). Modifications to *in situ* hybridization methods, using chromosomes at a stage in cell division (interphase) when they are less compact, increase map resolution to around 100 kb. Further banding refinement might allow chromosomal bands to be associated with specific amplified DNA fragments, an improvement that could be useful in analyzing observable physical traits associated with chromosomal abnormalities. Cosmids make good probes for FISH and can quickly be mapped to a chromosome region in this way.

Partial sequencing to facilitate mapping. Correlating mapping data from different laboratories has been a problem because of differences in generating, isolating, and mapping DNA fragments. A common reference system designed to meet these challenges uses partially sequenced unique regions (200–500 bp) to identify clones, contigs, and long stretches of sequence. Called sequence tagged sites, these short sequences have become standard markers for physical mapping (Hudson *et al.*, 1995).

Because coding sequences of genes represent most of the potentially useful information content of the genome but are only a fraction of the total DNA, partial sequencing of cDNAs instead of random genomic DNA has been used to produce expressed sequence tags (Adams *et al.*, 1991).

Sequence-tagged connectors and restriction digest mapping of BACs have been proposed as a rapid way to complete the human genome sequence (Mahairas *et al.*, 1999). This route has been selected by Celera Genomics who aim to complete the human genome sequence before the publicly funded programs.

Sequencing technology. Achieving the goals of the Human Genome Project required substantial improvements in the rate, efficiency, and reliability of standard sequencing procedures as well as the development of novel approaches. The Human Genome Project has provided a focus for the development of automated sequencing technology. Specific goals include the development of sequencing and detection schemes that are faster and more sensitive, accurate, and economical. Many novel sequencing technologies are now being explored, and the most promising ones will eventually be optimized for widespread use. Dideoxy sequencing has been adapted from a manual, radioactive procedure using slab gels to an automated fluorescent method using very thin gels or capillaries.

Developing technologies include direct reading of the base sequence on a DNA strand with the use of scanning tunneling or atomic force microscopes

(Shlyakhtenko *et al.*, 1999), mass spectrometric analysis of DNA sequence (Little *et al.*, 1997), sequencing by hybridization to short panels of nucleotides of known sequence (Hacia *et al.*, 1997) and microfabrication of gel and capillary devices (Schmalzing *et al.*, 1998). Despite these promising developments the genome sequence will be derived using large-scale application of the Sanger di-deoxy method.

References

Adams, M.D., Kelley, J.M., Gocayne, J.D., *et al.* (1991) Complementary-DNA sequencing – expressed sequence tags and human genome project. *Science* **252**: 1651–1656.

Ahrendt, S.A., Halachmi, S., Chow, J.T., Wu, L., Halachmi, N., Yang, S.C., Wehage, S., Jen, J. and Sidransky, D. (1999) Rapid p53 sequence analysis in primary lung cancer using an oligonucleotide probe array. *Proc. Natl Acad. Sci. USA* **96**: 7382–7387.

Barnes, W.M. (1994) PCR amplification of up to 35-kb DNA with high fidelity and high yield from lambda bacteriophage templates. *Proc. Natl Acad. Sci. USA* **91**: 2216–2220.

Chumakov, I.M., Rigault, P., LeGall, I., *et al.* (1995) A YAC contig map of the human genome. *Nature* **377** (Suppl.):175–297.

Cotton, R.G.H. (1997) *Mutation Detection*. Oxford University Press, Oxford.

Cotton, R.G.H., Rodrigues, N.R. and Campbell, R.D. (1998) Reactivity of cytosine and thymine in single base-pair mismatches with hydroxylamine and osmium tetroxide and its application to the study of mutations. *Proc. Natl Acad. Sci. USA* **85**: 4397–4401.

Crick, F. (1970) Central dogma of molecular biology. *Nature* **227**: 561–563.

Day, I.N.M., Odell, S.D., Cash, I.D., Humphries, S.E. and Weavind, G.P. (1995) Electrophoresis for genotyping – temporal thermal-gradient gel-electrophoresis for profiling of oligonucleotide dissociation. *Nucl. Acids Res.* **23**: 2404–2412.

Dib, C., Faure, S., Fizames, C., *et al.* (1996) A comprehensive genetic map of the human genome based on 5,264 microsatellites [see comments]. *Nature* **380**: 152–154.

Donis-Keller, H., Green, P., Helms, C., *et al.* (1987) A genetic linkage map of the human genome. *Cell* **51**: 319–337.

Freson, K., Peerlinck, K., Aguirre, T., *et al.* (1998) Fluorescent chemical cleavage of mismatches for efficient screening of the factor VIII gene. *Hum. Mutation* **11**: 470–479.

Goldrick, M.M., Kimbal, G.R., Liu, Q., Martin, L.A., Sommer, S.S. and Tseng, J.Y.H. (1996) NIRCA – a rapid robust method for screening for unknown point mutations. *Biotechniques* **21**: 106–112.

Gyapay, G., Schmitt, K., Fizames, C., *et al.* (1996) A radiation hybrid map of the human genome. *Hum. Mol. Genet.* **5**: 339–346.

Hacia, J.G., Edgemon, K., Hunt, N., Fodor, S.P., Brody, L.C. and Collins, F.S. (1997) Mutation screening and phylogenetic analysis of hereditary breast cancer genes using high density oligonucleotide arrays. *Am. J. Hum. Genet.* **61**: 361.

Hacia, J.G., Makalowski, W., Edgemon, K. *et al.* (1998) Evolutionary sequence comparisons using high-density oligonucleotide arrays. *Nature Genet.* **18**(2): 155–158.

Hansen, L.L., Justesen, J. and Kruse, T.A. (1996) Sensitive and fast mutation detection by solid phase chemical cleavage. *Hum. Mutation* **7**: 256–263.

Hudson, T.J., Stein, L.D., Gerety, S.S., *et al.* (1995) An STS-based map of the human genome. *Science* **270**: 1945–1954.

Jones, A.C., Austin, J., Hansen, N., *et al.* (1999) Optimal temperature selection for mutation detection by denaturing HPLC and comparison to single-stranded conformation polymorphism and heteroduplex analysis. *Clin. Chem.* **45**: 1133–1140.

Kim, U.J., Birren, B.W., Slepak, T., *et al.* (1996) Construction and characterization of a human bacterial artificial chromosome library. *Genomics* **34**: 213–218.

Korenberg, J.R., Yangfeng, T., Schreck, R. and Chen, X.N. (1992) Using fluorescence in situ hybridization (FISH) in genome mapping. *Trends Biotechnol.* **10**: 27–32.

Lai, E., Riley, J., Purvis, I. and Roses, A. (1998) A 4-Mb high-density single nucleotide polymorphism-based map around human APOE. *Genomics* **54**: 31–38.
Lewin, B. (1999) *Genes VI*. Oxford University Press, Oxford.
Little, D.P., Cornish, T.J., ODonnell, M.J., Braun, A., Cotter, R.J. and Koster, H. (1997) MALDI on a chip: Analysis of arrays of low femtomole to subfemtomole quantities of synthetic oligonucleotides and DNA diagnostic products dispensed by a piezoelectric pipet. *Anal. Chem.* **69**: 4540–4546.
Liu, Q., Feng, J., Buzin, C., *et al.* (1999) Detection of virtually all mutations – SSCP (DOVAM-S): A rapid method for mutation scanning with virtually 100% sensitivity. *Biotechniques* **26**: 932–936.
Livak, K.J. (1999) Allelic discrimination using fluorogenic probes and the 5′ nuclease assay. *Genet. Anal.* **14**: 143–149.
Mahairas, G.G., Wallace, J.C., Smith, K., *et al.* (1999) Sequence-tagged connectors: A sequence approach to mapping and scanning the human genome. *Proc. Natl Acad. Sci. USA* **96**: 9739–9744.
Myers, R.M., Larin, Z. and Maniatis, T. (1985) Detection of single base substitutions by ribonuclease cleavage at mismatches in rna-dna duplexes. *Science* **230**: 1242–1246.
Myers, R.M., Maniatis, T. and Lerman, L.S. (1987) Detection and localization of single base changes by denaturing gradient gel-electrophoresis. *Methods Enzymol.* **155**: 501–527.
Nickerson, D.A., Whitehurst, C., Boysen, C., Charmley, P., Kaiser, R. and Hood, L. (1992) Identification of clusters of biallelic polymorphic sequence-tagged sites (PSTSS) that generate highly informative and automatable markers for genetic-linkage mapping. *Genomics* **12**: 377–387.
Orita, M., Iwahana, H., Kanazawa, H., Hayashi, K. and Sekiya, T. (1989) Detection of polymorphisms of human DNA by gel electrophoresis as single-strand conformation polymorphisms. *Proc. Natl Acad. Sci. USA* **86**: 2766–2770.
Pastinen, T., Kurg, A., Metspalu, A., Peltonen, L. and Syvanen, A.C. (1997) Minisequencing: A specific tool for DNA analysis and diagnostics on oligonucleotide arrays. *Genome Res.* **7**: 606–614.
Schena, M. (1999) *DNA Microarrays: A practical approach*. Oxford University Press, Oxford.
Schmalzing, D., Adourian, A., Koutny, L., Ziaugra, L., Matsudaira, P. and Ehrlich, D. (1998) DNA sequencing on microfabricated electrophoretic devices. *Anal. Chem.* **70**: 2303–2310.
Schwartz, D.C. and Cantor, C.R. (1984) Separation of yeast chromosome-sized DNAs by pulsed field gradient gel-electrophoresis. *Cell* **37**: 67–75.
Sheffield, V.C., Beck, J.S., Kwitek, A.E., Sandstrom, D.W. and Stone, EM. (1993) The sensitivity of single-strand conformation polymorphism analysis for the detection of single base substitutions. *Genomics* **16**: 325–332.
Shlyakhtenko, L.S., Gall, A.A., Weimer, J.J., Hawn, D.D. and Lyubchenko, Y.L. (1999) Atomic force microscopy imaging of DNA covalently immobilized on a functionalized mica substrate. *Biophys. J.* **77**: 568–576.
Sokolov, B.P. (1990) Primer extension technique for the detection of single nucleotide in genomic DNA. *Nucl. Acids Res.* **18**: 3671.
Strachan, T. and Read, A.P. (1999) *Human Molecular Genetics*, 2nd edn. BIOS, Oxford.
Taylor, G.R. (1997) *Laboratory Methods for the Detection of Mutations and Polymorphisms in DNA*. CRC Press, London.
Tyagi, S., Bratu, D.P. and Kramer, F.R. (1998) Multicolor molecular beacons for allele discrimination. *Nature Biotechnol.* **16**: 49–53.
Underhill, P.A., Jin, L., Lin, A.A., *et al.* (1997) Detection of numerous Y chromosome biallelic polymorphisms by denaturing high-performance liquid chromatography. *Genome Res.* **7**: 996–1005.
Vogel, F. and Motulsky, A.G. (1986) *Human Genetics*, 2nd edn. Springer-Verlag, New York.
Weber, J.L. and May, P.E. (1989) Abundant class of human DNA polymorphisms which can be typed using the polymerase chain reaction. *Am. J. Hum. Genet.* **44**: 388–396.
Wu, Y., Hayes, V.M., Osinga, J., *et al.* (1998) Improvement of fragment and primer selection for mutation detection by denaturing gradient gel electrophoresis. *Nucl. Acids Res.* **26**: 5432–5440.
Youil, R., Kemper, B.W. and Cotton, R.G.H. (1995) Screening for mutations by enzyme mismatch cleavage with T4 endonuclease-7. *Proc. Natl Acad. Sci. USA* **92**: 87–91.

2

Methods for mapping complex disease traits

G. Mark Lathrop and Daniel E. Weeks

1. Introduction

Genomics has provided powerful tools for genetic studies that have led to the identification of many disease-related genes. Most genetic studies are based on the analysis of DNA variants that can be used to trace inheritance within families (linkage), and to compare allele and genotype frequencies in patient and control populations (association). The complete sequence of the human genome should soon be completely known, and much attention is being given to how this knowledge can improve the study of DNA variation. Within the next few years, catalogs of hundreds of thousands of single nucleotide polymorphisms (SNPs) will become available to complement existing microsatellite markers for genetic studies.

Although several hundred disease-related genes have been mapped and/or identified via genetic approaches, most of these are related to monogenic disorders with simple patterns of inheritance in families. Despite the improvement in molecular tools, identification of the genes implicated in common multifactorial diseases, such as heart attack, hypertension, diabetes or asthma, remains a challenging task because susceptibility to most of these disorders is caused by combinations of genes and environmental factors, potentially interacting with each other. However, the motivation for undertaking genetic investigations of multifactorial diseases is strong due to the anticipated consequences for public health, treatment and increased understanding of the etiological basis of disease. In tackling such problems, careful attention to study design, clinical resources and statistical analysis is required to maximize the chances of success. The purpose of this chapter is to provide an introduction to some of the approaches and statistical methodologies that are routinely applied to genetic studies of such disorders.

2. Sib-pair methods for qualitative traits

Over the last 15 years or so, the most powerful approach for identifying genes involved in monogenic diseases has been via linkage studies in which a relatively

Analysis of Multifactorial Disease, edited by T. Bishop and P. Sham.

small chromosome region containing the gene of interest is identified through the analysis of co-segregation of genetic markers with disease in families. For common multifactorial disorders, linkage studies are also undertaken, often in nuclear families and sometimes with the inclusion of additional close relatives such as aunts, uncles or cousins. Such families are relatively easy to collect, and they minimize the genetic heterogeneity that might be caused from multiple entries of susceptibility alleles in more extended pedigrees. Conceptually, the simplest situation arises when studying pairs of affected siblings. The methodology for analysing affected sib-pair data is well-established and has been used extensively first starting in the 1970s to investigate linkage between autoimmune disease and human leucocyte antigen (HLA) (Day and Simons, 1976; de Vries *et al.*, 1976; Penrose, 1935, 1946, 1953; Suarez *et al.* 1978).

In contrast to traditional lod score analysis, which assumes a genetic model for familial resemblance of a trait, the affected sib-pair method is 'nonparametric', in that no explicit assumptions are made about the mode of inheritance. Rather than assaying for recombination events between disease and marker, the tests of linkage are based on whether affected sib pairs share more than the expected number of marker alleles identical-by-descent (IBD) (two alleles are IBD if they are inherited from the same ancestor). The observed IBD distribution is compared to the expectation under independent segregation of the marker and trait. For independent segregation, each parental allele is transmitted with equal probability, and siblings inherit the same marker alleles 50% of time independently of affection status. When the data consist of independent affected sib pairs (i.e. each from a different family) with complete information on marker inheritance, the test for linkage can be made by a simple chi-square test comparing observed and expected IBD. In practice, a wide variety of approaches has been proposed for affected sib-pair tests of linkage (for examples, see references cited in Davis and Weeks, 1997). These utilize different methods for taking into account multiple affected siblings within a family, incomplete IBD information, different assumptions about the probabilities of IBD sharing in the presence of linkage, and different statistics for evaluating linkage.

Davis and Weeks (1997) and Abel *et al.* (1998) performed statistical comparison of the performance of several techniques and different computer implementations for single locus affected sib-pair analyses. Overall one of the best performances for both type I and II statistical error was obtained by analyzing affected sib-pair families as if they were segregating for a rare recessive disease with complete penetrance, while treating unaffected siblings and parents as having unknown disease phenotype irrespective of true disease status (Alcais *et al.*, 1999; Goring and Terwilliger, 2000a, 2000b, 2000c, 2000d; Satsangi *et al.*, 1996; Terwilliger and Goring, 2000; Trembath *et al.*, 1997). In this approach, the test statistic is simply the maximum logarithm of odds ratio (lod) score (MLS) found by estimating a recombination rate between the marker and the 'disease' locus [called a 'pseudo-marker' by Goring and Terwilliger (Goring and Terwilliger, 2000a, 2000b 2000c, 2000d, Terwilliger and Goring, 2000)]. However, here the recombination rate has no biological significance; it is simply a parameter that adjusts the IBD probabilities for affected siblings. Aside from its robust behavior, the pseudo-markers approach can be implemented easily in many of the existing programs for classical linkage analysis. Note that with this approach, there is no need to decide about weights for sibships with different numbers of affected offspring, as this is accounted for in the linkage analysis. Modifications have been proposed for multiple marker loci, study designs other than affected sib-pairs, and quantitative traits (Alcais and Abel, 1999; Alcais *et al.*, 1999; Goring and Terwilliger, 2000a, 2000b, 2000c, 2000d).

Whatever method is applied, the evidence for linkage of a specific marker in a set of affected sib pairs is most often assessed by a MLS statistic, which is analogous to a classical lod score in that its value is the $\log_{10}$ of the odds in favor of linkage (Risch, 1990). Traditionally, a MLS of 3.0 (odds of 1000 : 1 in favor of linkage) has been assumed as the threshold for declaring significant linkage for monogenetic disorders, in part because this corresponds to a genome-wide significance level, taking into account the testing of multiple markers throughout the whole genome, of about $P = 0.05$. A detailed discussion of critical limits for genome-wide significance of linkage under different study designs is provided by Lander and Kruglyak (1995). For complex diseases, it has become standard practice to designate regions of potential linkages with less stringent criteria of significance. The idea is that if one is carrying out genome-wide search, then it is appropriate to use a lower threshold (a MLS of perhaps 2.0) to indicate which areas of the genome should be investigated more thoroughly by typing more markers in the area and/or adding more sib pairs to the original study, or attempting replication in a different study (Brown *et al.*, 1994; Elston, 1992; Ghosh and Majumder, 2000; Guo and Elston, 2000a, 2000b). The former strategy is known as staged searching, where an initial genomic scan (Stage 1) is carried out at a low marker density and then those regions that contain elevated statistics are investigated by typing more markers (Stage 2).

A number of other practical issues arise when undertaking affected sib-pair linkage studies. First, it is very important to genotype the parents or unaffected siblings when possible. If parental genotypes are completely or partially absent, marker allele frequencies may have a profound effect on the test statistic; for example, as the IBD probability for related individuals that share a single rare allele is high, erroneously assuming that all marker alleles are equally frequent will generally inflate the test statistic in favor of linkage, if one or more of the alleles is more common in the general population (Ott, 1992). Therefore, marker frequencies should be estimated either from the data or from a matched control population prior to linkage analysis.

A second practical issue in applying is verification of putative relationships prior to linkage analysis. In general, any undetected nonpaternity or other pedigree errors reduces the power to detect linkage. There is a high probability of any errors being detected if parents, or three or more siblings have been genotyped in genome-wide linkage studies; computer programs such as PedCheck (O'Connell and Weeks, 1997) scan pedigree data for such inconsistencies, and provide information to aid in the correction of the putative family structures prior to analysis. However, for sib pairs, an erroneous relationship cannot be detected with certainty, unless genotypes on parents or other relatives are available, and a statistical framework must be applied to the totality of the marker data in order to evaluate if the relationship is correct or if the putative siblings are half-sibs or unrelated. Methods and computer programs are available for this, and should be applied to the data prior to linkage analysis (Boehnke and Cox, 1997; Douglas *et al.*, 2000; Goring and Ott, 1997; McPeek and Sun, 2000).

It is common practice to apply a variety of affected sib-pair methods for linkage analysis, including both multilocus and single locus techniques. Efficient algorithms for calculating multilocus IBD statistics in small families are now widely available, as for example, in the GeneHunter program (Kong and Cox, 1997; Kruglyak, 1996). Multilocus analysis overcomes to some extent lack of complete knowledge on IBD when parental genotypes are unknown, or when individual markers are not fully informative. Because multilocus methods are sensitive to genotype errors, and different methods have varying power to detect linkage, it is desirable to apply both single marker locus and multiple marker locus methods to all data sets. A general framework that allows incorporation of multilocus data and marker error under the 'pseudo-marker' approach cited above has been developed by Goring and Terwilliger (2000a, 2000b, 2000c, 2000d).

3. Power to detect linkage

Genome-wide linkage studies of affected sib pairs or nuclear families with multiple affected offspring have been widely adopted in an attempt to map susceptibility loci involved in complex disease with mixed success to date. In many instances, such as in studies of insulin-dependent diabetes mellitus (IDDM) and multiple sclerosis, the results from any single study have not provided strong evidence of linkage outside of the major histocompatibity complex region, and comparison of different studies has not pointed to clear replication of regions of potential linkage. In other diseases, the results have provided more compeling evidence for the localization of susceptibility loci. For example, a sib-pair study of obesity has led to strong evidence for linkage to human chromosome 10 (Hager *et al.*, 1998), whereas for inflammatory bowel disease high MLS has been found for several loci that have also been replicated in different studies (see below). What are the factors that may determine the success or failure of such investigations?

The ability of the affected sib-pair method to detect a disease susceptibility locus depends on the contribution the locus makes to family resemblance, which is often measured in terms of the increased risk to relatives of an affected proband as compared to the population prevalence (Bishop and Williamson, 1990; Risch, 1990b). For affected sib-pair studies, this can be measured by the sibling risk ratio λ_S of the risk to a sib of an affected proband versus the population prevalence (Risch, 1990b) [(e.g. IDDM, λ_S is about 15 (6%/0.4%) (Davies *et al.*, 1994)]. This λ_S is an overall risk ratio that summarizes the collective effect of all the disease loci plus any other nongenetic familial resemblance; higher λ_S indicates stronger familial effects. However, the power is a function of how many genes there are and how they interact with each other. If a complex disease is due to a number of susceptibility genes with alleles that act multiplicatively on the risk as described for example by Risch (1990b), and N loci are involved, with λ_i a risk ratio for the ith locus (i.e. the increased risk attributable to the probability that siblings are IBD at the ith susceptibility locus), then the overall $\lambda_S = \lambda_1\lambda_2\lambda_3\ldots\lambda_N - 1\lambda_N$ (Risch, 1990a, 1990b).

In the literature, the power to detect ith locus by linkage is then often measured as a function of the (unknown) magnitude of λ_i. For many diseases of interest in genetics, epidemiological data suggest that the overall sibling risk ratio will not be more than 2 or 3, and this is distributed over an unknown number of loci. Moreover, most estimated sibling recurrence risks reflect the effects of both genetic and common environmental factors that will weaken the power to detect linkage. It is easily deduced from power calculations in the literature that the required sample sizes to achieve reasonable power are usually much larger than 100 or so affected sib pairs (as contained in many published studies). Investigators should consider collecting 1000 or more affected sib pairs when tackling the genetics of common diseases. Although the task is daunting, sample sizes in this range can be obtained in many instances if networks of clinicians and scientists collaborate, as is common in other areas of research requiring large-scale epidemiological investigations.

Several comments are in order about the evaluation of power. Although power is often measured in terms of λ_i as described, it is important to realize that the underlying calculations are model dependent: the power is a direct function of the expected IBD probabilities at each susceptibility loci, and the relationship between the IBD probabilities and λ_i depends upon the assumptions about the action of the susceptibility alleles at and between the loci. Rybicki and Elston (2000) show how the power can vary dramatically for the same λ_i under different, biologically reasonable hypotheses regarding allele action. Several statistical methods have been developed that explicitly take interaction between putative susceptibility loci into account (Cordell *et al.*, 1995, 1999, 2000; Farrall, 1997; Lathrop and Ott, 1990), and these could increase power. Stratification of the sample based on linkage at an identified locus is often attempted but does not necessarily increase power, in addition to requiring cautious interpretation given the large number of potential subdivision of the data (see Leal and Ott, 2000 for a discussion).

Many of the power estimates that are contained in the literature provide optimistic assessments by assuming that the marker and the disease susceptibility locus are completely linked, and that markers are completely informative (*i.e.* that the IBD status of the siblings' alleles can be determined). Violation of these assumptions reduces power (or increases the minimum λ_S we can hope to detect) (Risch, 1990c); for example, Risch (1990b) found that the power to detect linkage using the MLS approach on a sample of 300 sib pairs decreased by about 50% at a recombination fraction of approximately 0.10. The effects of recombination can be reduced by staged screening as described above and analysis of multiple markers (Fulker and Cardon, 1994; Fulker *et al.*, 1995; Goldgar, 1990; Olson, 1995). Inclusion of parents and other siblings increases the power by providing more accurate estimates of IBD, as well as having other benefits such as those described above. It is very useful to examine candidate genes because there is, by definition, no recombination to take into account. The frequency of false-positive linkages increases when markers are not fully informative; this is because the dependence between observations at linked markers is lower when they are informative for partially overlapping family subsets. Thus, in real data where the information content is usually less than 100%, it would be appropriate to use a higher threshold of significance than the theoretical values based on fully informative markers as calculated in the literature.

Finally, another factor to take into account when evaluating the probability to detect linkage is the number of unlinked susceptibility loci that contribute to the overall sibling recurrence risk. Obviously in a genome-wide linkage study, the power to detect at least one of several loci each contributing λ to the overall sibling risk will be greater than the probability of detecting a specific locus with this contribution (the assumption in most power calculations). On the other hand, the expected number of families required to reproduce the detection of a specific locus is much greater than the number required for initial detection for one of many equally contributing loci (Suarez *et al.*, 1994). In other words, it should not be surprising to find many instances where statistically significant true-positive linkages found in one study are not replicated in a second study of comparable size, even when the affected sib pairs are drawn from the same population to reduce heterogeneity. Of course, clinical or population heterogeneity may also have a major impact on replication.

4. Inflammatory bowel disease

Inflammatory bowel diseases serve as interesting example of the power and limits of the present affected sib-pair study designs. These disorders are divided into Crohn's disease (CD) and ulcerative colitis (UC), which are distinguished on the basis of clinical features. Although the two diseases sometimes occur in the same family, suggesting some common genetic factors, there is sufficient distinction to make a reasonable subdivision, which has greatly aided genetic analysis. Prevalence is 100–200 per 100 000 in populations of European origin with λ_S of about 35 for CD and 15 for UC. Association with HLA alleles has been reported, and linkage has been confirmed in several studies although the λ_S attributed to the locus is relatively small. The combination of high recurrence risk and the small contribution of the HLA locus suggest that inflammatory bowel diseases are good candidates for a search for other genetic susceptibility factors, in contrast to some autoimmune disorders for which HLA appears to account for the majority of the recurrence risk.

CD affected sib-pair families were used for a genome-wide linkage study by Hugot *et al.* (1996) who found evidence of linkage to chromosome 16 at a locus designated *IBD1* (MLS = 3.2). A second affected sib-pair study by Satsangi *et al.* (1996) found significant linkage to chromosome 12 (*IBD2*; MLS = 5.5), and suggestive evidence of linkage on chromosomes 3 and 7, using a two-staged screening approach. The difference in results between the two studies could be explained in part by the inclusion of both UC and CD patients in the latter, and by the variable coverage of the genome provided by different marker panels. When markers from the region of *IBD1* were specifically selected, the CD families from the second study also gave evidence for linkage at *IBD1*, while the heterogeneity between results for CD and UC was statistically significant (Parkes *et al.*, 1996). Similar results were obtained by others on a separate family panel (Ohmen *et al.*, 1996). Additional studies have also confirmed linkage to *IBD1* principally in CD (Brant *et al.*, 1998; Cavanaugh *et al.*, 1998; Cho *et al.*, 1998; Curran *et al.*, 1998; Hampe *et al.*, 1999), but sometimes also for UC (Annese *et al.*, 1999; Miza *et al.*, 1998). Linkage to *IBD2* has also been reported in at least four family cohorts in addition to that of the first report (Curran *et al.*, 1998; Duerr *et al.*, 1998; Hampe *et al.*, 1999). *IBD1* and *IBD2* are now considered in the literature as confirmed linkages, although some studies did not report evidence of linkage to either one or both of these loci (Annese *et al.*, 1999; Brant *et al.*, 1998; Cho *et al.*, 1998; Duerr *et al.*, 2000; Rioux *et al.*, 1998).

The λ_S estimated to be contributed by *IBD1*, *IBD2* and *IBD3* (the designation for the HLA linkage) does not explain the totality of the sibling recurrence risk for inflammatory bowel disease, motivating continued searches for other susceptibility loci. Thus, Duerr *et al.* (2000) undertook a linkage study of 127 affected relative pairs (94 full sibling pairs, plus a number half-sibs, avuncular and other relative pairs) from 62 families with a very high density of microsatellite markers (751 markers with average spacing of 5 cM). They found evidence of linkage to chromosome 14 (MLS = 3.6) in a region that had previously been reported with suggestive evidence of linkage in an independent study of UC (MLS = 2.8) (Ma *et al.*, 1999). Because of this confirmation, this locus has been designated *IBD4*. Suggestive evidence of linkage has been found as reported for many other loci in the studies referenced above.

Inflammatory bowel diseases have proved a good target for linkage studies as shown by the findings of significant evidence of linkage with replication in different cohorts. The detection of loci has been aided by the division of patients into CD and UC groups on clinical grounds. Despite the relative success, the results of different studies show considerable heterogeneity as would be expected based on limited sample sizes, all being less than 200 affected sib pairs. Linkage studies of much larger cohorts are undoubtedly necessary to clarify the presence of different susceptibility loci. A sufficiently large cohort could be obtained by pooling the resources available in the different centers that have contributed to the work cited above, to the obvious benefit of all.

5. Analysis of extended pedigrees

Extended pedigrees provide the most powerful means for localization of monogenic disease, or traits that exhibit locus heterogeneity. Given some of the limitations for detecting linkage with affected sibs or small families, it can be useful also to study multifactorial traits in extended pedigrees in which the segregation pattern suggests the effect of a single locus, or the action of a small number of loci. This has been successful for mapping and identification of genes involved in a number of different diseases, such as maturity onset diabetes of the young (Winter *et al.*, 1999), and metabolic syndrome with insulin resistance associated with diabetes and hypertension (Bamoso *et al.*, 1999). Knowledge of the susceptibility genes involved in such traits often provides remarkable biological insight and important information for medical application even when the genes are not implicated in more common forms of these disorders.

One approach to exploiting extended pedigrees when single-gene segregation cannot be established with certainty is the affected-pedigree-member (APM) method of linkage analysis (Weeks and Lange, 1988, 1992), which measures marker similarity between all affected individuals that are genotyped by identity-by-state (i.e. similar alleles) rather than in terms of IBD. Thus, the APM method requires genotyping of only the affected individuals, and has proven to be quite useful in studying diseases of late onset, such as Alzheimer's disease (Pericak-Vance *et al.*, 1990, 1991). With the advent of Markov Chain Monte Carlo (MCMC) methods, multilocus estimates of IBD probabilities in very large extended pedigrees can now be achieved even with a large number of consanguinity loops (an important consideration in genetic studies of isolates), taking account of all the marker information available on both affected and nonaffected individuals (Heath, 1997; Kong *et al.*, 1993; Sobel and Weeks, 1998). In addition to the use of MCMC techniques for implementing classical linkage analysis (i.e. under a specific genetic model), several statistics have been proposed that compare functions of the estimated IBD to expectations under the absence of trait linkage in a model-free manner (Davis *et al.*, 1996).

With these statistics, theoretical large-sample *P* values are usually not appropriate as the asymptotic approximations may only be true as the number of independent families becomes large (usually not the case). A similar situation is found for quantitative trait loci, for example, in the all-relative-pairs test of Olson and Wijsman (Olson, 1994; Olson and Wijsman, 1993) and for the weighted least-squares robust sib-pair linkage test in samples of fewer than 300 sib pairs (Amos *et al.*, 1989). A solution to this common problem is to compute 'empirical' *P* values by simulation. An empirical *P* value is determined by simulating a large number of sets of data under the assumption of no linkage, calculating a statistic for each set, and then determining where in the distribution of simulated statistics the original statistic falls. In general, we recommend the use of empirical *P* values for all test statistics whenever possible, particularly when samples are small.

6. Quantitative traits

Quantitative traits have several advantages in genetic studies compared to a simple classification based on the presence or absence of a disease. Many diseases can be defined in terms of an underlying quantitative liability scale, such as blood pressure for hypertension or glucose tolerance for diabetes. Such traits provide a measure of disease severity, but can also be studied in unaffected family members or in individuals with extreme values of the phenotype but who are not clinically diagnosed with disease. For example, young adults who have extreme blood pressure for their age and sex may be useful in linkage analysis although they may not yet have been diagnosed with hypertension. Another advantage is that a specific quantitative trait may be influenced by a small number of factors, whereas a disease classification often represents a genetically and environmentally more heterogeneous phenotype. In these instances, the quantitative variable will be correlated more strongly with polymorphisms of an underlying gene, and therefore provide a better phenotype both for initial mapping purposes, and also for identifying the causative susceptibility variants at a locus.

A quantitative trait may be used as a selection criterion to increase power to detect linkage in an affected or discordant sib-pair strategy. For example, selecting affected individuals in the upper 5% or 1% extreme of the blood pressure distribution serves to increase the sibling risk ratio compared to a clinical definition of hypertension, which may be diagnosed in 10% of the general population of middle-age and older adults. A similar approach in which a stringent criterion (e.g. 1% extreme) is used as a criterion to select probands from clinics, whereas their relatives may be considered as affected with a less extreme phenotype (e.g. 5% extreme) also increases power, and has practical advantages for sample collection.

Considerable attention has been given to the power of discordant sib-pair analysis, in which siblings are chosen in opposite extremes of the distribution of a quantitative trait; this sampling strategy, although often difficult to implement in practice, can in theory reduce dramatically the number of sibling pairs or families that need to be genotyped in order to obtain significant evidence of linkage (Eaves, 1994; Risch and Zhang, 1995). In general, when studying a quantitative phenotype in an existing family panel it is useful to undertake simulation studies to determine which families provide most of the evidence for linkage and to concentrate on characterization of these. However, selection of informative families will be difficult if several quantitative phenotypes that are only partially correlated are being investigated simultaneously in the same families.

Direct linkage analysis of the quantitative variable is an alternative to its transformation into a qualitative phenotype. Selection for families in which some individuals are towards the extremes of the quantitative distribution is still necessary to achieve power; for unselected families the number of families necessary to detect even loci accounting for a high proportion of the trait variance (say 10–20%) is several thousand (Page *et al.*, 1998). One approach to the analysis of quantitative linkage data is to apply the so-called 'mixed model' which incorporates a major gene, usually under the assumption of two alleles; maximum likelihood estimates are obtained for three genotype means, a polygenic or other familial component of variance independent of the major gene, and residual environmental or error variance, recombination rates and potentially other parameters (Lalonel *et al.*, 1983) with programs such as REGRESS (Bonney *et al.*, 1998) and FINESSE (O'Connell *et al.*, 1998); ascertainment corrections can be incorporated to adjust for selection on the trait. For example, such approaches have been successfully used to map a major locus interacting with the β-globin gene to determine levels of fetal hemoglobin in adults (Craig *et al.*, 1996; Garner *et al.*, 1998).

Several methods have been developed that involve variance component analysis without incorporating a parametric model for the hypothesized major gene (Almasy *et al.*, 1996; Amos, 1994, Fulker and Cardon, 1994). Essentially, these methods involve first obtaining estimates of IBD statistics for relative pairs, and then estimating correlations between relatives that are a function of the IBD. Although this requires no hypothesis about the number of alleles at the major gene locus, the estimates and test statistics are usually obtained under the assumption of a multivariate normal distribution of the trait within families; care is needed as multivariate normality can be violated by a number of factors, in particular by ascertainment on the trait values. The assumption of mutivariate normality can be relaxed by using a multivariate *t* distribution (see Allison *et al.*, 1999 for a discussion of some of the issues involved). An interesting application of variance component analysis to detect linkage to quantitative levels of leptin is given in Comuzzie *et al.* (1997).

A popular model-free framework for sib-pair linkage analysis of quantitative traits is based on the Haseman–Elston approach (Elston, 2000; Haseman and Elston, 1972). This consists of regression of the square of the trait difference on the IBD at the marker locus for sibling pairs; the slope of the regression line is expected to be zero in the absence of linkage whereas it will be negative in the presence of linkage (because siblings are then more similar in phenotype when they have inherited the same alleles from their parents). A nonparametric approach based on the relationship between IBD and the ranks of the sib phenotype differences, rather than the squared difference, has been developed and implemented in the program MAPMAKER/SIBS (Kruglyak and Lander, 1995).

7. Association and linkage disequilibrium

The comparison of allele or genotype frequencies between patients and controls (association) is a second strategy for detection of disease susceptibility genes (Nöthen *et al.*, 1993; Owen and McGuffin, 1993). Association can be found either with functional genetic variants that have biological consequences related to disease, or with other variants that are in linkage disequilibrium with these (Parsian *et al.*, 1991). The advent of technologies that allow rapid discovery of genetic variants, such as SNPs, has greatly expanded the potential for such investigations. Considerable attention has focused recently on scope and feasibility for both genome-wide association and linkage studies using SNPs as a source of markers (see Risch and Teng, 1998; Terwilliger and Weiss, 1998; Xiong and Jin 1999). Although it is easy to find situations in which association will have greater power than linkage to detect susceptibility loci (for example when the susceptibility is due to a frequent allelic variant that is assayed directly), in practice the relative power depends very much on assumptions, such as the number of susceptibility variants at a locus, the magnitude of their linkage disequilibrium with other markers, and the comparative frequencies of the marker and susceptibility variants. It is more realistic to think of linkage and association as complementary approaches; indeed, association studies are often attempted as an approach to isolate a susceptibility gene in a region that has been identified by linkage.

Association arises either from the direct influence of a DNA variant on a phenotype, or from linkage disequilibrium creating combinations of alleles that occur more frequently than would be expected under random association. Linkage disequilibrium in turn can be due to a variety of causes, including recent admixture of populations with different allele frequencies, selection in favor of a specific allele (so that the haplotypes containing that allele increase in frequency), genetic drift or population bottlenecks, and new mutations (which create disequilibrium with closely linked markers). Linkage disequilibrium decays with time in proportion to the recombination fraction between the loci. Disequilibrium due to admixture, selection or drift between unlinked loci decays very rapidly, whereas disequilibrium between closely linked loci decays much more slowly. Thus, it is hoped that linkage disequilibrium between a marker and a trait locus will lead to the identification of a disease susceptibility gene in the near vicinity of the marker.

For single gene disorders, mapping by linkage disequilibrium is often applied to genetically isolated populations, since then it is much more likely that there is only one (or few) ancestral haplotype on which the disease mutation arose, and, if the population is relatively new, linkage disequilibrium may extend over greater distances than in an older population (Hastbacka *et al.*, 1992, 1994; Lander and Schork, 1994). There is now a wide variety of published methods for statistical analysis of linkage disequilibrium and haplotypes to determine the precise location of rare disease mutations (Graham and Thompson, 1997; Lam *et al.*, 2000; McPeek and Strahs, 1999; Terwilliger, 1995). For multifactorial traits, the methods must be modified to deal with complications such as common susceptibility variants, multiple susceptibility variants at the same locus, epistatic interaction between different genes, etc.

Population stratification either due to admixture or inappropriate matching of patients and controls is the major confounding factor in association studies. One approach to avoid the problem is by haplotype relative risk (HRR), where comparison is made between alleles that were or were not transmitted to affected offspring from their parents (Falk and Rubenstein, 1987; Ott, 1989; Schaid and Sommer, 1993; Spielman *et al.*, 1993; Terwilliger and Ott, 1992). HRR ensures that the control and case populations are well-matched (Falk and Rubenstein, 1987). Such family based association methods have been subject to numerous theoretical developments in recent years, many of which extending HRR or the transmission disequilibrium test (see below) to take into account such issues as incomplete information on parental genotypes.

From a practical viewpoint it is necessary to weigh the difficulties of obtaining the samples needed for family based association studies (for example, affected individuals with living parents may be extremely difficult to find) against the potential risk of heterogeneity in case–control studies, where the sample sizes can be made much larger. Probands for family based association studies may be unrepresentative of the general patient population, particularly for diseases of late onset if they have living parents. In the context of these difficulties, Devlin and Roeder (Bacanu *et al.*, 2000; Devlin and Roeder, 1999) have proposed a Baysesian ('genomic control') method of particular interest in which an internal control for potential population heterogeneity can be obtained from a case–control cohort. They suggest the calculation of an association statistic for markers throughout the genome under the assumption that any population heterogeneity should affect all markers. With appropriate statistics, this overall distribution can serve as a control against which disease associated markers are detected as outliers. The genomic control method does not require a Bonferroni correction for the number of markers tested, and it can be used also for candidate gene analysis.

In some instances linkage can be detected only when taking linkage disequilibrium simultaneously into account. For example, IDDM was shown to be associated with polymorphisms of the insulin gene region, but no evidence of significant linkage could be found as measured by identity-by-descent sharing among affected sib pairs. However, Julier *et al.* (1991) showed that many of the parents were homozygous for the markers that were associated with susceptibility, and thus were presumably noninformative in linkage analysis. Strong evidence for linkage was revealed by examining only the meioses from heterozygous parents, for which affected offspring exhibited IBD significantly greater than the expectation of 0.5 under the absence of linkage. This approach to linkage which takes into account the presence of linkage disequilibrium has been formalized as the transmission–disequilibrium test (TDT) (Spielman *et al.*, 1993, 1994), and applied to a number of other loci. When considering only parent–offspring trios, TDT is similar to HRR. However, it is important to note that when applied to affected sibling TDT provides a test of linkage, not of association. Thus, for example, sib-pair linkage and TDT results should not be presented as independent tests of linkage and association for the same set of families.

In general, it would be desirable to provide an integrated approach to model-free analysis that would allow simultaneous evaluation of linkage and association while incorporating data from different types of study designs such as affected sib pairs, parent–offspring trios and case–control cohorts. Such a framework was introduced in Trembath *et al.* (1997) and subsequently developed by Goring and Terwilleger (2000a, 2000b, 2000c, 2000d). The approach involves an extension of the 'pseudo-marker' linkage test described above for affected sib pairs. As described previously, the linkage component can be evaluated by calculating a lod score for a disease locus with two alleles, although the test itself is not model dependent. Association is incorporated by allowing for linkage disequilibrium between the marker and the alleles (denoted D and d) at the putative disease locus. For a marker locus with n alleles, under the hypothesis of association and linkage (H_2), we maximize the likelihood over the recombination rate, the frequencies of n–1 haplotypes associated with D and the frequencies of the n–1 haplotypes associated with d (n–1 in each case because the overall frequencies of D and d are constrained; this gives a total of $2n$–1 parameters). For the hypothesis of linkage without association (H_1), the likelihood is maximized over the recombination rate and the marker allele frequencies, assuming linkage equilibrium with D and d which have fixed allele frequencies (a total of n parameters). Finally, under the hypothesis of absence of linkage (H_0), the likelihood is maximized over the marker allele frequencies only, with recombination fixed at 0.5 (n–1 parameters).

The likelihood ratio statistics to test for linkage and association are $-2 \ln (L_1/L_0)$ and $-2 \ln (L_2/L_1)$ respectively, where L_i is the likelihood under the ith hypothesis. The significance of these test statistics can be judged by comparison to chi-square distributions with n–1 and 1 degrees of freedom, respectively. One of the advantages of the approach is that it can be implemented readily with minor modifications to existing programs for linkage analysis. An example of its application for testing linkage and association of susceptibility to psoriasis to the HLA locus with the LINKAGE programs is given in Trembath *et al.* (1997).

A related area that has stimulated extensive methodological investigation recently is linkage disequilibrium mapping with quantitative trait loci. This includes methods that examine the direct relationships between haplotypes or genotype combinations with phenotypes, assuming complete identification of causative variants (Boerwinkle *et al.*, 1986), and TDT extended to quantitative traits (Allison, 1997). Techniques are also available that take into account simultaneously linkage and partial linkage disequilibrium with polymorphisms of a gene in either parametric (Keavney *et al.*, 1998), or variance-components approaches (Abecasis *et al.*, 2000; Fulker *et al.*, 1999) similar to those discussed above. The variance component method of Fulker *et al.* (1999) and Abecasis *et al.* (2000) provides a means of controling for population stratification that may give rise to apparent association independently of linkage. This is done by partitioning the mean effect for the test locus into two terms; one is a nuisance parameter that absorbs the effects of population stratification, while the other depends only on the linkage disequilibrium at the test locus. The variance partition includes a component that enters into the between-relative covariance as a function of IBD at the test locus; this component of variance is nonzero if the linkage disequilibrium between the causative gene variants and the markers that have been tested is incomplete.

8. Example of the DCP1 gene

Leaving aside the question of family-based methods that are specific to genetic studies, the investigation of association in multifactorial diseases are in most ways similar to classical epidemiological investigations. They are subject to parallel needs of careful study design and large sample sizes in order to detect relatively modest differences of risk and potential interactions. As cited above, one advantage of genetic studies is the possibility to investigate intermediate phenotypes that may be related more directly to gene variation than a more complex disease phenotype. Studies of the relationship of the DCP1 gene, which encodes the angiotensin-I converting enzyme (ACE), with myocardial infarction (MI) and quantitative levels of plasma ACE illustrate these themes.

The example of DCP1 and MI shows the need for careful control of effects of subset selection and publication bias, and the potential for meta-analysis of multiple data sets. A relationship between DCP1 and MI was first postulated based on results from a case–control study of the ECTIM (Etude Cas-Tèmoin de l'Infarctus du Myocarde) which found an association with an insertion–deletion polymorphism (I/D) in intron 16 by comparison of genotype frequencies in 610 male patients aged 65 years or less who had survived MI for at least 3 months, and 733 control individuals without a history of coronary disease (Cambien *et al.*, 1992). In Caucasians, the D allele and the DD genotype occur with frequencies of 0.50–0.55 and 0.25–0.30 respectively. The ECTIM study found a modest increase in the risk ratio (RR) associated with DD genotype: 1.34 [95% confidence interval (CI) 1.05–1.70) overall, and a particularly strong association (RR = 3.2; 95% CI, 1.7–5.9) in one retrospectively defined subgroup of patients otherwise at low risk of disease (body mass index < 26 kg m^{-2} and apolipoprotein B concentrations < 1.25 g l^{-1}). Subsequently, almost 50 studies have been published involving the investigation of this polymorphism specifically in Caucasian MI patients and controls, and many other cardiovascular diseases and other populations (Keavney *et al.*, 2000; Stoessen *et al.*, 1997). The results of these investigations are contradictory, ranging from evidence that strongly supports association to its absence.

Meta-analysis, in which the results of different studies are combined in an appropriate manner, provides key information for interpreting this heterogeneity (Keavney *et al.*, 2000). For MI in Caucasians, the results of 35 smaller studies (200 or fewer MI patients) give a combined RR of 1.57 (99% CI, 1.38–1.78), whereas in 14 larger studies (>200 MI patients) the combined RR is 0.99 (99% CI, 0.90–1.08) suggesting a strong publication basis favoring positive results in small samples (*Figure 1*). It should also be noted that genotyping with knowledge of case–control status is more likely to result in a positive bias, particularly in studies involving a small number of samples; it is desirable that this should be controled by blind genotyping. Most of the published studies have been too small to test the hypothesis that the DD genotype had a particularly strong effect in a low risk group. However, a recent association study (Keavney *et al.*, 2000) has addressed some of the questions raised by the above results by examining the I/D polymorphism in a cohort of approximately 5000 MI patients and 6000 controls from ISIS-3 (Third International Study of Infarct Survival), including a low risk subgroup consisting of 836 patients and 3253 controls defined by the same criteria as used in ECTIM (with 10-fold more patients for the subgroup analysis than in the original report). The RR in ISIS was 1.10 (95% CI, 1.00–1.21) overall, and 1.06 (95% CI, 0.94–1.18) in the low risk subgroup. Thus, we can conclude that the data from large studies of MI are compatible with only a very modest increase in risk associated with the DD genotype, and that it is unlikely that DCP1 has a stronger effect in the low risk subgroup.

In contrast with the difficulties of interpreting disease association, the relationship of DCP1 with quantitative measures of plasma ACE levels can be established without ambiguity (McKenzie *et al.*, 1995; Tiret *et al.*, 1992). This

Published studies by effective size	No of MI cases	% DD genotype Cases	Controls
Original study: ECTIM	610	32%	27%
35 small studies (< 200 cases each)	3578	34%	26%
14 larger studies (200+ cases each)	6863	27%	27%
Largest study: ISIS	4629	29%	28%

Risk Ratio and 99% CI

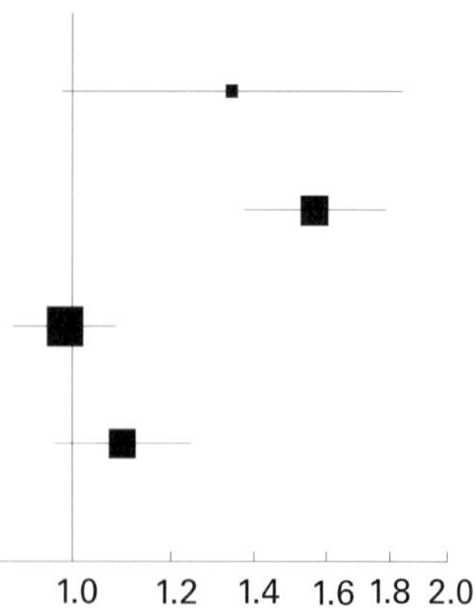

Figure 1. Meta-analysis of association between the DD genotype of the DCP1 gene and MI. Reprinted from Keavney (2000) *Nature* **23**: 270–271, by permission of Nature Publishing Group, New York.

Polymorphism (Rieder *et al.*, 1999)	(Villard *et al.*, 1996)	Haplotype A-1	A-2	B	C-1	C-2
–2851	T-5491C	T	T	C	T	T
–2826	A-5466C	A	A	C	A	A
–1252	T-3892C	T	T	C	T	T
2400	A-240T	A	A	T	A	A
2547	T-93C	T	T	C	T	T
3872		A	A	G	A	A
4504		G	G	C	G	G
6435		G	G	A	G	G
7831		A	G	G	G	G
8128	T1237C	C	C	T	C	T
8968		C	C	T	T	T
12257	G2215A	G	G	A	A	A
13145		C	C	A	A	A
13230		A	A	G	G	G
13336		A	A	T	T	T
13338		T	T	G	G	G
14094	I/D	I	I	D	D	D
14521	G2350A	A	A	G	G	G
15214		A	A	G	G	G
18222		A	A	G	G	G
20833		C	C	T	T	T
22251		C	C	T	T	T
23945	$4656(CT)_{3/2}$	3	3	2	2	2

Figure 2. Representative variant sites for common Caucasian haplotypes of the DCP1 gene showing clades A, B and C. Positions of the variant sites are shown with reference to the numbering of both Rieder *et al.* (1999) and Villard *et al.* (1996). Reprinted from Farrall *et al.* (1999) *Nature* **23**: 270–271, by permission of Nature Publishing Group, New York.

relationship provides a powerful tool for determining variants of the gene that could have a functional effect on plasma ACE as it is likely that any association with the I/D polymorphism in intron 16 is due to linkage disequilibrium rather than a direct casual effect. Statistical analysis shows an additive effect of the D allele for increasing serum ACE levels, and the genotype accounts for more than 35% of the variance of quantitative measurements of ACE after adjustment for age and sex. A catalog of frequent variants across a genomic region of 24 kb containing DCP1 has been obtained by sequencing of multiple individuals, and more than 37 SNPs have been described that are in strong linkage disequilibrium with the D allele (Reider *et al.*, 1999; Villard *et al.*, 1996). Based on genetic analysis, it is likely that this catalog of SNPs includes the principal functional variant or variants that have an effect on plasma ACE, and by implication, the most interesting variants to investigate for potential consequences in disease.

How can genetics help to refine the localization of the functional variants? In Caucasians, DCP1 variants fall into two principal haplotypes because of the strong linkage disequilibrium (Farrall *et al.*, 1999; Keavney *et al.*, 1998). One of

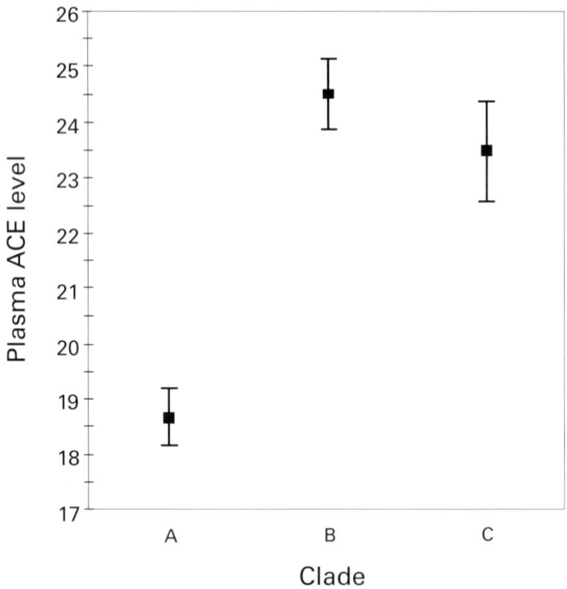

Figure 3. Means and standard deviations for standardized serum ACE levels associated with DCP1 haplotypes for clades A, B and C. ACE levels have been adjusted for age and sex. Units of ACE are arbitrary after adjustment. The results are from combined linkage and segregation analysis with the methods described by Keavney (1998). Reprinted from Keavney *et al.* (2000) *Lancet* **355**: 434–442, by permission of The Lancet.

these (denoted here as 'clade A') carries the I allele of the I/D polymorphism and other variants associated with low levels of ACE, whereas the other (clade B) carries the D allele and other variants associated with low levels of ACE. Together clades A and B have more than 55% frequency in Caucasians, whereas the third most common Caucasian haplotype (clade C) has a frequency of approximately 8%. The clade C haplotype may have originally been produced by a recombination event between clade A and clade B, as it contains an interesting combination of DCP1 variants: from the 5′ region to base 6435 of the gene it is largely identical to clade A, and from base 8968 through the 3′ region it is largely identical to clade B (and thus carries the D allele in intron 16), with a region of ambiguity between (*Figure 2*).

To aid in determining where the functional variants of the gene reside, it would be helpful to know how clade C haplotypes behave with respect to plasma ACE levels. The answer can be found by estimating the effects of the three haplotypes under an additive genetic model using approaches that have been described for quantitative linkage and association analysis above. The results show clearly that clade C haplotypes are associated with high levels of ACE that are statistically indistinguishable from the levels associated with clade B haplotypes with present data; the effects of both clade B and clade C are significantly different from those of clade A haplotypes (*Figure 3*). Thus, this refined haplotype analysis suggests that the major functional variants in DCP1 reside in the 3′ region. Further analysis linkage disequilibrium analysis in other populations is expected to refine further the region that is likely to contain the principal variant or variants of DCP1 that are functionally related to ACE.

9. Conclusion

The technology and information necessary for effective approaches to the genetics of multifactorial disease are rapidly being put into place. It can be anticipated that systematic exploration of many thousands of genes will be possible over the next few years, and the potential of genetics to reveal novel relationships between gene and phenotype will grow accordingly. A wide range of statistical methodologies is also available for genetic studies of multifactorial disease. Important factors for success in applying these approaches will be the collection of good clinical and phenotype information, and the adoption of study designs that are appropriate to the goals of genetic epidemiological investigations. In this regard, we can expect growing and fruitful contacts between epidemiologists and geneticists as genetics becomes a key element in epidemiological investigations and clinical trials.

References

Abecasis, G.R., Cardon, L.R. and Cookson, W.O. (2000) A general test of association for quantitative traits in nuclear families. *Am. J. Hum. Genet.* **66**: 279–292.

Abel, L., Alcais A. and Mallet, A. (1998) Comparison of four sib-pair linkage methods for analyzing sibships with more than two affecteds: interest of the binomial maximum likelihood approach. *Genet. Epidemiol.* **15**: 371–390.

Alcais, A. and Abel, L. (1999) Maximum-likelihood-binomial method for genetic model-free linkage analysis of quantitative traits in sibships. *Genet. Epidemiol.* **17**: 102–117.

Alcais, A., Philippi, A. and Abel, L. (1999) Genetic model-free linkage analysis using the maximum-likelihood-binomial method for categorical traits. *Genet. Epidemiol.* **17(suppl 1)**: S467–S472.

Allison, D.B. (1997) Transmission-disequilibrium tests for quantitative traits. *Am. J. Hum. Genet.* **60**: 676–690.

Allison, D.B., Neale, M.C., Zannolli, R., Schork, N.J., Amos, C.I. and Blangero, J. (1999) Testing the robustness of the likelihood-ratio test in a variance- component quantitative-trait loci-mapping procedure. *Am. J. Hum. Genet.* **65**: 531–544.

Almasy, L., Hixson, J.E., VandBerg, J.L., MacCluer, J.W. and Blangero, J. (1996) Multipoint quantitative trait linkage in general pedigrees. *Am. J. Hum. Genet.* **59**: A173.

Amos, C.I. (1994) Robust variance-components approach for assessing genetic linkage in pedigrees. *Am. J. Hum. Genet.* **54**: 535–543.

Amos, C.I., Elston, R.C., Wilson, A.F. and Bailey-Wilson, J.E. (1989) A more powerful robust sib-pair test of linkage for quantitative traits. *Genet Epidemiol* **6**: 435–449.

Annese, V., Latiano, A., Bovio, P. *et al.* (1999) Genetic analysis in Italian families with inflammatory bowel disease supports linkage to the IBD1 locus – a GISC study. *Eur. J. Hum. Genet.* **7**: 567–573.

Bacanu, S.A., Devlin, B. and Roeder, K. (2000) The power of genomic control. *Am. J. Hum. Genet.* **66**: 1933–1944.

Barroso, I., Gurnell, M., Crowley, V.E. *et al.* (1999) Dominant negative mutations in human PPARgamma associated with severe insulin resistance, diabetes mellitus and hypertension [see Comments]. *Nature* **402**: 880–883.

Bishop, D.T. and Williamson, J.A. (1990) The power of identity-by-state methods for linkage analysis. *Am. J. Hum. Genet.* **46**: 254–265.

Boehnke, M. and Cox, N.J. (1997) Accurate inference of relationships in sib-pair linkage studies. *Am. J. Hum. Genet.* **61**: 423–429.

Boerwinkle, E., Chakraborty, R. and Sing, C.F. (1986) The use of measured genotype information in the analysis of quantitative phenotypes in man. I. Models and analytical methods. *Ann. Hum. Genet.* **50**: 181–194.

Bonney, G.E., Lathrop, G.M. and Lalouel, J.M. (1988) Combined linkage and segregation analysis using regressive models. *Am. J. Hum. Genet.* **43**: 29–37.

Brant, S.R., Fu Y., Fields, C.T. *et al.* (1998) American families with Crohn's disease have strong evidence for linkage to chromosome 16 but not chromosome 12 [see Comments]. *Gastroenterology* **115**: 1056–1061.

Brown, D.L., Gorin, M.B. and Weeks, D.E. (1994) Efficient strategies for genomic searching using the affected-pedigree-member method of linkage analysis. *Am. J. Hum. Genet.*, **54**: 544–552.

Cambien, F., Poirier O., Lecerf, L. *et al.* (1992) Deletion polymorphism in the gene for angiotensin-converting enzyme is a potent risk factor for myocardial infarction. *Nature* **359**: 641–644.

Cavanaugh, J.A., Callen, D.F., Wilson, S.R. *et al.* (1998) Analysis of Australian Crohn's disease pedigrees refines the localization for susceptibility to inflammatory bowel disease on chromosome 16. *Ann. Hum. Genet.* **62**: 291–298.

Cho, J.H., Nicolae, D.L., Gold, L.H. *et al.* (1998) Identification of novel susceptibility loci for inflammatory bowel disease on chromosomes 1p, 3q, and 4q: evidence for epistasis between 1p and IBD1. *Proc. Natl Acad. Sci. USA* **95**: 7502–7507.

Comuzzie, A.G., Hixson, J.E., Almasy, L., Mitchell, B.D., Mahoney, M.C., Dyer, T.D., Stern, M.P., MacCluer, J.W. and Blangero, J. (1997) A major quantitative trait locus determining serum leptin levels and fat mass is located on human chromosome 2. *Nature Genet.* **15**: 273–276.

Cordell, H.J., Todd, J.A., Bennett, S.T., Kawaguchi, Y. and Farrall, M. (1995) Two-locus maximum lod score analysis of a multifactorial trait: joint consideration of IDDM2 and IDDM4 with IDDM1 in type 1 diabetes. *Am. J. Hum. Genet.* **57**: 920–934.

Cordell, H.J., Jacobs, K.B., Wedig, G.C. and Elston, R.C. (1999) Improving the power for disease locus detection in affected-sib-pair studies by using two-locus analysis and multiple regression methods. *Genet Epidemiol* **17(suppl 1)**: S521–S526.

Cordell, H.J., Wedig, G.C., Jacobs, K.B. and Elston, R.C. (2000) Multilocus linkage tests based on affected relative pairs. *Am. J. Hum. Genet.* **66**: 1273–1286.

Craig, J.E., Rochette, J., Fisher, C.A., Weatherall, D.J., Marc, S., Lathrop, G.M., Demenais, F. and Thein, S. (1996) Dissecting the loci controlling fetal haemoglobin production on chromosomes 11p and 6q by the regressive approach. *Nature Genet.* **12**: 58–64.

Curran, M.E., Lau, K.F., Hampe, J. (1998) Genetic analysis of inflammatory bowel disease in a large European cohort supports linkage to chromosomes 12 and 16 [see Comments]. *Gastroenterology* **115**: 1066–1071.

Davies, J.L., Kawaguchi, Y., Bennett, S.T. *et al.* (1994) A genome-wide search for human type 1 diabetes susceptibility genes. *Nature* **371**: 130–136.

Davis, S. and Weeks, D.E. (1997) Comparison of nonparametric statistics for detection of linkage in nuclear families: Single-marker evaluation. *Am. J. Hum. Genet.* **61**: 1431–1444.

Davis, S., Schroeder, M. Goldin, L.R. and Weeks, D.E. (1996) Nonparametric simulation-based statistics for detecting linkage in general pedigrees. *Am. J. Hum. Genet.* **58**: 867–880.

Day, N.E. and Simons, M.J. (1976) Disease susceptibility genes – their identification by multiple case family studies. *Tissue Antigens* **8**: 109–119.

de Vries, R.R.P., Fat, R.F.M., Lai, A., Nijenhuis, L.E. and Van Rood, J.J.J. (1976) HLA-linked genetic control of host response to *Mycobacterium leprae*. *Lancet* **ii**: 1328–1330.

Devlin, B. and Roeder, K. (1999) Genomic control for association studies. *Biometrics* 55: 997–1004.

Douglas, J.A., Boehnke, M. and Lange, K. (2000) A multipoint method for detecting genotyping errors and mutations in sibling-pair linkage data. *Am. J. Hum. Genet.* **66**: 1287–1297.

Duerr, R.H., Barmada, M.M., Zhang, L., Pfutzer, R. and Weeks, D.E. (1998) Linkage and association between inflammatory bowel disease and a locus on chromosome 12. *Am. J. Hum. Genet.* **63**: 95–100.

Duerr, R.H., Barmada, M.M., Zhang, L. (2000) High-density genome scan in Crohn disease shows confirmed linkage to chromosome 14q11–12. *Am. J. Hum. Genet.* **66**: 1857–1862.

Eaves, L.J. (1994) Effect of genetic architecture on the power of human linkage studies to resolve the contribution of quantitative trait loci. *Heredity* **72**: 175–192.

Elston, R.C. (1992) Designs for the global search of the human genome by linkage analysis. In: *XVIth International Biometric conference*. Hamilton, New Zealand.

Elston, R.C. (2000) Statistical Analysis for Genetics Epidemiology (http: //darwin.cwru.edu/pub/sage.html).

Falk, C.T. and Rubinstein, P. (1987) Haplotype relative risks: an easy reliable way to construct a proper control sample for risk calculations. *Ann. Hum. Genet.* **51**: 227–233.

Farrall, M. (1997) Affected sibpair linkage tests for multiple linked susceptibility genes. *Genet Epidemiol* **14**: 103–115.

Farrall, M., Keavney, B., McKenzie, C., Delephine, M., Matsuda, F. and Lathrop, G.M. (1999) Fine-mapping of an ancestral recombination breakpoint in DCP1 [letter]. *Nature Genet.* **23**: 270–271.

Fulker, D.W. and Cardon, L.R. (1994) A sib-pair approach to interval mapping of quantitative trait loci. *Am. J. Hum. Genet.* **54**: 1092–1103.

Fulker, D.W., Cherny, S.S. and Cardon, L.R. (1995) Multipoint interval mapping of quantitative trait loci, using sib pairs. *Am. J. Hum. Genet.* **56**: 1224–1233.

Fulker, D.W., Cherny, S.S., Sham, P.C. and Hewitt, J.K. (1999) Combined linkage and association sib-pair analysis for quantitative traits. *Am. J. Hum. Genet.* **64**: 259–267.

Garner, C., Mitchell, J., Hatzis, T., Reittie, J., Farrall, M. and Thein, S.L. (1998) Haplotype mapping of a major quantitative-trait locus for fetal hemoglobin production, on chromosome 6q23. *Am. J. Hum. Genet.* **62**: 1468–1474.

Ghosh, S. and Majumder, P.P. (2000) A two-stage variable-stringency semiparametric method for mapping quantitative-trait loci with the use of genomewide-scan data on sib pairs. *Am. J. Hum. Genet.* **66**: 1046–1061.

Goldgar, D.E. (1990) Multipoint analysis of human quantitative genetic variation. *Am. J. Hum. Genet.* **47**: 957–967.

Goring, H.H. and Ott, J. (1997) Relationship estimation in affected sib pair analysis of late-onset diseases. *Eur. J. Hum. Genet.* 5: 69–77.

Goring, H.H. and Terwilliger, J.D. (2000a) Linkage analysis in the presence of errors I: complex-valued recombination fractions and complex phenotypes [In Process Citation]. *Am. J. Hum. Genet.* **66**: 1095–1106.

Goring, H.H. and Terwilliger, J.D. (2000b) Linkage analysis in the presence of errors II: marker-locus genotyping errors modeled with hypercomplex recombination fractions [In Process Citation]. *Am. J. Hum. Genet.* **66**: 1107–1118.

Goring, H.H. and Terwilliger, J.D. (2000c) Linkage analysis in the presence of errors III: marker loci and their map as nuisance parameters. *Am. J. Hum. Genet.* **66**: 1298–1309.

Goring, H.H. and Terwilliger, J.D. (2000d) Linkage analysis in the presence of errors IV: joint pseudomarker analysis of linkage and/or linkage disequilibrium on a mixture of pedigrees and singletons when the mode of inheritance cannot be accurately specified. *Am. J. Hum. Genet.* **66**: 1310–1327.

Graham, J. and Thompson, E.A. (1997) Disequilibrium mapping via coalescent models of gene ancestry. *Am. J. Hum. Genet.* **61**: A199.

Guo, X. and Elston, R.C. (2000a) Two-stage global search designs for linkage analysis I: use of the mean statistic for affected sib pairs. *Genet. Epidemiol.* **18**: 97–110.

Guo, X. and Elston, R.C. (2000b) Two-stage global search designs for linkage analysis II: including discordant relative pairs in the study. *Genet. Epidemiol.* **18**: 111–127.

Hager, J., Dina, C., Francke, S. *et al.* (1998) A genome-wide scan for human obesity genes reveals a major susceptibility locus on chromosome 10. *Nat. Genet.* **20**: 304–308.

Hampe, J., Schreiber, S., Shaw, S.H. *et al.* (1999) A genomewide analysis provides evidence for novel linkages in inflammatory bowel disease in a large European cohort. *Am. J. Hum. Genet.* **64**: 808–816.

Haseman, J.K. and Elston, R.C. (1972) The investigation of linkage between a quantitative trait and a marker locus. *Behav. Genet.* **2**: 3–19.

Hastbacka, J., de la Chapelle, A., Kaitila, I., Sistonen, P., Weaver, A. and Lander, E. (1992) Linkage disequilibrium mapping in isolated founder populations: diastrophic dysplasia in Finland. *Nature. Genet.* **2**: 204–211.

Hastbacka, J., de la Chapelle, A., Mahtani, M.M. *et al.* (1994) The diastrophic dysplasia gene encodes a novel sulfate transporter: positional cloning by fine-structure linkage disequilibrium mapping. *Cell* **78**: 1073–1087.

Heath, S.C. (1997) Markov chain Monte Carlo segregation and linkage analysis for oligogenic models. *Am. J. Hum. Genet.* **61**: 748–760.

Hugot, J.P., Laurent-Puig, P., Gower-Rousseau, C. *et al.* (1996) Mapping of a susceptibility locus for Crohn's disease on chromosome 16. *Nature* **379**: 821–823.

Julier, C., Hyer, R.N., Davies, J. *et al.* (1991) Insulin-IGF2 region on chromosome 11p encodes a gene implicated in HLA-DR4-dependent diabetes susceptibility. *Nature* **354**: 155–159.

Keavney, B., McKenzie, C.A., Connell, J.M., Julier, C., Ratcliffe, P.J., Sobel, E., Lathrop, M. and Farrall, M. (1998) Measured haplotype analysis of the angiotensin-I converting enzyme gene. *Hum. Mol. Genet.* **7**: 1745–1751.

Keavney, B., McKenzie, C., Parish, S. *et al.* (2000) Lareg-scale test of hypothesised associations between the anitotensin-converting-enzyme insertion/deletion polymorphism and myocardial infarction in about 5000 cases and 6000 controls. *Lancet* **355**: 434–442.

Kong, A. and Cox, N.J. (1997) Allele-sharing models: LOD scores and accurate linkage tests. *Am. J. Hum. Genet.* **61**: 1179–1188.

Kong, A., Cox, N., Frigge, M. and Irwin, M. (1993) Sequential imputation and multipoint linkage analysis. *Genet. Epidemiol.* **10**: 483–488.

Kruglyak, L. and Lander, E.S. (1995) Complete multipoint sib-pair analysis of qualitative and quantitative traits. *Am. J. Hum. Genet.* **57**: 439–454.

Kruglyak, L., Daly, M.J., Reeve-Daly, M.P. and Lander, E.S. (1996) Parametric and nonparametric linkage analysis: a unified multipoint approach. *Am. J. Hum. Genet.* **58**: 1347–1363.

Lalouel, J.M., Rao, D.C., Morton, N.E. and Elston, R.C. (1983) A unified model for complex segregation analysis. *Am. J. Hum. Genet.* **35**: 816–826.

Lam, J.C., Roeder, K. and Devlin, B. (2000) Haplotype fine mapping by evolutionary trees. *Am. J. Hum. Genet.* **66**: 659–673.

Lander, E. and Kruglyak, L. (1995) Genetic dissection of complex traits: guidelines for interpreting and reporting linkage results. *Nat. Genet.* **11**: 241–247.

Lander, E.S. and Schork, N.J. (1994) Genetic dissection of complex traits. *Science* **265**: 2037–2048.

Lathrop, G.M. and Ott, J. (1990) Analysis of complex diseases under oliogenic models and intrafamilial heterogeneity by the LINKAGE programs. *Am. J. Hum. Genet.* **47**: A188.

Leal, S.M. and Ott, J. (2000) Effects of stratification in the analysis of affected-sib-pair data: benefits and costs. *Am. J. Hum. Genet.* **66**: 567–575.

Ma, Y., Ohmen, J.D., Li, Z. *et al.* (1999) A genome-wide search identifies potential new susceptibility loci for Crohn's disease. *Inflamm. Bowel Dis.* 5: 271–278.

McKenzie, C.A., Julier, C., Forrester, T., McFarlane-Anderson, N., Keavney, B., Lathrop, G.M., Ratcliffe, P.J. and Farrall, M. (1995) Segregation and linkage analysis of serum angiotensin I converting enzyme levels: evidence for two quantitative trait loci. *Am. J. Hum. Genet.* **57**: 1426–1435.

McPeek, M.S. and Strahs, A. (1999) Assessment of linkage disequilibrium by the decay of haplotype sharing, with application to fine-scale genetic mapping. *Am. J. Hum. Genet.* **65**: 858–875.

McPeek, M.S. and Sun, L. (2000) Statistical tests for detection of misspecified relationships by use of genome-screen data. *Am. J. Hum. Genet.* **66**: 1076–1094.

Mirza, M.M., Lee, J., Teare, D. *et al.* (1998) Evidence of linkage of the inflammatory bowel disease susceptibility locus on chromosome 16 (IBD1) to ulcerative colitis. *J. Med. Genet.* **35**: 218–221.

Nöthen, M.M., Propping, P. and Fimmers, R. (1993) Association versus linkage studies in psychosis genetics. *J. Med. Genet.* **30**: 634–637.

O'Connell, J.R. and Weeks, D.E. (1998) PedCheck: A program for identifying genotype incompatibilities in linkage analysis. *Am. J. Hum. Genet.* **63**: 259–266.

O'Connell, J.R., Margetic, N., Farrall, M. *et al.* (1998) Dissecting complex diseases with FINESSE. In: International Genetic Epidemiology Society. Arcachon, France.

Ohmen, J.D., Yang, H.Y., Yamamoto, K.K. *et al.* (1996) Susceptibility locus for inflammatory bowel disease on chromosome 16 has a role in Crohn's disease, but not in ulcerative colitis. *Hum. Mol. Genet.* 5: 1679–1683.

Olson, J.M. (1994) Some empirical properties of an all-relative-pairs linkage test. *Genetic. Epidem.* **11**: 41–49.

Olson, J.M. (1995) Multipoint linkage analysis using sib pairs: an interval mapping approach for dichotomous outcomes. *Am. J. Hum. Genet* **56**: 788–798.

Olson, J.M. and Wijsman, E.M. (1993) Linkage between quantitative trait and marker loci: methods using all relative pairs. *Genet. Epidemiol.* **10**: 87–102.

Ott, J. (1989) Statistical properties of the haplotype relative risk. *Genet Epidemiol* **6**: 127–130.

Ott, J. (1992) Strategies for characterizing highly polymorphic markers in human gene mapping. *Am. J. Hum. Genet.* **51**: 283–290.

Owen, M.J. and McGuffin, P. (1993) Association and linkage: complementary strategies for complex disorders. *J. Med. Genet.* **30**: 638–639.

Page, G.P., Amos, C.I. and Boerwinkle, E. (1998) The quantitative LOD score: test statistic and sample size for exclusion and linkage of quantitative traits in human sibships. *Am. J. Hum. Genet.* **62**: 962–968.

Parkes, M., Satsangi, J., Lathrop, G.M., Bell, J.I. and Jewell, D.P. (1996) Susceptibility loci in inflammatory bowel disease [letter]. *Lancet* **348**: 1588.

Parsian, A., Todd, R.D., Devor, E.J., O'Malley, K.L., Suarez, B.K., Reich, T. and Cloninger, C.R. (1991) Alcoholism and alleles of the human D2 dopamine receptor locus: studies of association and linkage. *Arch. Gen. Psychiatry.* **48**: 655–663.

Penrose, L.S. (1935) The detection of autosomal linkage in data which consist of pairs of brothers and sisters of unspecified parentage. *Ann. Eugen.* **6**: 133–138.

Penrose, L.S. (1946) A further note on the sib-pair linkage method. *Ann. Eugen.* **13**: 25–29.

Penrose, L.S. (1953) The general purpose sib-pair linkage test. *Ann. Eugen.* **18**: 120–124.

Pericak-Vance, M.A., Bebout, J.L., Yamaoka, L.A. *et al.* (1990) Linkage studies in familial Alzheimer's disease, application of the affected pedigree member (APM) method of linkage analysis. In Proceedings of the International Symposium on Dementia: Molecular Biology and Genetics of Alzheimer's Disease (eds T. Miyatake, D.J. Selkoe and Y. Ihara). Elsevier, Amsterdam, pp. 215–228.

Pericak-Vance, M.A., Bebout, J.L., Gaskell, P.C. Jr. *et al.* (1991) Linkage studies in familial Alzheimer disease: evidence for chromosome 19 linkage. *Am. J. Hum. Genet.* **48**: 1034–1050.

Rieder, M.J., Taylor, S.L., Clark, A.G. and Nickerson, D.A. (1999). Sequence variation in the human angiotensin converting enzyme. *Nature Genet.* **22**: 59–62.

Rioux, J.D., Daly, M.J., Green, T. *et al.* (1998) Absence of linkage between inflammatory bowel disease and selected loci on chromosomes 3, 7, 12, and 16 [see Comments]. *Gastroenterology* **115**: 1062–1065.

Risch, N. (1990a) Linkage strategies for genetically complex traits. I. Multilocus models. *Am. J. Hum. Genet.* **46**: 222–228.

Risch, N. (1990b) Linkage strategies for genetically complex traits. II. The power of affected relative pairs. *Am. J. Hum. Genet.* **46**: 229–241.

Risch, N. (1990c) Linkage strategies for genetically complex traits. III. The effect of marker polymorphism on analysis of affected relative pairs. *Am. J. Hum. Genet.* **46**: 242–253.

Risch, N. and Teng, J. (1998) The relative power of family-based and case–control designs for linkage disequilibrium studies of complex human diseases I. DNA pooling. *Genome Res* **8**: 1273–1288.

Risch, N. and Zhang, H. (1995) Extreme discordant sib pairs for mapping quantitative trait loci in humans. *Science* **268**: 1584–1589.

Rybicki, B.A. and Elston, R.C. (2000) The relationship between the sibling recurrence-risk ratio and genotype relative risk. *Am. J. Hum. Genet.* **66**: 593–604.

Satsangi, J., Parkes, M., Louis, E. *et al.* (1996) Two stage genome-wide search in inflammatory bowel disease provides evidence for susceptibility loci on chromosomes 3, 7 and 12. *Nature Genet.* **14**: 199–202.

Schaid, D.J. and Sommer, S.S. (1993) Genotype relative risks: methods for design and analysis of candidate-gene association studies. *Am. J. Hum. Genet.* **53**: 1114–1126.

Sobel, E. and Weeks, D.E. (1998) Multipoint IBD estimation at arbitrary locations using general pedigrees. *Am. J. Hum. Genet.* **63**: A309.

Spielman, R.S., McGinnis, R.E. and Ewens, W.J. (1993) Transmission test for linkage disequilibrium: the insulin gene region and insulin-dependent diabetes mellitus (IDDM). *Am. J. Hum. Genet.* 52: 506–516.

Spielman, R.S., McGinnis, R.E. and Ewens, W.J. (1994) The transmission/disequilibrium test detects cosegregation and linkage. *Am. J. Hum. Genet* **54**: 559–560.

Staessen, J.A., Wang, J.G., Ginocchio, G., Petrov, V., Saavedra, A.P., Soubrier. F., Vlietinck, R. and

Fagard, R. (1997) The deletion/insertion polymorphism of the angiotensin converting enzyme gene and cardiovascular-renal risk. *J. Hypertens.* **15**: 1579–1592.
Suarez, B.K., Rice, J.P. and Reich, T. (1978) The generalized sib pair IBD distribution: its use in the detection of linkage. *Ann. Hum. Genet.* **42**: 87–94.
Suarez, B.K., Hampe, C.L. and Van Eerdewegh, P. (1994) Problems of replicating linkage claims in psychiatry. In: *Genetic Approaches to Mental Disorder* (eds E.S. Gershon and C.R. Cloninger). American Psychiatric Press, Washington, DC.
Terwilliger, J.D. (1995) A powerful likelihood method for the analysis of linkage disequilibrium between trait loci and one or more polymorphic marker loci. *Am. J. Hum. Genet.* **56**: 777–787.
Terwilliger, J.D. and Goring, H.H. (2000) Gene mapping in the 20th and 21st centuries: statistical methods, data analysis, and experimental design. *Hum. Biol.* **72**: 63–132.
Terwilliger, J.D. and Ott, J. (1992) A haplotype-based 'haplotype relative risk' approach to detecting allelic associations. *Hum. Hered.* **42**: 337–346.
Terwilliger, J.D. and Weiss, K.M. (1998) Linkage disequilibrium mapping of complex disease: fantasy or reality? *Curr. Opin. Biotechnol.* **9**: 578–594.
Tiret, L., Rigat, B., Visvikis, S., Breda, C., Corvol, P., Cambien, F. and Soubrier, F. (1992) Evidence, from combined segregation and linkage analysis, that a variant of the angiotensin I-converting enzyme (ACE) gene controls plasma ACE levels. *Am. J. Hum. Genet.* **51**: 197–205.
Trembath, R.C., Clough, R.L., Rosbotham, J.L. *et al.* (1997) Identification of a major susceptibility locus on chromosome 6p and evidence for further disease loci revealed by a two stage genome-wide search in psoriasis. *Hum. Mol. Genet.* **6**: 813–820.
Villard, E., Tirek, L., Visvikis, S., Rakotovao, R., Cambien, F. and Soubrier, F. (1996) Identification of new polymorphisms of the angiotensin I-converting enzyme (ACE) gene, and study of their relationship to plasma ACE levels by two-QTL segregation-linkage analysis. *Am. J. Hum. Genet.* **58**: 1268–1278.
Weeks, D.E. and Lange, K. (1988) The affected-pedigree-member method of linkage analysis. *Am. J. Hum. Genet.* **42**: 315–326.
Weeks, D.E. and Lange, K. (1992) A multilocus extension of the affected-pedigree-member method of linkage analysis. *Am. J. Hum. Genet.* **50**: 859–868.
Winter, W.E., Nakamura, M. and House, D.V. (1999) Monogenic diabetes mellitus in youth. The MODY syndromes. *Endocrinol. Metab. Clin. North. Am.* **28**: 765–785.
Xiong, M. and Jin, L. (1999) Comparison of the power and accuracy of biallelic and microsatellite markers in population-based gene-mapping methods. *Am. J. Hum. Genet.* **64**: 629–640.

3

The design of studies for investigating linkage and association

John P. Rice, Nancy L. Saccone and Brian K. Suarez

1. Introduction

Recent advancements in marker development and the sequencing of the human genome have enabled the mapping and identification of genes for single locus disorders to proceed at a rapid pace. Attention is now focused increasingly on how to identify genes for common, complex diseases. Twin, adoption, and family studies support the hypothesis of strong (but not complete) genetic components for disorders such as schizophrenia, manic depression and noninsulin dependent diabetes mellitus. Nonetheless, segregation analysis and linkage analysis have not yet proven successful in delineating homogeneous subtypes or in identifying specific genes which contribute to susceptibility in the majority of cases.

A basic question is: What is the optimal design for the detection of genes for a complex trait? Unfortunately, there is no clear answer to this question. General approaches include: (i) genome screens using highly polymorphic repeat polymorphisms; (ii) testing candidate genes using association methods; (iii) using animal models; (iv) focusing on related biological measures (endophenotypes) to identify candidate susceptibility genes; and (v) genome screens or association studies using single nucleotide polymorphisms (SNPs), often based on linkage disequilibrium in special populations. Within each of these approaches there are choices regarding ascertainment, sampling, marker selection and density, phenotype definition and analytic method. We review the implications of these approaches and choices in the context of several genetic models.

2. The generalized single locus model

We begin with the single trait gene setting. While the design considerations under the assumption of a single major locus are clearly relevant for studies of Mendelian traits, issues that affect power in this setting are important for studies

Analysis of Multifactorial Disease, edited by T. Bishop and P. Sham.

of complex traits as well. Clinical and genetic heterogeneity, reduced penetrance, and incomplete marker information will complicate the genetic study of both simple and complex traits. It is also useful to begin with the single gene model because complex diseases sometimes exhibit subtypes that follow Mendelian inheritance. Identification of trait genes for Mendelian subtypes may then improve understanding of the underlying biology of the common, complex trait as well.

2.1 Description

In the Single Major Locus Model (SML) it is assumed there is single trait locus. The trait may be either continuous or polychotomous. We consider the case of a two allele locus for a dichotomous phenotype. Denoting the two alleles by *A* and *a*, let p be the frequency of *A* and $q = 1 - p$ be the frequency of *a*. Under the assumption of random mating, the probabilities for the three genotypes *AA*, *Aa*, *aa* are p^2, $2pq$, q^2, respectively. For a continuous phenotype, it is typically assumed that the distribution within a genotype is normal, so that parameters consist of the three genotypic means and a common variance. For a dichotomous trait, the penetrances f_{AA}, f_{Aa}, f_{aa} are defined as the probability that individuals of genotype *AA*, *Aa*, *aa*, respectively, are affected. Accordingly, the SML can be described in terms of the four parameters $p, f_{AA}, f_{Aa}, f_{aa}$. We further assume that all familial resemblance is due to the single locus. Thus, the phenotype of an individual depends only on his/her genotype. That is, if $(X_1, \ldots, X_m)$ denotes the phenotypes in a family of size m, and $(g_1, \ldots, g_m)$ denotes their genotypes, then $P(X_i | g_1, \ldots, g_m) = P(X_i | g_i)$ and $P(X_i, X_j | g_i, g_j) = P(X_i | g_i)\, P(X_j | g_j)$, for individuals i and j, where *P*( ) denotes the probability of the event in parentheses.

These assumptions enable the efficient computation of the likelihood of large pedigrees using the Elston and Stewart algorithm (Elstron and Stewart, 1971) as implemented in the LINKAGE package (Lathrop *et al.*, 1984). The SML has served as the 'standard' analytic model for Mendelian traits.

2.2 Design issues for the SML

The goal of linkage analysis is to identify an (unknown) disease locus by identifying a known marker locus linked to the disease. Consider a dominant trait locus with alleles *D* and *d*. If *D* is dominant over *d*, then *Dd* and *DD* individuals express the same phenotype. Consider also a second locus with alleles *A* and *a*. Four combinations (or haplotypes) are produced: *DA*, *Da*, *dA*, and *da*. In the process of meiosis during gamete formation, crossing over between homologous chromosomes can occur, but if loci are close physically on the same chromosome, they will tend to be transmitted as a unit. That is, if a parent is *DA*/*da* (indicating that *D* and *A* are on one of the two homologous chromosomes, and *d* and *a* are on the other), then *DA* and *da* gametes will be overrepresented. The haplotypes *Da* and *dA* are termed recombinants, and the probability that a parent will produce a recombinant is called the recombination fraction and denoted θ. If two loci are on different chromosomes, they are unlinked and $\theta = 0.5$.

These considerations lead to a few principles which guide design issues. If a parent has haplotypes *Da*/*da*, then there is no way to distinguish recombinant from nonrecombinant gametes. That is:

P1: Parents must be doubly heterozygous to be informative for linkage

Secondly, if there is only one child, we do not have enough information to test for linkage (in the absence of association or inbreeding) because we do not know the phase of the parent. That is, we do not know which two alleles occur together on the same chromosome, we do not know whether the parent is *DA*/*da*, or *Da*/*dA*, so we cannot know whether or not the child is a recombinant. Thus:

P2: More than one child is needed to be informative for linkage (in the absence of association)

In deciding whether there is evidence for linkage in a data set, investigators usually rely on the logarithm of the odds ratio (lod) score Z, with a value of Z above 3 taken as evidence for linkage, and a value below -2 taken as evidence against linkage (Morton, 1955; Ott, 1991). A value of 3 corresponds to odds in favor of linkage of 1000 : 1, and corresponds to a significance level of $\alpha = 0.0001$.

In the context of a genome screen, multiple, nonindependent statistical tests are performed. With an infinitely dense map, the genomewide criterion of a lod score of 3.6 corresponds to a genome-wide significance level of 0.05 (Lander and Kruglyak, 1995). That is, in every 20 genome screens, we would expect one lod score of 3.6 due to chance. Even when a 10 cM screen is performed, areas of interest defined by moderate lod scores are often fine mapped, so the lod score of 3.6 has replaced the lod score of 3.0 as the level of support for linkage in many recent studies.

In the simplest case, that of the child of a phase known heterozygous parent *DA*/*da* and a doubly homozygous parent *da*/*da*, a nonrecombinant child will contribute 0.3 to the lod score when $\theta = 0$. In general, the power of a sample depends on the information content of the markers used, the spacing of the markers, the penetrances of the disease locus and the method of analysis. We illustrate this in the case of a dominant disorder.

We consider a single family consisting of two parents and 15 children. We assume the father and six of the children are affected, and the mother and nine children are unaffected. We assume the disease is a (rare) fully penetrant dominant, so the affected individuals are *Dd* and the unaffected individuals are *dd*.

Impact of incomplete marker information. We used the program SIMLINK (Boehnke, 1986; Ploughman and Boehnke, 1989) to simulate marker data with two, three or four equally frequent alleles and recombination fractions of $\theta = 0.0$, 0.05 and 0.1. Recall from principle P1 that a parent must be heterozygous at the marker to be informative for linkage. With four equally frequent alleles, the heterozygosity is 75%, and the power decreases as the marker heterozygosity decreases. The results are displayed in *Table 1*.

Note the decrease of 0.3 in the phase unknown case since one child determines phase at $\theta = 0.0$. In Table 1, the power is the percentage of time the lod score of the sample is above 3. In general, if SNPs are used in a genome screen, they will have to be densely spaced to achieve the same information as a single highly polymorphic marker.

Table 1. Simulation results for incomplete marker information

Phase	θ	Number of Alleles	Mean lod score	Power
Known	0.0	Infinite	4.52	100%
Unknown	0.0	Infinite	4.22	100%
Unknown	0.0	4	2.92	64%
Unknown	0.0	3	2.60	56%
Unknown	0.0	2	1.67	27%
Unknown	0.05	4	2.36	56%
Unknown	0.05	3	2.95	46%
Unknown	0.05	2	1.19	17%
Unknown	0.1	4	2.02	43%
Unknown	0.1	3	1.49	26%
Unknown	0.1	2	0.84	12%

Impact of reduced penetrance. The above results assumed a completely penetrant dominant model, so there was a one-to-one correspondence between genotype and phenotype. This means that all affected people were *Dd* and all unaffected people were *dd*. Indeed, an unaffected child receiving *d* from the affected parent was as informative as an affected child receiving *D* from that parent. For diseases with a variable age of onset, an unaffected person may carry the *D* allele, but not show the phenotype because they are only part way through the risk period. Individuals with a *dd* genotype may be diagnosed with the phenotype because of another reason or simply because of diagnostic error. These situations lead to incomplete penetrances where the *f* values are between 0 and 1. We simulated phenotypes in the above family so that all affected individuals are *Dd* and the unaffected parent is *dd*, but reduced the penetrance of the *Dd* genotype. For example, if $f_{Dd} = 0.8$, then 20% of the children who are of genotype *Dd* will have the unaffected phenotype. Thus, there is ambiguity whether an unaffected person has genotype *dd* and is a recombinant, or has genotype *Dd* and is a nonpenetrant nonrecombinant. The results are displayed in *Table 2*.

Note how the power to detect linkage is reduced as the penetrance is decreased. We have assumed also that $f_{dd} = 0$. If we had allowed this 'phenocopy' rate to increase, there would be an even more dramatic decrease in power.

In a linkage study, in addition to the power to detect linkage, it is also important to be able to exclude areas of the chromosome where there is no trait locus. The last column of *Table 2* shows the average lod score when the true θ is 1/2 (i.e. no linkage) in each simulation, but the lod score is computed at θ = 0.0. Note that reduced penetrance also affects the power to exclude linkage.

One important consideration in study design is to minimize the impact of reduced penetrance. For many diseases, an unaffected person contributes little to linkage information, so that one strategy is to genotype only affected individuals and use them in analysis. We restate this as Principle 3.

> P3: Reduced penetrance decreases the power to detect and exclude linkage, and should be considered in a particular sampling strategy

In some circumstances, however, unaffected individuals can enhance an analysis. For instance, as discussed below, genotypic data on unaffected individuals

Table 2. Simulation results for reduced penetrance

	$\theta = 0.0$		$\theta = 0.5$
Penetrance	Mean lod score	Power	Mean lod score
1.0	2.92	64%	−16.34
0.9	2.08	19%	−8.34
0.8	1.75	14%	−6.65
0.7	1.41	3%	−6.60
0.6	1.26	0%	−7.17
0.5	1.18	0%	−5.79

can protect against false inference of linkage due to allele frequency misspecification.

False detection of linkage. There are three causes of false-positive linkage reports. The first is significance due to chance. Although the LOD score of 3 criterion corresponds to an α of 0.0001 when testing a single marker, a genome screen leads to multiple nonindependent tests and increases the genomewide significance level (Lander and Kruglyak, 1995). Moreover, for many disorders the definition of the phenotype is problematic and investigators perform multiple genome screens with different definitions of who is affected. In the context of the SML, the penetrances may not be known, and investigators may analyze under multiple models (e.g. dominant, recessive and intermediate), so that it is not uncommon to find reports with nine genome screens (three phenotypic definitions each analyzed under three different penetrance models). In addition, other analytic methods may be used, and all analyses may be repeated on subsets of the data. Although these approaches may be justified under the rationale that the researchers do not want to miss true signals (i.e. increased power), this is done at the expense of increasing type I error in a way which cannot be deduced. This has led to conflicting reports which are difficult to interpret in a rigorous way (Moldin, 1997). Thus:

> P4:A clear analytic plan which allows calculation of the *P*-value of the lod score sould be drafted prior to analysis

Another major difficulty can occur if the phenotypes are developed with knowledge of the genotypic marker data, or if the marker data are cleaned with knowledge of the phenotypes. Similarly, if the analytic results are used to change phenotypes or genotypes, then the probabilities associated with lod scores are no longer valid. Thus,

> P5: Phenotypic definition and genotyping must be done in a blind fashion to each other and to the analytic results

In most circumstances, errors in model specifications or data preparation will reduce the power of linkage analysis. There is, however, one important exception which affects lod score methods as well as 'nonparametric' sib-pair methods. This is incorrect specification of marker allele frequencies, or even underlying heterogeneity of marker frequencies within strata of families (Ott, 1991; Hauser *et al.*, 1996). This occurs because the likelihood of a pedigree depends on the allele frequencies when the parents are not genotyped. (If parents are genotyped, then this bias does not arise). To illustrate this principle, consider a population where 50% of the families only have allele 1 and 50% only have allele 2 for a two allele marker. Consider a sample of affected sib pairs drawn from this population and assume the parents are not genotyped. Thus, all sib pairs will be (11,11) in the first subpopulation and (22,22) in the second. The allele frequencies are 0.5, so likelihood computations would infer that 50% of the parents are heterozygous 12 and transmitted the same allele to their offspring. This creates false evidence for linkage.

> P6: Incorrect marker allele frequencies or population admixture can lead to spurious evidence for linkage when parents are not genotyped

2.3 Summary for single locus traits

For some complex diseases discussed elsewhere in this volume, there are Mendelian subforms such as mature onset diabetes of the young or the rare single gene forms of Alzheimer's disease. Given the overall heterogeneity, large pedigrees segregating these genes have been used to localize major loci. To conduct a linkage study is a straightforward application of the above principles if we in fact have a single locus subform. Power calculations can be done with SIMLINK (Boehnke, 1986) or SLINK (Weeks *et al.*, 1990) on the available family material prior to genotyping and can model incomplete penetrance and incomplete marker information.

Selection of families can be guided by the above principles. For example, in the case of a fully penetrant (rare) dominant, an affected parent is heterozygous, whereas in the case of a recessive disorder, an affected parent is homozygous. Knowledge of the mode of transmission is critical in deciding on ascertainment prior to power calculations. Unfortunately, this knowledge is incomplete for complex phenotypes.

Problems associated with marker allele frequencies may be difficult to deal with in design. Genotyping of parents protects against this problem, but may not be feasible for diseases of late onset. The inclusion of unaffected individuals offers a partial solution. In the above example of population admixture, discordant pairs of siblings will also have the same genotype and provide evidence against linkage. For a known stratification such as ethnic group, allele frequencies may be estimated separately within strata and combined in analysis by distinguishing the alleles within each group. That is, if, for example, there were 10 alleles in group 1 and 12 alleles in group 2, a total of 22 alleles (with frequencies summing to 2) could be used in a combined analysis.

A critical question is the impact of performing power calculations and analyses under the SML for a trait which is *not* controlled by a single locus or which is heterogeneous. The above guidelines are meaningful for a trait for which the SML assumptions hold in the sample ascertained. These methods have been remarkably successful when this is true, and have not been for complex traits when analyzed under the SML. This in part reflects the low (unknown) power, so that results have yielded false-positive linkage signals which have not replicated. Another difficulty is that an incorrect trait model leads to bias in the estimate of θ, so that localization of susceptibility loci for complex traits would be problematic.

3. Affected sib-pair methods

The lack of success of using the SML for complex traits moved the field to proceed from the approach of first using segregation analysis to estimate penetrances and then applying the SML linkage model, to sib-pair methods using techniques introduced by Kruglyak and Lander (1995) and Hinds and Risch (ASPEX, personal communication). Affected sib-pair methods do not require explicit specification of a genetic model for the disease and thus may appear preferable for studying diseases of uncertain, and complex, genetic etiology. However, this apparent advantage may be only partly true. Equivalencies between affected sib-pair tests and parametric tests have been demonstrated in particular cases (Knapp *et al.*, 1994). Furthermore, implicit model assumptions exist for affected sib-pair tests, and the power of such tests is thus influenced by the appropriateness of these assumptions (Whittemore, 1996), although it is also argued that nonparametric tests have been designed to be robust, with reasonable power to detect an effect under a range of other models. Also relevant, maximization of parametric LOD scores over more than one model has been shown to have greater power than a nonparametric approach in many cases of a simulated complex or heterogeneous trait (Abreu *et al.*, 1999).

3.1 Parameterization of the SML

One key in understanding the resurgence of these methods first introduced by Penrose (1953) deals with use of λ to parameterize the SML

As noted above, the SML is defined in terms of four parameters $[q, f_{AA}, f_{Aa}, f_{aa}]$. The prevalence of the trait, K, is given by $K = p^2 f_{AA} + 2pqf_{Aa} + q^2 f_{aa}$. Thus an alternative parameterization is given by $[K, f_{AA}, f_{Aa}, f_{aa}]$. This can be useful when the value of K is known. Another parameterization (Kempthorne, 1957) is in terms of the additive and dominance variance; $[K, V_A, V_D, q]$ where:

$$K = p^2 f_{AA} + 2pqf_{Aa} + q^2 f_{aa}$$

$$V_A = 2pq\alpha^2$$

$$V_D = p^2 q^2 \delta^2$$

$$\alpha = p(f_{Aa} - f_{AA}) + q(f_{aa} - f_{Aa})$$

$$\delta = f_{aa} - 2f_{Aa} + f_{AA}$$

James (1971) used this parameterization to show that only three parameters of the SML may be estimated using proband-relative pairs, no matter how many different types of relatives are studied. Moreover, he notes that K_0 and K_S, the risk to offspring and siblings of a proband, respectively, are given by:

$$K_0 = K + \frac{\frac{1}{2}V_A}{K}$$

$$K_s = K + \frac{\frac{1}{2}V_A + \frac{1}{4}V_D}{K}$$

Risch (1990a, 1990b, 1990c) used a SML parameterization in terms of λ_R, where $\lambda_R = K_R/K$, and R denotes a type of relative.

The importance of these latter two parameterizations can be seen in the distribution of identity by descent (IBD) in sibling pairs conditioned on their phenotypes (Suarez *et al.*, 1978). Indeed this distribution does not depend on the parameter q, but only on $[K, V_A, V_D]$ or $[K, \lambda_s, \lambda_0]$. In fact, the distribution in affected sibling pairs depends only on λ_s and λ_0 (and on the recombination fraction θ). This is a key to understanding the sib-pair methods described below.

3.2 Power calculations for affected sib-pair designs

Earlier sib-pair methodology considered IBD sharing at a marker locus. Current methods parameterize sharing in terms of $Z = (Z_0, Z_1, Z_2)$ where Z_i is the probability that an affected sib-pair shares i genes identical by descent at the disease locus. Risch (1990b) showed that $Z_0 = 0.25/\lambda_s$, $Z_1 = 0.5\ \lambda_0/\lambda_s$, with $Z_2 = 1 - Z_0 - Z_1$. If the locus is not related to disease then $\lambda_0 = \lambda_s = 1$ and Z = (1/4, 1/2, 1/4). Thus multipoint marker data enable the estimation of Z as well as disease gene location as implemented in the programs ASPEX (Hinds and Risch, personal communication) and MAPMAKER/SIBS (Kruglyak and Lander, 1995) using the 'possible triangle' method of Holmans (1993). The lod score is maximized over Z to provide a test of linkage.

These methods all allow exclusion mapping (analogous to having a parametric lod score less than -2). This is done for a particular value of λ, so that areas of the genome may be excluded when the lod score is <-2 for this λ. The exclusion curves can be plotted with ASPEX or MAPMAKER/SIBS for specified λ.

The distribution of the maximum lod score is not known under the null hypothesis as the parameter space is degenerate. Hauser *et al.* (1996) provide simulation results in terms of $\lambda = \lambda_o = \lambda_s$, where λ is the locus-specific recurrence risk ratio. They assume the penetrance $f_{Aa} = (f_{AA} + f_{aa}) / 2$, so that $V_D = 0$ and $\lambda_s = \lambda_0$. For $\lambda = 1.2$, the power is 0.00 for $N = 100$ sib pairs and 12% for $N = 800$ with a lod score cut-off of 3. For $\lambda = 2.0$, the power is 15% for $N = 100$ and 94% with $N = 400$. These figures are based on markers with four equally frequent alleles and an intermarker distance of 10 cM with the disease locus in the middle of the interval. The power calculations depend on the true λ for a specific locus, the value of which is not known for most complex disorders. It is clear from Table V of Hauser *et al.* (1996) that the power associated with 100 affected sib pairs is relatively low (15% for $\lambda = 2.0$ and 45% for $\lambda = 3.0$), so it is perhaps not surprising that there are contradictory findings among the many reports which use 100 or fewer sib pairs.

A more subtle issue is that of replication of findings. If there were six unlinked disease loci for a disorder, and the power to detect any one of them is, for example, 20%, then the power to detect at least one is $1 - (0.8)^6 = 74\%$. That is, there may be a high probability of correctly detecting a susceptibility locus, but in a replication study looking at the specific locus, the power is low (20% in our example). This presents a problem unless the replication sample is much larger (Suarez *et al.*, 1994).

Replicability of results can also be affected if different studies use different ascertainment schemes. Depending on the underlying (unknown) genetic architecture of the trait, certain pedigree structures will be more powerful than others in a sib-pair analysis. McCarthy *et al.* (1998) compared the power of different ascertainment schemes for sib-pair studies, for a range of one- and two-locus disease models. They assumed an infinitely dense, completely polymorphic map and sampled one affected sib pair per sibship for analysis. They concluded that while random affected sib-pair (RAS) ascertainment (collection of affected sib pairs without regard to the disease status of other nuclear family members) would be less powerful than the 'optimal' family structure for a given set of model parameters, the loss of power is often relatively modest. While certain additional restrictions may increase power slightly in many circumstances, for certain underlying models, such an approach could fail (e.g. under certain models, the only pairs that contribute to excess allele sharing are those from densely affected sibships, and linkage would be missed if only sparsely affected sibships were ascertained). Thus RAS ascertainment may represent a 'reasonable hedge' that tends to be robust under a wide variety of assumptions. Attempting a narrow ascertainment strategy to find 'optimal' families is often risky in contrast, given that the 'true' model is unknown. Even though observable epidemiological data may restrict the model parameters by specifying population prevalence and risk to sibs or other relatives, these parameters still would have a range of allowable values and would not necessarily determine an optimal family structure.

Other heuristic guidelines outlined by McCarthy *et al.* include the observation that with rare exceptions, nuclear families with two affected parents contribute little power. Intuitively, such a structure could result in affected sibs inheriting non-IBD susceptibility alleles at a trait locus (or non-IBD alleles at a tightly linked marker locus), from different parents.

Further design issues regarding family structure for studies of oligogenic traits are discussed in section 4 below.

3.3 Mean IBD sharing

An alternative to examining the full IBD distribution in sib-pairs is to examine the mean distribution $0 \times Z_0 + 1 \times Z_1 + 2 \times Z_2$. Under the null hypothesis, this is $0 \times 1/4 + 1 \times 1/2 + 2 \times 1/4 = 1$. Thus the average sharing with respect to a specific parent is 50%. Deviations in sharings may be tested using the Haseman–Elston (Haseman and Elston, 1972) statistic as suggested by Suarez and VanEerdewegh (1984) and Blackwelder and Elston (1985). We performed a set of simulations to examine the average sharing under various oligogenic models and the power to detect one of the trait loci.

We used the models of Suarez *et al.* (1994) to generate oligogenic traits. We consider N independent diallelic trait loci with alleles (A_i, a_i), i = 1, ... N, with each A_i having the same frequency p_N. We define the liability X as the number of capital 'A' alleles; an individual is affected if $X \geqslant N$. For a given population prevalence K, the frequency p_N is determined to yield that K. Thus the heritability of the trait is 100% with each quantitative trait locus (QTL) contributing $1/N$ of the total variance of X. The mean of X is $\mu = 2Np_N$ and the variance is $\sigma_G^2 = 2Np_N(1 - p_N)$.

In addition, we allow for a reduced heritability h^2 by sampling from a normal random variable with mean 0 and variance of $\sigma_G^2(1-h^2)/h^2$ and adding this perturbation to the genotypic score defined above. The contribution of each QTL is then h^2/N of the total variance in liability.

For each n-locus model, genotypes were simulated for n fully informative, completely linked ($\theta = 0$) markers. Samples were ascertained by requiring at least two affected sibs in the bottom generation; all families simulated had 10 members (four grandparents, two parents, and four offspring).

For each disease model, the mean IBD at one locus was calculated from samples of 40 000 observations of sharing with respect to a parent at the locus (20 000 observations in the case of the single locus model). The perturbation in mean IBD score P from the *a priori* expected value of 0.5 gives an indication of the ability of this study design to detect linkage under these models.

If N affected sib pairs are sampled, there are $M = 2{\star}N$ observations of IBD sharing of sib pairs with respect to a parent. If M_1 of these observations have an IBD score of 1 and $M-M_1$ have a score of 0, then for $M > 20$,

$$W = \frac{2M_1 - M}{\sqrt{M}}$$

is approximately a unit normal random variable under the null hypothesis H_0 of no linkage ($P = 0.5$), and is expected to be positive when linkage is present. That is, under H_0, the observed statistic M_1 is a binomial random variable and asymptotically follows the distribution $N(M/2, M/4)$. The null hypothesis H_0 is rejected at significance level α if $W > Z_\alpha$, where Z_α is the one-sided α critical value of the unit normal random variable.

Under the alternative hypothesis H_1 ($P > 0.5$), W is approximately normal with distribution

$$N[(2P-1)\sqrt{M}, 4P(1-P)]$$

The (minimum) sample size $N = M/2$ of affected sib pairs required to detect linkage at a significance level α with power $1-\beta$ is given by:

$$\frac{Z_\alpha - (2P-1)\sqrt{M}}{2\sqrt{P(1-P)}} \le Z_{1-\beta}$$

So

$$N \ge \frac{1}{2}\left[\frac{Z_\alpha - 2\sqrt{P(1-P)}\,Z_{1-\beta}}{2P-1}\right]^2$$

Required sample sizes corresponding to a significance level of $\alpha = 0.0001$ (as would correspond to a lod of 3) and power of 80% from the above asymptotic formulae are given in *Table 3* for prevalences of 1% and 10%. Note that these numbers correspond to nonparametric (model-free) sib-pair analyses.

The above example uses oligogenic models that assume equal action of the multiple loci, but it serves as a useful illustration. Results indicate that in some circumstances, mean IBD sharing at a single locus is elevated enough to make nonparametric affected sib-pair analyses feasible. In particular, if the population prevalence is relatively low ($K = 0.01$), we see that for these particular models, the sample sizes required are realistic if the disease is at least 50% heritable and no more than four to six loci are involved. For the $K = 0.1$ case, the sample sizes

Table 3. Simulation results for oligogenic models

	Heritability (h^2)							
	100%		75%		50%		25%	
Number of loci	*P*	*N*	*P*	*N*	*P*	*N*	*P*	*N*
10% Prevalence								
1	70.7%	59	70.8%	59	69.0%	70	64.0%	131
2	62.7%	159	60.2%	251	58.2%	389	55.1%	983
4	56.2%	669	54.8%	1106	53.8%	1815	52.5%	4112
6	54.1%	1510	53.2%	2592	52.0%	6264	51.7%	9048
8	52.9%	3094	52.6%	3740	51.4%	12502	51.1%	20360
10	52.3%	4926	51.8%	7645	51.1%	20075	50.6%	63078
1% Prevalence								
1	74.4%	42	74.3%	42	74.4%	42	74.2%	43
2	70.2%	62	62.7%	160	62.9%	154	62.6%	163
4	61.6%	192	58.4%	369	57.5%	466	55.5%	870
6	57.6%	449	56.2%	685	55.6%	834	53.5%	2072
8	56.0%	711	54.8%	1149	53.8%	1802	52.4%	4438
10	54.4%	1357	53.9%	1720	53.1%	2624	51.9%	7118

Mean IBD sharing P at 1 locus and sample size N (= number of sib pairs) required to detect linkage at a significance level of $\alpha = 0.0001$ and 80% power, calculated from mean sharing carried to the fifth decimal place, for polygenic models

rapidly become demanding beyond four loci even if the trait is 100% heritable, and are prohibitive for multilocus reduced heritability cases.

Note that the use of a parametric dominant model for analysis of the one locus, 100% heritable, lod score of about 3 at a tightly linked marker, as expected for a rare dominant disease, in contrast with the 42 pairs listed in Table 3. In this setting, nonparametric analysis in effect adds noise to the data via the unaffected parent and is less powerful than model based pedigree analysis.

3.4 Selection of genetic markers and the utility of multi-stage genome screen designs

The power to detect a trait locus is clearly affected by the density of the marker map used in a genome screen. The number of polymorphic markers available has increased greatly and allows very dense coverage of the genome. Efficiency of a study, in addition to its power, must also be considered. Genotyping the full sample of sib pairs or families with a very dense grid of markers may be beyond the resources of a project, and a strategy that retains good power while reducing genotyping costs may be desired.

Sequential designs, which perform the genome screening in two (or more) stages, are worth considering for the above purposes. Two (complementary) approaches to improve efficiency of a genome screen are sometimes referred to as 'grid-tightening' and 'sample splitting.' In grid tightening, an initial screen is performed with a coarse grid of markers. Regions meeting a relaxed test criterion in the initial screen are targeted, and a tighter grid of markers is genotyped in these regions for the second stage. Signals above a more stringent criterion in the second screen are then reported as positive evidence for linkage. In 'sample splitting', initial genotyping is carried out for a subset of the full sample. Again, regions with modest evidence are targeted with a denser grid in the full sample, and regions meeting a more stringent criterion are reported. These strategies may be combined.

Elston (1992) determined the optimal sample size and spacing between markers when the sampling units are pairs of affected relatives. He minimized the total cost, as determined by the cost per subject recruited and the cost per marker typing. In the first stage of sampling, more widely spaced markers and a larger significance level are used, followed by a second stage of more narrowly spaced markers around those significant from the first stage. He provides formulae to determine the optimal spacing, sample size, and stage one cut-off to achieve a given power in terms of the risk ratio λ for that type of affected relative pair. These algorithms are available in the DESPAIR computer program.

Holmans and Craddock (1997) examined combinations of strategies. For simulated samples of 200 sib pairs with parents, they compared the power and number of genotypings required for sib-pair studies under different designs. For uniform stage two LOD score criteria, LOD score criteria for stage one were chosen to fix the genomewide type I error rate at 0.05.

Results showed that both grid tightening and sample splitting are important tools for increasing the efficiency of a genome scan. In general, strategies with an initial grid of 20 cM were more efficient than those with tighter initial grids. If an emphasis of the study is to minimize genotyping, their findings recommend typing half the pairs without parents on a 20 cM grid in stage 1, and all pairs plus parents with a 10 cM grid in stage 2. If an emphasis on power is preferred, the recommended strategy would be to type all 200 pairs without parents on a 20 cM grid in stage 1, followed by a 5 cM grid for all pairs and parents in stage 2.

Holmans and Craddock (1997) also evaluated the utility of genotyping parents of the sib pairs at stage 1 (note that they examined only strategies that genotyped all pairs and parents at stage 2, and that the disease status of parents was not used in the linkage analyses). They found that under the Hardy–Weinberg equilibrium, in many cases the power gained by typing parents did not offset the increased genotyping cost. However, in simulations of a stratified sample with allele frequency heterogeneity, genotyping of parents is desirable. Clearly, genotyping of parents in the second stage can control type I error. Furthermore, in the presence of stratification, genotyping parents in the first stage decreased the overall amount of genotyping required, due to follow-up of fewer false positives from stage one.

Hauser *et al.* (1996) also investigated the efficiency of genotyping and marker selection strategies. Their conclusions were similar to those of Holmans and Craddock, and recommended typing of only the affected pairs on a 10 cM to 20 cM grid in the initial screen.

3.5 Summary of sib-pair design issues

Sib-pair methods are model free in that they do not depend on the assumptions of the SML model or on knowledge of trait penetrances or gene frequencies. They do, however, depend on marker allele frequencies as discussed already. Most studies of complex traits have relied on affected sib-pair methods. The Haseman and Elston (1972) method analyzes all types of sib pairs (both affected, discordant and both unaffected) and performs a single test of overall linkage.

The sample sizes required to achieve adequate power as indicated in *Table 3* can be sobering. Moreover, these sizes are to detect linkage, and much larger samples would be needed to localize the genes once linkage is detected.

It is important to note the improved performance of these methods for a 1% trait versus a 10% trait. Thus the use of a strictly defined phenotype may be a key when designing studies relying on sib-pair analysis. The importance of phenotype definition is discussed further in section 6.

The large sample sizes required, the difficulties of gene localization, and problems associated with replication for complex diseases may indicate the need for other approaches. If the λ (and corresponding elevation in percent sharing) is high, then this design should detect such a locus. However, phenotypic heterogeneity or epistatic effects of many loci of small effect will make it difficult to detect disease genes for many traits currently being studied using these methods.

4. Family structure designs

In this section we address the question of whether there is an optimal sampling design in terms of family structure for linkage analysis. This is a contentious issue with some workers arguing for the largest possible sampling frame (which includes not only large pedigrees but, at the extreme, the entire breeding population), while other workers argue for small sampling units such as sib pairs (which can allow greater ascertainment stringency than larger units). Under the rubric 'family structure' we include not only the size of the sampling unit but also the distribution of phenotypes in the unit. Our position is that there is no optimal family structure but rather many optima, depending on specifics of the trait or disease under study and the analytical method chosen for analysis. This point has also been made by McCarthy *et al.* (1998) with regard to one- and two-locus models (see discussion in section 3.2).

Clearly not all families are equal with respect to the information they bring to a linkage analysis. By judiciously choosing the types of families we accept to study, we can dramatically lower or increase the probability of detecting linkage. The objective is simple; we wish to maximize the number of informative (i.e. segregating) meioses at the loci that affect susceptibility. Unfortunately, we do not know which loci are susceptibility loci in advance of a linkage study. And we are unable to arrange intercrosses or backcrosses at these loci experimentally except inferentially through a 'progeny test'.

It is obvious that certain sampling units, by virtue of their structure, have advantages over other sampling units. For instance, a strength of three-generation families is that they provide direct information about phase. Among two-generation families, larger sibships ought to be better than smaller sibships, because they allow more accurate phase inferences. Except when parents are genetically related, families with a single offspring provide no linkage information. Of course, there are many other factors in addition to family structure that will affect experimental design – not the least of which is the cost of genotyping and, importantly, the availability and expense of ascertaining the sample. For fully penetrant Mendelian traits the choice of families is straightforward. Notwithstanding a new mutation or substantial variation in age of onset or expressivity, families segregating simple dominant traits will perforce contain at least one "affected" parent. Families with two affected parents are less informative. For rare recessive traits, virtually all informative families will contain two unaffected parents. Families containing one parent with the recessive phenotype provide only half the information available from families without an affected parent while families with both parents affected provide no information.

For complex non-Mendelian traits or diseases, the optimal family structure for mapping susceptibility loci depends on the true mode of transmission which, in general, is unknown to a researcher. Indeed, for any particular mode of transmission, the optimal family structure can even depend on such 'nuisance' parameters as the allele frequency at the susceptibility loci. This is easily demonstrated by again considering the simple additive six-locus QTL model discussed in section 3.3 and the fixed family structure shown in Figure 1. As before, we suppose that all six diallelic susceptibility loci are unlinked to one another and that the frequency of the 'A' allele (i.e. the allele that increases risk) is the same at each QTL. Let the probability that a person is affected be 1 if the sum of the '*A*' alleles over all six loci is 6 or more and zero otherwise. As the frequency of each '*A*' allele is assumed to be the same at each QTL, it is a simple matter to solve for the allele frequency that gives rise to a particular population prevalence (K) of the trait. Here we are interested in two population prevalences; K=0.01 and K=0.10. For K=0.01, p_i=0.17481 and for K=0.10, p_i=0.28817. We randomly simulated families with the structure shown in *Figure 1*, separately for K=1% and K=10%, and retained for linkage analysis those families that contain two or more affected offspring in the third generation. We simulated a completely informative marker linked to each of the six QTLs at a recombination fraction of zero. Details of the model can be found in Suarez *et al.* (1994).

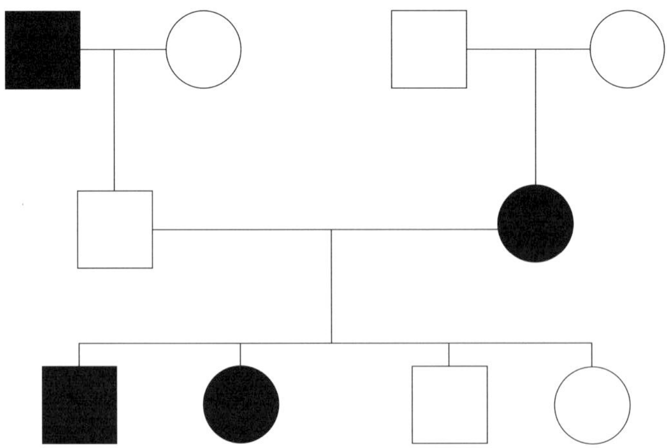

Figure 1. Ten member family structure used in simulations. This family is denoted 2/1(O)/1 to indicate two affected children and an affected parent and grandparent on opposite sides of the pedigree.

To analyze these families for linkage we chose two statistics: the t_2 statistic of Blackwelder and Elston (1985) which uses information from the entire sibship – not just the affected members – and, for contrast, the nonparametric linkage (NPL) Z-score statistic (Whittemore and Halpern, 1994a, 1994b) as implemented in GENEHUNTER (Kruglyak *et al.*, 1996). The NPL Z-score uses genotypic information from the entire pedigree but only affected pedigree members contribute to the Z-score.

With respect to the distribution of affected family members, there are 63 unique, nonredundant configurations for the family structure shown in *Figure 1*. We determined the relative frequency of each configuration by simulating families at random until 100 000 were ascertained. (Some configurations never occurred among the simulated families and others were exceedingly rare.) For each value of K, the configurations were ranked and the 10 most frequent were determined. Families were then generated randomly until 9000 ascertained pedigrees of each configuration were obtained (except for the three rarest configurations for K=1%, where fewer families were available).

Pedigrees with the same configuration were assembled into groups of 250 families for linkage analysis. For each group of 250 families, all six QTLs were analyzed with both methods (since all QTLs are equivalent) and the t_2 and NPL Z-scores were recorded. For most of the configurations, the 9000 simulated families allowed 36 replications for each of the six QTLs in N = 250 families. *Table 4* reports the 10 most frequent configurations, and various summary statistics including the mean, standard deviation, rank and relative efficiency for the t_2 scores and the NPL Z-scores. With respect to the reported rank, we first performed a simple ranking based on the mean value of the linkage statistic, then tested for significant differences, grouping together configurations with means that are not significantly different from one another at $P < 0.05$. For instance, the NPL Z-score analysis for K = 1% resulted in seven ranked clusters: the mean NPL Z-score of configuration 2/0/1 and 3/0/0 are not different (rank = 2.5) nor are the NPL Z-scores for configurations 3/1/0, 4/1/0 and 2/1(O)/1, all of which receive a rank of 8.

The results from these simulations provide a number of interesting insights that may be of help in designing linkage studies for complex phenotypes. Despite the fact that the ascertainment rule and the generating model for each value of K is exactly the same (except for the frequency of the 'A' allele at each QTL), the relative abundance of the various configurations are very different. For K = 1%, the 10 most frequent configurations account for 94.9% of all ascertained families whereas the 10 most frequent configurations for K = 10% account for only 71.8% of all ascertained families. Moreover, only seven of the top 10 configurations are common for both population prevalences. Perhaps more disturbing for the purpose of experimental design, however, is that the 'best' family structure is not the most common family structure for either of the linkage statistics investigated here. Thus, for a six QTL additive oligogenic model with a population prevalence of 1%, the optimal family consists of three affected offspring and no affected parents or grandparents when the t_2 statistic is evaluated, whereas the optimal family consists of three affected children, no affected parents and one affected grandparent when the NPL Z-score is evaluated. This underscores the notion

Table 4. Linkage analyses of the 10 most frequent family structures for a six QTL trait with a population prevalence of 1% or 10%.

Family structure[a]	Relative frequency	Mean t_2 score	SD	Rank	RE	Mean NPL-Z score	SD	Rank	RE
1% Prevalence									
2 / 0 / 0	0.405	5.461	0.98	2.5	0.81	5.206	0.90	4.0	0.40
2 / 1 / 0	0.193	3.743	0.98	8.5	0.38	3.138	0.93	10.0	0.15
2 / 1(S) / 1	0.088	3.731	1.13	8.5	0.38	4.521	1.19	6.0	0.30
2 / 0 / 1	0.074	4.817	1.02	4.0	0.63	6.763	1.20	2.5	0.68
3 / 1 / 0	0.066	4.123	1.11	6.0	0.46	3.803	1.15	8.0	0.21
3 / 0 / 0	0.055	6.069	1.14	1.0	1.00	6.891	1.13	2.5	0.70
3 / 1(S) / 1	0.035	4.065	1.04	6.0	0.45	4.896	1.14	5.0	0.36
3 / 0 / 1	0.013	5.569	1.21	2.5	0.84	8.217	1.39	1.0	1.00
4 / 1 / 0	0.010	4.034	1.12	6.0	0.44	3.901	1.16	8.0	0.23
2 / 1(O) / 1	0.010	2.731	0.95	10.0	0.20	3.665	1.07	8.0	0.20
10% Prevalence									
2 / 1(S) / 1	0.157	3.100	1.02	6.0	0.44	3.295	1.07	5.5	0.36
2 / 0 / 0	0.129	4.697	1.00	1.5	1.00	4.386	0.98	2.0	0.64
2 / 1 / 0	0.107	3.166	0.88	6.0	0.45	2.341	0.94	9.0	0.18
2 / 0 / 1	0.100	4.523	0.99	1.5	0.93	5.464	1.12	1.0	1.00
3 / 1(S) / 1	0.077	3.412	1.03	4.0	0.53	3.600	1.16	3.5	0.43
3 / 1 / 0	0.050	3.762	1.10	3.0	0.64	2.981	1.12	7.5	0.30
2 / 1 / 2(O)	0.037	2.499	1.07	9.5	0.28	3.341	1.12	5.5	0.37
2 / 1(O) / 1	0.029	2.598	1.07	9.5	0.31	2.983	1.10	7.5	0.30
3 / 1 / 2(O)	0.027	3.049	1.04	6.0	0.42	3.720	1.10	3.5	0.46
2 / 1(S) / 2(S)	0.025	2.884	1.09	8.0	0.38	1.949	1.06	10.0	0.13

A completely informative marker is linked with a recombination fraction of zero to each QTL. Thirty-six samples of 250 families with each configuration were analyzed except for configurations 3/0/1, 4/1/0 and 2/1(O)/1 where 27, 24 and 20 replicates, respectively, were analyzed. RE, Relative efficiency compared to the 'best' configurations.

[a]n1/n2/n3, Number of affected offspring/number of affected parents/number of affected grandparents. (S): Affected parent is child of affected grandparent; when applied to grandparents (S) denotes that the two affected grandparents are spouses. (O) denotes 'opposite', when applied to an affected parent, (O) denotes that the affected grandparent is on the opposite side of the family; when applied to grandparents (O) denotes that the two affected grandparents are not spouses.

that power is not simply a property of the family's structure, but also depends on the analytic approach. For the configurations reported in *Table 4*, the NPL *Z*-score outperforms the t_2 statistic in six of the 10 family configurations for both $K = 1\%$ and $K = 10\%$.

As both t_2 and the NPL statistic are asymptotically standard normal variables, the relative efficiency of any two configurations can be obtained by computing the ratio of the squared scores. *Table 4* reports these relative efficiencies by comparing the various configurations to the configuration that gives the greatest evidence for linkage. For example, for $K=1\%$, analyzed with the t_2 statistic, the 3/0/0 configuration yields the largest mean score of 6.069 for 250 families. The relative efficiency of the next 'best' family structure (3/0/1) is 0.84. Taking the inverse of this ratio indicates that 19% more families with the structure 3/0/1 are required to yield the same linkage information that the 3/0/0 structure contains (when analyzed with the t_2 statistic).

It is generally recognized that for diseases with a simple genetic architecture, bilineal families (i.e. families containing affected ancestors on both the mother's and father's side), offer less linkage information than unilineal families. This also appears to be true for an additive oligogenic phenotype of the sort considered here. *Table 4* shows that for a threshold trait with a 1% prevalence, only one of the 10 most frequently ascertained family configurations is bilineal and, when analyzed with the t_2 statistic, provides the least power to detect QTLs. When analyzed with the NPL Z-score approach, the bilineal configuration ranks in the second least informative cluster. For a disorder with 10% prevalence, bilineal pedigrees account for a larger proportion of the ascertained families but still tend to offer less linkage information than other structures. A notable exception, however, is the 3/1/2(O) structure – especially when analyzed with the NPL Z-score statistic. Of course this configuration also contains six affected members, which increases the usefulness of the NPL Z-score approach.

Table 4 also shows that among nonbilineal pedigrees, families with an affected parent are, on average, less efficient than families where neither parent is affected. For instance, the relative mean efficiency of the four configurations without affected parents for the $K=1\%$ phenotype is 82% and 69.5% compared with 42.2% and 23.8% for the five configurations with an affected parent for the t_2 and NPL Z-score statistic, respectively. In other words, for the oligogenic model considered here, approximately twice as many families with an affected parent are required to attain the same t_2 score as obtained when neither parent is affected. Using the NPL Z-score, the loss in relative efficiency for families with an affected parent is approximately three-fold. These ratios are the same for the $K = 10\%$ phenotype. This is a reasonable finding as our ascertainment rule required at least two affected offspring. Unaffected parents capable of having a minimum of two of four affected children must have a liability close to the threshold and must, therefore, be segregating at many of the QTLs. It is our impression that multiplex sibships with unaffected parents contain, on average, more linkage information than other family structures for every additive QTL model we have simulated.

Another principle can be deduced from comparing the results for the two population prevalences considered here. Even after excluding families with a bilineal structure, the mean of all t_2 scores, as well as the mean of all NPL Z-scores, for the $K = 1\%$ trait, are 27% and 53% larger, respectively, than that obtained for the $K = 10\%$ trait, even though the generating model is exactly the same in both cases. Accordingly, for any quasi-continuous phenotype that conforms to the assumptions of the additive QTL model, the preferred sampling design is to ascertain families with the more 'deviant' phenotype.

The t_2 statistic of Blackwelder and Elston (1985) uses only the marker and phenotypic information contained in the sibship. Thus, although we were able to 'bin' the families according to their configuration for the purpose of this analysis, the linkage test we used ignores this information. Nonetheless, such phenotypic information is often available to a researcher (even if the ancestors' DNA is not) and should be used to 'prioritize' family selection.

So we return to the question asked at the beginning of this section: Is there an optimal family structure for linkage analysis? The results of the simulations reported here suggest that, at least for a simple additive QTL model, there is no optimal 'one-size-fits-all' approach to experimental design. There are, however, some general sampling principles that ought to improve the prospect of identifying susceptibility loci for complex phenotypes. These principles include: (i) the avoidance of bilineal families, (ii) the preferential ascertainment of multiplex sibships where neither parent is affected; and (iii) employing a 'narrow' definition of the phenotype.

5. Association studies

An alternative to linkage and allele-sharing methods is association-based analysis. Rather than relying on cosegregation of a susceptibility gene and linked markers within families (with possibly different markers linked in different families), association studies hope to detect allele variants that are associated with susceptibility across larger populations. That is, such studies seek to detect linkage disequilibrium due to linkage between a marker locus and a locus bearing a disease-causing mutation.

Until recently, association studies have generally been candidate gene studies. (Prospects for genomewide association studies are discussed in section 5.1.) Subjects are genotyped at loci thought to be reasonable candidates for involvement in the disorder. Examples include: for psychiatric disorders, polymorphisms in structural genes for neurotransmitter receptors or enzymes that break down neurotransmitters; for cardiovascular disease, polymorphisms in genes coding for apolipoproteins; for diabetes, polymorphisms in the insulin gene region.

In the past, case–control studies were common. Under this design, a group of cases is collected, along with a group of matched controls. Allele frequencies at candidate loci are compared for significant differences between case and control groups. This design does not require the ascertainment of family units and hence has appeal due to the relative ease of implementation. However, a well known disadvantage of this approach is that allelic association between a disease and locus can arise as an artifact of population admixture, rather than due to chromosomal proximity of the marker and a trait-causing mutation. Differences in allele frequencies in distinct strata, together with different affection rates for the strata, can lead to spurious association. As an exaggerated example with two strata, if at the locus being tested, a particular allele is observed only in stratum one, any trait having much higher prevalence in stratum one would likely appear to be associated with this allele (subject to sampling variation).

This problem can be addressed by using 'family based controls,' as popularized by the transmission disequilibrium test (TDT) (Spielman *et al.*, 1993), which has similarities to tests proposed by others (e.g. Falk and Rubenstein, 1987; Field *et al.*, 1986). These tests are based on the premise that a parent who is heterozygous with one associated allele and one nonassociated allele at the locus under study will transmit the associated allele to affected offspring more often than expected by chance. The TDT in effect uses the nontransmitted alleles as controls which are well matched for the ancestry of the parent: as the control and disease alleles for this observation are contributed by a single individual (the parent), stratification is reduced or eliminated. The TDT is a test of linkage of the marker locus and disease locus, but is able to detect this linkage only if association is also present.

The TDT has been used to show significant association due to linkage of the 5′ flanking polymorphism adjacent to the insulin gene, and insulin-dependent diabetes mellitus (IDDM) (Spielman *et al.*, 1993). Previously, this locus had demonstrated strong association with but no linkage to IDDM in most studies.

Unfortunately in some circumstances the TDT is not feasible (e.g. in a study of a late onset disease such as Alzheimer's). When the TDT design is a viable option, the primary concerns in using the TDT over a case–control design are power and efficiency. As pointed out by Morton and Collins (1998), the TDT is less efficient than a standard case–control study. For instance, when parent–child pairs are used for the TDT, each informative parent–child pair provides two observed alleles (one transmitted and one non-transmitted), while in a case–control study, each fully informative case–control pair provides four alleles for observation.

Morton and Collins also note that the use of multiplex sibships and the inclusion of normal sibs within a TDT study increases efficiency of the TDT, representing a possible strategy to balance the concerns of efficiency and potential population admixture. In theory, a risk allele should be transmitted to an unaffected child less often than expected by chance, complementing the observed transmissions in affected offspring. However, this strategy brings the additional concern that 'normal' sibs may not yet be through the risk period and thus are subject to censoring; at the same time, they are likely to have elevated risk as well, compared with that of the general population. Depending on the trait under study, requiring unaffected sibs or relatives to act as controls may limit effective choices in terms of phenotyping certainty. Confidence in the phenotype of controls is important for an association study; Morton and Collins also make the related observation that a case–control design using 'hypernormal' controls increases efficiency compared with using 'normal' controls. The role of appropriate phenotype definition is discussed further in section 6.

The concerns requiring attention when performing association studies are well illustrated by the story of the dopamine D2 receptor (DRD2) gene and alcoholism. The dopamine system has long been considered a candidate for involvement in alcoholism. An association between alcoholism and the Taq1-A1 polymorphism in DRD2 was first reported by Blum *et al.* (1990), in a sample of 35 alcoholics and 35 controls. This polymorphism has since been studied by many groups, with conflicting results from association studies. Several linkage studies between the region of DRD2 and alcoholism found negative results. More recently, family based analyses using the TDT and related tests (Edenberg *et al.*, 1998) showed no evidence for linkage or association of the DRD2 Taq1-A1 allele with alcohol dependence. Note that the frequency of the Taq1-A1 allele varies among populations, ranging from 0.09 to 0.80 according to different reports. This highlights the need for either careful matching of cases and controls or the use of family based analyses.

Another concern is that association studies may be unable to detect disease genes if multiple disease-causing mutations, which arose independently on different haplotype backgrounds, are represented in the sample under study. Association is hence commonly studied in young, genetically isolated populations. For such populations, linkage disequilibrium is expected to extend over greater distances, and there are likely to be fewer distinct disease-causing alleles present, than in older, outbred populations.

5.1 Association based genomewide studies

An alternative to genomewide linkage studies and candidate gene studies is the use of genomewide association (linkage disequilibrium) studies. The power expected from such approaches has been well illustrated in studies by Risch and Merikangas (1996, 1997) and by Camp (1997, 1999).

Using a multiplicative single locus two-allele model, Risch and Merikangas demonstrate that for certain parameter settings, the increase in mean IBD sharing for affected sib-pairs is not substantial enough to keep sample sizes within reach for nonparametric sib-pair analysis, analogous to the above demonstration in section 3.3 for certain multilocus disease models. In addition, they show that the corresponding sample sizes for TDT association studies are significantly more favorable.

To see their parameterization, note that for the case where the recombination fraction $\theta = 0$ and marker heterozygosity = 1, the expected proportion P of alleles shared by an affected sib pair can be parameterized in terms of the genotypic risk ratio (GRR, the increased chance that an individual with a particular genotype has the disease) γ and the allele frequency p of the risk increasing allele A. Specifically, they define the GRR so that the risk of individuals with genotype Aa is γ times the risk for those with aa, and the risk for AA is γ^2 times that of aa (multiplicative model). Using formulas from Risch (1987; 1990b), they show that the expected mean IBD sharing P can be written as

$$P = \frac{1 + w}{2 + w} \quad \text{where } w = \frac{pq(\gamma - 1)^2}{(p\gamma + q)^2}$$

They then compute required sample sizes for 80% power at a significance level of 0.0001, just as in section 3.3, for GRRs of $\gamma = 4$, 2, and 1.5, and for allele frequencies $p = 0.01, 0.1, 0.50, 0.80$. Except in the cases where $\gamma = 4$ and $p = 0.1$ or 0.5, sample sizes range in the thousands.

To compare these sample sizes with those required for TDT association studies, Risch and Merikangas (1996) take note of a limitation of association studies: the actual polymorphism within the disease gene (or a polymorphism in strong disequilibrium) must be available and chosen for analysis. [This point has been discussed further by Muller-Myhsok and Abel (1997).] Allowing for a screen of 1 000 000 independent association tests (corresponding to 10 diallelic polymorphisms in each of 100 000 genes), Risch and Merikangas point out that for the same genomewide significance level as a nonparametric linkage screen using 500 markers and a point significance level 0.0001, an association study must require a point significance level of $\alpha = 5 \times 10^{-8}$. Results are presented for family units of parents plus one affected offspring and for parents plus two affected offspring. Furthermore, their sample sizes are calculated taking into account that out of this full sample, only a proportion of the parents will be heterozygous at a given marker and thus contribute to the TDT analysis at that marker. Under this framework, even with the smaller significance level, the required sample sizes for TDT analysis under the corresponding γ and p are invariably smaller than those for nonparametric sib-pair analysis, in some cases by orders of magnitude.

The study by Camp (1997) extends the above investigation to additional modes of inheritance (additive, common recessive, common dominant) in addition to multiplicative, for a single locus two-allele model. Using formulae from Suarez *et al.* (1978), the expected proportion P of alleles shared IBD by an affected sib pair can be written in terms of the allele frequency p of the susceptibility allele A and the penetrances f_0, f_1 and f_2 for the genotypes *aa*, *Aa* and *AA* respectively. Then, using the corresponding penetrance functions (which can be written in terms of GRRs for *AA* and *Aa* with respect to *aa*) for the different modes of inheritance and various allele frequencies p, Camp computes sample sizes required for nonparametric affected sibling pair analysis and contrasts this with sample sizes required for TDT association analyses. The pointwise significance levels used correspond to a genomewide significance of 0.05 with an infinitely dense marker map in the case of affected sib-pair analysis [a more stringent criterion as suggested by Lander and Kruglyak (1995)], and with 1 000 000 independent tests in the case of the TDT. Again, results indicate that the sample sizes (in family units) required for the TDT are smaller than those needed for nonparametric linkage analysis.

Although nonparametric linkage analysis will certainly continue to be a productive tool for complex disease analysis, the above discussions have generated much interest in the future feasibility of genomewide association scans. The current state of technology (in terms of functional polymorphisms available for testing) generally limits association studies to the testing of candidate genes, as undirected scans of the maps available would be inefficient, with low prospects for success. However, it has been pointed out that the large number of gene polymorphisms desired for an association scan, while not yet readily available, may be promised by the advent of SNP markers. Recently, Halushka and colleagues (1999) reported on 874 candidate SNPs in 75 candidate genes for blood pressure homeostasis and hypertension; this indicates the feasibility of this approach.

5.2 SNP markers: their expected utility for genome screens

SNPs have several advantageous properties as DNA markers. They are abundant in the genome: on average, any two copies of a chromosome are expected to differ in approximately one nucleotide position every 1000 bases. There has also been promising development of efficient methods for assaying/typing SNPs (Landegren *et al.*, 1998).

SNP markers are biallelic with maximum heterozygosity of 50%; hence it is expected that a genome scanning set of SNP markers must be denser to be as informative as the highly polymorphic microsatellite marker sets currently in use (see section 2.2). Kruglyak (1997) performed simulations and demonstrated that a map of 700–900 moderately polymorphic biallelic markers is equivalent to typical 300–400 microsatellite marker maps. SNP maps containing on the order of thousands of markers are expected to become available, thus promising highly informative maps for future linkage analyses.

An additional question is whether SNP markers are sufficiently dense in the genome to serve as maps for linkage disequilibrium (LD) and association mapping. To address this concern, simulation studies by Kruglyak (1999) estimate the extent of LD surrounding common gene variants in the general human population and also in population isolates. The results suggest that on average, a useful level of LD is unlikely to extend beyond an average distance of roughly 3 kb in the general population, implying that approximately 500 000 SNPs will be necessary for genomewide association scans using SNPs. This number is theoretically within reach when compared to the number of common SNPs expected to exist in the human genome. Another interesting finding from these simulations is that in many cases, the extent of LD in isolated populations may be similar to that in the general population, contrary to traditional expectations.

Kruglyak points out some caveats to his analysis, noting that human population history is more complex than assumed for these simulations. He also notes that linkage disequilibrium may be enhanced in some 'hotspot' regions of the genome. Some empirical studies of SNP markers in various populations support predictions of the limited extent of linkage disequilibrium (Carlson and Cox, 1998; Clark *et al.*, 1998). However, Taillon-Miller *et al.* (2000) have found extended regions of linkage disequilibrium on the X chromosome, with significant disequilibrium among SNP markers up to hundreds of kilobases apart, in more than one population. Further investigations will shed light on the future utility and necessary density and composition of SNP based maps for whole genome association studies.

6. Phenotype definition

Success in identifying susceptibility loci will depend on phenotypic definition and appropriate data analytic techniques. Although there is hope that studies will lead to a molecular basis for disease definition, clinical uncertainty may preclude the discovery of linkage.

6.1 Alternative phenotypic definitions

We have focused in this chapter on design issues related to a dichotomous disease phenotype. We have seen in section 3 that a strictly defined phenotype can aid detection of linkage. An alternative to the use of a clinical phenotype is to study an underlying neurobiological or neurobehavioral phenotype associated with the phenotype of interest. These intermediate phenotypes may be more closely linked to gene expression, may be measurable in unaffected individuals, and may lead to candidate genes which will not be detected through a linkage study of the clinical phenotype. Often these endophenotypes are quantitative measures with minimal measurement error associated with them. Examples would include the study of cholesterol rather than 'heart attack' (Bodzioch *et al*, 1999; Brooks-Wilson *et al.*, 1999), or the study of event-related brain potential abnormalities rather than alcoholism (Begleiter and Porjesz, 1999). This strategy should be considered when appropriate endophenotypes are available. Similarly, animal models offer an alternative to linkage or association studies of the clinical phenotype.

6.2 Impact of diagnostic error

We close by examining the impact of diagnostic error on the heritability and risk

Table 5. Impact of diagnostic error on h2 and λ

Sensitivity (s)	Specificity (t)	Prevalence (K^*)	h^2	λ
1.00	1.00	0.100	100%	3.24
0.95	1.00	0.095	97%	3.24
0.90	1.00	0.090	94%	3.24
1.00	0.95	0.145	68%	2.01
0.95	0.95	0.140	65%	1.98
0.90	0.95	0.135	61%	1.94
1.00	0.90	0.190	50%	1.56
0.95	0.90	0.185	46%	1.53
0.90	0.90	0.180	43%	1.50

ratio λ of a dichotomous polygenic trait. We assume that a trait is determined by an underlying liability scale which in turn is determined by the additive effects of many genes with individuals above a threshold being affected. We assume the threshold yields a prevalence K of 10% and the heritability is 100%. The risk to first degree relatives (K_1) is 32% (Rice, 1993; Rice *et al.*, 1987).

Let s be the sensitivity of diagnosis (the probability of correctly diagnosing a true case) and let t be the specificity (the probability of correctly diagnosing a noncase). The observed prevalence of the trait is given by $K^* = sK + (1 - t)(1 - K)$ (*Table 5*).

The values of heritability (h^2) and risk ratio ($\lambda = K_1/K^*$) are given for various values of s and t for this trait with 10% prevalence. Note that there is only a minor impact of reduced sensitivity. However, the inclusion of false positives (reduced specificity) has a major impact. A specificity of 90% reduces h^2 and λ by one-half or more and would have a dramatic effect in reducing power compared with the situation of no diagnostic error. This underscores the necessity of making clinical diagnoses as carefully as possible and the potential benefit of repeated measures to reduce error.

7. Conclusions

The recent progress in molecular genetics, together with the prospect in the near future of having complete human genome sequence, a catalog of all human genes, and a dense map of SNPs in coding regions, heralds a new era in the identification of disease genes for complex diseases. The optimal approach to utilize these technologies remains unclear, and will probably evolve over the next few years.

Against this background, we have addressed several design issues. We have emphasized basic genetic principles in the context of the simple SML model, but these apply equally well to multi-locus situations. The major question is: What are the effect sizes for the genes we wish to detect? As noted in *Table 3*, required sib-pair sample sizes may range from the hundreds to the tens of thousands, and we do not know the underlying deviation in sharing above 50%. It is clear that having a phenotype with a lower prevalence reduces the sample sizes required. Moreover, diagnostic error (see *Table 5*, section 6.2) can have a devastating effect on power.

Alternative approaches to traditional linkage analysis such as studying a different phenotype (related endophenotypes) or a different organism (animal studies) depend on their relationship to the disease phenotype. Similarly, the successful use of SNPs depends on disequilibrium and the existence of a relatively small set of disease gene polymorphisms. The conclusion at this point in time is that no one approach is optimal and comprehensive, and success may depend on a combination of strategies and on the epidemiologic characteristics of the disease of interest.

Acknowledgments

This work was supported in part by NIH grants MH37685, MH31302, AA12239 and MH17104 (NLS), and by a CaP CURE award and an award from the Urological Research Foundation. Special thanks to J. Lin for programming assistance and to C. Roark for preparation of the manuscript.

References

Abreu, P.C., Greenberg, D.A. and Hodge, S.E. (1999) Direct power comparisons between simple LOD scores and NPL scores for linkage analysis in complex diseases. *Am. J. Hum. Genet.* **65**: 847–857.

Begleiter, H. and Porjesz, B. (1999) What is inherited in the predisposition toward alcoholism? A proposed model. *Alcoholism: Clin. Exp. Res.* **23**: 1125–1135.

Blackwelder, W.C. and Elston, R.C. (1985) A comparison of sib-pair linkage tests for disease susceptibility loci. *Genet. Epidemiol.* **2**: 85–97.

Blum, K., Noble, E.P., Sheridan, P.J., Montgomery, A., Ritchie, T., Jagadeeswaran, P., Nogami, H., Briggs, A.H. and Cohn, J.B. (1990) Allelic association of the human dopamine D2 receptor gene in alcoholism. *JAMA* **263**: 2055–2060.

Bodzioch, M., Orso, E., Klucken, J. *et al.* (1999) The gene encoding ATP-binding cassette transporter 1 is mutated in Tangier disease. *Nature Genet.* **22**: 347–351.

Boehnke, M. (1986) Estimating the power of a proposed linkage study: A practical computer simulation of approach. *Am. J. Hum. Genet.* **39**: 513–527.

Brooks-Wilson, A., Marcil, M., Clee, S. *et al.* (1999) Mutations in *ABC1* Tangier disease and familial high-density lipoprotein deficiency. *Nature Genet.* **22**: 336–345.

Camp, N.J. (1997) Genomewide transmission/disequilibrium testing – consideration of the genotypic relative risks at disease loci. *Am. J. Hum. Genet.* **61**: 1424–1430.

Camp, N.J. (1999) Genomewide transmission/disequilibrium testing: a correction. *Am. J. Hum. Genet.* **64**: 1485–1487.

Carlson, C.S. and Cox, D.R. (1998) Linkage disequilibrium of SNPs on human chromosome 21. *Am. J. Hum. Genet.* **63**: A284.

Clark, A.G., Weiss, K.M., Nickerson, D.A. *et al.* (1998) Haplotype structure and population genetic inferences from nucleotide-sequence variation in human lipoprotein lipase. *Am. J. Hum. Genet.* **63**: 595–612.

Edenberg, H.J., Foroud, T., Koller, D.L. *et. al.* (1998) A family-based analysis of the association of the dopamine D2 receptor (DRD2) with alcoholism. *Alcoholism: Clin. Exp. Res.* **22**: 505–512.

Elston, R.C. (1992) Designs for the global search of the human genome by linkage analysis. *Proceedings of the XVIth International Biometric Conference*. Hamilton, New Zealand.

Elston, R.C. and Stewart, J. (1971) A general model for the analysis of pedigree data. *Hum. Hered.* **21**: 523–542.

Falk, C.T. and Rubinstein, P. (1987) Haplotype relative risks: an easy reliable way to construct a proper control sample for risk calculations. *Ann. Hum. Genet.* **51**: 227–233.

Field, L.L., Fothergill-Payne, C., Bertrams, J. and Baur, M.P. (1986) HLA-DR effects in a large German IDDM data set. *Genet. Epidemiol. Suppl.* **1**: 323–328.

Halushka, M., Fan, J., Bentley, K., Hsie, L., Shen, N., Weber, A., Cooper, R., Lipshutz, R. and Chakravarti, A. (1999) Patterns of single-nucleotide polymorphisms in candidate genes for blood-pressure homeostasis. *Nature Genet.* **22**: 239–247.

Haseman, J.K. and Elston, R.C. (1972) The investigation of linkage between a quantitative trait and a marker locus. *Behav. Genet.* **2**: 3–19.

Hauser, L., Boehnke, M., Guo, S. and Risch, N. (1996) Affected-sib-pair interval mapping and exclusion for complex genetic traits: sampling considerations. *Genet. Epidemiol.* **13**: 117–137.

Holmans, P. (1993) Asymptotic properties of affected-sib-pair linkage analysis. *Am. J. Hum. Genet.* 52: 362–374.

Holmans, P. and Craddock, N. (1997) Efficient strategies for genome scanning using maximum-likelihood affected-sib-pair analysis. *Am. J. Hum. Genet.* **60**: 657–666.

James, J.W. (1971) Frequency in relatives for an all-or-none trait. *Ann Hum Genet* **35**: 47–48.

Kempthorne, O. (1957) *An Introduction to Genetic Statistics*. John Wiley & Sons, New York.

Knapp, M., Seuchter, S.A. and Baur, M.P. (1994) Linkage analysis in nuclear families 2: Relationship between affected sib-pair tests and lod score analysis. *Hum. Hered.* **44**: 44–51.

Kruglyak, L. (1997) The use of a genetic map of biallelic markers in linkage studies. *Nature Genet.* **17**: 21–24.

Kruglyak, L. (1999) Prospects for whole-genome linkage disequilibrium mapping of common disease genes. *Nature Genet.* **22**: 139–144.

Kruglyak, L. and Lander, E.S. (1995) Complete multipoint sib-pair analysis of qualitative and quantitative traits. *Am. J. Hum. Genet.* **57**: 439–454.

Kruglyak, L., Daly, M.J., Reeve-Daly, M.P. and Lander, E.S. (1996) Parametric and nonparametric linkage analysis: a unified multipoint approach. *Am. J. Hum. Genet.* **58**: 1347–1363.

Landagren, U., Nilsson, M., and Kwok, P.-Y. (1998) Reading bits of genetic information: methods for single nucleotide polymorphism analysis. *Genome Res.* **8**: 769–776.

Lander, E. and Kruglyak, L. (1995) Genetic dissection of complex traits: guidelines for interpreting and reporting linkage results. *Nature Genet.* **11**: 241–247.

Lathrop, G.M., Hooper, A.B., Huntsman, J.W. and Ward, R.H. (1984) Strategies for multilocus linkage analysis in humans. *Proc. Natl Acad. Sci. USA* **81**: 3443–3446.

McCarthy, M.I., Kruglyak, L. and Lander, E.S. (1998) Sib-pair collection strategies for complex diseases. *Genet. Epidemiol.* **15**: 317–340.

Moldin, S. (1997) The maddening hunt for madness genes. *Nature Genet.* **17**(2) : 127–129.

Morton, N.E. (1955) Sequential tests for the detection of linkage. *Am. J. Hum. Genet.* **7**: 277–318.

Morton, N.E. and Collins, A. (1998) Tests and estimates of allelic association in complex inheritance. *Proc. Natl. Acad. Sci. USA* **95**: 11389–11393.

Muller-Myhsok, B. and Abel, L. (1997) Genetic analysis of complex diseases. *Science* **275**: 1328–1329.

Ott, J. (1991) *Analysis of Human Genetic Linkage*. The Johns Hopkins University Press, Baltimore.

Penrose, L.S. (1953) The general-purpose sib-pair linkage test. *Ann. Eugenics* **18**: 120–124.

Ploughman, L.M. and Boehnke, M. (1989) Estimating the power of a proposed linkage study for a complex genetic trait. *Am. J. Hum. Genet.* **44**: 543–551.

Rice, J. (1993) Phenotype definition for genetic studies. *Eur. Arch. Psychiat. Clin. Neurosci.* **243**: 158–163.

Rice, J., Endicott, J., Knesevich, M.A. and Rochberg, N. (1987) The estimation of diagnostic sensitivity using stability data: an application to major depressive disorder. *J. Psychiatr. Res.* **21**: 337–345.

Risch, N. (1990a) Linkage strategies for genetically complex traits. I. Multilocus models. *Am. J. Hum. Genet.* **46**: 222–228.

Risch, N (1990b) Linkage strategies for genetically complex traits. II. The power of affected relative pairs. *Am. J. Hum. Genet.* **46**: 229–241.

Risch, N. (1990c) Linkage strategies for genetically complex traits. III. The effect of marker polymorphism on analysis of affected relative pairs. *Am. J. Hum. Genet.* **46**: 242–253.

Risch, N. (1992) Corrections to "Linkage Strategies for Genetically Complex Traits. III. The Effect of Marker Polymorphism on Analysis of Affected Relative Pairs" (*Am. J. Hum. Genet.* **46**: 242–253, 1990). *Am. J. Hum. Genet.* **51**: 673–675.

Risch, N. and Merikangas, K. (1996) The future of genetic studies of complex human diseases. *Science* **273**: 1516–1517.

Risch, N. and Merikangas, K. (1997) Genetic analysis of complex diseases. *Science* **275**: 1329–1330.

Spielman, R.S., McGinnis, R.E. and Ewens, W.J. (1993) Transmission test for linkage disequilibrium: the insulin gene region and insulin-dependent diabetes mellitus (IDDM). *Am. J. Hum. Genet.* **52**: 506–516.

Suarez, B.K. and Van Eerdewegh, P. (1984) A comparison of three affected-sib-pair scoring methods to detect HLA-linked disease susceptibility genes. *Am. J. Hum. Genet.* **18**: 135–146.

Suarez, B.K., Rice, J. and Reich, T. (1978) The generalized sib pair IBD distribution: its use in the detection of linkage. *Ann. Hum. Genet.* **42**: 87–94.

Suarez, B.K., Hampe, C.L. and Van Eerdewegh, P. (1994) Problems of replicating linkage claims in psychiatry. In: *Genetic Approaches to Mental Disorders* (eds E. S. Gershon and C. R. Cloninger). American Psychiatric Press Inc., Washington, DC, pp. 23–46.

Taillon-Miller, P., Bauer-Sardiña, I., Saccone, N.L. *et al.* (2000) Juxtaposed regions of extensive and minimal linkage disequilibrium in human Xq25 and Xq28. *Nature Genet.* **25**: 324–328.

Weeks, D.E., Ott, J. and Lathrop, G.M. (1990) SLINK: A general simulation program for linkage analysis. *Am. J. Hum. Genet.* **47**: A204 (abstract.).

Whittemore, A.S. (1996) Genome scanning for linkage: an overview. *Am. J. Hum. Genet.* **59**: 704–716.

Whittemore, A.S. and Halpern, J. (1994a) Probability of gene identity by descent: computation and applications. *Biometrics* **50**: 109–117.

Whittemore, A.S. and Halpern, J. (1994b) A class of tests for linkage using affected pedigree members. *Biometrics* **50**: 118–120.

4

Genetic susceptibility to common cancers: a model for genetics of a complex trait

David E. Goldgar

1. Introduction

The last decade has seen significant progress in the localization, cloning and characterization of genes involved in susceptibility to cancer. This success is not only true of relatively rare tumors characteristic of specific Mendelian syndromes, but has now been extended to many more common cancer sites (though still largely in the familial context). As in other common diseases, the early successes in cancer gene localization were found in the relatively rare cancer-associated syndromes such as familial retinoblastoma (Sparkes *et al.*, 1983), multiple endocrine neoplasia types 1 and 2 (Larsson *et al.*, 1988; Simpson *et al.*, 1987), and neurofibromatosis I and II (Barker *et al.*, 1987; Rouleau *et al.*, 1987). These syndromes are characterized by nearly full penetrance and by the presence of relatively rare (in the overall population) types of tumors, together with numerous other distinguishing characteristics [e.g., café-au-lait spots in neurofibromatosis type I (NFI)]. For more common cancers, the successes were also in the context of clearly identifiable syndromes such as childhood tumors in Li-Fraumeni syndrome (Li and Fraumeni, 1969; Malkin *et al.*, 1990), and presence of large numbers of polyps in familial adenomatous polyps (FAP)/Gardners syndrome (Gardner, 1951). In these syndromes there would often be an increased incidence of more frequent cancers such as breast cancer in young women and colorectal cancer in FAP, but with markedly early onset. The only real difficulties one faces in mapping these quasi-Mendelian syndromes, is that for the more rare syndromes, the gathering of sufficient family material to detect linkage may be problematic, as was the case for some of the more recent localizations of tumor associated syndromes (e.g., familial Wilms tumor (Rahman *et al.*, 1996; McDonald *et al.*, 1998), thyroid cancer in conjunction with multinodular goiter

Analysis of Multifactorial Disease, edited by T. Bishop and P. Sham.

(Bignell *et al.*, 1997). Another factor that allowed the early localization and identification of these syndromes was the lack of locus heterogeneity. Nearly all of the above-mentioned syndromes are characterized by a single locus which accounts for all (or most) of the recognized families with these syndromes.

Among the most common cancers, the identification of susceptibility loci has been most successful for cancers of the breast and ovary, bowel and melanoma. In being asked to write this chapter, the first question that comes to mind is what sets cancer apart from other complex phenotypes? That cancer genetics has proven to be successful (or more accurately, perceived to have made great progress) when compared to other disciplines, is due in my opinion, to several factors as detailed below.

(i) *Diagnostic accuracy.* Although there are still a few 'gray areas' in diagnosis, particularly with the classification of *in-situ* cancers, reliable data is usually available from medical records, path reports and cancer registry. This is in contrast to many of the other diseases and human traits described in this book such as asthma, the psychiatric disorders and others.

(ii) *Population based cancer registries.* In many countries worldwide, and individual states/regions within other countries, reporting all cases of cancers diagnosed to the registries is required by law. Often, this registration includes histology, clinical staging and survival data. This ensures complete accurate diagnosis of all cancers in a region and allows for computer-based verification of family histories supplied by a family proband. Currently, cancer incidence data on 183 populations from 50 countries are available as part of the International Agency for Research on Cancer (IARC) network of cancer registries which must meet basic requirements of data quality and completeness (IARC, 1997). More importantly for genetic studies, in several areas covered by such registries, there are also large computerized genealogies which have been computer-linked to the cancer registries such as those found in Utah, Iceland and Sweden. This allows for easy family ascertainment and for the assessment of familial risk and clusters of associated tumor sites.

(iii) *Phenotypically defined subsets, cancer syndromes.* Although cancer is in itself a very common disease, a number of rare syndromes have been recognized for a long time. These cancer syndromes, which include retinoblastoma, Li–Fraumeni syndrome, familial adenomatous polyposis, and the multiple endocrine neoplasias (Types 1 and 2), appear to segregate in an autosomal dominant fashion with nearly full penetrance. Under the two-hit model proposed by Knudson (1971) in the context of retinoblastoma, the expectation for cancer cases arising from inherited predisposition are (i) earlier onset and (ii) more frequent bilaterality (and/or multiple primary cancers). These features allow easier recognition of probands and families most likely to be segregating a susceptibility allele.

(iv) *Availability of tissue specimens.* One feature that is perhaps unique to the genetics of cancer is given by the very nature of the disease itself, that is the presence of abnormal collections of cells. We now realize that these tumors have undergone a series of genetic changes that have removed the natural controls on growth of damaged DNA. In practical terms, the fact that these tumors are removed surgically and archived as standard practice, provides, in many cases, a source of normal DNA from the surrounding tissue to allow genotyping. In particular, this provides a source of DNA from deceased individuals of a pedigree that would otherwise be unavailable. Unfortunately, DNA extracted from this sort of material is not generally suitable for routine large-scale genotyping; however, for testing of candidate regions, it can be extremely useful when lymphocyte DNA is not available. In addition, a comparison of normal and tumor tissue provides clues to regions that, at least at the somatic level, can be important in the development of a particular cancer. This includes such techniques as loss of heterozygosity (LOH) studies, comparative genome hybridization (CGH), serial analysis of gene expression (SAGE), representational difference analysis (RDA) and gene expression oligonucleotide high-density arrays (Gene-Chips). Finally, tumors can, in certain circumstances provide additional linkage evidence if genotypes of both normal and tumor tissue can be obtained. Under the two-hit model first proposed by Knudson (1971) in the context of retinoblastoma, cancer is a result of inactivation of both alleles at a tumor suppressor locus (in addition to other somatic changes). Thus, for many cancers, an individual inherits one 'hit' in the form of a mutated allele and acquires the other in a tumor cell. One observes then, consistent loss of the wild type (non-linked) allele in tumors of patients in a linked family. This is most useful for classifying small families as linked or not to a particular candidate locus, and can be also useful in the context of a genomic search for testing suggestive regions of linkage. Methods for including tumor loss data into formal linkage analysis to increase power have been proposed by several authors. (e.g. Rebbeck *et al.*, 1994).

2. Evidence for familial aggregation in common cancers

Evidence of familial aggregation of common cancers comes from a variety of sources. These include:

(i) the observation of specific large multigenerational pedigrees in which it would be difficult, if not impossible, to ascribe to chance clustering;
(ii) analysis of reported family histories of probands ascertained for the presence of a particular cancer compared to family histories of control subjects;
(iii) systematic studies of familial risk based upon analyses of family/genealogical databases linked to population-based cancer registries;
(iv) formal segregation analysis of family data ascertained through one or more index cases.

The results of almost all population-based investigations of cancer show evidence of an elevated familial risk for nearly all cancer sites. For the more common cancer sites such as breast, prostate and colon, there appears to be a two-fold excess of cancer among first degree relatives of cancer probands, while certain other cancers which are less frequent, such as testicular cancer and thyroid cancer, are characterized by an eight-fold familial risk (Goldgar *et al.*, 1994a; Hemminki and Dong, 2000; Ron *et al.*, 1991). When only cancers with relatively early onset are considered (e.g., premenopausal breast cancer), in general the familial risks are increased, consistent with the familial clustering being genetic in origin rather than a primarily shared environment, although the latter cannot be ruled out. More specific modeling using complex segregation analyses have also been useful in characterizing hypothesized susceptibility loci for many common cancers including breast (Claus *et al.*, 1990), prostate (Carter *et al.*, 1992) and colorectal (Houlston *et al.*, 1995). In all three cases, these analyses have favored autosomal dominant inheritance of a susceptibility locus conferring lifetime risks of about 80%.

3. Breast cancer

3.1 Large pedigrees

The first published report of a cancer-prone family was that of Pierre Broca, who in 1866 published a treatise on 'traits des tumeurs' which contained a descriptive report of a family in which there were more than 10 cases of early onset breast cancer, as well as five cases of liver cancer (Broca, 1866). It must be remembered that breast cancer incidence would have been considerably lower 100 years ago than today, making this family even more remarkable. Gardner and Stephens (1950) later reported a large Utah kindred (family 107) with nine cases of breast cancer in three generations. This family was updated several times (Bishop *et al.*, 1988; Easton *et al.*, 1997) and now contains over 36 cases of female breast cancer, three male breast cancer cases, and five ovarian cancers. A mutation in *BRCA2* was found in this family shortly after the gene was cloned in 1995.

3.2 Linkage studies and positional cloning

Although there had been a number of early reports of linkage of familial breast cancer to classical blood group and enzyme markers such as glutamate-pyruvate, transaminase (GPT; King *et al.*, 1980) and the ABO blood group (in a single large Utah pedigree; Skolnick *et al.*, 1984), subsequent events proved both these linkages incorrect. In the latter instance it was the subsequent identification of a single recombinant individual that resulted in the evidence for linkage to ABO to almost completely disappear (Skolnick *et al.*, 1984). Given a disease as common as breast cancer and the fact that we now know that even the inherited forms are genetically heterogeneous, it would have been surprising indeed if linkage had been identified using only the 20 or so blood group, protein and enzyme polymorphisms available until the mid 1980s. In retrospect, the localization of *BRCA1* to chromosome 17q by the group of Mary-Claire King in 1990 (Hall *et al.*, 1990) is an interesting case study of linkage of common cancers in the familial setting in the pre-microsatellite era. It is especially noteworthy that the sample included primarily site-specific breast cancer families, rather than families containing both breast and ovarian cancer. At the time of these investigations, the genetic markers available for typing consisted of primarily restriction fragment length polymorphism (RFLP) markers usually with two or three alleles. In addition, there were a set of highly variable VNTR minisatellite probes available, but they were not randomly spaced throughout the genome and reading them on the gels posed difficulties in many cases. This was at the time when the first polymerase chain reaction (PCR)-based microsatellite markers were being developed. In the initial paper in *Science* a total of 23 pedigrees were described. The maximum total lod score across all 23 families was 2.35 with the VNTR marker D17S74 at a recombination fraction of 0.20. When heterogeneity was allowed for, the lod score rose to 3.28 at a distance of 0.0014 from D17S74 with an estimated 40% of families linked. When the authors explained the heterogeneity by age at onset of the cases in the pedigrees, they obtained an overall lod score of 5.98 in the seven pedigrees with the lowest mean age at diagnosis. After *BRCA1* was cloned and mutations could be examined, mutations in *BRCA1* were found in six of the seven families with the lowest age of onset. Interestingly, the location of the marker D17S74 lies approximately 25 cM from *BRCA1*, outside the interval suggested by Hall *et al.* (1990). This demonstrates the effect a single informative marker can have on linkage analysis.

After the initial report which appeared in December 1990, Narod *et al.* (1991) demonstrated that a large fraction of families with breast and ovarian cancer were linked to *BRCA1*. After analysis of over 214 families, the location of the *BRCA1* gene had moved considerably proximal to the original location. This was due to the availability of more highly informative short tandem repeat (STR) markers, and the use of a model which better incorporated age of onset, as well as the larger data set. A key to the eventual cloning of the *BRCA1* gene by the consortium led by Mark Skolnick at the University of Utah and Myriad Genetics was the availability of large extended families. The gene was shown to be clearly linked and by virtue of the size of the family, had sufficient opportunities for recombination within the linked region in individuals who had high likelihood of being mutation carriers based on their phenotype. This is exemplified by kindred 2082, a large Utah kindred comprising several hundred descendants of a Utah pioneer who settled in the Utah valley in 1847. The family was ascertained from the Utah population Database (UPDB) in 1991 based on a cluster of four ovarian cancers and two early onset breast cancers found in related individuals. The pedigree was then expanded to contain a total of 25 breast and 20 ovarian cancers with a lod score for linkage to genetic markers on chromosome 17q of 9.0 (Goldgar *et al.*, 1994b). Subsequent fine mapping identified a branch of the family in which a recombination event occurred within the linked region, moving the distal boundary some 2 cM from its previous position. Similarly, a large family collected in the Midwestern USA by Henry Lynch and studied by Steven Narod (unpublished), showed a recombinant which moved the proximal boundary about 1 cM distal. In the end, the Utah/Myriad group were left with a region of 650 kbp, which facilitated the search for and testing of candidate cDNAs in the region. These large families also provided a useful resource for studies of penetrance, modifying factors associated with a single mutation.

For the localization of *BRCA2*, a key factor was the recognition that families in which there was a case of breast cancer in a male as well as early onset female breast cancer and/or ovarian cancer, gave convincing linkage evidence against the involvement of *BRCA1* (Stratton *et al.*, 1994). It is important to remember that this was before the *BRCA1* gene was cloned and one could stratify families by mutation status. The ability to select families which were largely not due to *BRCA1* and likely to be due to the same predisposition gene were key factors in the localization of *BRCA2* to chromosome 13 in 1994 (Wooster *et al.*, 1994). Here again, two large families, Utah K107 described above, and B186, a large multi-generational Irish kindred which also had a case of male breast cancer, played key roles in establishing definitive linkage. Although the clear majority of families with *BRCA2* mutations did not have a case of male breast cancer, it was the ability of being relatively certain that those that did, would be due to *BRCA2* which added additional power to the linkage analysis by reducing locus heterogeneity. Interestingly, although much effort was taken to reduce the critical region containing the gene through analysis of key recombination events as with *BRCA1*, the cloning of *BRCA2* was greatly facilitated by analysis of a pancreatic tumor in which a small homozygous deletion was identified (Schutte *et al.*, 1995) which narrowed the region to allow the testing of a reasonable number of small candidate loci, a portion of one showing clear truncating mutations in several families (Wooster *et al.*, 1995). Shortly afterwards, the full coding sequence of the very large *BRCA2* gene (~3400 amino acids) was identified and additional mutations described (Tavtigian *et al.*, 1996).

3.3 Functions of BRCA1 *and* BRCA2

Although at the time of writing all the functions of the proteins coded for by these two genes are unknown, it is clear that both play a role in the repair of DNA double strand breaks, and *BRCA1* is thought to play a role in transcriptional activation of a number of other genes, though most of these have not yet been identified. The current state of knowledge about the function of *BRCA1* and *BRCA2* is given in a review article by Scully (in press). Mutations have been identified throughout both genes, most of which lead to truncated protein products. There is some evidence that the risk of breast cancer conferred by these mutations depends somewhat on the position of the mutation within the gene (i.e., the length of the predicted protein) (Gayther *et al.*, 1995, 1997). These results must be interpreted with caution, however, since the effect, while statistically significant, explains only a small proportion of the phenotypic variance, and because of messenger RNA (mRNA) instability through nonsense-mediated RNA decay, the position of the protein does not necessarily correspond to the length or presence of the BRCA protein.

4. Colon cancer

4.1. Adenomatous polyposis coli (APC)

The first gene predisposing to colorectal cancer to be localized and then identified was associated with the relatively rare syndrome FAP. FAP is characterized by the appearance of large numbers of colonic polyps appearing in childhood with an associated high risk of developing cancer of the colon. Penetrance is very high. Following a report of a constitutional interstitial deletion in chromosome 5q, traditional linkage analysis of several large pedigrees using RFLP markers localized the gene responsible for FAP to chromosome 5q21 (Bodmer *et al.*, 1987; Leppert *et al.*, 1987). The specific gene for FAP, *APC* was cloned in 1991 (Groden *et al.*,

1991; Kinzler *et al.*, 1991). This large gene, which has also been shown to be associated with desmoid tumors (Gardners syndrome), and Turcot syndrome, is involved in the regulation of β-catenin and may suppress tumorigenesis by inhibiting catenin-related transcription of growth-promoting genes such as *c-Myc* and *cyclin-D1*, leading to apoptosis. In addition to its role in FAP, as many as 80% of sporadic colorectal cancers have been found to harbor somatic *APC* mutations (reviewed by Goss and Groden, 2000). Based on studies of adenomas, *APC* mutations are thought to be a very early event in colon tumorigenesis.

4.2 Hereditary non-polyposis coli

The first documented instance of what we now call hereditary nonpolyposis coli (HNPCC) was a kindred seen in 1895 by Warthin (family G) and published in 1913. This family had an unusual constellation of endometrial, gastric and colorectal cancer with gastric being the most frequently observed tumor. It was not until the 1960s before this became recognized as a familial cancer syndrome with families being ascertained and characterized worldwide. Although first denoted cancer family syndrome or Lynch syndrome type II after Dr Henry Lynch who was the first to ascertain a collection of families with this syndrome (Lynch *et al.*, 1977), the term HNPCC was given to distinguish the syndrome from FAP described above. As a result of population-wide cancer registry, good genealogical characterization, and an apparent relatively high prevalence of the syndrome, researchers in Finland were able to collect a large number of families, and were also able to characterize the genetic epidemiology of the syndrome (Aaltonen *et al.*, 1994).

4.3 The replication error (RER) phenotype

The year 1993 was really the banner year in understanding the genetics of HNPCC. In May of that year, the Finnish group headed by Albert de la Chapelle reported the linkage of an HNPCC phenotype to markers on chromosome 2p15–16 in two Finnish families (Peltomaki *et al.*, 1993). Later that year a Swedish group found linkage to chromosome 3p in families not linked to chromosome 2 (Lindblom *et al.*, 1993). Concomitant with the linkage findings, the observation was made that a fraction of all colorectal cancers in general, and almost all tumors from HNPCC patients exhibited a phenomenon known as microsatellite instability (MIN). Tumors exhibiting this phenotype have been denoted replication error positive (RER+). This observation provided key clues to the nature of the genes responsible, as it indicated that good candidates would be genes involved in repair of mismatches due to misalignment of repeat sequences occurring during DNA replication. Following on from this lead, Richard Kolodner and colleagues who had been working on mismatch repair (MMR) genes in yeast identified the human homolog (*hMSH2*) to the yeast mismatch repair protein Mut-S; this turned out to be the gene responsible for the families linked to chromosome 2 (Fishel *et al.*, 1994; Leach *et al.*, 1993). This provided obvious candidates for other genes involved in the HNPCC syndrome. Subsequently, the human homolog of the yeast Mut-L gene, *hMLH1*, was found to be responsible for the chromosome

3-linked families (Bronner *et al.*, 1994). Since then, mutations in at least three other genes involved in mismatch repair have been identified in a smaller number of HNPCC families. Of these, *MSH6* appears to be the most important (Kolodner *et al.*, 1999) and may be associated to a lesser degree with MIN. Although 15% of all colorectal tumors have the RER+ phenotype, only a small fraction of these are due to germline mutations in the (known) mismatch repair genes. The majority of such RER+ tumors appear to be the result of genomic instability caused by a combination of somatic mutation and silencing by methylation. A useful review of the current knowledge of the genetics of colorectal cancer is provided by Lynch and de la Chapelle (1999).

5. Melanoma

The first studies of familial melanoma focused on a syndrome first recognized clinically by Lynch *et al.* (1978) called familial atypical multiple mole melanoma (FAMMM) and by Clark *et al.* (1978), who denoted this clustering as B-K mole syndrome after the surnames of the two probands. Later, the more general term of dysplastic nevus syndrome (DNS) became more widely used. The phenotype is characterized by one or more of the following characteristics: large numbers of moles, very large moles, or moles which have specific characteristics of dysplasia when viewed histopathologically. There is, however, some controversy about whether the key feature is the number of moles, the underlying pathology (i.e., should a person with a single dysplastic nevus be counted as affected), and even the reliability of the histopathological features themselves. These families are also characterized by a high incidence of melanoma. There is evidence that these so-called dysplastic nevi represent a true precursor lesion for the subsequent transformation to malignancy, akin to the polyp in colon cancer (Reimer *et al.*, 1978). Certainly the number and/or size of moles is a strong risk factor for melanoma in epidemiological studies. In addition, segregation studies (Goldgar *et al.*, 1991) have shown that a phenotype based on total number and size of nevi, adjusted for body size was most consistent with Mendelian segregation from among the series of models tested. Earlier, using a set of families with the DNS phenotype, Greene *et al.* (1983) presented evidence of linkage of a locus predisposing to melanoma and DNS was linked to the Rh blood group locus on chromosome 1. Later, Bale *et al.* (1989) further refined the location to between two RFLP markers on chromosome 1p, PND and D1S47. The maximum location score in these kindreds was over 5 at a position in between these two markers. Subsequently, two reports, one from the Netherlands and the other from Utah, failed to replicate this finding. The Utah study cited the diagnostic uncertainty in the definition of DNS as a possible cause in the failure to replicate as well as genetic heterogeneity. Other groups also failed to replicate this finding (Gruis *et al.*, 1990; Kefford *et al.*, 1991; Nancarrow *et al.*, 1992). In 1992, Cannon-Albright *et al.* reported linkage to chromosome 9p in a series of 11 kindreds including three quite extensive families containing 14, 16 and 22 cases of melanoma, respectively. Using malignant melanoma (MLM) exclusively as the definition of the affected phenotype, a lod score of > 10 was achieved with two markers on chromosome 9p with no evidence of heterogeneity reported, although several of the families were uninformative for the tested markers. The linkage to chromosome 9p was quickly replicated by an Australian group who also found linkage to this region (Nancarrow *et al.*, 1993), again with no significant evidence for locus heterogeneity. Even the group reporting the original 1p linkage found significant evidence for a 9p locus in their set of families, although some evidence still remained for a locus on chromosome 1p.

Through a combined analysis of tumors and through identification of recombinants in three large clearly linked families, the region on chromosome 9p was narrowed down to 2 cM (Cannon-Albright *et al.*, 1994a). From there, an analysis of a large number of tumors at multiple sites showed evidence of both hemizygous and homozygous deletions in this region of chromosome 9p detected using representational difference analysis. By examining the minimum region of overlap of these key homozygous deletions, a set of candidate cDNAs was derived through exon trapping and screening of cDNA libraries. One of these candidate cDNAs showed evidence of mutations in the 9p-linked kindreds. Comparison of the sequence of this gene with the GenBank database showed this gene to be a previously identified cell cycle control gene, *p16* (Okamoto *et al.*, 1994). More specifically, p16 acts as an inhibitor of a cyclin-dependent kinase 4 (CDK4) inhibitor which controls the progression from G2 to S phase in the cell cycle allowing stoppage for the repair of DNA. Mutations in *p16* allow cells to proceed through the cell cycle and replicate without undergoing DNA repair and therefore acquire mutations that confer a growth advantage leading to malignancy. In spite of the linkage data which showed no evidence for locus heterogeneity, germ-line mutations in the coding regions could be identified only in about 50% of high-risk melanoma families, and not all families linked to 9p, indicating the presence of noncoding mutations or large genomic rearrangements or deletions. Because p16 acts as an inhibitor of CDK4, this might be considered as another candidate gene for familial melanoma. Indeed, not long after the discovery of the role of *p16* in melanoma susceptibility, Zuo *et al.* (1996) reported two families with identical mutations in *CDK4* which affect binding by p16. Subsequently, another mutation has been identified in this gene, but it appears to be a rare event in familial melanoma in general, leaving open the possibility of other melanoma susceptible loci.

Although no supporting evidence from other data sets has been forthcoming on the 1p locus, this could be a result of the focus of many other studies on malignant melanoma, while the 1p locus may be largely or solely implicated in the DNS phenotype. There is some evidence that the nevus related phenotypes may not be totally explained by germline mutations in the *p16* gene, even in families where the melanoma susceptibility is due to such mutations (Cannon-Albright *et al.*, 1994b; Hussussian *et al.*, 1994). The problem, however, arises in that there is no clear definition of DNS, although there are certainly individuals who have such a florid nevus phenotype that they could be diagnosed across the room. However, in less dramatic cases, the diagnosis has sometimes depended on a single nevus displaying a dysplastic histopathology. Given that there seems to be little agreement among pathologists about just what constitutes a 'dysplastic nevus', it is no surprise that there have been difficulties in replicating the linkage. Whether or not there exists a locus on chromosome 1p which contributes either to a nevus phenotype or to malignant melanoma, will only be resolved when the gene is identified and mutations are found. It has recently been shown that in a series of Australian twin pairs not selected for any melanoma or nevus phenotypes, the phenotypes defined by number of nevi and by total nevus density exhibited excess sharing of markers in the *p16* region on chromosome 9p (Zhu *et al.*, 1999). This raises the possibility that polymorphic variation in the *p16* gene may control normal variation in nevus occurrence.

6. Prostate cancer

Compared with, for example, breast and colon cancer, our understanding of the underlying genetic basis for prostate cancer is relatively less developed. This is in spite of the fact that prostate cancer is, at least in most developed countries, more frequent than either breast or colorectal cancer, and in population studies is associated with a higher familial risk (Goldgar *et al.*, 1994a). As in the case of breast cancer, population-based segregation analyses have shown evidence for autosomal dominant inheritance (Carter *et al.*, 1992; Grönberg *et al.*, 1997; Schaid *et al.*, 1998), although some studies have shown support for X-linked or recessive inheritance (Hayes *et al.*, 1995; Monroe *et al.*, 1995; Narod *et al.*, 1995; Woolf 1960), based on differences in risk to brothers compared to fathers. In the last three years or so, there have been a number of efforts to map susceptibility loci for prostate cancer using a genomic search of STR markers in families with multiple cases of prostate cancer. Linkages have been reported for loci on 1q24–25, 1q42, Xq, and on 1p36, 20q, and with weaker evidence, a region of 16q. In all cases however, as detailed below, the evidence has been far less than convincing and studies of replication have had mixed results. The first linkage was to chromosome 1q24–25 by a large NIH/Swedish consortium (Smith *et al.*, 1996). The genetic model assumed for the parametric linkage analyses was that of Carter *et al.* (1992) which yielded a maximal total lod score of 3.65. However, statistically significant evidence of locus heterogeneity was present and a lod score assuming heterogeneity (hlod) was 5.43 with an estimated 34% of the families linked to this locus, which was denoted HPC1 (hereditary prostate cancer 1). Almost all of the evidence for linkage to this region came from a subset of families characterized by many (>4) cases of prostate cancer and with relatively early onset (mean age < 65 years) of the disease. The next reported location of a prostate cancer susceptibility locus was the X chromosome (HPCX; Xu *et al.*, 1998). Again, there was significant evidence of linkage (multipoint lod of 3.85) and locus heterogeneity with an estimated 16% of the total data set of 360 families being attributed to HPCX. In contrast to the 1q24–25 locus, the Xq linkage evidence was not confined to early onset cancer, and in fact, may play more of a role in cancer diagnosed later in life. The localization of a prostate cancer susceptibility locus on the X chromosome is consistent with reports of higher familial risk to brothers of probands than to fathers, though this could also be attributed to recessive effects. This localization was soon followed by a linkage to a more distal region on chromosome 1q than HPC1, 1q42, by a collaborative study of French and German prostate cancer families (Berthon *et al.*, 1998). In addition, a report of linkage to chromosome 1p36 was reported (Gibbs *et al.*, 1999a), although the effect was modest and largely confined to families with brain and prostate cancer, designated CAPB (cancer of prostate and brain). More recently, there has been a report of linkage of a prostate cancer susceptibility locus on chromosome 20 based on a genome-wide search of 162 families with at least three cases of prostate cancer (Berry *et al.*, 2000a), again with evidence of heterogeneity.

Since the original reports, a number of studies have been undertaken to replicate these loci. In the case of the 1q24–25 locus, some studies have found evidence largely supporting the original report (Cooney *et al.*, 1997; Hsieh *et al.*, 1997; Neuhausen *et al.*, 1999), while three others (Berthon *et al.*, 1998; Eeles *et al.*, 1998; McIndoe *et al.*, 1997) failed to find evidence of linkage to this region. However, due to the relatively small size of these studies, the authors could not rule out a 1q24 locus accounting for less than 30% of all HPC families. In a further study conducted recently by the Swedish group which contributed to the original report, a further 28 Swedish prostate cancer families were analyzed, with only the nine families with average age of onset lower than 65 years, providing any support for this region. A recent international collaborative study pooling 772 families from Australia, North America, the United Kingdom and Scandinavia, estimated that overall, 6% of all families could be attributed to the 1q24–25 (HPC1) locus. However, if one considered only the 491 families which showed evidence of male-to-male transmission, 11% of families were estimated to be linked with an overall heterogeneity lod score of 2.56 ($P<0.001$).

Attempts to replicate the 1q42 locus have consistently been negative (Eeles *et al.*, 1998; Gibbs *et al.*, 1999b; Whittemore *et al.*, 1999), and the only attempts at replication of the 1p36 linkage did not find evidence to support the association with brain cancer (Badzioch *et al.*, in press; Berry *et al.*, 2000b). However, the latter study did have some small evidence in favor of linkage in families with early onset. To date, only one study has replicated the HPCX locus (Lange *et al.*, 1999) with positive, but not significant lod scores found in the set of families without evidence of male-to-male transmission.

6.1 Potential complications in prostate cancer linkage

One might certainly wonder why there has been more difficulty in localizing prostate cancer loci (at least with any reasonable degree of certainty) compared with the other cancers described in this chapter. Given the advances in sequence information, physical mapping resources, etc., it is also of interest that, to date, no prostate susceptibility loci (other than perhaps *BRCA1* and *BRCA2*) have been cloned. Both prostate and breast cancer are sex-limited diseases which generally results in less informative families for linkage, and both occur at relatively equal frequencies in the respective sex. Both also have similar patterns of familial relative risk, and segregation analyses have suggested similar allele frequencies and penetrances of underlying susceptibility locus or loci. However, prostate cancer studies present several problems that might explain some of the inconsistencies in linkage studies as well as the failure to positionally clone any susceptibility loci from the regions suggested by these linkage analyses. The most obvious difference is in the later age of onset of prostate cancer in both sporadic and familial cancer, being on average about 10–15 years later than breast cancer. This reduces the informativeness of families for study. In addition, men, as a rule, are less willing to participate in research studies than women which also reduces the total amount of linkage information available. Another complication is that due to the recent advent (at least in the USA) of prostate screening by prostate specific antigen (PSA), cases are detected much earlier than criteria based on purely clinical criteria. Many of these cases, if left untreated, would likely resolve on their own and never be diagnosed clinically. This has led to over-diagnosis of prostate cancer, particularly in men who have one or more affected relatives. This in turn could result in an increased rate of phenocopies in prostate cancer families, further confounding linkage studies. We must also consider the hypothesis that prostate cancer is more genetically heterogeneous than breast cancer and that each locus makes a roughly equal contribution in terms of the proportion of families attributable to that locus. Finally, aside from perhaps age at diagnosis, there so far seems to be no specific phenotypic feature in either the cases or tumors themselves, or in the families to allow a stratification of families. The only exception may be the linkage to chromosome 1p, which at least in the initial report, seemed to be primarily in families which had one or more cases of brain cancer in addition to prostate cancer (Gibbs *et al.*, 1999a), but this finding has not been replicated. As previously discussed, the presence of ovarian cancer in *BRCA1* families, male breast cancer in *BRCA2*, and endometrial cancer and the presence of the RER phenotype in colon tumors to delineate families likely due to mismatch repair mutations were key features in the ability to localize and/or identify the genes involved in those cancer families.

7. Evidence for additional cancer susceptibility loci for breast, colorectal, and melanoma, and strategies to find them

From studies conducted by a number of investigators, it is clear that in addition to *BRCA1* and *BRCA2*, that there are additional genes that predispose to breast cancer (Ford *et al.*, 1998; Schubert *et al.*, 1997; Serova *et al.*, 1997). Although evidence suggests that these may be slightly less penetrant than *BRCA1* and *BRCA2*, these loci are likely to still cause substantial familial aggregation. Although there have been reports of linkages, specifically to a region of chromosome 6q which includes the estrogen receptor locus (Zuppan *et al.*, 1991), and also to chromosome 8p (Kerangueven *et al.*, 1995; Seitz *et al.*, 1997) to a region which is frequently lost in breast tumors. However, an analysis by Rahman *et al.* (in press) of a large collection of families showed that the proposed region on chromosome 8p, even if real, does not account for the vast majority of non-*BRCA1*/*BRCA2* inherited breast cancers. Similarly, a study by Shugart *et al.* (1999) ruled out involvement of the tumor suppressor gene *PTEN*, which is associated with early onset breast cancer in the context of Cowden's syndrome in familial breast cancer. In order to identify these loci, studies are ongoing in both high-risk families and in affected sister pairs. Preliminary results from at least one of the family studies indicates that there are likely to be multiple loci involved acting either additively or multiplicatively which account for these families, making the task of identifying any of them more difficult. Unlike breast cancer, however, there is little evidence of susceptibility loci for ovarian cancer other than *BRCA1/2* and perhaps to a lesser degree the mismatch repair genes (Gayther *et al.*, 1999).

The evidence for genetic susceptibility to colon cancer aside from the mismatch repair genes discussed earlier in this chapter seems to be less than that for breast cancer. Certainly there have been reports of families with multiple cases of cancer which, based on the tumor phenotype, are not due to any of the known (or unknown) mismatch repair genes (Lewis *et al.*, 1996). However, these families are not so numerous and the age at onset not so remarkable, that chance or environmental clustering cannot be ruled out in their etiology. It is likely that loci not associated with mismatch repair defects are of lower penetrance, and therefore a sib-pair strategy may be useful in localizing these loci.

For melanoma, there appears to be at least one other major susceptibility locus other than the *p16* locus on chromosome 9p. Most studies have found about half of all families with clearly hereditary malignant melanoma to be due to this locus. A large collaborative Australian family study has potentially identified one such additional locus (G. Mann, personal communication) but the evidence is not yet at a sufficient level to be considered a definitive finding.

8. Role of low penetrance loci in common cancer

Thus far, this chapter has focused on the role of genes that confer relatively high cancer risks to individuals who are heterozygous for a deleterious mutation. However, these loci cannot explain all the familial risk in these cancers and it is likely that there are other loci which are involved but which individually do not give rise to detectable familial clustering. These loci will be difficult, if not impossible to detect using a linkage-based approach. More likely, these loci which may be associated with a two or three-fold increased risk of cancer (or even less), are more amenable to examination using either population-based or family-based case–control studies. Many of the advantages of cancer as a phenotype, which have already been discussed at the beginning of this chapter in reference to analysis of cancer in families, apply to the case–control situation as well, most notably the existence of cancer registries. Because for the association-based approach the goal is to detect linkage disequilibrium between a susceptibility locus and a genetic marker (although in some cases the susceptibility and the tested polymorphism may be identical), rather than cosegregation in families, the marker must be very close to the disease susceptibility locus. Although the distance will vary depending on the age of the susceptibility allele, the genomic sequence surrounding both the marker and allele, the mutation rate at the marker locus, the structure of the population and other factors, a general rule of thumb is that the marker locus should be within about 100 kbp of the disease allele. In order for a genome-wide search for a novel susceptibility locus it is estimated that the number of random single nucleotide polymorphisms will need to be in the order of 50 000–100 000. At the present time there are an insufficient number of mapped single nucleotide polymorphisms (SNPs) and the technology to analyze them is prohibitive. For this reason, almost all such studies have focused on polymorphisms within genes which biologically may be important in the cancer under study, or which are associated with basic cellular or physiological processes. In most cancer studies, these have focused on polymorphisms located within or near:

(i) genes coding for enzymes involved in the metabolism and detoxification of carcinogens including the cytochrome P450 family and glutathione-S-transferase (GSTs);
(ii) genes involved in the repair of DNA damage;
(iii) genes related to cell growth and differentiation or steroid hormone pathways;
(iv) known high-risk genes such as *p16* or *BRCA1*.

It is likely that interpretation of results from such studies will be difficult, if not impossible, without consideration of environmental exposures and lifestyle factors that interact with the genetic susceptibility polymorphism. Fortunately, there are a large number of both case–control studies and cohort studies whose focus is on the environmental (diet, lifestyle, occupational exposures, etc.) aspects (e.g., Riboli and Kaaks, 1997), but for whom biological samples are available. These can be used either for genotyping of candidate susceptibility polymorphisms or a genome-wide search for novel susceptibility loci.

A comprehensive literature survey of all such studies that have been done in cancer would fill an entire volume, but I will try to illustrate some of the work that has been done to demonstrate some of the problems inherent in such studies. For the interested reader, a useful review of genetic polymorphisms in breast cancer can be found in an article by Dunning *et al.* (1999) which reports results of a meta-analysis of 46 published case–control studies which examined polymorphisms in 18 different genes, including all of the types listed above. In these studies a total of 12 statistically significant associations were found. However, no two studies provided independent confirmation of any of these associations. Many of the studies had limited power to detect odds ratios (OR) less than 2.5; for this reason a meta-analysis was performed, after which three polymorphisms remained significant, with the CYP19 $(TTTA)_n$ polymorphism showing the strongest effect (OR=2.33, P=0.002). For most polymorphisms, the combined analysis ruled out effects larger than 1.5. However, a more recent study by these same authors (Healey *et al.*, 2000) found no effect of polymorphisms in this gene in a series of approximately 600 breast cancer cases and 400 controls from the British population. Similarly, a recent review of the role of GST-M and GST-T in colorectal cancer emphasized the difficulties inherent in such studies (Cotton *et al.*, 2000).

9. Summary

As detailed in this chapter, the search for genes which, when altered, cause markedly increased susceptibility to common cancers of the breast, ovary, bowel, prostate and skin, has found a certain degree of success in the last decade. These successes have come about for a number of reasons notably: candidate loci based on homology to a lower organism and motivated by observations of a specific tumor phenotype; a combination of linkage and the observation of a homozygous deletion in a related tumor; and by standard linkage/recombinant mapping/positional cloning.

However, there are several commonalities shared by all these successes. First there is the recognition of specific constellations of cancers characterized by early age of onset compared to non-familial cancer; second, the ability to avoid complications caused by misclassifications and misdiagnosis. Finally, one would have to include the magnitude of the prize, mandated by the importance of these diseases to the general population and the overemphasis in the press of the importance of genetic susceptibility to these disorders. This has in some sense insured the funding by both government and commercial sources of a large-scale effort to attack these problems. It is noteworthy that almost all of the localizations to date in cancer genetics have been achieved using traditional linkage analysis in which a genetic model is specified, rather than the model-free methods which are commonly applied to other complex traits. This is due, in large part, to population-based segregation analyses and/or studies of familial risk which provide a reasonable genetic model for use in model-based methods.

9.1 Future prospects

The study of genetic susceptibility to cancer has been aided in part by good diagnostic criteria, excellent case registration, often population-based, and by the presence of tumors from which candidate loci or regions can be suggested through analyses of mechanistic genetic pathways in key cellular processes or studies of differences in tumor and normal tissue. These are features which should continue to help in developing a true understanding of the genetic pathways which lead to cancer both at the germline level as susceptibility and the somatic alterations which predispose an individual cell to become malignant and spread throughout the body.

Although the successes in cancer genetics have been many, there still remain a number of challenges to be met in the coming years. Perhaps the greatest disappointment is that no definitive prostate cancer susceptibility locus has been identified either by a candidate gene approach, or by positional cloning (although by the time this book is printed, this will likely no longer hold). There has also been unexpected difficulty in localizing additional breast cancer susceptibility loci in spite of two large international consortia and despite the evidence that approximately half of site-specific female breast cancer families are unexplained by *BRCA1* and *BRCA2*. This likely reflects the problem of substantial locus heterogeneity or perhaps less likely, a heretofore unrecognized strong environmental component or viral agent causing the observed familial clustering. As in all complex diseases, the most daunting challenge will lie in the localization and identification of lower risk loci, which may be more common in the population, and therefore account for a larger fraction of cancer than the relatively rare, high-risk genes. The recent localization of a testicular cancer predisposition gene to the X chromosome (Rapley *et al.*, 2000) using an affected relative pair approach holds promise for similar success for low risk susceptibility loci in the common cancers. It should be noted, however, that the success in testicular cancer came after several years of intensive effort and was realized only when families were stratified by a clinical phenotype, presence of bilateral disease.

In addition, one must also consider susceptibility polymorphisms which cause very modest increased risks of cancer (relative risks on the order of 1.5). In the end, it is likely that these effects will only be definitively characterized by considering specific relevant environmental exposures as well as other genetic effects. Given the advantages of studying cancer as a phenotype outlined in this chapter, if the combination of genetics and epidemiology are going to be useful in untangling the complex interactions that surely govern the etiology of complex traits, it will likely be in the area of neoplastic disease. The long-term hope is, of course, that this greater understanding will translate into improved cancer prevention and/or treatment strategies and consequently reduce the all-too-high worldwide burden of this disease.

Acknowledgments

The author would like to thank Ms Yvette Granjard for her assistance in assembling the reference material and Prof Tim Bishop for not being overly strict about deadlines.

References

Aaltonen, L.A., Sankila, R., Mecklin, J.P., Jarvinen, H., Pukkala, E., Peltomaki, P. and de la Chapelle, A. (1994) A novel approach to estimate the proportion of hereditary nonpolyposis colorectal cancer of total colorectal cancer burden. *Cancer Detect. Prev.* **18**: 57–63.

Badzioch, M., Eeles, R., Leblanc, G., *et al.* (2000) Suggestive evidence for a site-specific early-onset prostate cancer gene on chromosome 1p36. *J. Med. Genet.* (in press).

Bale, S.J., Dracopoli, N.C., Tucker, M.A., Clark, W.H., Jr., Fraser, M.C., Stanger, B.Z., Green, P., Donis-Keller, H., Housman, D.E. and Greene, M.H. (1989) Mapping the gene for hereditary cutaneous malignant melanoma-dysplastic nevus to chromosome 1p. *New Engl. J. Med.* **320**: 1367–1372.

Barker, D., Wright, E., Nguyen, K., *et al.* (1987) Gene for von Recklinghausen neurofibromatosis is in the pericentromeric region of chromosome 17. *Science* **236**: 1100–1102.

Berry, R., Schroeder, J.J., French, A.J., McDonnell, S.K., Peterson, B.J., Cunningham, J.M., Thibodeau, S.N. and Schaid, D.J. (2000a) Evidence for a prostate cancer-susceptibility locus on chromosome 20. *Am. J. Hum. Genet.* **67**: 82–91.

Berry, R., Schaid, D.J., Smith, J.R., *et al.* (2000b) Linkage analyses at the chromosome 1 loci 1q24–25 (HPC1), 1q42.2–43 (PCAP), and 1p36 (CAPB) in families with hereditary prostate cancer. *Am. J. Hum. Genet.* **66**: 539–546.

Berthon, P., Valeri, A., Cohen-Akenine, A., *et al.* (1998) Predisposing gene for early-onset prostate cancer, localized on chromosome 1q42.2–43. *Am. J. Hum. Genet.* **62**: 1416–1424.

Bignell, G.R., Canzian, F., Shayeghi, M., Stark, M., *et al.* (1997) Familial nontoxic multinodular thyroid goiter locus maps to chromosome 14q but does not account for familial nonmedullary thyroid cancer. *Am. J. Hum. Genet.* **61**: 1123–1130.

Bishop, D.T., Cannon-Albright, L., McLellan, T., Gardner, E.J. and Skolnick, M.H. (1988) Segregation and linkage analysis of nine Utah breast cancer pedigrees. *Genet. Epidemiol.* 5: 151–169.

Bodmer, W.F., Bailey, C.J., Bodmer, J., *et al.* (1987) Localization of the gene for familial adenomatous polyposis on chromosome 5. *Nature* **328**: 614–616.

Broca, P. (1866) Influence héréditaire – *Traité des Tumeurs – Tome premier: Des tumeurs en général.* (ed. P. Asselin), Paris. pp. 149–154.

Bronner, C.E., Baker, S.M., Morrison, P.T., *et al.* (1994) Mutation in the DNA mismatch repair

gene homologue hMLH1 is associated with hereditary non-polyposis colon cancer. *Nature* **368**: 258–261.
Cannon-Albright, L.A. Goldgar, D.E., Meyer, L.J., *et al.* (1992) Assignment of a locus for familial melanoma, MLM, to chromosome 9p13-p22. *Science* **258**: 1148–1152.
Cannon-Albright, L.A., Goldgar, D.E., Neuhausen, S., *et al.* (1994a) Localization of the 9p melanoma susceptibility locus (MLM) to a 2-cM region between D9S736 and D9S171. *Genomics* **23**: 265–268.
Cannon-Albright, L.A., Meyer, L.J., Goldgar, D.E., Lewis, C.M., McWhorter, W.P., Jost, M., Harrison, D., Anderson, .D.E, Zone, J.J. and Skolnick, M.H. (1994b) Penetrance and expressivity of the chromosome 9p melanoma susceptibility locus. *Cancer Res.* **54**: 6041–6044.
Carter, B.S., Beaty, T.H., Steinberg, G.D., Childs, B. and Walsh, P.C. (1992) Mendelian inheritance of familial prostate cancer. *Proc. Natl Acad. Sci. USA* **89**: 3367–3371.
Clark, W.H., Reimer, R.R., Greene, M., Ainsworth, A.M. and Mastrangelo, M.J. (1978) Origin of familial malignant melanomas from heritable melanocyte lesions: the B-K mole syndrome. Arch. Dermatol. **114**: 732–738.
Claus, E.B., Risch, N.J. and Thompson, W.D. (1990) Age at onset as an indicator of familial risk of breast cancer. *Am. J. Epidemiol.* **131**: 961–972.
Cooney, K.A., McCarthy, J.D., Lange, E., Huang, L., Miesfeldt, S., Montie, J.E., Oesterling, J.E., Sandler, H.M. and Lange, K. (1997) Prostate cancer susceptibility locus on chromosome 1q: a confirmatory study. *J. Natl Cancer Inst.* **89**: 955–959.
Cotton, S.C., Sharp, L., Little, J. and Brockton, N. (2000) Glutathione S-transferase polymorphisms and colorectal cancer: a HuGE review. *Am. J. Epidemiol.* **151**: 7–32.
Dunning, A.M., Healey, C.S., Pharoah, P.D., Teare, M.D., Ponder, B.A. and Easton, D.F. (1999) A systematic review of genetic polymorphisms and breast cancer risk. *Cancer Epidemiol. Biomarkers Prev.* **8**: 843–854.
Easton, D.F., Steele, L., Fields, P., *et al.* (1997) Cancer risks in two large breast cancer families linked to BRCA2 on chromosome 13q12–13. *Am. J. Hum. Genet.* **61**: 120–127.
Eeles, R.A., Durocher, F., Edwards, S., *et al.* (1998) Linkage analysis of chromosome 1q markers in 136 prostate cancer families. The Cancer Research Campaign/British Prostate Group U.K. Familial Prostate Cancer Study Collaborators. *Am. J. Hum. Genet.* **62**: 653–658.
Fishel, R., Lescoe, M.K., Rao, M.R., Copeland, N.G., Jenkins, N.A., Garber, J., Kane, M. and Kolodner, R. (1994) The human mutator gene homolog MSH2 and its association with hereditary nonpolyposis colon cancer. *Cell* **77**: 167.
Ford, D., Easton, D.F., Stratton, M., *et al.* (1998) Genetic heterogeneity and penetrance analysis of the *BRCA1* and *BRCA2* genes in breast cancer families. *Am. J. Hum. Genet.* **62**: 676–689.
Gardner, E.J. (1951) A genetic and clinical study of intestinal polyposis, a predisposing factor for carcinoma of the colon and rectum. *Am. J. Hum. Genet.* **3**: 167–176.
Gardner, E.J. and Stephens, F.E. (1950) Breast cancer in one family group. *Am. J. Hum. Genet.* **2**: 30–40.
Gayther, S.A., Warren, W., Mazoyer, S., *et al.* (1995) Germline mutations of the *BRCA1* gene in breast and ovarian cancer families provide evidence for a genotype–phenotype correlation. *Nat. Genet.* **11**: 428–433.
Gayther, S.A., Mangion, J., Russell, P., Seal, S., Barfoot, R., Ponder, B.A., Stratton, M.R. and Easton, D. (1997) Variation of risks of breast and ovarian cancer associated with different germline mutations of the *BRCA2* gene. *Nat. Genet.* **15**: 103–105.
Gayther, S.A., Russell, P., Harrington, P., Antoniou, A.C., Easton, D.F. and Ponder, B.A. (1999) The contribution of germline *BRCA1* and *BRCA2* mutations to familial ovarian cancer: no evidence for other ovarian cancer-susceptibility genes. *Am. J. Hum. Genet.* **65**: 1021–1029.
Gibbs, M., Stanford, J.L., McIndoe, R.A., *et al.* (1999a) Evidence for a rare prostate cancer-susceptibility locus at chromosome 1p36. *Am. J. Hum. Genet.* **64**: 776–787.
Gibbs, M., Chakrabarti, L., Stanford, J.L., *et al.* (1999b) Analysis of chromosome 1q42.2–43 in 152 families with high risk of prostate cancer. *Am. J. Hum. Genet.*, **64**: 1087–1095.
Goldgar, D.E., Cannon-Albright, L.A., Meyer, L.J., Piepkorn, M.W., Zone, J.J. and Skolnick, M.H. (1991) Inheritance of nevus number and size in melanoma and dysplastic nevus syndrome kindreds. *J. Natl Cancer Inst.* **83**: 1726–1733.
Goldgar, D.E., Easton, D.F., Cannon-Albright, L.A. and Skolnick, M.H. (1994a) A systematic

population-based assessment of cancer risk in first degree relative of cancer probands. *J. Natl Cancer Inst.* **86**: 1600–1608.

Goldgar, D.E., Fields, P., Lewis, C.M., Tran, T.D., Cannon-Albright, L.A., Ward, J.H., Swensen, J. and Skolnick, M.H. (1994b) A large kindred with 17q-linked breast and ovarian cancer: genetic, phenotypic and genealogical analysis. *J. Natl Cancer Inst.* **86**: 200–209.

Goss, K.H. and Groden, J. (2000) Biology of the adenomatous polyposis coli tumor suppressor. *J Clin Oncol.* **18**: 1967–1979.

Greene, M.H., Goldin, L.R., Clark, W.H., Lovrien, E., Kraemer, K.H., Tucker, M.A., Elder, E.D., Fraser, M.C. and Rowe, S. (1983) Familial cutaneous malignant melanoma: autosomal dominant trait possibly linked to the Rh locus. *Proc. Natl Acad. Sci. USA* **80**: 6071–6075.

Groden, J., Thliveris, A., Samowitz, W., *et al.* (1991) Identification and characterization of the familial adenomatous polyposis coli gene. *Cell* **66**: 589–600.

Grönberg, H., Damber, L., Damber, J.E. and Iselius, L. (1997a) Segregation analysis of prostate cancer in Sweden: support for dominant inheritance. *Am. J. Epidemiol.* **146**: 552–557.

Grönberg, H., Isaacs, S.D., Smith, J.R., *et al.* (1997b) Characteristics of prostate cancer in families potentially linked to the hereditary prostate cancer 1 (HPC1) locus. *JAMA* **278**: 1251–1255.

Grönberg, H., Xu, J., Smith, J.R., *et al.* (1997c) Early age at diagnosis in families providing evidence of linkage to the hereditary prostate cancer locus (HPC1) on chromosome 1. *Cancer Res.* **57**: 4707–4709.

Gruis, N.A., Bergman, W. and Frants, R.R. (1990) Locus for susceptibility to melanoma on chromosome 1p. *New Engl. J. Med.* **322**: 853–854.

Hall, J.M., Lee, M.K., Newman, B., Morrow, J.E., Anderson, L.A., Huey, B. and King, M.C. (1990) Linkage of early-onset familial breast cancer to chromosome 17q21. *Science* **250**: 1684–1689.

Hayes, R.B., Liff, J.M., Pottern, L.M., *et al.* (1995) Prostate cancer risks in US blacks and whites with a family history of cancer. *Int. J. Cancer* **60**: 361–364.

Healey, C.S., Dunning, A.M., Durocher, F., Teare, D., Pharoah, P.D., Luben, R.N., Easton, D.F. and Ponder, B.A. (2000) Polymorphisms in the human aromatase cytochrome P450 gene (*CYP19*) and brease cancer risk. Carcinogenesis **21**: 189–193.

Houlston, R.S., Collins, A., Kee, F., Collins, B.J., Shields, D.C. and Morton, N.E. (1995) Segregation analysis of colorectal cancer in Northern Ireland. *Hum. Hered.* **45**: 41–48.

Hsieh, C.L., Oakley-Girvan, I., Gallagher, R.P., Wu, A.H., Kolonel, L.N., Teh, C.Z., Halpern, J., West, D.W., Paffenbarger, R.S. Jr and Whittemore, A.S. (1997) Prostate cancer susceptibility locus on chromosome 1q: a confirmatory study. *J. Natl Cancer Inst.* **89**: 1893–1894.

Hemminki, K. and Dong, C. (2000) Familial relationships in thyroid cancer by histopathological type. *Int. J. Cancer* **85**: 201–205.

Hussussian, C.J., Struewing, J.P., Goldstein, A.M., Higgins, P.A., Ally, D.S., Sheahan, M.D., Clark, W.H. Jr, Tucker, M.A. and Dracopoli N.C. (1994) Germline p16 mutations in familial melanoma. *Nat. Genet.* **8**: 15–21.

International Agency for Research on Cancer (1997) Cancer incidence in five continents. (eds D.M. Parkin, S.L. Whelan, J. Ferlay, L. Raymond, and J. Young). IARC Sci. Pub. No 143.

Kefford, R.F., Salmon, J., Shaw, H.M., Donald, J.A. and McCarthy, W.H. (1991) Hereditary melanoma in Australia: variable association with dysplastic naevi and absence of genetic linkage to chromosome 1p. *Cancer Genet. Cytogenet.* **51**: 45–55.

Kerangueven, F., Essioux, L., Dib, A., Noguchi, T., Allione, F., Geneix, J., Longy, M., Lidereau, R., Eisinger, F. and Pebusque, M.J. (1995). Loss of heterozygosity and linkage analysis in breast carcinoma: indication for a putative third susceptibility gene on the short arm of chromosome 8. *Oncogene* **10**: 1023–1026.

King, M.C., Go, R.C., Elston, R.C., Lynch, H.T. and Petrakis, N.L. (1980) Allele increasing susceptibility to human breast cancer may be linked to the glutamate–pyruvate transaminase locus. *Science* **208**: 406–408.

Kinzler, K.W., Nilbert, M.C., Su, L.K., *et al.* (1991) Identification of FAP locus genes from chromosome 5q21. *Science* **253**: 661–665.

Kolodner, R.D., Tytell, J.D., Schmeits, J.L., *et al.* (1999) Germline *msh6* mutations in colorectal cancer families. *Cancer Res.* **59**: 5068–5074.

Knudson, A.G. (1971) Mutation and cancer: statistical study of retinoblastoma. *Proc. Natl Acad. Sci. USA* **68**: 820–823.

Lange, E.M., Chen, H., Brierley, K., Perrone, E.E., Bock, C.H., Gillanders, E., Ray, M.E. and

Coone, K.A. (1999) Linkage analysis of 153 prostate cancer families over a 30-cM region containing the putative susceptibility locus HPCX. *Clin. Cancer Res.* 5: 4013–4020.

Larsson, C., Skogseid, B., Oberg, K., Nakamura, Y. and Nordenskjold, M. (1988) Multiple endocrine neoplasia type 1 gene maps to chromosome 11 and is lost in insulinoma. *Nature* **332**: 85–87.

Leach, F.S., Nicolaides, N.C., Papadopoulos, N., *et al.* (1993) Mutations of a *mutS* homolog in hereditary nonpolyposis colorectal cancer. *Cell.* **75**: 1215–1225.

Leppert, M., Dobbs, M., Scrambler, P., *et al.* (1987) The gene for familial polyposis coli maps to the long arm of chromosome 5. *Science* **238**: 1411–1413.

Lewis, C.M., Neuhausen, S.L., Daley, D., Black, F.J., Swensen, J., Burt, R.W., Cannon-Albright, L.A. and Skolnick, M.H. (1996) Genetic heterogeneity and unmapped genes for colorectal cancer. *Cancer Res.* **56**: 1382–1388.

Li, F.P. and Fraumeni, J.F. (1969) Rhabdomyosarcoma in children: an epidemiologic study and identification of a familial cancer syndrome. *J. Natl Cancer Inst.* **43**: 1364–1373.

Lindblom, A., Tannergard, P., Werelius, B. and Nordenskjold, M. (1993) Genetic mapping of a second locus predisposing to hereditary non-polyposis colon cancer. *Nat. Genet.* 5: 279–282.

Lynch, H.T. and de la Chapelle, A. (1999) Genetic susceptibility to non-polyposis colorectal cancer. *J. Med. Genet.* **36**: 801–818.

Lynch, H.T., Lynch, P.M. and Harris, R.E. (1977) Proximal colon cancer in familial carcinoma of the colon exclusive of familial polyposis coli. *Lancet* **1**: 1306–1307.

Lynch, H.T., Frichot, B.C. and Lynch, J.F. (1978) Familial atypical multiple mole melanoma syndrome. *J. Med. Genet.* **15**: 352–356.

Malkin, D., Li, F.P., Strong, L.C., *et al.* (1990) Germ line *p53* mutations in a familial syndrome of breast cancer, sarcomas, and other neoplasms. *Science* **250**: 1233–1238.

McDonald, J.M., Douglass, E.C., Fisher, R., Geiser, C.F., Krill, C.E., Strong, L.C., Virshup, D. and Huff, V. (1998) Linkage of familial Wilms' tumor predisposition to chromosome 19 and a two-locus model for the etiology of familial tumors. *Cancer Res.* **58**: 1387–1390.

McIndoe, R.A., Stanford, J.L., Gibbs, M., *et al.* (1997) Linkage analysis of 49 high-risk families does not support a common familial prostate cancer-susceptibility gene at 1q24–25. *Am. J. Hum. Genet.* **61**: 347–353.

Monroe, K.R., Yu, M.C., Kolonel, L.N., Coetzee, G.A., Wilkens, L.R., Ross, R.K. and Henderson, B.E. (1995) Evidence of an X-linked or recessive genetic component to prostate cancer risk. *Nat. Med.* **1**: 827–829.

Nancarrow, D.J., Palmer, J.M., Walters, M.K., Kerr, B.M., Hafner, G.J., Garske, L., McLeod, G.R. and Hayward, N.K. (1992) Exclusion of the familial melanoma locus (MLM) from the PND/D1S47 and LMYC 1 regions of chromosome arm 1p in 7 Australian pedigrees. *Genomics* **12**: 18–25.

Nancarrow, D.J., Mann, G.J., Holland, E.A., *et al.* (1993) Confirmation of chromosome 9p linkage in familial melanoma. *Am. J. Hum. Genet.* **53**: 936–942.

Narod, S.A., Feunteun, J., Lynch, H.T., Watson, P., Conway, T., Lynch, J. and Lenoir, G.M. (1991) Familial breast-ovarian cancer locus on chromosome 17q12-q23. *Lancet* **33**: 82–83.

Narod, S.A., Dupont, A., Cusan, L., Diamond, P., Gomez, J.L., Suburu, R. and Labrie, F. (1995) The impact of family history on early detection of prostate cancer. *Nat. Med.* **1**: 99–101.

Neuhausen, S.L., Farnham, J.M., Kort, E., Tavtigian, S.V., Skolnick, M.H. and Cannon-Albright, L.A. (1999) Prostate cancer susceptibility locus HPC1 in Utah high-risk pedigrees. *Hum. Mol. Genet.* 8: 2437–2442.

Okamoto, A., Demetrick, D.J., Spillare, E.A., *et al.* (1994) Mutations and altered expression of p16INK4 in human cancer. *Proc. Natl Acad. Sci. USA* **91**: 11045–11049.

Peltomaki, P., Aaltonen, L.A., Sistonen, P., *et al.* (1993) Genetic mapping of a locus predisposing to human colorectal cancer. *Science* **260**: 810–812.

Rahman, N., Arbour, L., Tonin, P., Renshaw, J., Pelletier, J., Baruchel, S., Pritchard-Jones, K., Stratton, M.R. and Narod, S.A. (1996) Evidence for a familial Wilms' tumour gene (FWT1) on chromosome 17q12-q21. *Nat. Genet.* **13**: 461–463.

Rahman, N., Teare, D., Seal, S. *et al.* (2000) Absence of evidence for a familial breast cancer gene on chromosome 8p12–p22. *Oncogene*, in press.

Rapley, E.A., Crockford, G.P., Teare, D., *et al.* (2000) Localization to Xq27 of a susceptibility gene for testicular germ-cell tumours. *Nat. Genet.* **24**: 197–200.

Rebbeck, T.R., Lustbader, E.D. and Buetow, K.H. (1994) Somatic allele loss in genetic linkage analysis of cancer. *Genet. Epidemiol.* **11**: 419–429.
Reimer, R.R., Clark, W.H. Jr, Greene, M.H., Ainsworth, A.M. and Fraumeni, J.F. Jr. (1978) Precursor lesions in familial melanoma. A new genetic preneoplastic syndrome. *JAMA* **239**: 744–746.
Riboli, E. and Kaaks, R. (1997) The EPIC project: rationale and study design. *Int. J. Epidemiol.* **26**(Suppl. 1): S6–S14.
Ron, E., Kleinerman, R.A., LiVolsi, V.A. and Fraumeni, J.F. Jr. (1991) Familial nonmedullary thyroid cancer. *Oncology* **48**: 309–311.
Rouleau, G.A., Wertelecki, W., Haines, J.L., Hobbs, W.J., Trofatter, J.A., Seizinger, B.R., Martuza, R.L., Superneau, D.W., Conneally, P.M. and Gusella, J.F. (1987) Genetic linkage of bilateral acoustic neurofibromatosis to a DNA marker on chromosome 22. *Nature* **329**: 246–248.
Schaid, D.J., McDonnell, S.K., Blute, M.L. and Thibodeau, S.N. (1998) Evidence for autosomal dominant inheritance of prostate cancer. *Am. J. Hum. Genet.* **62**: 1425–1438.
Schubert, E.L., Lee, M.K., Mefford, H.C., Argonza, R.H., Morrow, J.E., Hull, J., Dann, J.L. and King, M.C. (1997) *BRCA2* in American families with four or more cases of breast or ovarian cancer: recurrent and novel mutations, variable expression, penetrance, and the possibility of families whose cancer is not attributable to *BRCA1* or *BRCA2*. *Am. J. Hum. Genet.* **60**: 1031–1040.
Schutte, M., Rozenblum, E., Moskaluk, C.A., Guan, X., Hoque, A.T., Hahn, S.A., da Costa, L.T., de Jong, P.J. and Kern, S.E. (1995) An integrated high-resolution physical map of the *DPC/BRCA2* region at chromosome13q12. *Cancer Res.* 55: 4570–4574.
Scully, R. (2000) Role of *BRCA* gene dysfunction in breast and ovarian cancer predisposition. *Breast Cancer Res.* in press.
Seitz, S., Rohde, K., Bender, E., Nothnagel, A., Kolble, K., Schlag, P.M. and Scherneck, S. (1997) *Oncogene* **14**: 741–743.
Serova, O., Mazoyer, S., Puget, N., Dubois, V., Tonin, P., Shugart, Y., Goldgar, D., Narod, S.A., Lynch, H.T. and Lenoir, G.M. (1997) Mutations in *BRCA1* and *BRCA2* in breast cancer families: are there more breast cancer susceptibility genes? *Am. J. Hum. Genet.* **60**: 486–495.
Shugart, Y.Y., Cour, C., Renard, H., *et al.* (1999) Linkage analysis of 56 multiplex families exclude the Cowden disease gene *PTEN* as a major contributor to familial breast cancer. *J. Med. Genet.* **36**: 720–721.
Simpson, N.E., Kidd, K.K., Goodfellow, P.J., *et al.* (1987) Assignment of multiple endocrine neoplasia type 2A to chromosome 10 by linkage. *Nature* **328**: 528–530.
Skolnick, M.H., Thompson, E.A., Bishop, D.T. and Cannon L.A. (1984) Possible linkage of a breast cancer-susceptibility locus to the ABO locus: sensitivity of LOD scores to a single new recombinant observation. *Genet. Epidemiol.* **1**: 363–373.
Smith, J.R., Freije, D., Carpten, J.D., *et al.* (1996) Major susceptibility locus for prostate cancer on chromosome 1 suggested by a genome-wide search. *Science* **274**: 1371–1374.
Smith, T.M., Lee, M.K., Szabo, C.I., Jerome, N., McEuen, M., Taylor, M., Hood, L. and King, M.C. (1996) Complete genomic sequence and analysis of 117 kb of human DNA containing the gene *BRCA1*. *Genome Res.* **6**: 1029–1049.
Sparkes, R.S., Murphree, A.L., Lingua, R.W., Sparkes, M.C., Field, L.L., Funderburk, S.J., and Benedict, W.F. (1983) Gene for hereditary retinoblastoma assigned to human chromosome 13 by linkage to esterase D. *Science* **219**: 971–973.
Stratton, M.R., Ford, D., Neuhausen, S., *et al.* (1994) Familial male breast cancer is not linked to *BRCA1* locus on chromosome 17q. *Nat. Genet.* **7**: 103–106.
Tavtigian, S.V., Simard, J., Rommens, J., *et al.* (1996) The complete *BRCA2* gene and mutations in chromosome 13q-linked kindreds. *Nat. Genet.* **12**: 333–337.
Warthin, A.S. (1913) Heredity with reference to carcinoma. *Arch. Intern. Med.* **12**: 546–555.
Whittemore, A.S., Lin, I.G., Oakley-Girvan, I., Gallagher, R.P., Halpern, J., Kolonel, L.N., Wu, A.H. and Hsieh, C.L. (1999) No evidence of linkage for chromosome 1q42.2–43 in prostate cancer. *Am. J. Hum. Genet.* **65**: 254–256.
Woolf, C.M. (1960) An investigation of the familial aspects of carcinoma of the prostate. *Cancer* **13**: 739–744.
Wooster, R., Neuhausen, S., Mangion, J., *et al.* (1994) Localization of a breast cancer susceptibility gene (*BRCA2*) to chromosome 13q 12–13. *Science* **265**: 2088–2090.

Wooster, R., Bignell, G., Lancaster, J., *et al.* (1995) Identification of the breast cancer susceptibility gene *BRCA2*. *Nature* **378**: 789–792.

Xu, J., Meyers, D., Freije, D., *et al.* (1998) Evidence for a prostate cancer susceptibility locus on the X chromosome. *Nat. Genet.* **20**: 175–179.

Zhu, G., Duffy, D.L., Eldridge, A., *et al.* (1999) A major quantitative-trait locus for mole density is linked to the familial melanoma gene *CDKN2A*: a maximum-likelihood combined linkage and association analysis in twins and their sibs. *Am. J. Hum. Genet.* **65**: 483–492.

Zuo, L., Weger, J., Yang, Q., Goldstein, A.M., Tucker, M.A., Walker, G.J., Hayward, N. and Dracopoli, N.C. (1996) Germline mutations in the p16INK4a binding domain of CDK4 in familial melanoma. *Nat. Genet.* **12**: 97–99.

Zuppan, P., Hall, J.M., Lee, M.K., Ponglikitmongkol, M. and King, M.C. (1991) Possible linkage of the estrogen receptor gene to breast cancer in a family with late-onset disease. *Am. J. Hum. Genet.* **48**: 1065–1068.

5

Genetic epidemiology of Alzheimer's disease

G. Roks and C.M. van Duijn

1. Introduction

Dementia is a major health problem in the elderly. It is a syndrome characterized by impairment in intellectual functioning resulting in a distressing condition both for the patient and caregiver. Alzheimer's disease (AD) is the most common cause of dementia in Western society. AD is clinically characterized by an insidous onset of decline in memory and at least one other area of cognition. Additional characteristics are a gradually progressive course, a preserved level of consciousness, and absence of other conditions able to cause these symptoms. The pathological hallmark in brains of AD patients are extracellular plaques composed mainly of the amyloid-β peptide and intracellular neurofibrillary tangles containing hyperphosphorylated tau protein (Braak and Braak, 1991).

In this chapter a review of the current knowledge on the genetic epidemiology of AD is given. First the diagnostic issues in AD will be discussed briefly. Secondly, the epidemiology of AD and the familial aggregation are discussed. Thirdly, known AD genes are described together with their estimated contribution to the occurrence of the disease in the general population. Finally, efforts to identify novel genes involved in AD are discussed in light of the developments in molecular biology research.

2. AD diagnosis and subtypes

Although improved considerably, the clinical and pathological diagnosis of AD is still ambiguous. At the neuropathological level none of the characteristics, including the amyloid plaques and tau pathology, are unique as they can also be found in nondemented subjects or patients with other neurodegenerative disorders.

The second most frequent cause of dementia is vascular dementia (Ott *et al.*, 1995). Although both AD and vascular dementia have widely accepted criteria (McKhann *et al.*, 1984; Roman *et al.*, 1993), the distinction in clinical practice and

Analysis of Multifactorial Disease, edited by T. Bishop and P. Sham.

even neuropathological examination is often very difficult. Typically, vascular dementia is more acute in onset and has a stepwise progression, contrasting the gradually slow progression in AD. However, a considerable number of patients have features of both AD and vascular dementia. Subjects developing a slowly progressive cognitive decline after a vascular event, or subjects in whom the gradually cognitive decline is complicated by cerebrovascular disease, are difficult to classify exclusively as AD or vascular dementia. The situation is even more complicated by the fact that vascular pathology and its determinants may be risk factors for AD (Hofman *et al.*, 1997).

Another disease showing substantial overlap with AD is Parkinson's disease. About 33% of patients with Parkinson's disease suffer from dementia while AD patients often develop Parkinsonian symptoms (Perl *et al.*, 1998). Further, the familial clustering of both diseases points to the overlap in clinic, pathology, and aetiology (Hofman *et al.*, 1998). At the level of pathology the distinction from idiopathic Parkinson's disease is clear with cortical plaques and tangles in AD and Lewy bodies in the substantia nigra in Parkinson's disease. However, there is ongoing debate on the existence of a distinct form of dementia, which is pathologically characterized by Lewy bodies but clinically presents as an AD-like disorder (Perl *et al.*, 1998).

In addition to problems in the differential diagnosis, genetic epidemiological research is further hampered by the fact that AD itself is a heterogeneous disorder. A subgroup of AD, which is often distinguished in clinical as well as research settings, is the early-onset form of AD. There is no agreement on the cut-off point to define early-onset. Although age 65 years is classically used, others prefer 50 or 70 years as a cut-off. Whatever cut-off is used, no consistent differences in clinical or pathological features between patients with early- and late-onset AD have been found. Another frequently used subdivision is autosomal-dominant, familial, and sporadic AD. The autosomal-dominant form of AD is rare, probably accounting for less than 5% in early- and late-onset AD (Rao *et al.*, 1994). The familial cases, i.e. cases with at least one affected first-degree relative, may explain about 30% of early-onset patients and about 20% of the late-onset (Tol *et al.*, 1999). Thus, for both early- and late-onset AD, sporadic cases are most common. There are no clear differences in clinical or neuropathological presentation between dominant, familial, and sporadic AD, other than that autosomal-dominant and familial AD more often present at an earlier age than the sporadic form (van Duijn *et al.*, 1991).

3. Epidemiology of AD

3.1 General epidemiology

The number of patients affected with AD (prevalence of disease) increases strongly with age. In *Figure 1*, data on the prevalence of AD in 12 European studies have been pooled and re-analysed (Rocca *et al.*, 1991). Below age 70, less than 1% of the subjects in the general population suffer from AD. However, the prevalence increases exponentially with advancing age (see *Figure 1*). By age 90, up to 30% of subjects are affected with AD. There are only minor differences in

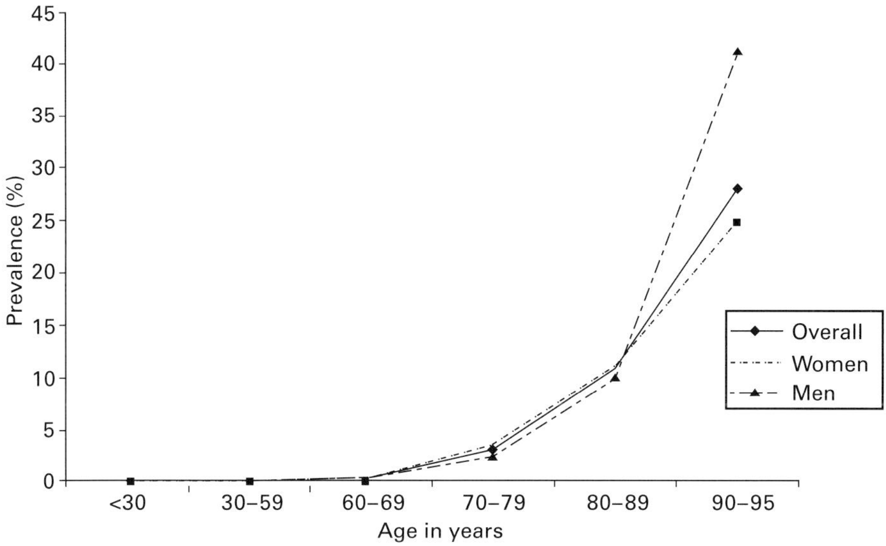

Figure 1. Age-specific prevalences of Alzheimer's disease from six pooled European studies for both sexes combined, men, and women (Rocca *et al.*, 1991).

prevalence between men and women, with a slightly higher prevalence in women before age 90 years (*Figure 1*).

Studies comparing geographical variation have yielded important clues for the aetiology of common disorders such as cardiovascular disease, cancer, and osteoporosis. Since differences in prevalence of a chronic disorder such as AD may be determined for a large part by regional differences in mortality, it is preferred to compare the number of newly diagnosed patients (incidence of disease). There is considerable variation in incidence of AD between populations, in particular after age 75 years (van Duijn, 1996). However, there is no evidence for a geographical trend in incidence rates, e.g. North–South trends in Europe that have been observed in cardiovascular disease and cancer (van Duijn, 1996). The differences observed in AD are most likely due to methodological problems, including a low number of subjects, non-response and competing mortality in the elderly, and co-morbidity complicating the diagnosis of AD at old age.

There is some evidence for a difference between the relative proportion of AD and vascular dementia between different populations (Jorm, 1991). In Caucasian populations AD is by far the most frequent type of dementia while in Asian populations vascular dementia has been diagnosed in up to 60% of the dementia patients. Given the similarity in incidence of AD in Asian and Caucasian populations, there may be a higher frequency of genetic or environmental determinants for vascular dementia in Asian populations. However, an alternative explanation may be that the differences in subtypes of dementia are difficult to interpret due to the lack of biologic markers and unique clinical features for these subtypes.

Putative risk factors for AD include thyroid disease, depression, vascular disease and its risk factors, aluminium and head trauma (van Duijn, 1996). Potential protective factors include hormone replacement therapy and anti-inflammatory medication (van Duijn, 1996). For most of these factors the evidence is not convincing up to date as findings of studies have continued to be contradicting (van Duijn, 1996). A major limitation of research of non-genetic factors to date is the lack of longitudinal follow-up studies of AD (van Duijn, 1996). This type of study has proven to be extremely valuable in cardiovascular and cancer research.

Perhaps the most important development in research of the epidemiology of AD concerns the role of vascular factors. In recent years, there has been growing evidence for a role of various vascular factors associated with a modest (up to two-fold) increase in risk for AD. Vascular factors that have been implicated in the risk of AD include hypertension, diabetes mellitus, atherosclerosis, estrogen replacement therapy, and smoking (Brenner *et al.*, 1994; Elias *et al.*, 1993; Hofman *et al.*, 1997; Kawas *et al.*, 1997; Lee, 1994; Leibson *et al.*, 1997; Merchant *et al.*, 1999; Ott *et al.*, 1998, 1999; Skoog *et al.*, 1996; Tang *et al.*, 1996). Although few studies have addressed these factors in a long-term follow-up approach, the hypothesis of vascular pathology modifying the risk of AD offers a novel framework to study risk factors for and perhaps also prevention of AD. The underlying biological model through which vascular factors may be involved in AD remains to be elucidated. These factors by themselves may have (small) additive effects. Together their effects may result in a load of vascular pathology, which influences the expression of AD. Indeed the risk of AD appears to be associated with the number of vascular (risk) factors (Hofman *et al.*, 1997). How the risk of AD is modified by vascular pathology is unclear. Vascular pathology may merely exaggerate already ongoing AD pathology in brain. The additional vascular pathology may compromise the brain further, leading to an earlier onset (and subsequently higher risk) of disease. However, it may be speculated that vascular pathology can be a primary cause of AD in a subset of patients (Hofman *et al.*, 1997).

3.2 Familial aggregation

Several twin studies have been performed in the past. A study summarizing the data of three twin-studies showed concordance rates in monozygotic twin pairs of 80% and 35% in dizygotic twins, resulting in a heritability estimate of 0.84 (Plassman and Breitner, 1996). However, twin studies of AD are not straightforward in interpretation. A major problem is the AD diagnosis, which requires extensive clinical examinations. Differences may exist between medical centres in particular when twins were diagnosed in the past when the diagnosis was not yet standardized. Given the late and variable onset of AD, differences in onset age between twin pairs may lead to false disconcordant findings.

In support of twin research, epidemiological studies of risk in first-degree relatives also point to a strong familial clustering of the disease. However, also in these studies problems related to the diagnosis and age-related expression are encountered. In a re-analysis of seven case-control studies first-degree relatives of AD patients had a 3.5-fold increase in risk for developing AD (van Duijn *et al.*, 1991). This relative risk decreased with age at onset of the proband but even by age 80 years a statistically significant 2.6-fold increase in risk for first-degree relatives was observed. From these relative risks it can be calculated that at the population level up to 34% of the occurrence of AD may be explained by hereditary factors (Tol *et al.*, 1999). For clinical genetic practice, absolute risks for relatives of patients are relevant, which have been adjusted for censored observations given the age-dependent expression. The lifetime cumulative risk for first-degree relatives of early- and late-onset AD patients was found to be 39% compared to about 12% in first-degree relatives of controls (Lautenschlager *et al.*, 1996; van Duijn *et al.*, 1993).

With regard to the mode of inheritance, numerous extended families with multiple AD cases have been published (St George-Hyslop *et al.*, 1989). However, most families concern cases with an early-onset AD. The inheritance patterns in these families are predominantly autosomal-dominant. It is important to realize that these families are an atypical subgroup also within the early-onset AD group. Segregation analysis in 198 early-onset AD families indicated that patterns of familial clustering can be explained by transmission of a major autosomal-dominant gene in less than 1% of the families (van Duijn *et al.*, 1993). Even in early-onset AD, the pattern of familial aggregation in the majority of the families fits a dominant gene with a multifactorial component (21%). In late-onset AD the situation is technically more complex because of censoring of at-risk individuals, small families, and incomplete family histories. In a segregation analysis comprising 400 late-onset families there was evidence for a single major gene with a population frequency of 1.5% (Rao *et al.*, 1994). However, due to lack of power no distinction could be made between dominant, recessive, or additive models.

In conclusion, there is a substantial contribution of genetic factors to the aetiology of AD. However, heritability is difficult to assess in AD. Besides some early-onset AD families, no conclusion about the mode of inheritance can be drawn. In most cases, the disease appears to be of multifactorial origin with several genetic and environmental factors involved.

4. Genes involved in Alzheimer's disease

4.1 Autosomal-dominant mutations

Early genetic research focused on early-onset families segregating AD in an autosomal-dominant pattern. Given the rare occurrence of the disease before age 65 (see *Figure 1*) there is a high probability that there is a single gene underlying AD in these cases. Initial genetic research targeted chromosome 21 as a candidate chromosome because of the observation that trisomy 21 patients develop pathology identical to that observed in AD patients (Wisniewski *et al.*, 1985). Using linkage analysis, early-onset AD was found to be linked to the q11.2-q21.2 region on chromosome 21 (St George-Hyslop *et al.*, 1987) , which contains the gene encoding the amyloid precursor protein (APP) (Tanzi *et al.*, 1987). Yet the first mutation in APP responsible for AD was not found earlier than 1991 (Goate *et al.*, 1991). Since then, eight different mutations have been found that cause autosomal-dominant forms of AD with an early onset (van Broeckhoven *et al.*, 1985). The APP gene consists of 18 exons but the mutations are all located in exon 16 and 17 that code for the βA4 amyloid, suggesting that either this is the functional region or mutations in the other regions are not compatible with life.

The most frequently described mutation is a valine to isoleucine change at codon 717. This mutation is found in 11 families while the others are described in one family only (AD mutation database: http: //molgen-www.uia.ac.be/ADMutations/). APP mutations are not uniquely associated with AD. A mutation in codon 693 is associated with cerebral haemorrhages due to vascular depositions of amyloid (Levy *et al.*, 1990), while one single mutation in codon 692 has been shown to present with either a cerebral haemorrhage or AD (Hendriks *et al.*, 1992).

Given the limited impact of APP mutations (see also *Section 4.3*) in terms of families explained by mutations in this gene, several genome-wide searches were conducted. In 1992 findings of the first genomic screen of autosomal-dominant AD were published suggesting linkage to chromosome 14 (Schellenberg *et al.*, 1992). Three independent groups confirmed these findings (Mullan *et al.*, 1992; St George-Hyslop *et al.*, 1992; Van Broeckhoven *et al.*, 1992). The gene was isolated in 1995 (Sherrington *et al.*, 1995). This gene was identified in patients with early-onset AD and designated presenilin-1 (PSEN-1). As APP, PSEN-1 is a transmembrane protein, with seven transmembrane domains and one hydrophilic loop. Mutations associated with AD are clustered in the transmembrane domains, especially domain 2 and the membrane-associated region of the hydrophilic loop (Cruts and Van Broeckhoven, 1998). Up to date, 56 mutations in seven different exons of the PSEN-1 gene have been found (AD mutation database: http: //molgen-www.uia.ac.be/ADMutations/). The function of the presenilin protein is unclear, but mutations in the gene lead to altered APP processing *in vivo*, cell cultures, and transgenic animals (Borchelt *et al.*, 1996; Duff *et al.*, 1996; Scheuner *et al.*, 1996).

The third gene harboring dominant mutations was found based on its homology with PSEN-1 (67% amino acid sequence homology concentrated in the transmembrane domains) (Levy-Lahad *et al.*, 1995). Two missense mutations in this second presenilin gene (PSEN-2) on chromosome 1 (q31-q42) were identified in two families with early-onset AD (Rogaev *et al.*, 1995). Until now only four mutations in PSEN-2 are described of which three are localized transmembranic (AD mutation database: http://molgen-www.uia.ac.be/ADMutations/). The functional relationship of these mutations with AD remains unclear.

4.2 Genetic susceptibility

In 1991, linkage to a region on chromosome 19 in families with late-onset AD was reported (Pericak-Vance *et al.*, 1991). Albeit that evidence was weak and based solely on an affected only analysis. The apolipoprotein E gene (APOE) on chromosome 19 was tested as a candidate because of the presence of apolipoprotein E in senile plaques and its affinity to amyloid (Strittmatter *et al.*, 1993). The APOE gene has three common alleles (the wild-type APOE*3, APOE*2 and APOE*4) coding for three different isoforms of the protein (Mahley, 1988). The frequency of APOE*4 was found to be significantly increased in familial AD patients (Strittmatter *et al.*, 1993). The number of studies that confirmed the association between APOE*4 and AD is overwhelming (Farrer *et al.*, 1997). In a meta-analysis of 40 studies conducted up to 1995 (Farrer *et al.*, 1997), both early- and late-onset AD, sporadic and familial forms, were associated with APOE*4. Early findings of a decreased frequency of APOE*2 in AD patients (Chartier-Harlin *et* *al.*, 1994; Corder *et al.*, 1994) were not confirmed by others (Sorbi *et al.*, 1994; van Duijn *et al.*, 1995). APOE*4 is not exclusively associated with AD. A number of studies also found association with vascular dementia (Slooter *et al.*, 1997), or Lewy body dementia (Lippa *et al.*, 1995), which makes this gene unsuitable for clinical diagnostics (Slooter *et al.*, 1996). Risk may differ across subgroups. There is some evidence that APOE interacts with other genetic and environmental risk factors. Vascular risk factors in generalized atherosclerosis and serum cholesterol levels may modulate the APOE*4-associated risk of AD (Hofman *et al.*, 1997; Jarvik *et al.*, 1995). Other factors include head injury (Mayeux *et al.*, 1995), herpes simplex virus (Itzhaki *et al.*, 1997), smoking (Ott *et al.*, 1998), and estrogen use (Tang *et al.*, 1996). However, identification of gene–gene and gene–environment interaction requires large population-based studies of which there are few available at present. This makes confirmation of the findings on gene–gene and gene–environment interaction difficult.

4.3 Population attribution

The frequency of major mutations depends for a large part on the ascertainment of families. Families ascertained for linkage studies (Schellenberg *et al.*, 1992) or genetic counselling (Campion *et al.*, 1999; Finckh *et al.*, 2000) have shown high mutation frequencies. When considering a population-based sample of families with an autosomal-dominant form of early-onset AD, 18% was explained by mutations in known genes, of which PSEN-1 mutations are most common (Cruts *et al.*, 1998). When considering all early-onset AD patients in a population-based sample, 0.5% of the patients carry APP mutations, 6% carry PSEN-1 mutations, and 1% carry PSEN-2 mutations (Cruts *et al.*, 1998; van Duijn *et al.*, 1991). Frequencies of these mutations in controls approach zero implying that the relative risk associated with these mutations is very high. The absence of asymptomatical carriers after age 70 years in families segregating APP or PSEN-1 mutations suggest a high penetrance approaching 100%. From a clinical genetic perspective, this makes these genes useful for risk prediction and diagnosis of AD. Nevertheless, the mutations causing AD remain unknown in 82% of the families with early-onset AD this concerns also extended pedigrees segregating AD.

In the general population early-onset AD is only a minor fraction of all AD (<1%) (Ott *et al.*, 1995). Therefore, on the level of the general population the attribution of these mutations to the occurrence of disease is estimated to be very small. Together, the three dominant genes do not appear to explain more than 0.1% of the occurrence of all AD; (Tol *et al.*, 1999).

At the population level, APOE is a more important determinant of AD. Although the relative risk of AD is moderately (1.3 to 3.2 times) increased for carriers of one APOE*4 allele (APOE24, APOE34), the population-attributable risk suggests that the APOE*4 allele may explain about 10 to 17% of the disease in the general population (Tol *et al.*, 1999). The substantial contribution to the frequency of AD in the population is for a large part explained by the large number of subjects heterozygous for the APOE*4 allele (±25% in Caucasians). For homozygotes the situation is different. Despite the high risk for AD (6 to 15 times increased), the contribution to AD in the general population is limited (<2%) because of the very low frequency of this genotype (Tol *et al.*, 1999). *APOE* is a susceptibility gene that increases risk but does not always lead to AD (Meyer *et al.*, 1998). Given its modest increase in risk, its role in risk prediction in the clinical genetic practice and diagnosis in the neurological practice is limited (Slooter *et al.*, 1996).

5. Efforts to identify new AD genes

It is clear that research on the genetics of AD is far from completed. Several strategies to identify novel genes have been followed including candidate gene approach and genome screening. None of these approaches yielded equivocal results so far.

The candidate gene approach has been successful in identifying AD genes. However, these studies were based on strong evidence towards region of interest, e.g. observations of the occurrence of amyloid pathology in Down syndrome patients pointing to chromosome 21 (APP), linkage studies flagging chromosome 19 (APOE) or chromosome 14 (PSEN-1), or homology to a known gene (PSEN-2). Recent candidate gene studies have been less successful despite the fact that for each of them there is a biological justification (*Table 1*). A large number of genes was studied but none of the studies of the candidate genes listed in *Table 1* have been replicated consistently. Methodology in some studies may explain part of the inconsistencies. In a number of studies genotype frequencies in control samples were not in Hardy Weinberg equilibrium, suggesting the presence of bias (Cavalli-Sforza and Bodmer, 1971). This bias might have occurred during control ascertainment or at laboratory assessment. An important issue in case-control studies is the source population of controls. Hospital-based controls are especially prone to selection bias. Inclusion of subjects with unknown pathology related to the genetic factor under study may result in false positive findings. Furthermore, errors in the genotype classification due to reading or genotyping mistakes may cause deviation from Hardy Weinberg equilibrium.

Although candidate gene studies have also been disappointing in other fields than AD, it is only fair to say that these studies in AD are far from their limits. Critical evaluation of studies in terms of possible false positive findings due to ascertainment and stratification has been limited both by investigators and referees, evidenced by findings based on control data not in Hardy Weinberg equilibrium. On the other hand, there is an issue of small studies underlying false negative results, which is often neglected. To overcome the problem of false negative results in small studies pooling valid studies in a meta-analysis may lead to a more intelligible interpretation of scientific evidence.

With regard to the candidate gene approach there are opportunities for progress in studies of the promotor areas of known genes (*APOE*, *APP*, *PSEN-1*) (Bullido *et al.*, 1998; Roks *et al.*, 1998; van Duijn *et al.*, 1999) in that more detailed studies on the functionality of these polymorphisms is needed. There have been limited efforts to increase the validity of candidate gene studies using other designs, for example the transmission disequilibrium test (TDT) or sibpair design. Nor was studying samples for the presence of population admixture, potentially a source of bias (Lander and Schork, 1994).

Table 1. Candidate genes for Alzheimer's disease

Candidate gene	Possible role	References
1α-antichymotrypsin	Binds to β amyloid	Haines *et al.,* 1996; Kamboh *et al.*, 1995
VLDL receptor	Binds to APOE	Chung *et al.*, 1996; Okuizumi *et al.*, 1995
LDL receptor-related protein (LRP)	APOE receptor in brain	Lambert *et al.*, 1998; Woodward *et al.*, 1998
α2-macroglobuline	Binds to β amyloid	Blacker *et al.*, 1998; Rudrasingham *et al.*, 1999
PSEN-1 intron 8	Alteration in known AD gene	Wragg *et al.*, 1996; Yasuda *et al.*, 1999
PSEN-1 promotor	Promotor known AD gene	van Duijn *et al.*, 1999
ACE	Gene involved in vascular pathology	Kehoe *et al.*, 1999
APOE promotor	Regulation of APOE expression	Bullido *et al.*, 1998; Roks *et al.*, 1998
Butyrylcholinesterase	Associated with plaques and tangles	Brindle *et al.*, 1998; Lehman *et al.*, 1997
α-synuclein	Component of plaques	Hellman *et al.*, 1998; Xia *et al.*, 1996
HLA	Inflammation	Payami *et al.*, 1991; Small *et al.*, 1991
Bleomycin hydrolase	APP processing	Farrer *et al.*, 1998; Montoya *et al.*, 1998
Tau	Main protein in tangles	Lilius *et al.*, 1998; Roks *et al.*, 1999
Nitric oxide synthase	Dilatation of small vessels	Dahyat *et al.*, 1999

VLDL, very low density lipoprotein; LRP, low density lipoprotein receptor-related protein; ACE, angiotensin converting enzyme; HLA, human leucocyte antigen

Successful candidate gene studies in AD up to date distinguish themselves from unsuccessful ones by the fact that they were supported by positive results in linkage analysis. Unfortunately genome screening for new AD genes has not yielded consistent results. Several searches have been conducted, targeting the most common late-onset form of AD. Different approaches have been followed including the parametric lod score method (Pericak-Vance *et al.*, 1997) and the non-parametric sibpair method (Kehoe *et al.*, 1999; Pericak-Vance *et al.*, 1997). Chromosome 12 was identified in two genomic screens (Kehoe *et al.*, 1999; Pericak-Vance *et al.*, 1997), although one may argue that the regions with significant lod scores do not overlap. In this region the candidate gene alpha-2 macroglobuline (A2M) has been put forward as an AD gene. This gene has been extensively studied without yielding conclusive findings (Blacker *et al.*, 1998; Chen *et al.*, 1998; Crawford *et al.*, 1999; Dow *et al.*, 1999; Hu *et al.*, 1999; Korovaitseva *et al.*, 1999; Kovacs *et al.*, 1999; Liao *et al.*, 1998; Myllykangas *et al.*, 1999; Rogaeva *et al.*, 1999; Rudrasingham *et al.*, 1999; Shibata *et al.*, 1999; Singleton *et al.*, 1999; Wavrant-De Vrieze *et al.*, 1999). *Figure 2* shows a meta-analysis of the studies conducted to date. The figure and the overall odds ratios show that it is unlikely that A2M is indeed the gene explaining linkage of AD to chromosome 12 (Koster *et al.*, 2000).

Main problems encountered in genome screens, in particular in late-onset AD, are phenocopies limiting the statistical power of linkage and sibpair studies. The

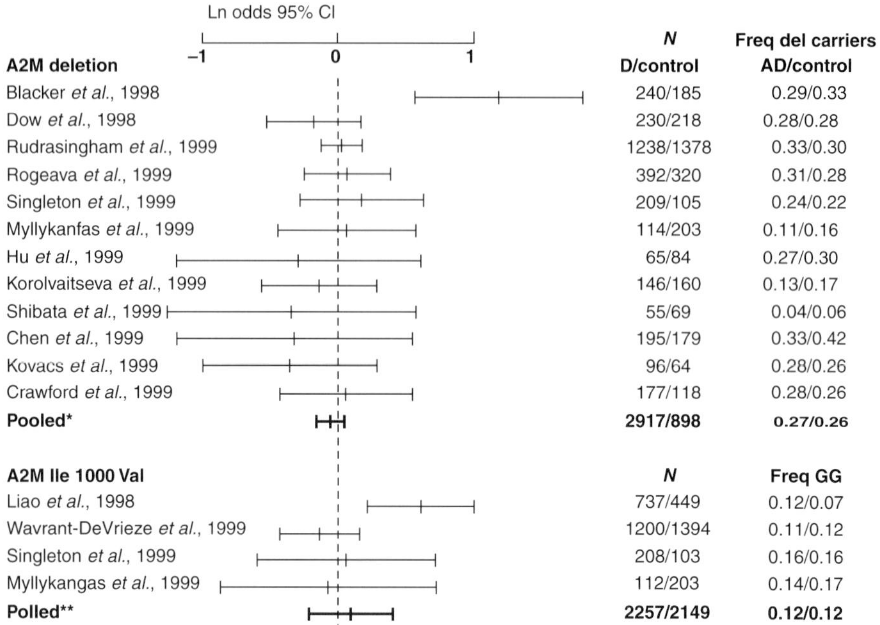

Figure 2. Meta-analysis of the A2M deletion and Ile 1000 Val polymorphism of all published case-control studies so far (Ln odds ratios and 95% CI).
* Odds ratio 0.99 (0.88–1.12). ** Odds ratio 1.07 (0.76–1.49).

distinction of phenocopies from the genetic form of the disease is clinically impossible at present. When assessing the statistical power, the lack of knowledge on the number of different genes and pathways involved is a major limitation. Although the heritability for AD is high it cannot be excluded that there are a large number of genes with modest effect involved. These genes may be implicated in AD through complex interactions. This has serious implications for the power of linkage and sib-pair studies, but to what extent cannot yet be overseen.

An opportunity not explored fully in screening the genome are association studies in genetically isolated populations (Lander and Schork, 1994). Chances of success may be better when genetic studies are performed in founder populations which are genetically homogeneous (Lander and Schork, 1994). This approach has not been taken to its limits in the field of AD. Within these populations, AD genetics may be less complex in terms of number of genes involved due to founder effects and genetic drift (Khoury *et al.*, 1993; Lander and Schork, 1994). An obvious disadvantage of localizing genes in such a population may be the limited relevance in the general population. On the other hand, any gene identified may yield important clues for proteins involved in the AD pathology.

6. Concluding remarks

Considerable progress in AD genetics has been made in the last decade. Although the autosomal-dominant mutations do not explain a large proportion of the disease, in particular the discovery of the APP and PSEN-1 genes have contributed extensively to our knowledge of the pathogenesis of AD and provided a target for long-awaited drug development. Further, identification of these genes has opened opportunities for clinical genetic counselling in families with autosomal-dominant forms of early-onset AD.

From a clinical perspective, findings on APOE have not yielded major implications. Nevertheless, APOE explains a substantial part of the occurrence of disease in the general population. Due to the large proportion of AD explained by this polymorphism, APOE may offer new opportunities for prevention in the future. However, development of effective strategies awaits progress in epidemiological research on environmental factors involved. Particular findings on vascular risk factors, which may be preventable, offer opportunities in this respect.

The challenge for the near future will be to identify new genes involved in the aetiology of AD. There is evidence for an AD locus on chromosome 12, which yields a start for further molecular genetic studies. Although candidate and genome searches have not reached their limits in terms of statistical power and methodology, novel approaches to identify genes are on the way. A possibility to identify new genes is screening all genes identified in the human genome with single nucleotide polymorphisms (SNPs) (Risch and Merikangas, 1996). This approach may be followed in association and sibpair studies. At present, this approach awaits major progress in technology and development of informative SNPs in exons. Once these technical problems are conquered, the next step will be ascertaining AD series large enough for a powerful analysis.

As reviewed AD is a heterogeneous disorder. Development in the clinical diagnosis, in particular of disease phenotypes may enable further progress in unravelling AD genetics. The dissection of late-onset AD into more homogeneous subgroups may lead to progress in genetic epidemiologic research. Although our knowledge of the clinical course of AD is limited, there is growing evidence that the aetiology of AD is strongly linked with that of other neurodegenerative diseases, including vascular disease, tau disorders, and Lewy body pathology. Studies of the genetics of these disorders, in particular brain-related atherosclerosis, may open opportunities for the genetic dissection of AD.

The genetic aetiology of AD is not fully understood. As in other complex disorders, major difficulties are encountered in research aiming to further unravel genetics. Methodological and technical developments in the near future may facilitate new breakthroughs.

Acknowledgment

This work was supported by the Netherlands Organisation for Scientific Research (NWO) and the Netherlands Institute for Health Sciences (NIHES).

References

Blacker, D., Wilcox, M.A., Laird, N.M., Rodes, L., Horvath, S.M., Go, R.C., *et al.* (1998) Alpha-2 macroglobulin is genetically associated with Alzheimer disease. *Nat. Genet.* **19**: 357–360.

Borchelt, D.R., Thinakaran, G., Eckman, C.B., Lee, M.K., Davenport, F., Ratovitsky, T., *et al.*

(1996) Familial Alzheimer's disease-linked presenilin 1 variants elevate Abeta1–42/1–40 ratio *in vitro* and *in vivo*. *Neuron* **17**: 1005–1013.

Braak, H. and Braak, E. (1991) Neuropathological stageing of Alzheimer-related changes. *Acta Neuropathol.* **82**: 239–259.

Brenner, D.E., Kukull, W.A., Stergachis, A., van Belle, G., Bowen, J.D., McCormick, W.C., *et al.* (1994) Postmenopausal estrogen replacement therapy and the risk of Alzheimer's disease: a population-based case-control study. *Am. J. Epidemiol.* **140**: 262–267.

Brindle, N., Song, Y., Rogaeva, E., Premkumar, S., Levesque, G., Yu, G., *et al.* (1998) Analysis of the butyrylcholinesterase gene and nearby chromosome 3 markers in Alzheimer disease. *Hum. Mol. Genet.* **7**: 933–935.

Bullido, M.J., Artiga, M.J., Recuero, M., Sastre, I., Garcia, M.A., Aldudo, J., *et al.* (1998) A polymorphism in the regulatory region of APOE associated with risk for Alzheimer's dementia. *Nat. Genet.* **18**: 69–71.

Campion, D., Dumanchin, C., Hannequin, D., Dubois, B., Belliard, S., Puel, M., *et al.* (1999) Early-onset autosomal dominant Alzheimer disease: prevalence, genetic heterogeneity, and mutation spectrum. *Am. J. Hum. Genet.* **65**: 664–670.

Cavalli-Sforza, L.L. and Bodmer, W.F. (1971) The genetics of human populations (eds Kennedy, D. and Park, R.B.). San Francisco, W.H. Freeman and Company.

Chartier-Harlin, M.C., Parfitt, M., Legrain, S., Perez-Tur, J., Brousseau, T., Evans, A., *et al.* (1994) Apolipoprotein E, epsilon 4 allele as a major risk factor for sporadic early and late-onset forms of Alzheimer's disease: analysis of the 19q13.2 chromosomal region. *Hum. Mol. Genet.* **3**: 569–574.

Chen, L., Baum, L., Ng, H.K., Chan, L.Y., Sastre, I., Artiga, M.J., *et al.* (1999) Apolipoprotein E promoter and alpha2-macroglobulin polymorphisms are not genetically associated with Chinese late onset Alzheimer's disease. *Neurosci. Lett.* 269: 173–177.

Chung, H., Roberts, C.T., Greenberg, S., Rebeck, G.W., Christie, R., Wallace, R., *et al.* (1996) Lack of association of trinucleotide repeat polymorphisms in very-low-density lipoprotein receptor gene with Alzheimer's disease. *Ann. Neurol.* **39**: 800–803.

Corder, E.H., Saunders, A.M., Risch, N.J., Strittmatter, W.J., Schmechel, D.E., Gaskell, P.C., Jr., *et al.* (1994) Protective effect of apolipoprotein E type 2 allele for late onset Alzheimer disease. *Nat. Genet.* 7: 180–184.

Crawford, F., Town, T., Freeman, M., Schinka, J., Gold, M., Duara, R., *et al.* (1999) The alpha-2 macroglobulin gene is not associated with Alzheimer's disease in a case-control sample. *Neurosci. Lett.* **270**: 133–136.

Cruts, M. and Van Broeckhoven, C. (1998) Presenilin mutations in Alzheimer's disease. *Hum. Mutat.* **11**: 183–190.

Cruts, M., van Duijn, C.M., Backhovens, H., Van den Broeck, M., Wehnert, A., Serneels, S., *et al.* (1998) Estimation of the genetic contribution of presenilin-1 and -2 mutations in a population-based study of presenile Alzheimer disease. *Hum. Mol. Genet.* **7**: 43–51.

Dahiyat, M., Cumming, A., Harrington, C., Wischik, C., Xuereb, J., Corrigan, F., *et al.* (1999) Association between Alzheimer's disease and the NOS3 gene. *Ann. Neurol.* **46**: 664–667.

Dow, D.J., Lindsey, N., Cairns, N.J., Brayne, C., Robinson, D., Huppert, F.A., *et al.* (1999) Alpha-2 macroglobulin polymorphism and Alzheimer disease risk in the UK. *Nat. Genet.* **22**: 16–17; discussion 21–12.

Duff, K., Eckman, C., Zehr, C., Yu, X., Prada, C.M., Perez-tur, J., *et al.* (1996) Increased amyloid-beta42(43) in brains of mice expressing mutant presenilin 1. *Nature* **383**: 710–713.

Elias, M.F., Wolf, P.A., D'Agostino, R.B., Cobb, J. and White, L.R. (1993) Untreated blood pressure level is inversely related to cognitive functioning: the Framingham Study. *Am. J. Epidemiol.* **138**: 353–364.

Farrer, L.A., Cupples, L.A., Haines, J.L., Hyman, B., Kukull, W.A., Mayeux, R., *et al.* (1997) Effects of age, sex, and ethnicity on the association between apolipoprotein E genotype and Alzheimer disease. A meta-analysis. APOE and Alzheimer Disease Meta Analysis Consortium. *JAMA* **278**: 1349–1356.

Farrer, L.A., Abraham, C.R., Haines, J.L., Rogaeva, E.A., Song, Y., McGraw, W.T., *et al.* (1998) Association between bleomycin hydrolase and Alzheimer's disease in caucasians. *Ann. Neurol.* **44**: 808–811.

Finckh, U., Muller-Thomsen, T., Mann, U., Eggers, C., Marksteiner, J., Meins, W., *et al.* (2000) High prevalence of pathogenic mutations in patients with early-onset dementia detected by sequence analyses of four different genes. *Am. J. Hum. Genet.* **66**: 110–117.

Goate, A., Chartier-Harlin, M.C., Mullan, M., Brown, J., Crawford, F., Fidani, L., *et al.* (1991) Segregation of a missense mutation in the amyloid precursor protein gene with familial Alzheimer's disease. *Nature* **349**: 704–706.

Haines, J.L., Pritchard, M.L., Saunders, A.M., Schildkraut, J.M., Growdon, J.H., Gaskell, P.C., *et al.* (1996) No genetic effect of alpha1-antichymotrypsin in Alzheimer disease. *Genomics* **33**: 53–56.

Hellman, N.E., Grant, E.A. and Goate, A.M. (1998) Failure to replicate a protective effect of allele 2 of NACP/alpha-synuclein polymorphism in Alzheimer's disease: an association study. *Ann. Neurol.* **44**: 278–281.

Hendriks, L., van Duijn, C.M., Cras, P., Cruts, M., Van Hul, W., van Harskamp, F., *et al.* (1992) Presenile dementia and cerebral haemorrhage linked to a mutation at codon 692 of the beta-amyloid precursor protein gene. *Nat. Genet.* **1**: 218–221.

Hofman, A., Schulte, W., Tanja, T.A., van Duijn, C.M., Haaxma, R., Lameris, A.J., *et al.* (1989) History of dementia and Parkinson's disease in 1st-degree relatives of patients with Alzheimer's disease. *Neurology* **39**: 1589–1592.

Hofman, A., Ott, A., Breteler, M.M., Bots, M.L., Slooter, A.J., van Harskamp, F., *et al.* (1997) Atherosclerosis, apolipoprotein E, and prevalence of dementia and Alzheimer's disease in the Rotterdam Study. *Lancet* **349**: 151–154.

Hu, C.J., Sung, S.M., Liu, H.C., Lee, K.Y., Hsu, W.C., Wong, W.K., *et al.* (1999) No association of alpha-2 macroglobulin gene five-nucleotide deletion with AD in Taiwan Chinese. *Neurology* **53**: 642–643.

Itzhaki, R.F., Lin, W.R., Shang, D., Wilcock, G.K., Faragher, B. and Jamieson, G.A. (1997) Herpes simplex virus type 1 in brain and risk of Alzheimer's disease. *Lancet* **349**: 241–244.

Jarvik, G.P., Wijsman, E.M., Kukull, W.A., Schellenberg, G.D., Yu, C. and Larson, E.B. (1995) Interactions of apolipoprotein E genotype, total cholesterol level, age, and sex in prediction of Alzheimer's disease: a case-control study. *Neurology* **45**: 1092–1096.

Jorm, A.F. (1991) Cross-national comparisons of the occurrence of Alzheimer's and vascular dementias. *Eur. Arc.h Psychiatry Clin. Neurosci.* **240**: 218–222.

Kamboh, M.I., Sanghera, D.K., Ferrell, R.E. and DeKosky, S.T. (1995) APOE*4-associated Alzheimer's disease risk is modified by alpha 1-antichymotrypsin polymorphism. *Nat. Genet.* **10**: 486–488.

Kawas, C., Resnick, S., Morrison, A., Brookmeyer, R., Corrada, M., Zonderman, A., *et al.* (1997) A prospective study of estrogen replacement therapy and the risk of developing Alzheimer's disease: the Baltimore Longitudinal Study of Aging. *Neurology* **48**: 1517–1521.

Kehoe, P.G., Russ, C., McIlory, S., Williams, H., Holmans, P., Holmes, C., *et al.* (1999) Variation in DCP1, encoding ACE, is associated with susceptibility to Alzheimer disease. *Nat. Genet.* **21**: 71–72.

Kehoe, P., Wavrant-De Vrieze, F., Crook, R., Wu, W.S., Holmans, P., Fenton, I., *et al.* (1999) A full genome scan for late onset Alzheimer's disease. *Hum. Mol. Genet.* **8**: 237–245.

Khoury, M.J., Beaty, T.H. and Cohen, B.H. (1993) Fundamentals of Genetic Epidemiology. Oxford University Press, New York.

Korovaitseva, G.I., Premkumar, S., Grigorenko, A., Molyaka, Y., Galimbet, V., Selezneva, N., *et al.* (1999) Alpha-2 macroglobulin gene in early- and late-onset Alzheimer disease. *Neurosci. Lett.* **271**: 129–131.

Koster, M.N., Dermaut, B., Cruts, M., Houwing-Duistermaat, J.J., Roks, G., Tol, J. *et al.* (2000) The alpha-2 macroglobin gene in AD: a population-based study on meta-analysis. *Neurology* (in press).

Kovacs, T., Cairns, N.J., Lantos, P.L. (1999) Alpha-2-macroglobulin intronic polymorphism is not associated with autopsy-confirmed late-onset Alzheimer's disease. *Neurosci. Lett.* **273**: 61–83.

Lambert, J.C., Wavrant-De Vrieze, F., Amouyel, P. and Chartier-Harlin, M.C. (1998) Association at LRP gene locus with sporadic late-onset Alzheimer's disease. *Lancet* 351: 1787–1788.

Lander, E.S. and Schork, N.J. (1994) Genetic dissection of complex traits. *Science* **265**: 2037–2048.

Lautenschlager, N.T., Cupples, L.A., Rao, V.S., Auerbach, S.A., Becker, R., Burke, J., *et al.* (1996) Risk of dementia among relatives of Alzheimer's disease patients in the MIRAGE study: What is in store for the oldest old? *Neurology* **46**: 641–650.

Lee, P.N. (1994) Smoking and Alzheimer's disease: a review of the epidemiological evidence. *Neuroepidemiology* **13**: 131–144.

Leibson, C.L., Rocca, W.A., Hanson, V.A., Cha, R., Kokmen, E., O'Brien, P.C., *et al.* (1997) Risk of dementia among persons with diabetes mellitus: a population-based cohort study. *Am. J. Epidemiol.* **145**: 301–308.

Lehmann, D.J., Johnston, C. and Smith, A.D. (1997) Synergy between the genes for butyrylcholinesterase K variant and apolipoprotein E4 in late-onset confirmed Alzheimer's disease. *Hum. Mol. Genet.* **6**: 1933–1936.

Levy, E., Carman, M.D., Fernandez-Madrid, I.J., Power, M.D., Lieberburg, I., van Duinen, S.G., *et al.* (1990) Mutation of the Alzheimer's disease amyloid gene in hereditary cerebral hemorrhage, Dutch type. *Science* **248**: 1124–1126.

Levy-Lahad, E., Wasco, W., Poorkaj, P., Romano, D.M., Oshima, J., Pettingell, W.H., *et al.* (1995) Candidate gene for the chromosome 1 familial Alzheimer's disease locus. *Science* **269**: 973–977.

Lilius, L., Forsell, C., Axelman, K., Matitila, K.M., Andreadis, A., Viitanen, M., *et al.* (1998) Tau gene polymorphism confers susceptibility to Alzheimer's disease. *Neurobiol. Aging* **19**: S68.

Liao, A., Nitsch, R.M., Greenberg, S.M., Finckh, U., Blacker, D., Albert, M., *et al.* (1998) Genetic association of an alpha2-macroglobulin (Val1000lle) polymorphism and Alzheimer's disease. *Hum. Mol. Genet.* **7**: 1953–1956.

Lippa, C.F., Smith, T.W., Saunders, A.M., Crook, R., Pulaski-Salo, D., Davies, P., *et al.* (1995) Apolipoprotein E genotype and Lewy body disease. *Neurology* **45**: 97–103.

McKhann, G., Drachman, D., Folstein, M., Katzman, R., Price, D. and Stadlan, E.M. (1984) Clinical diagnosis of Alzheimer's disease: report of the NINCDS-ADRDA Work Group under the auspices of Department of Health and Human Services Task Force on Alzheimer's Disease. *Neurology* **34**: 939–944.

Mahley, R.W. (1988) Apolipoprotein E: cholesterol transport protein with expanding role in cell biology. *Science* **240**: 622–630.

Mayeux, R., Ottman, R., Maestre, G., Ngai, C., Tang, M.X., Ginsberg, H., *et al.* (1995) Synergistic effects of traumatic head injury and apolipoprotein-epsilon 4 in patients with Alzheimer's disease. *Neurology* **45**: 555–557.

Merchant, C., Tang, M.X., Albert, S., Manly, J., Stern, Y. and Mayeux, R. (1999) The influence of smoking on the risk of Alzheimer's disease. *Neurology* **52**: 1408–1412.

Meyer, M.R., Tschanz, J.T., Norton, M.C., Welsh-Bohmer, K.A., Steffens, D.C., Wyse, B.W., *et al.* (1998) APOE genotype predicts when–not whether–one is predisposed to develop Alzheimer disease. *Nat. Genet.* **19**: 321–322.

Montoya, S.E., Aston, C.E., DeKosky, S.T., Kamboh, M.I., Lazo, J.S. and Ferrell, R.E. (1998) Bleomycin hydrolase is associated with risk of sporadic Alzheimer's disease. *Nat. Genet.* **18**: 211–212.

Mullan, M., Houlden, H., Windelspecht, M., Fidani, L., Lombardi, C., Diaz, P., *et al.* (1992) A locus for familial early-onset Alzheimer's disease on the long arm of chromosome 14, proximal to the alpha 1-antichymotrypsin gene. *Nat. Genet.* **2**: 340–342.

Myllykangas, L., Polvikoski, T., Sulkava, R., Verkkoniemi, A., Crook, R., Tienari, P.J., *et al.* (1999) Genetic association of alpha2-macroglobulin with Alzheimer's disease in a Finnish elderly population. *Ann. Neurol.* **46**: 382–390.

Okuizumi, K., Onodera, O., Namba, Y., Ikeda, K., Yamamoto, T., Seki, K., *et al.* (1995) Genetic association of the very low density lipoprotein (V_LDL) receptor gene with sporadic Alzheimer's disease. *Nat. Genet.* **11**: 207–209.

Ott, A., Breteler, M.M., van Harskamp, F., Claus, J.J., van der Cammen, T.J., Grobbee, D.E., *et al.* (1995) Prevalence of Alzheimer's disease and vascular dementia: association with education. The Rotterdam study. *BMJ* **310**: 970–973.

Ott, A., Slooter, A.J., Hofman, A., van Harskamp, F., Witteman, J.C., Van Broeckhoven, C., *et al.* (1998) Smoking and risk of dementia and Alzheimer's disease in a population-based cohort study: the Rotterdam Study. *Lancet* **351**: 1840–1843.

Ott, A., Stolk, R.P., van Harskamp, F., Pols, H.A., Hofman, A. and Breteler, M.M. (1999) Diabetes mellitus and the risk of dementia: The Rotterdam Study. *Neurology* **53**: 1937–1942.

Payami, H., Kaye, J., Becker, W., Norman, D. and Wetzsteon, P. (1991) HLA-A2, or a closely linked gene, confers susceptibility to early-onset sporadic Alzheimer's disease in men. *Neurology* **41**: 1544–1548.

Pericak-Vance, M.A., Bebout, J.L., Gaskell, P.C., Jr., Yamaoka, L.H., Hung, W.Y., Alberts, M.J., *et al.* (1991) Linkage studies in familial Alzheimer disease: evidence for chromosome 19 linkage. *Am. J. Hum. Genet.* **48**: 1034–1050.

Pericak-Vance, M.A., Bass, M.P., Yamaoka, L.H., Gaskell, P.C., Scott, W.K., Terwedow, H.A., *et al.* (1997) Complete genomic screen in late-onset familial Alzheimer disease. Evidence for a new locus on chromosome 12 [see comments]. *JAMA* **278**: 1237–1241.

Perl, D.P., Olanow, C.W. and Calne, D. (1998) Alzheimer's disease and Parkinson's disease: distinct entities or extremes of a spectrum of neurodegeneration? *Ann. Neurol.* **44**: S19–31.

Plassman, B.L. and Breitner, J.C. (1996) Recent advances in the genetics of Alzheimer's disease and vascular dementia with an emphasis on gene-environment interactions. *J. Am. Geriatr. Soc.* **44**: 1242–1250.

Rao, V.S., van Duijn, C.M., Connor-Lacke, L., Cupples, L.A., Growdon, J.H. and Farrer, L.A. (1994) Multiple etiologies for Alzheimer disease are revealed by segregation analysis. *Am. J. Hum. Genet.* 55: 991–1000.

Risch, N. and Merikangas, K. (1996) The future of genetic studies of complex human diseases. *Science* **273**: 1516–1517.

Rocca, W.A., Hofman, A., Brayne, C., Breteler, M.M., Clarke, M., Copeland, J.R., *et al.* (1991) Frequency and distribution of Alzheimer's disease in Europe: a collaborative study of 1980–1990 prevalence findings. The EURODEM-Prevalence Research Group. *Ann. Neurol.* **30**: 381–390.

Rogaev, E.I., Sherrington, R., Rogaeva, E.A., Levesque, G., Ikeda, M., Liang, Y., *et al.* (1995) Familial Alzheimer's disease in kindreds with missense mutations in a gene on chromosome 1 related to the Alzheimer's disease type 3 gene. *Nature* **376**: 775–778.

Rogaeva, E.A., Premkumar, S., Grubber, J., Serneels, L., Scott, W.K., Kawarai, T., *et al.* (1999) An alpha-2-macroglobulin insertion-deletion polymorphism in Alzheimer disease. *Nat. Genet.* **22**: 19–22.

Roks, G., Cruts, M., Bullido, M.J., Backhovens, H., Artiga, M.J., Hofman, A., *et al.* (1998) The –491 A/T polymorphism in the regulatory region of the apolipoprotein E gene and early-onset Alzheimer's disease. *Neurosci. Lett.* **258**: 65–68.

Roks, G., Dermaut, B., Heutink, P., Julliams, A., Backhovens, H., Van de Broeck, M., *et al.* (1999) Mutation screening of the tau gene in patients with early-onset Alzheimer's disease. *Neurosci. Lett.* **277**: 137–139.

Roman, G.C., Tatemichi, T.K., Erkinjuntti, T., Cummings, J.L., Masdeu, J.C., Garcia, J.H., *et al.* (1993) Vascular dementia: diagnostic criteria for research studies. Report of the NINDS-AIREN International Workshop. *Neurology* **43**: 250–260.

Rudrasingham, V., Wavrant-De Vrieze, F., Lambert, J.C., Chakraverty, S., Kehoe, P., Crook, R., *et al.* (1999) Alpha-2 macroglobulin gene and Alzheimer disease. *Nat. Genet.* **22**: 17–19; discussion 21–12.

St George-Hyslop, P.H., Tanzi, R.E., Polinsky, R.J., Haines, J.L., Nee, L., Watkins, P.C., *et al.* (1987) The genetic defect causing familial Alzheimer's disease maps on chromosome 21. *Science* **235**: 885–890.

St George-Hyslop, P.H., Myers, R.H., Haines, J.L., Farrer, L.A., Tanzi, R.E., Abe, K., *et al.* (1989) Familial Alzheimer's disease: progress and problems. *Neurobiol Aging* **10**: 417–425.

St George-Hyslop, P., Haines, J., Rogaev, E., Mortilla, M., Vaula, G., Pericak-Vance, M., *et al.* (1992) Genetic evidence for a novel familial Alzheimer's disease locus on chromosome 14. *Nat. Genet.* **2**: 330–334.

Schellenberg, G.D., Bird, T.D., Wijsman, E.M., Orr, H.T., Anderson, L., Nemens, E., *et al.* (1992) Genetic linkage evidence for a familial Alzheimer's disease locus on chromosome 14. *Science* **258**: 668–671.

Scheuner, D., Eckman, C., Jensen, M., Song, X., Citron, M., Suzuki, N., *et al.* (1996) Secreted amyloid beta-protein similar to that in the senile plaques of Alzheimer's disease is increased *in vivo* by the presenilin 1 and 2 and APP mutations linked to familial Alzheimer's disease. *Nat. Med.* **2**: 864–870.

Sherrington, R., Rogaev, E.I., Liang, Y., Rogaeva, E.A., Levesque, G., Ikeda, M., *et al.* (1995) Cloning of a gene bearing missense mutations in early-onset familial Alzheimer's disease. *Nature* **375**: 754–760.

Shibata, N., Ohnuma, T., Takahashi, T., Ohtsuka, E., Ueki, A. and Arai, H. (1999) Genetic association between alpha-2 macroglobulin and Japanese sporadic Alzheimer's disease. *Neurosci. Lett.* **271**: 132–134.

Singleton, A.B., Gibson, A.M., McKeith, I.G., Ballard, C.A., Perry, R.H., Ince, P.G., *et al.* (1999) Alpha2-macroglobulin polymorphisms in Alzheimer's disease and dementia with Lewy bodies. *NeuroReport* **10**: 1507–1510.

Skoog, I., Lernfelt, B., Landahl, S., Palmertz, B., Andreasson, L.A., Nilsson, L., *et al.* (1996) 15-year longitudinal study of blood pressure and dementia. *Lancet* **347**: 1141–1145.

Slooter, A.J., Breteler, M.B., Ott, A., Van Broeckhoven, C. and van Duijn, C.M. (1996) APOE genotyping in differential diagnosis of Alzheimer's disease. *Lancet* **348**: 334.

Slooter, A.J., Tang, M.X., van Duijn, C.M., Stern, Y., Ott, A., Bell, K., *et al.* (1997) Apolipoprotein E epsilon4 and the risk of dementia with stroke. A population-based investigation. *JAMA* **277**: 818–821.

Small, G.W., Ebeling, S.C., Matsuyama, S.S., Heyman, A., Reisner, E.G., Renvoize, E.B., *et al.* (1991) Variable association of HLA-A2 in men with early-onset Alzheimer disease. *Neurobiol. Aging* **12**: 375–377.

Sorbi, S., Nacmias, B., Forleo, P., Latorraca, S., Gobbini, I., Bracco, L., *et al.* (1994) ApoE allele frequencies in Italian sporadic and familial Alzheimer's disease. *Neurosci. Lett.* **177**: 100–102.

Strittmatter, W.J., Saunders, A.M., Schmechel, D., Pericak-Vance, M., Enghild, J., Salvesen, G.S., *et al.* (1993) Apolipoprotein E: high-avidity binding to beta-amyloid and increased frequency of type 4 allele in late-onset familial Alzheimer disease. *Proc. Natl Acad. Sci. USA* **90**: 1977–1981.

Tang, M.X., Jacobs, D., Stern, Y., Marder, K., Schofield, P., Gurland, B., *et al.* (1996) Effect of oestrogen during menopause on risk and age at onset of Alzheimer's disease. *Lancet* **348**: 429–432.

Tanzi, R.E., Gusella, J.F., Watkins, P.C., Bruns, G.A., St George-Hyslop, P., Van Keuren, M.L., *et al.* (1987) Amyloid beta protein gene: cDNA, mRNA distribution, and genetic linkage near the Alzheimer locus. *Science* **235**: 880–884.

Tol, J., Slooter, A.J.C. and Van Duijn, C.M. (1999) Genetic factors in early- and late-onset Alzheimer's disease. In: *Research and Perspectives in Alzheimer's Disease* (ed. R. Mayeux) Springer-Verlag, Heidelberg, pp. 33–40.

Van Broeckhoven, C.L. (1995) Molecular genetics of Alzheimer disease: identification of genes and gene mutations. *Eur. Neurol.* **35**: 8–19.

Van Broeckhoven, C., Backhovens, H., Cruts, M., De Winter, G., Bruyland, M., Cras, P., *et al.* (1992) Mapping of a gene predisposing to early-onset Alzheimer's disease to chromosome 14q24.3. *Nat. Genet.* **2**: 335–339.

van Duijn, C.M. (1996) Epidemiology of the dementias: recent developments and new approaches. *J. Neurol. Neurosurg. Psychiatry* **60**: 478–488.

van Duijn, C.M., Hendriks, L., Cruts, M., Hardy, J.A., Hofman, A. and Van Broeckhoven, C. (1991a) Amyloid precursor protein gene mutation in early-onset Alzheimer's disease. *Lancet* **337**: 978.

van Duijn, C.M., Clayton, D., Chandra, V., Fratiglioni, L., Graves, A.B., Heyman, A., *et al.* (1991b) Familial aggregation of Alzheimer's disease and related disorders: a collaborative re-analysis of case-control studies. EURODEM Risk Factors Research Group. *Int. J. Epidemiol.* **20**: S13–20.

van Duijn, C.M., Farrer, L.A., Cupples, L.A. and Hofman, A. (1993) Genetic transmission of Alzheimer's disease among families in a Dutch population based study. *J Med Genet* **30**: 640–646.

van Duijn, C.M., de Knijff, P., Wehnert, A., De Voecht, J., Bronzova, J.B., Havekes, L.M., *et al.* (1995) The apolipoprotein E epsilon 2 allele is associated with an increased risk of early-onset Alzheimer's disease and a reduced survival. *Ann. Neurol.* **37**: 605–610.

van Duijn, C.M., Cruts, M., Theuns, J., Van Gassen, G., Backhovens, H., Van den Broeck, M., *et al.* (1999) Genetic association of the presenilin-1 regulatory region with early-onset Alzheimer's disease in a population-based sample. *Eur. J. Hum. Gen.* **7**: 801–806.

Wavrant-DeVrieze, F., Rudrasingham, V., Lambert, J.C., Chakraverty, S., Kehoe, P., Crook, R., *et al.* (1999) No association between the alpha-2 macroglobulin I1000V polymorphism and Alzheimer's disease. *Neurosci. Lett.* **262**: 137–139.

Wisniewski, K.E., Wisniewski, H.M. and Wen, G.Y. (1985) Occurrence of neuropathological changes and dementia of Alzheimer's disease in Down's syndrome. *Ann. Neurol.* **17**: 278–282.

Woodward, R., Singleton, A.B., Gibson, A.M., Edwardson, J.A. and Morris, C.M. (1998) LRP gene and late-onset Alzheimer's disease. *Lancet* **352**: 239–240.

Wragg, M., Hutton, M., Talbot, C. (1996) Genetic association between intronic polymorphism in presenilin-1 gene and late-onset Alzheimer's disease. Alzheimer's Disease Collaborative Group. *Lancet* **347**: 509–512.

Xia, Y., Rohan de Silva, H.A., Rosi, B.L., Yamaoka, L.H., Rimmler, J.B., Pericak-Vance, M.A., *et al.* (1996) Genetic studies in Alzheimer's disease with an NACP/alpha-synuclein polymorphism. *Ann. Neurol.* **40**: 207–215.

Yasuda, M., Hirono, N., Maeda, K., Imamura, T., Mori, E. and Tanaka, C. (1999) Case-control study of presenilin-1 intronic polymorphism in sporadic early and late onset Alzheimer's disease. *J. Neurol. Neurosurg. Psychiatry* **66**: 722–726.

Molecular genetic studies of the idiopathic epilepsies

Andrew Makoff, Lina Nashef and Philip Asherson

1. Introduction

Epilepsy is a common disorder with a prevalence of between 0.5 and 1% and an overall incidence of 0.05%. Epilepsy may occur alone or as a symptom of another disease process. It consists of a heterogeneous mix of overlapping syndromes with many causes. Genetic factors are particularly important and are thought to make a significant contribution in approximately 40% of all affected individuals (Anderson and Hauser, 1993; Elmslie and Gardiner, 1995). As in many other common disorders familial segregation is complex, with single gene inheritance patterns observed in a minority of cases and accounting for no more than 1–2% of all epilepsies (Gardiner, 1990). Family and twin studies clearly demonstrate the importance of genetic factors in common forms of epilepsy and among these the highest familial risks occur in idiopathic epilepsies. Furthermore, there is overlap between epilepsies that are considered to be primarily of genetic etiology and those that are secondary to environmental or acquired factors (Degen *et al.*, 1996; Shorvon, 1991). One important observation is that epilepsy in diverse etiologies is influenced by age at presentation (Commission on Classification and Terminology of the International League against Epilepsy, 1989; Shinnar and Moshe, 1991) and the effect of environmental triggers probably differ depending on age of exposure.

In this chapter we set out to review the major advances in our understanding of the genetic basis of the idiopathic epilepsies and speculate on likely approaches and further advances in the coming years with emphasis on the common forms of idiopathic generalized epilepsies. The most exciting development is the finding that mutations within ion channel genes cause certain forms of idiopathic epilepsy in both humans and mice. However, it is not yet proven that variation in such genes will confer susceptibility to the more common and genetically complex forms of idiopathic epilepsy.

Analysis of Multifactorial Disease, edited by T. Bishop and P. Sham.

1.1 Classification of epilepsies

Epilepsy was initially classified according to seizure type but emphasis has shifted to a syndromic classification: one which takes into account not only seizures but also the overall clinical picture, investigation results, likely pathological substrate and prognosis (see *Table 1*). Two themes which are useful, despite areas of difficulty and overlap, underpin the classification: (i) the distinction between partial and generalized epilepsies and (ii) the distinction between idiopathic and symptomatic epilepsies. These are likely to survive the current review by the International League Against Epilepsy (ILAE) Task Force for Classification.

There are many inherited conditions where epilepsy is only one of several abnormalities. Some may be of interest if co-segregation occurs between a gene for idiopathic epilepsy and a gene for a different inherited disorder, a so-called contiguous gene syndrome. Other disorders such as tuberous sclerosis or autism, both of which have high rates of accompanying epilepsy, may be brought about by shared molecular mechanisms. With recent developments in imaging techniques, there has been increasing interest in inherited epilepsies associated with cortical dysgenesis/dysplasia where a structural basis for the epilepsy has been demonstrated. A distinction is made between such secondary or symptomatic epilepsies and idiopathic epilepsies where no structural basis is present on routine imaging, although more subtle microscopic or functional imaging abnormalities may still occur.

There has been much confusion over the use of the term idiopathic in the epilepsy literature and indeed in clinical practice. In older epidemiological studies, the term was used for all epilepsies where no clear cause was apparent and was often based on history alone. In the past, when investigative facilities were limited, the term applied to the majority. As would be expected improved investigations, particularly imaging, resulted in a decrease in the proportion of so-called idiopathic epilepsies. This purely negative use of the term is not intended here. Idiopathic epilepsy, in the sense used by the ILAE commission on classification

Table 1. Classification of human epilepsies

1. Localization-related (partial)
   - Idiopathic (e.g. benign childhood epilepsy with centrotemporal spikes)
   - Symptomatic
   - Cryptogenic
2. Generalized
   - Idiopathic (e.g. childhood absence, juvenile absence, juvenile myoclonic, epilepsy with *grand mal* seizures on awakening, photosensitive epilepsy)
   - Cryptogenic or symptomatic (e.g. West syndrome, Lennox–Gastaut)
   - Symptomatic: subdivided into those due to nonspecific or to specific aetiologies
3. Undetermined whether focal or generalized
   - With both generalized and focal features
   - Without unequivocal generalized or focal features
4. Special syndromes
   - Situation related seizures, e.g. febrile convulsions

Adapted from Commission on Classification and Terminology of the International League Against Epilepsy (1989).

Table 2. Casewise concordance for some major epilepsy syndromes in monozygous (MZ) and dizygous (DZ) twin pairs (adapted from Berkovic *et al.*, 1998)

Syndrome	Twin pairs	MZ	DZ
Idiopathic generalized	41	0.76 (± 0.08)	0.33 (± 0.12)
Idiopathic partial	10	0.33 (± 0.25)	0.33 (± 0.25)
Cryptogenic partial	30	0.55 (± 0.14)	0
Febrile seizures	82	0.58 (± 0.10)	0.14 (± 0.06)

and in this chapter, is a positive diagnosis. Lack of evidence for structural abnormality on imaging (where performed) is only one of many features. A cryptogenic category, on the other hand, includes those where no positive features allow classification within either the symptomatic or idiopathic category, with the assumption that an as yet unidentified etiology is present. A positive family history of epilepsy is found more often in idiopathic generalized epilepsy (IGE) than in epilepsy as a whole with higher concordance found among monozygotic (MZ) as compared with (DZ) dizygotic twins (*Table 2*) (Berkovic *et al.*, 1990; Berkovic *et al.*, 1998; Gedda and Tatarelli, 1971). However, the most recent large study, using up to date clinical assessment and investigations, has also demonstrated a high concordance within the cryptogenic partial epilepsies (Berkovic *et al.*, 1998). This suggests that this category also includes unrecognized idiopathic cases (see *Table 2*). It is evident that the current classification has its limitations. Genetic studies need to proceed despite these acknowledged shortcomings and will no doubt help to refine and redefine future classifications.

1.2 The electroencephalograph (EEG) and idiopathic epilepsy

Electroencephalographic findings, many of which are age related, are relevant to the classification and genetics of idiopathic epilepsy. The subject has been reviewed by Doose (1997). There is considerable variation in EEG pattern in the normal population. This variation is inherited and is identical in maturation, frequency spectrum and alpha-rhythm asymmetries in MZ twins. Normal EEG variants confer a different susceptibility to epilepsy. For example, the low voltage EEG found in 4–10% of healthy adults and inherited as an autosomal dominant trait is considered protective against seizures. On the other hand the monomorphic generalized alpha EEG variant favors the generation of generalized spikes and waves and increases the expression of photosensitivity – the photoparoxysmal response (PPR).

Five important patterns are associated with epilepsy: 4–7 s rhythms, occipital 3–4 s rhythms, focal sharp waves, generalized spike/wave (GSW) and PPR. Doose reports that the less familiar 4–7 s rhythm (often spindle-shaped monomorphous bilaterally synchronous runs) is sometimes seen in those with febrile convulsions and early onset IGE. It is considered 'excitatory' and its persistence into adulthood is associated with intractability of complex partial seizures. Occipital intermittent rhythmic delta activity at 3–4 s, on the other hand, while also observed in children with febrile convulsions or absences, is considered 'inhibitory'. Among siblings of photosensitive children for example, a PPR and GSW is seen less frequently when this pattern is present than in siblings without this pattern. In childhood absence epilepsy (CAE), its presence is associated with the pure syndrome with typical 3 Hz GSW, good prognosis and a lower risk of developing later generalized tonic clonic seizures (GTCS) (Loiseau, 1992).

Focal sharp waves, sometimes with shifting laterality and not indicative of a fixed structural anomaly, are found in idiopathic partial epilepsies such as benign epilepsy of childhood with centrotemporal spikes (BECTS or rolandic epilepsy). The idiopathic partial epilepsies with complex inheritance may represent a spectrum including BECTS, pseudo-Lennox Gastaut, epilepsy with occipital paroxysms, Landau Kleffner syndrome and Electrical Status Epilepticus during sleep. Generalized EEG findings, a biological hallmark of IGE consist of paroxysms of bilateral synchronous and symmetrical discharges that interrupt the normal background rhythm consisting of generalized slow waves, generalized spike and wave, or generalized polyspike and wave occurring at rest or during hyperventilation. These reflect diffuse cortical hyperexcitability with probable genetic heterogeneity. The PPR segregates separately and is considered to be an independent marker. Despite overlap, some patterns emerge. Examples include the typical 3 Hz GSW pattern at rest and activated by hyperventilation usually seen in CAE, and the polyspike and wave pattern, often with photosensitivity, associated with juvenile myoclonic epilepsy (JME). The situation, however, is by no means straightforward. A polyspike and wave pattern is sometimes seen in otherwise typical absences and conversely the 3 Hz GSW can be seen in JME. GSW is also found in approximately 2% of healthy children with a maximum incidence at 7–8 years of age. PPR (using a wide definition where the whole continuum of responses to photic stimulation is included as shown in *Table 3*) is observed in 8% of healthy children aged 5–15 years, with the maximum incidence in teenage girls. It is estimated that only 3% of children with PPR will develop epilepsy in childhood or adolescence. However, the generalized type IV pattern (*Table 3*) has a much closer association with epilepsy in general, and IGE in particular, supporting the notion that expression of the PPR is modified not only by age but also by other traits. Focal discharges (without a consistent source) are not infrequently observed in association with GSW but are transmitted in families independently of other genetically determined EEG patterns (Doose, 1997).

The potential of EEG findings in genetic studies has perhaps not been fully realized. Besides aiding in syndromic classification of affected probands, the EEG so far has been used in some studies to identify affected but asymptomatic cases. The various patterns may in themselves be worth studying in association studies as potentially useful phenotypes.

Table 3. Photoparoxysmal response (Waltz *et al.*, 1992)

Response type	EEG pattern
Type I	Occipital spikes within alpha rhythm (mostly adults)
Type II	Parieto-occipital spikes with biphasic slow waves
Type III	As Type II with frontal spread
Type IV	Generalized spike wave ('photoconvulsive response')

1.3 Approaches to gene mapping in idiopathic epilepsy

Traditional linkage and positional cloning approaches have been highly successful in identifying disease genes for Mendelian disorders in which a single major locus is both necessary and sufficient to cause the disorder. As we shall see, this has also been the case among the idiopathic epilepsies with genes identified for several well defined syndromes showing autosomal dominant inheritance. These important studies have identified causative genes encoding ion channels such as neuronal nicotinic acetylcholine receptor subunits, voltage gated potassium and sodium channels. However, these epilepsy syndromes account for a small minority of all idiopathic epilepsies and identifying the susceptibility genes for the common forms of idiopathic epilepsies with complex inheritance presents a far greater challenge.

The study of idiopathic epilepsies, like many other conditions with complex or polygenic inheritance, is complicated by genetic heterogeneity, incomplete penetrance, phenotypic variation and phenocopies. A further difficulty is uncertainty over inclusion and exclusion criteria for genetic studies. The misdiagnosis as epilepsy of other conditions causing blackouts is well recognized, and, as previously outlined by Gardiner (1990), the nature of the condition is such that 'reliable and accurate clinical data...is often difficult to obtain. The exact description and circumstances surrounding a seizure may never be available, and subjects may be unaware of seizures occurring in early childhood. Even if the seizures can be defined accurately, particular persons may defy categorization.... Not only may the particular variety of seizure...change with age but a mixture of different seizure types may occur in the same subject and within a particular pedigree.' The difficulty in defining suitable phenotypes for genetic studies is reflected in the wide range of different definitions that have been applied. For example, in JME different investigators have each adopted their own inclusion criteria which has led to difficulties in interpretation of linkage data (see section 2.2).

Several complementary approaches have been taken in linkage studies of the idiopathic epilepsies. The simplest and most successful approach has been to target rare Mendelian subtypes using conventional model-dependent linkage. More recently, investigators have extended this strategy to more common subtypes which are sometimes found to aggregate in large multigenerational pedigrees with a Mendelian appearance. This has had some initial success in large families with CAE, generalized epilepsy with febrile seizures plus (GEFS+) (see section 2.4) and some partial epilepsy syndromes (Ottman et al., 1995; Scheffer et al., 1998). A more general approach to the common subtypes has been to target those with relatively high familial risks for model free linkage, which does not require specification of the mode of transmission. In addition, sampling strategies such as the collection of affected sibling pairs, may be made more homogeneous by selecting only cases with restricted definitions of subtypes.

Despite the success of linkage studies, there are several limitations when it comes to the analysis of common idiopathic epilepsies and as a result some researchers have focused on the alternative strategy of association. At the clinical level the selection of familial cases results in the loss of clinical data both in terms of the numbers available for study and in terms of the breadth of the original clinical data which is no longer representative of the full range of idiopathic epilepsies. Furthermore, some susceptibility genes are likely to be shared between subtypes and unclassified cases, a view supported by the observation of increased familial rates and high MZ : DZ concordance ratios for the broad generalized group. Other genes may specifically predispose to certain clinical features or subtypes. A general approach designed to encompass these possibilities would be to carry out association studies initially using a broad category of IGE. Subtype analysis may also be performed based not only on subtype classification, but also on the original raw clinical data; for example different seizure-types and patterns, age of onset, EEG findings and response to individual drug treatments.

At the genetic level, it has not been possible to determine in advance the likely number of genes influencing risk to the common idiopathic epilepsies and the expected size of their effects. The expectation is that with the exception of rare mendelian subtypes, more common genetic influences will come from a few or many genes of moderate or minor effect. In this context association strategies are appropriate because compared with linkage there is far more power to detect genes of such effect size (Collins *et al.*, 1997; Risch and Merikangas, 1996).

To date the common IGEs that have received most attention are JME and CAE. This is probably because they are clinically well defined and high familial risks have been clearly established. A measure that is frequently used in evaluating the relative merits of linkage and association approaches is the ratio of sibling risk to population prevalence (γ_s). In JME for example, the risk to siblings of epilepsy with JME or related syndromes is 7% of which 30% are JME and the population prevalence is approximately 0.05%, giving $\gamma_s = 0.021/0.0005 = 42$ (Elmslie *et al.*, 1997). The value of γ_s at individual loci, the so-called locus-specific value (γ_g), will depend on the number of genes and the size of their individual effects. If, for the purpose of illustration, we assume that there are three loci of equal effect influencing JME then each locus would confer a γ_g of approximately 14; an effect size which is within the resolving power of sib-pair linkage approaches (Risch and Merikangas, 1996). It is for this reason that to date the major efforts in the analysis of JME have adopted linkage strategies, with an increasing focus on model free as well as parametric methods.

In contrast the γ_s across all idiopathic epilepsies is in the range 5–10 and it is likely that γ_g values for individual loci will be much smaller, perhaps in the range of 1.5–3. In their often cited paper Risch and Merikangas (1996) applied an extremely stringent test, power of 80% at $P < 5 \times 10^{-8}$, and estimated that a sample of 340 cases would detect a disease gene with frequency of 0.5 and γ_g of 2.0 compared with 2498 affected sib-pairs required for linkage analysis. Despite this, association approaches in complex disorders including idiopathic epilepsies are still in their infancy and in most cases have been applied only to the analysis of a few candidate genes.

The major advances in our current knowledge of the genetic basis of the idiopathic epilepsies will be reviewed in the next two sections. Findings in both humans and mice provide considerable support for the important role of ion channel genes and the genetic control of neuronal excitability. Together these provide, in addition to pharmacological and animal research, a large array of candidate genes, the relevance of which to the vast majority of idiopathic epilepsies with complex inheritance is yet to be defined.

2. Molecular findings in human idiopathic epilepsies

2.1 Mendelian idiopathic epilepsies

Autosomal dominant frontal lobe epilepsy (ADNFLE). ADNFLE is a rare idiopathic partial epilepsy characterized by nocturnal seizures which first occur in childhood but usually continue into adulthood (Scheffer *et al.*, 1995). It exhibits Mendelian segregation and was mapped to 20q13 in an Australian kindred (Phillips *et al.*, 1995). The disease-causing mutation was found to be a single base change in *CHRNA4* which results in a serine to phenylalanine amino acid substitution in the second transmembrane domain of the encoded nicotinic acetylcholine receptor α4 subunit (Steinlein *et al.*, 1995). Subsequently, a second ADNFLE family was shown to have a mutation in the same gene; this time a 3 nucleotide insertion resulting in a leucine amino acid insertion also in the second transmembrane domain (Steinlein *et al.*, 1997).

The nicotinic acetylcholine receptors, both of the neuronal and muscle type, are a group of multisubunit cation channels, all of which are permeable to sodium ions but which vary in their selectivity for calcium ions (McGehee and Role, 1995). They comprise five subunits, each of which has four transmembrane domains. The second transmembrane domain is known to line the ion pore and mutations in this region have been shown to affect the ion selectivity of the channel (Bertrand *et al.*, 1993; Cohen *et al.*, 1992; Galzi *et al.*, 1992). The α4 subunit forms functional ion channels when expressed in cultured cells or *Xenopus* oocytes but requires the co-expression of a β subunit. *In situ* hybridization and immunohistochemical studies have shown that the α4–β2 combination is very common in the central nervous system (CNS) and are found in both presynaptic and postsynaptic locations (McGehee and Role, 1995). When in a presynaptic location, nicotinic acetylcholine receptors act to facilitate neurotransmitter release (McGehee and Role, 1996). Co-expression of either α4 ADNFLE variant with the β2 subunit has enabled the physiological and pharmacological consequences of the mutations to be explored. This has shown that the Ser248Phe amino acid change increases the rate of desensitization of the receptor and reduces its calcium permeability (Kuryatov *et al.*, 1997; Weiland *et al.*, 1996). The leucine insertion variant also displayed reduced permeability to calcium (Steinlein *et al.*, 1997). Both mutations therefore appear to result in a decrease in receptor activity, but the overall effect depends on whether excitatory or inhibitory neurons are involved. Interestingly, the mutations appear to be clinically indistinguishable (Nakken *et al.*, 1999). A possible partial explanation for seizures in ADNFLE is that the reduced calcium permeability of receptors containing mutant α4 subunits compromises the facilitation of release of inhibitory neurotransmitters by presynaptic nicotinic receptors. However the molecular basis for the seizures in ADNFLE caused by these two mutations still remains to be explained.

There is genetic heterogeneity in ADNFLE and a locus at 15q24 has recently been implicated in another ADNFLE family (Phillips *et al.*, 1998). This region contains the *CHRNA3*, *CHRNA5*, *CHRNB4* cluster that encodes other nicotinic acetylcholine receptor subunits: α3, α5 and β4.

Benign familial neonatal convulsions (BFNC). BFNC also shows a mendelian autosomal dominant pattern of inheritance. Several families have been investigated. Like ADNFLE there is genetic heterogeneity and two loci have been mapped: at 20q13 and at 8q24 (Leppert *et al.*, 1989; Lewis *et al.*, 1993). The affected gene at 20q13 was found to be the gene for a novel potassium channel, *KCNQ2* which is situated approximately 30 kb from *CHRNA4*, implicated in ADNFLE (Biervert *et al.*, 1998; Singh *et al.*, 1998). The significance of two closely linked, but apparently unrelated genes mutated in two different idiopathic epilepsies is not clear. *KCNQ2* encodes a voltage gated potassium channel which is in the same potassium channel family as the cardiac *KVQT1* (renamed *KCNQ1*), mutations in which are responsible for long QT syndrome and Jervell and Lange–Nielsen cardio-auditory syndrome. The affected gene in BFNC families that map to 8q24 is *KCNQ3*, another voltage gated potassium channel from the same family (Charlier *et al.*, 1998).

Voltage-gated potassium channels form a large group of diverse proteins which share a common structure of six putative transmembrane domains (S1–S6) and a pore-forming loop (P) between S5 and S6 [for a review see Jan and Jan (1997)]. The most extensively studied is the *shaker* subgroup which is a group of multimeric proteins consisting of four α subunits (Doyle *et al.*, 1998; Mackinnon, 1991). The KCNQ2 encoded potassium channel appears to form heterotetramers with KCNQ3 channels. Expression of either subunit alone in oocytes produces a channel with a weak or no voltage dependent potassium current that is greatly stimulated by co-expression of both subunits (Wang *et al.*, 1998; Yang *et al.*, 1998). The potassium channel formed by the two subunits appears to define the M channel which is responsible for determining neuronal subthreshold excitability. Co-expression of different BFNC mutant forms of *KCNQ2* with wild type *KCNQ3* reduced the size of the current by about two-fold as did the converse co-expression with a *KCNQ3* BFNC mutant (Schroeder et al., 1998). When the heterozygous situation was simulated by co-expression of mutant and wild type subunit of one gene with wild type subunit of the other in the ratio of 1: 1: 2, the reduction of wild type current was approximately 20–40%. BFNC mutations are dominant but usually only produce seizures shortly after birth. This suggests that the *KCNQ2/KCNQ3* potassium channel has a very critical role in determining excitability in early postnatal development when even a small reduction in its hyperpolarizing current is potentially proconvulsive. Modest reductions in this potassium current appear to be better tolerated after this early developmental stage.

Benign familial infantile convulsions (BFIC). BFIC was originally described in Italian families and more recently outside Italy (Echenne *et al.*, 1994; Lee *et al.*, 1993; Vigevano *et al.*, 1992) and is autosomal dominant. It is defined by the presence of nonfebrile convulsions, which first occur between 3 and 12 months of age and, like BFNC, has a favorable outcome. A linkage study has been carried out in five Italian families (Guipponi *et al.*, 1997) which detected significant linkage on chromosome 19 and excluded several other regions of the genome including 8q24 and 20q13. Positive maximum likelihood lod (logarithm of the odds ratio) scores were obtained with neighboring markers *D19S49* (4.98), *D19S250*

(5.16), *D19S114* (6.36), *D19S249* (5.40), *D19S251* (4.73) and *D19S245* (5.21), and the maximum multipoint lod score > 8 was obtained for the interval *D19S250–D19S245*. A common disease associated haplotype was identified in each family, consistent with a founder effect with all five families. This region spanning approximately 8 Mb at 19q11–q13 is almost identical to the region linked to the GEFS+ syndrome in two apparently unrelated families from Australia, and has been shown to be due to a mutation in a sodium channel gene (see Section 2.4).

Another region that has been of interest in a number of similar epilepsy syndromes is chromosome 16p12–q12. A variant of BFIC called familial infantile convulsions and choreoathetosis (ICCA) has been described which combines the BFIC syndrome with the involuntary movement disorder, paroxysmal choreoathetosis (PC) (Szepetowski *et al.*, 1997). Four French families with ICCA were studied in a genome scan with 119 markers which specifically excluded 19p12 and other regions previously implicated in idiopathic epilepsies. Significant linkage under autosomal dominant inheritance was found on chromosome 16. Positive LOD scores were found with seven markers with a maximum of 6.76 for *D16S3133* at a recombination fraction of 0.0. Haplotype analysis identified a 10 Mb region between *D16S517* and *D16S401* on either side of the centromere (16p12–q12).

In another study by the same group, a Chinese family with a similar combination of infantile convulsions and paroxysmal dyskinesia was found to map to the same region of chromosome 16. There were however some clinical differences, the main one being that affected individuals within the Chinese family had generalized seizures whereas cases within the French pedigrees had partial seizures (Lee *et al.*, 1998). More recently, another combination of seizures with a dystonia was mapped to the same centromeric region of chromosome 16 (Guerrini *et al.*, 1999). This syndrome, which combines BECTS with paroxysmal exercise induced dystonia and writer's cramp, was mapped in a large inbred Italian family. Unlike ICCA, this disorder is inherited recessively. If these similar syndromes involve the same gene, it would provide an example of a gene in which different mutations can cause either generalized or partial seizures.

This region of chromosome 16 contains four candidate genes which encode nonvoltage gated sodium channel β and γ subunits, a sodium/glucose co-transporter and a calcium transporter SERCA1 (Callen *et al.*, 1991; Voilley *et al.*, 1995; Wells *et al.*, 1993), although none of these are primarily expressed within the CNS. Mutations in *ATP2A1*, the gene for SERCA1, have been implicated in an autosomal recessive form of Brody disease, a rare disorder involving exercise induced muscle cramps and stiffness (Odermatt *et al.*, 1996). The resemblance of Brody disease to some aspects of the above syndromes is intriguing. Yet SERCA1 and its mRNA have only been found in fast twitch skeletal muscle (Brandl *et al.*, 1986; Wu *et al.*, 1995). It is possible that on chromosome 16 near to *ATP2A1* there is a gene in which mutations can confer susceptibility to epilepsy. In such a scenario, the syndromes which combine an epilepsy with a dystonia may result either from a deletion affecting both genes or from a mutation in a common control region.

BFIC has also been reported to partially co-segregate with familial hemiplegic migraine (FHM) in a large Dutch–Canadian family (Terwindt *et al.*, 1997). The

19p13.1 region, which contains *CACNA1A* implicated in both FHM in humans and absence type seizures in mice (see section 3.1) was excluded from linkage to either phenotype in this family, as were the two BFNC linkage regions on 20q13.3 and 8q24 (Terwindt *et al.*, 1997).

Other mendelian idiopathic epilepsies. A number of other pedigrees with autosomal dominant partial epilepsies are under investigation including familial temporal lobe epilepsy (Berkovic *et al.*, 1996), partial epilepsy with variable foci (Scheffer *et al.*, 1998), familial multifocal epilepsy with pericentral spikes (Kinton *et al.*, 1998) and partial epilepsy with auditory features (Ottman *et al.*, 1995). It is likely that the genetic contribution to partial epilepsy has been unrecognized when presenting beyond childhood. Another mendelian syndrome, GEFS+, is discussed in detail in section 3.4.

2.2 Common forms of IGE

Common forms of IGE have a complex pattern of inheritance. Individuals with IGE have seizures of generalized types, including absences, myoclonic jerks, myoclonic absences or GTCS. Seizures may be precipitated by sleep deprivation and alcohol. They may also be triggered by photic stimulation and may demonstrate exacerbation during menstruation. Presentations are influenced by age. Seizures may follow a particular pattern. A florid onset, for example, is observed in CAE with many daily attacks. Morning myoclonus predominantly affecting the upper limbs is observed in JME. GTCS often occur on awakening. The role of environmental factors in precipitating clinical epilepsy in those susceptible has not been well studied and requires further investigation. There is an increased incidence in IGE of a prior history of febrile convulsions. Those affected have normal neurology, normal imaging and usually normal cognition unless the epilepsy is very severe and of early onset. EEG, as already discussed, shows GSW. A positive PPR may also be demonstrated. IGE is subdivided into various age related syndromes but individuals do not always fit into a subtype. These syndromes overlap, one syndrome may evolve into another and seizure types can be influenced by age and medication. More than one syndrome may be observed within pedigrees, which nevertheless is usually a generalized epilepsy syndrome. Of interest are Berkovic's findings that in concordant twins, DZ twins were usually concordant for the major group (partial or generalized) while MZ twins were also concordant for the subtype (S.F. Berkovic, personal communication). Sex differences are also noted with a higher incidence of certain syndromes in girls, and in relatives of female probands. As can be seen below, genetic studies have focused on CAE, JME and to a lesser extent on JAE.

Dravet *et al.* (1992) considered that the rarely reported benign infantile myoclonic epilepsy, occurring in the first or second year of life, may be the earliest expression of IGE. EEG in sleep shows brief bursts of GSW. Family history is often positive for epilepsy or convulsions. Prognosis and response to treatment is good, although GTCS may occur later in adolescence. More difficult is the nosology of two relatively rare and intractable childhood syndromes: severe myoclonic epilepsies of infancy (SMEI) which presents in the first year and (the idiopathic) myoclonic astatic epilepsy of early childhood (MAE) which presents in the first 5 years of life. Both have mostly normal development prior to onset and often a positive family history of early onset epilepsy or febrile convulsions among first degree relatives. Clinical observations of large series suggest polygenic rather than simple mendelian inheritance. A number of EEG abnormalities are observed including GSW, variable foci and early photosensitivity in both syndromes as well as febrile and afebrile seizures of different types. There are differences in seizure semiology between the two syndromes with partial seizures, for example, occurring in addition to generalized seizures in SMEI but not, as a rule in MAE. The development of the epilepsy in both syndromes affects normal development and is followed by cognitive and neurological deterioration. It is of interest that the proband in several of the GEFS+ families (see section 2.4) had MAE. It is certainly possible that susceptibility genes may be shared between these syndromes and other IGE subtypes.

Childhood absence epilepsy (CAE). The next age related syndrome is that of CAE which presents between the ages of 3 and 12 years with a peak at 6–7 years with frequent daily absences (for a review, see Loiseau, 1992). Absences usually respond to appropriate treatment and remit with age. Persistence is said to be relatively rare and is reported in only 6% of cases. In general, absences are less troublesome in older age groups; not only are they less frequent, but the clinical features associated with paroxysmal EEG changes are less pronounced, and indeed may only take the form of transient cognitive impairment. It is thus evident that the method of ascertainment will influence the reported frequency of the persistence of absences. GTCS occur in about 40% of cases and are more likely in boys, with an age of onset of absences of greater than 8 years, and in the presence of a PPR. They are less likely with early treatment and a rapid response and in the presence of occipital delta rhythms. Less than 10% of cases go on to develop myoclonic jerks. Loiseau (1992) considered as misdiagnosed those cases presenting with absences a few years before the first myoclonic jerks. Certainly cases appearing to evolve into JME are a minority. A positive family history has been reported in 15–40% of cases (Loiseau, 1992). The risk of developing epilepsy in siblings is about 10%, 25% of whom have absences.

A persistent form of CAE in a large Indian multi-generation family was investigated by both a parametric method assuming autosomal dominance and a model free method (Fong *et al.*, 1998). In this family a maximum lod score of 3.6 was obtained at a recombination fraction of 0.00 for the marker *D8S502* in the 8q24 region and significant *P* values for several markers in the same region using the nonparametric method. Five smaller families with the persistent form of childhood absence had an aggregate maximum lod score of 2.4 for the marker *D8S537* within 1 Mb of *D8S502* and a similar result with the model free method.

A haplotype analysis was performed in this large pedigree using 16 closely linked markers ordered according to data from Genethon and the Whitehead Institute. In this analysis, the CAE region appears to map to 8q24 and is very close to, but on the telomeric side of, *KCNQ3*. This is the gene implicated in one form of BFNC. However, uncertainties in the map order prevent firm conclusions being made about flanking markers. The map order used is significantly different from that given by the University of Southampton Linkage Database (LDB) which is based primarily on physical mapping data. Using the LDB marker to marker distances and map order, the distance between the linked region and *KCNQ3* appears to be less than 3 Mb and there is some overlap between the two regions. It therefore remains feasible that a causative gene for BFNC may also be a susceptibility gene for CAE.

Interestingly, a study of Italian families of IGE patients (Zara *et al.*, 1995) also maps to 8q24, albeit with only suggestive evidence for linkage. Again both parametric and nonparametric methods were used in the analysis. Assuming autosomal recessive inheritance, a maximum lod score of 1.96 at a recombination fraction of 0.05 was obtained for the marker *D8S256* which lies within the CAE region. A significant *P* value was obtained for the same marker in the nonparametric analysis. This IGE study which includes patients with absences may well have detected a CAE susceptibility gene in a more heterogeneous population. However, an attempt to replicate this finding with German IGE families was unsuccessful (Sander *et al.*, 1998a). Besides *KCNQ3*, two other genes in the 8q24 region have also excited interest: the human homolog of the mouse jerky gene (Morita *et al.*, 1998) and *GRINA* which encodes an *N*-methyl-D-aspartate (NMDA) associated subunit (Lewis *et al.*, 1996).

Juvenile absence epilepsy (JAE). JAE has a later age of onset than CAE occurring around puberty. It differs from CAE in that absences are less frequent and some 80% of cases have GTCS, often on awakening. There is therefore an overlap between this syndrome and epilepsy with *grand mal* seizures on awakening (EGMSA). Photosensitivity was reported to be less frequent in one study. Myoclonic jerks occur in a minority and are slightly more frequent than in CAE. Although linkage studies on JAE families have yet to be reported, JAE patients have been used as a subset in several linkage studies of IGE families (Durner *et al.*, 1999; Sander *et al.*, 1998a; Zara *et al.*, 1995). An association has, however, been reported between a sample of JAE probands and a tetranucleotide repeat polymorphism within *GRIK1*, the gene for the kainate type of glutamate receptor GluR5 (Sander *et al.*, 1997b). *GRIK1* maps to 21q22.1 near to the gene for the G-protein coupled inward rectifying potassium channel Kir3.2 (Ohira *et al.*, 1997) which has been implicated in the weaver mouse (see Section 3.4).

Juvenile myoclonic epilepsy (JME). JME is relatively common, accounting for 4–7% of all epilepsies (Janz *et al.*, 1992; Wolf, 1992), and about 15–20% of IGEs. The age of onset in 79% of cases is between 12 and 18 years, but much younger and older ages of onset of otherwise typical cases do occur. Two studies noted lower mean age of onset, also in the teens, in those with a positive PPR (Wolf, 1992). JME is characterized by myoclonic jerks and, in 95% of cases, also by GTCS especially on awakening (Jain *et al.*, 1996; Janz *et al.*, 1992; Janz, 1997). Approximately one-third of JME cases are photosensitive. JME shows considerable overlap with other forms of IGE, in particular with EGMSA. Between 8 and 37% of patients have absences, with estimates varying depending on whether polygraphic recordings are performed (Jain *et al.*, 1996; Janz *et al.*, 1992; Panayiotopoulos *et al.*, 1989). However, a pyknoleptic presentation of absences (daily frequent clinical absences) is rare, occurring in about 5% of JME patients. Between 5 and 13% of first degree relatives of JME probands have been found to have epilepsy (Jain *et al.*, 1996; Janz *et al.*, 1992). Of these, in one study, one-third had JME, one-third an absence epilepsy and one-third epilepsy with GTCS (Janz *et al.*, 1992). In another older study by Tsuboi in 1977 (quoted by Wolf, 1992) 14% had 'myoclonic petit mal' (JME), 11% nonpyknoleptic absences (JAE) and 3.4% pyknolepsy (CAE), and 85% exhibited GTCS. In addition, several clinically normal siblings of JME probands have abnormal EEGs (Delgado Escueta *et al.*, 1994; Greenberg *et al.*, 1992).

In recent years several groups have performed linkage studies in the search for major susceptibility loci for JME. The disorder does not, however, show simple Mendelian inheritance and as a consequence a number of different genetic models have been applied in these studies. Despite a number of positive linkage reports the accumulated data is not yet sufficient to indicate which of these are true linkages. There is a major problem in defining the phenotype and different groups have applied different criteria to define affected individuals within JME pedigrees. The first report of linkage came from a group in California who detected linkage of JME to the human leukocyte antigen (HLA) region on 6p21.3. They obtained a lod score of 3.03 with a definition of the phenotype which included asymptomatic family members with an abnormal EEG (Greenberg and Delgado Escueta, 1993; Greenberg *et al.*, 1988). In this study a negative lod score was obtained with the HLA marker if such individuals were classified as unaffected. A similar result was obtained by a German group who used an even broader definition of the phenotype in a set of German families, which included relatives of the JME proband with absences and GTCS (Durner *et al.*, 1991; Weissbecker *et al.*, 1991). They also obtained lod scores in excess of 3 with markers in the HLA region. In contrast with the first study, positive lod scores were maintained even after nonepileptic individuals with abnormal EEGs were re-classified as unaffected, but not if patients with GTCS and/or absence seizures were excluded from the phenotype. A further study by a British group using Caucasian families from the UK and Sweden, failed to confirm linkage on chromosome 6, although a small positive lod score of 1.38 was obtained for the HLA-DQA1/A2 marker at a recombination fraction of 0.1 (Whitehouse *et al.*, 1993). In this study, relatives of the proband with GTCS, or absences, or both were included in the phenotype but no EEGs were performed on unaffected individuals.

A later study by the American group focused mainly on South American families, including a large family originating from Belize. They continued to use a phenotype definition which included abnormal EEGs in otherwise asymptomatic individuals, but subdivided the families into two groups according to whether the families contained individuals with absence seizures. The 'classic' JME families with the narrower phenotype showed strong linkage to a large region of 6p but this was significantly centromeric to the HLA region, whereas the families including relatives with absences gave negative lod scores throughout 6p (Liu *et al.*, 1995). The highest lod scores in the former group of families were for the markers *D6S272*, *D6S294*, *D6S428* and *D6S257*, defining a 7 cM region in 6p12–p11. This region is approximately 15 Mb in length and is at least 12 Mb from the HLA complex, according to the University of Southampton LDB. In a follow-up paper, the two groups of families were combined and the dataset was extended with some new families; additional markers were then tested (Liu *et al.*, 1996). Linkage to the same region was reported with lod scores greater than 3, although these maximized at higher recombination fractions, perhaps due to an increase in heterogeneity within the set of pedigrees. A further study by the German group confirmed the original linkage to the HLA region with a maximum lod score of 3.08 (Sander *et al.*, 1997a). In this study, markers spanning the 6p12–p11 region identified in the South American families all gave negative lod scores. Finally, the British group performed a further study and again obtained negative lod scores, this time using markers spanning both putative linkage regions on 6p (Elmslie *et al.*, 1996). In this study they used a far more restricted definition of the phenotype, only counting JME individuals as affected. Moreover, families with fewer than two JME patients were excluded from the study.

Some limited evidence in support of the HLA locus has come from association studies. However in one study (Greenberg *et al.*, 1996) the effect is small and involved only 24 JME patients and 24 controls, while in another study of 66 JME patients and 77 controls a trend in the same direction failed to achieve significance (Sander *et al.*, 1997a). A candidate gene which has recently been mapped to 6p21.3, very close to the HLA locus, is the gene for $GABA_BR1$ (Peters *et al.*, 1998). This G-protein-coupled γ-amino butyrate (GABA) receptor is found in both pre- and postsynaptic locations. The postsynaptic receptors appear to be coupled to inwardly rectifying potassium channels (Kir3.1–3.4). Of these, Kir3.2 (formerly GIRK2) either when mutated in weaver mice or absent in gene knockout mice causes seizures (see section 3.4). Moreover, $GABA_B$ receptors have been implicated in mouse models of absence epilepsy (Caddick and Hosford, 1996; Hosford *et al.*, 1992). The frequencies of three polymorphic variants of $GABA_BR1$ were compared between JME patients, combined CAE and JAE patients and controls, but no significant differences were apparent (Sander *et al.*, 1999a). The JME patients in this study included patients from families showing positive linkage to the HLA locus, but no significant association with HLA-DR (Sander *et al.*, 1997a)

The British group who failed to find evidence for linkage in 6p, found strong evidence for linkage of JME to 15q13–q14, again using a very narrow definition of phenotype (Elmslie *et al.*, 1997). This was especially interesting as they used only candidate gene markers, close to the nicotinic acetylcholine receptor α subunit genes. They found a peak lod score of 4.42 and a nonparametric linkage score of 2.94 ($P = 0.00048$) very close to *CHRNA7*, the gene for the α7 subunit of the nicotinic acetylcholine receptor (Elmslie *et al.*, 1997). This finding was not replicated in two other linkage studies of JME families that had previously shown linkage to 6p (Durner *et al.*, 1998; Sander *et al.*, 1999b). Interestingly, this region has also shown linkage to the partial idiopathic epilepsy BECTS (see section 2.3) and to the sensory gating deficit phenotype of schizophrenia (Freedman *et al.*, 1997; Neubauer *et al.*, 1998).

Unlike the α4 receptor, α7 is able to form homomultimeric ion channels which predominantly have a presynaptic location. The receptor is also highly permeable to calcium ions which is consistent with a facilitatory presynaptic role (MacDermott *et al.*, 1999; McGehee and Role, 1996). A transgenic mouse has been constructed that lacks *CHRNA7* which exhibits continuous generalized hypersynchronous 4–7 Hz sharp wave activity on EEG (Orr Urtreger *et al.*, 1997). Independent evidence for its involvement in the sensory gating deficit associated with schizophrenia comes from pharmacological studies both in patients responding to nicotine and in animal models (Leonard *et al.*, 1996). Further studies with this gene are complicated by it being partly duplicated (Gault *et al.*, 1998). The duplicated region covers more than half the coding sequence (exons 6–10) and is attached to a gene of unknown function and with which it is transcribed.

It is possible therefore that different mutations in the same gene predispose to both schizophrenia and different idiopathic epilepsies. Of possible relevance are two related patients who have had GTCS and either bipolar affective disorder or schizoaffective disorder (Calzolari *et al.*, 1996). Both have a partial trisomy of chromosomes 15 (pter–q13.3) and 18 (q23–qter) derived from a familial balanced translocation which may include the *CHRNA7* region.

In summary, three different loci have been implicated in JME (6p12–p11, 6p21.3 and 15q13–q14), and although two of these regions are close to each other they appear to be distinct.

Other and unclassified IGEs. Attention has focused on the above categories as in their 'pure' forms they are easy to recognize. Other subtypes, such as EGMSA and photosensitive epilepsies overlap with these categories. There are also many IGE cases with both GTCS (without a specific pattern) and GSW on EEG, but these are not usually the subject of genetic studies except in large pedigrees. However, several genetic studies have included apparently unclassified IGE cases. Many of these come from families which include individuals with specific subtypes and thus do not represent unselected cases. Some of the JME studies which found linkage to 6p21.3 used a sufficiently broad definition of phenotype to fit into this category (Durner *et al.*, 1991; Sander *et al.*, 1997a; Weissbecker *et al.*, 1991). Two studies looked at the linkage of IGE to 6p21.3. Sander *et al.* (1995) used IGE families where the proband had an absence epilepsy and failed to find linkage to the HLA locus (Sander *et al.*, 1995). Zara *et al.* (1995) used IGE families which included both classified and unclassified IGEs as well as febrile convulsions. They also failed to detect linkage of IGE to 6p in Italian families but did find suggestive evidence for linkage to 8q24, near to *KCNQ3* and near to a linkage region for CAE (see *Childhood Absence Epilepsy*).

In a similar approach to that of Liu *et al.* (1995), Durner et al. (1999) subdivided their IGE families into two groups according to whether or not the families contained any individuals with JME. They investigated linkage to chromosome 8 and failed to find any evidence for linkage whether or not asymptomatic individuals with a GSW on EEG were classified as affected or unaffected. However if families with any members who had JME were excluded, linkage was found over a broad region of 8p near to the centromere with a maximum lod score of 3.24 when asymptomatic GSW individuals were included in the phenotype or 3.18 when excluded. They therefore also failed to support the 8q24 location. Because of multiple testing, this new region of chromosome 8 must be viewed with caution until independent replication. It contains *CHRNB3*, the gene for the β3 subunit of the nicotinic acetylcholine receptor.

An unusual EEG trace was found in 11 individuals from an Italian family of

IGE patients, of which four had seizures including two with bilateral spike wave discharges (Zara *et al.*, 1998). This EEG trait consisted of delta bursts with bilateral synchronous high voltage slow waves and appeared to segregate in an autosomal dominant manner. Linkage was investigated in a genomewide scan which detected linkage to 3p14.2–p12.1 giving a maximum lod score of 3.81 at a recombination fraction of 0.0. Near to this region is *CACNA1D* which encodes the voltage gated calcium channel α1D subunit which forms L-type calcium channels and is localized in the brain and endocrine systems. Involvement of mutations in another voltage gated calcium channel gene (*CACNA1A*) in seizures, at least in the mouse, makes this a strong candidate gene (see Section 3.1).

2.3 Common idiopathic partial epilepsies

Benign epilepsy of childhood with centrotemporal spikes (BECTS or rolandic epilepsy). BECTS is a common idiopathic partial epilepsy found in 15–20% of young people with epilepsy. As with many other idiopathic epilepsies, there is a higher incidence of febrile convulsions than in the general population with a possible association with headache/migraine. In close relatives there is a higher incidence of a history of febrile convulsions, partial or generalized seizures and both generalized and focal epileptiform EEG changes (Lerman, 1992). As already mentioned, the idiopathic partial epilepsies with complex inheritance may represent a spectrum which includes BECTS. The characteristic age dependent EEG trait is centrotemporal spikes often followed by slow waves. In a recent linkage study involving 22 families both BECTS patients and asymptomatic relatives with CTS were counted as affected. Both parametric and nonparametric analyses had strong evidence for linkage in 15q13–q14 in the vicinity of the α7 subunit of the nicotinic acetylcholine receptor (*CHRNA7*). In the parametric analysis a maximum lod score of 3.16 was obtained under an autosomal recessive mode of inheritance for the marker *D15S1010* at a recombination fraction of 0.02. Furthermore there was evidence for heterogeneity. These results suggest linkage to the same region as was found linked in some JME families (see section 2.2).

In another study, a family of BFNC patients was described in which two affected siblings exhibited generalized seizures within a few days of birth and went on to develop BECTS in early childhood. This condition subsequently resolved after the age of 10 years (Maihara *et al.*, 1999). Six family members had BFNC, all of which had their first seizure shortly after birth. It has been long recognized that about 10% of BFNC patients continue to exhibit seizures after the neonatal period, but the nature of the seizures has not been studied systematically. This family may represent an area of overlap between these two conditions. It will be interesting to establish whether this BFNC family, in common with other BFNC families has mutations in either *KCNQ2* or *KCNQ3*.

Finally, as described above, BECTS with paroxysmal exercise induced dystonia and writer's cramp was mapped to chromosome 16 (see section 2.1).

2.4 Febrile convulsions

Febrile convulsions are seizures associated with fever and are very common in childhood, affecting between 2 and 5% of children aged 3 months to 5 years (O'Donohoe, 1992). The majority are simple (brief and generalized occurring in otherwise normal children), while a minority are complex (prolonged, sometimes with partial features and transient or permanent neurological deficit). Febrile convulsions are a heterogeneous group of disorders comprising both idiopathic and symptomatic etiologies and their genetic analysis is therefore not straightforward. Complex febrile seizures may in some cases reflect pre-existing brain disease and they may also be causally related to the later development of partial epilepsies. However, the evidence from family and twin studies suggests that the majority of febrile seizures are likely to reflect a genetic predisposition to epilepsy. Although simple febrile convulsions usually have no consequences of their own, they are associated with an increased risk for later development of idiopathic epilepsy in general and IGE in particular (O'Donohoe, 1992). There is a considerable genetic influence in febrile seizures, with a three to four-fold increase in the risk to first degree relatives compared with the general population (Hauser *et al.*, 1985; Miller *et al.*, 1998). Twin studies (see *Table 2*) estimate heritability to be in the order of 80% (Berkovic *et al.*, 1998; T. Price and P. Asherson, unpublished data). There is evidence for considerable genetic heterogeneity in this common group of disorders with some relatively rare families displaying an autosomal dominant inheritance pattern, and others a complex pattern.

In one study, a large Australian family with 88 individuals from three generations was investigated (Wallace *et al.*, 1996). A total of 19 affected family members were defined by having at least one febrile seizure before the age of 6 years and included an obligate carrier who had seizures in adulthood. A genomewide scan using 275 markers was performed on these 19 individuals, plus their parents, which excluded all but four chromosomes. Further genotyping with additional markers excluded all except chromosome 8. After inclusion of all family members a two-point lod score of 2.25 was obtained for *D8S530* at a recombination fraction of 0. By making adjustments for the observed penetrance of 60% and estimating the phenocopy rate at 3%, a multipoint lod score of 3.4 was obtained between *D8S553* and *D8S543* (8q13–q21). Two candidate genes in this region are the genes for calbindin and corticotrophin releasing factor, both of which have been implicated in seizures in rat models. To date no other study has found evidence for linkage of febrile convulsions to this region.

In another study, 19 families of patients with febrile convulsions were investigated (Johnson *et al.*, 1998). Following a genomewide scan with 130 markers, only one large family gave a significant lod score which indicated possible linkage to chromosome 19. When other nearby markers were used, significant pairwise lod scores were obtained for *D19S216* (4.3), *D19S427* (4.1) and *D19S177* (4.5). Haplotype analysis identified a disease associated haplotype between flanking markers *D19S591* and *D19S395* which define a 1.5–2 cM region of 19p13.3. Within this region is the gene for Cayman ataxia and on the syntenic region of mouse chromosome 10 is jittery/hesitant (Kapfhamer *et al.*, 1996; Nystuen *et al.*, 1996).

A syndrome of GEFS+ was described in a very large family with an autosomal dominant pattern of inheritance (Scheffer and Berkovic, 1997). More than one phenotype was observed within the pedigree. The most common, denoted as febrile seizures plus (FS+), comprised febrile seizures continuing beyond 6 years or the subsequent development of afebrile seizures. Afebrile seizures were usually generalized and included absences, myoclonic seizures and atonic seizures. One individual, the proband, had MAE. Subsequently, nine other GEFS+ families were described, in seven of which at least the proband had MAE but this may reflect an ascertainment bias (Singh *et al.*, 1999). A linkage study was performed with 378 individuals from the largest of these more recent pedigrees, of which 42 had seizures including 26 with GEFS+ (Wallace *et al.*, 1998). A genomewide scan excluded all chromosomes except 19. Several markers on chromosome 19 gave a significant multipoint lod score over 19q13.1 with a peak score of 3.85 at *D19S414*. One of the candidate genes in this region is *SCN1B*, the gene for the β1 subunit of the voltage gated sodium channel. Mutation screening of *SCN1B* revealed a base substitution found only in affected family members and which results in a conserved cysteine residue being replaced by tryptophan, disrupting a disulphide bridge. The normal function of the β1 subunit is to modulate the activity of the α subunit, including reduction of the inactivation time after activation of the channel. Co-expression in *Xenopus* oocytes of the mutant β1 subunit with the α subunit resulted in slower inactivation kinetics of the sodium channel, similar to the kinetics with the α subunit alone. Such an effect is consistent with increased neuronal excitability. The involvement of sodium channels in seizures has long been recognized and many anti-epilepsy drugs target sodium channels. A linkage region overlapping this region was implicated in families with BIFC (see section 2.1).

The GEFS+ syndrome in this family behaves as a dominant disorder with low penetrance: of 18 people with this mutation seven were unaffected, indicating a penetrance of only 60%. All eight individuals with GEFS+ had the mutation, but six out of 12 who had uncomplicated febrile seizures had both copies of the wild type cysteine allele. It is likely that other genes influenced the clinical presentations within this family. Mutations in *SCN1B* have been sought in GEFS+ patients from other families and so far only one other apparently unrelated family has been found to have the same mutation.

Recently Baulac *et al.* (1999) reported a large French family with a phenotype similar to GEFS+. They excluded the 19q *SCN1B* gene and found significant linkage to 2q21–q33 with a maximum pairwise lod score of 3.00 at a recombination fraction of 0.00. Close to this linkage region are the genes for four sodium channel α subunits (*SCN1A*, *SCN2A1*, *SCN2A2*, *SCN3A*), all of which are expressed in the brain (Han *et al.*, 1991; Litt *et al.*, 1989; Lu *et al.*, 1992; Malo *et al.*, 1994a, 1994b). A fifth gene from the same α subunit family has been mapped to the homologous gene cluster on mouse chromosome 2 and which is also expressed in the brain (Beckers *et al.*, 1996). As the β1 subunit interacts with the α subunits, these genes represent highly plausible candidate genes for this syndrome.

3. Mouse epilepsies and their relation to human studies

There are several mouse strains which have a phenotype including epilepsy and many are used as experimental animal models for different types of epilepsy. Some of these segregate in a Mendelian and usually recessive fashion, presumably because they were found in inbred strains. By contrast, almost all of the mutations so far reported for human idiopathic epilepsies are dominant. We have reviewed some recent developments in this field and relate them to human disease.

3.1 Tottering (tg)

This is a recessive mutation which has a phenotype comprising mainly ataxia, motor seizures and absence type seizures. Fine mapping localized *tg* to a gene on mouse chromosome 8. This is the mouse homolog of the rat voltage gated α1a calcium subunit gene which encodes the pore forming subunit of P/Q type calcium channels (Fletcher *et al.*, 1996). The α subunits comprise four repeating units, each of which have a similar structure to a voltage gated potassium channel α subunit (see section 2.1). Each repeat contains six transmembrane domains (S1–S6) and a pore forming loop (P) between S5 and S6. Sequencing of the mouse gene from wild type and *tg* mice identified a base substitution in *tg* which results in a proline being replaced by a leucine in the P loop of the second repeat and which presumably affects the function of the ion pore. There are two other mouse mutations in the same gene: leaner (tg^{la}) and rolling (tg^{rol}). The leaner mouse has a much more severe phenotype than tottering, additionally having cerebellar neurodegeneration and often failing to survive beyond a few weeks. This mutation is a G to A base change at a splice site resulting in an altered carboxyl terminus in the calcium channel subunit. The phenotype of tg^{rol} is intermediate in severity between *tg* and tg^{lu} but the mutation has not yet been identified. The way in which these mutations in the α1a subunit result in the epileptic and ataxic phenotypes observed is not clear, especially as they appear to result in reduced P current activity in Purkinje cells (Lorenzon *et al.*, 1998; Wakamori *et al.*, 1998) and a reduction in glutamatergic but not GABAergic transmission in the somatosensory thalamus (Caddick *et al.*, 1999). Compensatory changes may well be occurring. Campbell and Hess (1999) demonstrated an upregulation of L-type calcium channels in tottering mice which appears to contribute to the motor seizure phenotype.

Several genetically inherited neurological conditions in humans are due to mutations in *CACNA1A* (the gene for the calcium channel α1a subunit): FHM, progressive ataxia, episodic ataxia type 2 (EA-2) and spinocerebellar ataxia type 6 (SCA6) (Ophoff *et al.*, 1998). Four different base substitution mutations leading to single amino acid changes at positions throughout the calcium channel subunit have been identified in FHM families, one of which is in the P loop of the second repeat and very close to the equivalent position of the *tg* mouse mutation (Ophoff *et al.*, 1996). Patients with this mutation also have progressive ataxia associated with FHM. Interestingly a family with progressive ataxia and associated migraine has a mutation in the P loop in the first repeat (Yue *et al.*, 1997). Four mutations in EA-2 families have been described. Two are very close to each other near to the beginning of the third repeat and both (a deletion and a splice site mutation) result in a frame shift and premature truncation of translation (Ophoff *et al.*, 1996). Recently, two new EA-2 families were found to have novel mutations (Eunson *et al.*, 1999). One of these families is of particular interest as the proband has an absence epilepsy and learning difficulties in addition to EA-2. The mutation in this family is in the S6 transmembrane domain of the fourth repeat and results in an arginine residue being replaced by a premature stop codon resulting in the absence of the carboxyl-terminal intracellular region. The mutation in the other family is in an intron near to the end of the coding region and may affect splicing. *CACNA1A* has a polymorphic series of CAG repeats which are translated to a polyglutamine tract in at least one of the many splice variants of the calcium channel subunit which are known to exist. In some SCA6 families and also in an EA-2 family there appears to be an increase in the number of these repeats (Jodice *et al.*, 1997; Zhuchenko *et al.*, 1997). Many neurological diseases are caused by expansions of trinucleotide repeats, but these are much larger expansions than the ones so far described in *CACNA1A* (Brice, 1998; Koshy and Zoghbi, 1997; Warren, 1996). Consequently, the relevance of these modest increases in the number of CAG repeat units to these ataxias remains uncertain.

The three mutations in the mouse α1a calcium channel gene all result in ataxia and seizures. The mutations in the human homolog have ataxias or FHM. So far, as described above, only one case of a human epilepsy being associated with a mutation in *CACN1A* has been reported. Other families have been reported which combine epilepsy with either FHM or ataxia. We have already described a family which combined FHM with BFIC and was not linked to *CACNA1A* (see section 2.1). Another study of three similar families, also unlinked to 19q, showed linkage to 1q21–q23 where the G-protein coupled inward rectifying potassium channel gene (Kir3.3, formerly GIRK3) is a good candidate gene (Ducros *et al.*, 1997). A family which combined spinocerebellar ataxia with motor seizures was also unlinked to *CACNA1A* and showed linkage to 22q13–qter (Matsuura *et al.*, 1999). Also of interest is the over-representation of epilepsy in families with episodic ataxia type 1, a rare disorder of ataxia associated with myokymia, due to point mutations in the human voltage gated potassium channel *Kv1.1* gene on 12p13 (Zuberi *et al.*, 1999).

One study has investigated *CACNA1A* in human epilepsy both by linkage and association using haplotype relative risk (Sander *et al.*, 1998b). The CAG repeat polymorphism was studied in IGE families and in two subsets defined by having at least one member with JME or CAE/JAE. The results were negative, although there was a trend for association with JME using a narrow definition which excluded nonepileptic relatives with GSW on EEG. However, the study involved small numbers (n = 55 for all IGE patients) which severely limited the power to detect susceptibility genes in a complex disorder and only one polymorphism. Firm conclusions await further studies with larger numbers and additional mutations.

3.2 Lethargic (lh)

These mice have a very similar phenotype to tottering, combining ataxia with motor seizures and absence type seizures. The *lh* mutation is also recessive and was mapped to the gene for the voltage gated calcium channel β4 subunit located on mouse chromosome 2 (Burgess *et al.*, 1997). This subunit interacts with the α1 subunits including α1a. The mutation is a 4 base insertion which disrupts a splice junction resulting in two altered transcripts at considerably lower levels. The resulting polypeptides are truncated and lack the region that is necessary for interaction with α1 subunits. No β4 subunit has been detected in lethargic mice, but the other three β subunits are all able to replace β4 in interacting with α1 subunits (McEnery *et al.*, 1998; Burgess *et al.*, 1999). As with tottering, the relationship between the mutation and the observed phenotype is likely to result from a complex series of changes. The gene for the human homolog of mouse β4 (*CACNB4*) has been mapped to 2q22–q23, but so far it has not been associated with human disease (Escayg *et al.*, 1998).

3.3 *Stargazer* (sg)

The stargazer mouse is another mutation with a phenotype which combines absence seizures with ataxia. The *sg* mutation was shown to be due to the insertion of a transposon into an intron of a novel gene which has homology to the gene for the voltage gated calcium channel γ subunit (Letts *et al.*, 1998). This subunit was previously identified in skeletal muscle and modulates L-type calcium channels by interacting with α1s subunits. Transcription of the novel gene in stargazer mice results in abnormal mRNA species with very little normal transcript. The novel protein also appears to have calcium channel modulating properties in P/Q-type calcium channels and is brain specific. It has been designated the γ2 subunit. It has been speculated that the phenotype in stargazer results from inappropriate calcium entry in the cerebellum as a consequence of inefficient regulation of P/Q calcium currents.

3.4 *Weaver* (wv)

The homozygous weaver mouse (*wv/wv*) has a complex and severe phenotype including severe ataxia, hyperactivity, tremor, seizures and male sterility and usually results in death within the first 3 weeks. The neuronal abnormalities are associated with neurodegeneration mainly in the cerebellum and substantia nigra pars compacta. The heterozygote (*wv*/+) is also affected but to a lesser degree and the phenotype is limited mainly to sporadic seizures. The mutation causing the weaver phenotype is a glycine to serine change in the pore-lining H5 domain of a G-protein coupled inward rectifying potassium channel subunit (Kir3.2, formerly GIRK2) (Patil *et al.*, 1995). These potassium channels are activated by a number of receptors including $GABA_BR1$ receptors (see section 2.2).

Although this mutation leads to a decrease in G-protein activated potassium current in cerebellar granule cells and hippocampal slices (Jarolimek *et al.*, 1998; Kofuji *et al.*, 1996; Surmeier *et al.*, 1996), loss of Kir3.2 function cannot explain completely the weaver phenotype because mice lacking Kir3.2 do not exhibit the same symptoms. Homozygous null mice (–/–) have seizures but are otherwise normal, whereas heterozygous null mice (+/–) are indistinguishable from wild type mice (Signorini *et al.*, 1997). When expressed in oocytes, Kir3.2 containing the *wv* mutation display loss of ion selectivity, being permeable to sodium ions (Slesinger *et al.*, 1996) and calcium ions (Silverman *et al.*, 1996). In cerebellar granule cells from *wv/wv* mice, significant sodium currents were detected in one study (Kofuji *et al.*, 1996) and elevated calcium levels in another (Womack *et al.*, 1998). These observations have led to the view that the weaver phenotype results from a gain of function due to leaking of inward sodium and/or calcium currents through ion channels containing Kir3.2 (*wv*) subunits. However, loss of normal function is also clearly involved. Furthermore, Kir3.1 (GIRK1) subunits which appear to form heteromultimeric ion channels with Kir3.2 (Jelacic *et al.*, 1999;

Wischmeyer *et al.*, 1997) are present at reduced levels in both –/– and *wv/wv* mice (Liao *et al.*, 1996; Signorini *et al.*, 1997).

4. Current status and future prospects

In this chapter we have reviewed the major findings in human genetic studies of the idiopathic epilepsies as well as several mouse epilepsy syndromes which may provide important clues about the identity of human epilepsy genes. Several genes have been implicated in rare Mendelian epilepsies and, as already noted, these have so far universally occurred within ion channel genes. Considerable progress has also been made in the analysis of more complex forms of idiopathic epilepsy, which is demonstrated by a number of encouraging reports of linkage to subtypes such as JME and CAE. Discrepancies between studies are in part due to differences in the clinical assessment and categorization of cases. It is also very likely that the genetic heterogeneity found within the Mendelian epilepsies will turn out to be present in the more common forms. These considerations add further complexity to the study of polygenic disorders. Discrepancies, in this context, are not surprising and we should remain encouraged by the identification of overlapping linkage regions in the studies that have been described so far.

Continued success in the analysis of common idiopathic epilepsies will almost certainly require the wider adoption of association approaches. The development of very dense marker maps and the technology to screen large numbers of genotypes will enable genomewide scans and a comprehensive analysis of candidate genes to be undertaken. In one recent report on large scale mapping of single nucleotide polymorphisms (SNPs), one SNP was detected on average every 1001 base pairs and a total of 3241 SNPs identified (Wang *et al.*, 1998). In parallel with identification of these markers, the group collaborated in the development of a prototype DNA micro-array, which enables simultaneous genotyping of 500 SNPs. A SNP consortium has now been established to identify and map 200 000 such markers over the next 2 years. At the same time, comprehensive surveys of candidate gene markers are now being published, which are likely to be the antecedents of comprehensive analysis of all identified genes once the human genome sequence has been completed. These and other technical developments such as the use of mass spectrometry for high throughput genotyping (Pieles *et al.*, 1993; Yip and Hutchens, 1992) are expected to facilitate the analysis of very large samples with very dense maps, greatly increasing the power to detect genes of small effect in complex disorders.

References

Anderson, V.E. and Hauser. W.A. (1993) Genetics. In: *A Textbook of Epilepsy* (eds Laidlow, J., Richens, A. and Chadwick, D.). Churchill Livingstone, Edinburgh, pp. 47–75.

Baulac, S., Gourfinkel-An, I., Picard, F., Rosenberg-Bourgin, M., Prudhomme, J.F., Baulac, M., Brice, A. and LeGuern, E. (1999) A second locus for familial generalized epilepsy with febrile seizures plus maps to chromosome 2q21-q33. *Am. J. Hum. Genet.* **65**: (in press).

Beckers, M.C., Ernst, E., Belcher, S., Howe, J., Levenson, R. and Gros, P. (1996) A new sodium channel alpha-subunit gene (Scn9a) from schwann cells maps to the Scn1a, Scn2a, Scn3a cluster of mouse chromosome 2. *Genomics* **36**: 202–205.

Berkovic, S.F., Howell, R.A., Hopper, J.L., Hay, D.A. and Andermann, E. (1990) A twin study of epilepsy. *Epilepsia* **31**: 813.

Berkovic, S.F., McIntosh, A., Howell, R.A., Mitchell, A., Sheffield, L.J. and Hopper, J.L. (1996) Familial temporal lobe epilepsy: a common disorder identified in twins. *Ann. Neurol.* **40**: 227–235.

Berkovic, S.F., Howell, R.A., Hay, D.A. and Hopper, J.L. (1998) Epilepsies in twins: genetics of the major epilepsy syndromes. *Ann. Neurol.* **43**: 435–445.

Bertrand, D., Galzi, J.L., Devillers Thiery, A., Bertrand, S. and Changeux, J.P. (1993) Mutations at 2 distinct sites within the channel domain m2 alter calcium permeability of neuronal alpha-7 nicotinic receptor. *Proc. Natl Acad. Sci. USA* **90**: 6971–6975.

Biervert, C., Schroeder, B.C., Kubisch, C., Berkovic, S.F., Propping, P., Jentsch, T.J. and Steinlein, O.K. (1998) A potassium channel mutation in neonatal human epilepsy. *Science* **279**: 403–406.

Brandl, C.J., Green, N.M., Korczak, B. and Maclennan, D.H. (1986) 2 Ca-2+ ATPase genes – homologies and mechanistic implications of deduced amino-acid-sequences. *Cell* **44**: 597–607.

Brice, A. (1998) Unstable mutations and neurodegenerative disorders. *J. Neurol.* **245**: 505–510.

Burgess, D.L., Jones, J.M., Meisler, M.H. and Noebels, J.L. (1997) Mutation of the Ca2+ channel beta subunit gene cchb4 is associated with ataxia and seizures in the lethargic (lh) mouse. *Cell* **88**: 385–392.

Burgess, D.L., Biddlecome, G.H., McDonough, S.I., Diaz, M.E., Zilinski, C.A., Bean, B.P., Campbell, K.P. and Noebels, J.L. (1999) Beta subunit reshuffling modifies N- and P/Q-type Ca2+ channel subunit compositions in lethargic mouse brain. *Mol. Cell. Neurosci.* **13**: 293–311.

Caddick, S.J. and Hosford, D.A. (1996) The role of GABA(B) mechanisms in animal models of absence seizures. *Mol. Neurobiol.* **13**: 23–32.

Caddick, S.J., Wang, C.S., Fletcher, C.F., Jenkins, N.A., Copeland, N.G. and Hosford, D.A. (1999) Excitatory but not inhibitory synaptic transmission is reduced in lethargic (cacnb4(1h)) and tottering (cacnala(tg)) mouse thalami. *J. Neurophysiol.* **81**: 2066–2074.

Callen, D.F., Baker, E., Lane, S., *et al.* (1991) Regional mapping of the Batten disease locus (cln3) to human chromosome-16p12. *Am. J. Hum. Genet.* **49**: 1372–1377.

Calzolari, E., Aiello, V., Palazzi, P., *et al.* (1996) Psychiatric disorder in a familial 15;18 translocation and sublocalization of myelin basic protein to 18q22.3. *Am. J. Med. Genet.* **67**: 154–161.

Campbell, D.B. and Hess, E.J. (1999) L-type calcium channels contribute to the tottering mouse dystonic episodes. *Mol. Pharmacol.* **55**: 23–31.

Charlier, C., Singh, N.A., Ryan, S.G., Lewis, T.B., Reus, B.E., Leach, R.J. and Leppert, M. (1998) A pore mutation in a novel KQT-like potassium channel gene in an idiopathic epilepsy family. *Nature Genet.* **18**: 53–55.

Cohen, B.N., Labarca, C., Davidson, N. and Lester, H.A. (1992) Mutations in m2 alter the selectivity of the mouse nicotinic acetylcholine-receptor for organic and alkali-metal cations. *J. Gen. Physiol.* **100**: 373–400.

Collins, F.S., Guyer, M.S. and Chakravarti, A. (1997) Variations on a theme: cataloging human DNA sequence variation. *Science* **278**: 1580–1581.

Commission on Classification and Terminology of the International League Against Epilepsy (1989) Proposal for revised classification of epilepsies and epilepsy syndromes. *Epilepsia* **30**: 389–399.

Degen, R., Degen, H.E. and Ahlemeyer, K. (1996) Contribution to the genetics of symptomatic generalized tonic-clonic seizures: waking and sleep EEGs in siblings. *Acta Neurol. Scand.* **93**: 9–13.

Delgado Escueta, A.V., Serratosa, J.M., Liu, A., Weissbecker, K., Medina, M.T., Gee, M., Treiman, L.J. and Sparkes, R.S. (1994) Progress in mapping human epilepsy genes. *Epilepsia* **35**: S29–S40.

Doose, H. (1997) Genetic EEG traits in the pathogenesis of the epilepsies. *J. Epilepsy* **10**: 97–110.

Doyle, D.A., Cabral, J.M., Pfuetzner, R.A., Kuo, A.L., Gulbis, J.M., Cohen, S.L., Chait, B.T. and Mackinnon, R. (1998) The structure of the potassium channel: molecular basis of K+ conduction and selectivity. *Science* **280**: 69–77.

Dravet, C., Bureau, M. and Roger, M. (1992) Benign myoclonic epilepsy in infants. In: *Epileptic Syndromes in Infancy, Childhood and Adolescence* (eds J. Roger, M. Bureau, C. Dravet, F.E. Dreifus, A. Perret and P. Wolf). John Libbey, London, Paris, Rome, pp. 67–74.

Ducros, A., Joutel, A., Vahedi, K., *et al.* (1997) Mapping of a second locus for familial hemiplegic migraine to 1q21-q23 and evidence of further heterogeneity. *Ann. Neurol.* **42**: 885–890.

Durner, M., Sander, T., Greenberg, D.A., Johnson, K., Beck Mannagetta, G. and Janz, D. (1991) Localization of idiopathic generalized epilepsy on chromosome 6p in families of juvenile myoclonic epilepsy patients. *Neurology* **41**: 1651–1655.

Durner, M., Greenberg, D.A., Shinnar, S., ***et al.*** (1998) No evidence for linkage of juvenile myoclonic epilepsy to chromosome 15. *Epilepsia* **39(suppl. 6)**: 145.

Durner, M., Zhou, C.L., Fu, D.Y., ***et al.*** (1999) Evidence for linkage of adolescent-onset idiopathic generalized epilepsies to chromosome 8 – and genetic heterogeneity. *Am. J. Hum. Genet.* **64**: 1411–1419.

Echenne, B., Humbertclaude, V., Rivier, F., Malafosse, A. and Cheminal, R. (1994) Benign infantile epilepsy with autosomal-dominant inheritance. *Brain Dev.* **16**: 108–111.

Elmslie, F. and Gardiner, M. (1995) Genetics of the epilepsies. *Curr. Opin. Neurol.* **8**: 126–129.

Elmslie, F.V., Williamson, M.P., Rees, M., ***et al.*** (1996) Linkage analysis of juvenile myoclonic epilepsy and microsatellite loci spanning 61 cM of human chromosome 6p in 19 nuclear pedigrees provides no evidence for a susceptibility locus in this region. *Am. J. Hum. Genet.* **59**: 653–663.

Elmslie, F.V., Rees, M., Williamson, M.P., ***et al.*** (1997) Genetic mapping of a major susceptibility locus for juvenile myoclonic epilepsy on chromosome 15q. *Hum. Mol. Genet.* **6**: 1329–1334.

Escayg, A., Jones, J.M., Kearney, J.A., Hitchcock, P.F. and Meisler, M.H. (1998) Calcium channel beta 4 (cacnb4): human ortholog of the mouse epilepsy gene lethargic. *Genomics* **50**: 14–22.

Eunson, L.H., Jouvenceau, A., Ramesh, V., Nairne, A., Hyman, N., Wood, N.W., Spauschus, A., Kullmann, D.M. and Hanna, M.G. (1999) New mutations in the voltage-gated calcium channel gene CACNA1A in British families with episodic ataxia type 2. *Am. J. Hum. Genet.* **65**: 1646.

Fletcher, C.F., Lutz, C.M., O'Sullivan, T.N., Shaughnessy, J.D., Hawkes, R., Frankel, W.N., Copeland, N.G. and Jenkins, N.A. (1996) Absence epilepsy in tottering mutant mice is associated with calcium-channel defects. *Cell* **87**: 607–617.

Fong, G.C.Y., Shah, P.U., Gee, M.N., ***et al.*** (1998) Childhood absence epilepsy with tonic-clonic seizures and electroencephalogram 3–4-Hz spike and multispike-slow wave complexes: linkage to chromosome 8q24. *Am. J. Hum. Genet.* **63**: 1117–1129.

Freedman, R., Coon, H., Myles Worsley, M., ***et al.*** (1997) Linkage of a neurophysiological deficit in schizophrenia to a chromosome 15 locus. *Proc. Natl Acad. Sci. USA* **94**: 587–592.

Galzi, J.L., Devillers Thiery, A., Hussy, N., Bertrand, S., Changeux, J.P. and Bertrand, D. (1992) Mutations in the channel domain of a neuronal nicotinic receptor convert ion selectivity from cationic to anionic. *Nature* **359**: 500–505.

Gardiner, R.M. (1990) Genes and epilepsy. *J. Med. Genet.* **27**: 537–544.

Gault, J., Robinson, M., Berger, R., ***et al.*** (1998) Genomic organization and partial duplication of the human alpha 7 neuronal nicotinic acetylcholine receptor gene (chrna7). *Genomics* **52**: 173–185.

Gedda, L. and Tatarelli, R. (1971) Essential isochronic epilepsy in MZ twin pairs. *Acta Genet. Med. Gemello. Roma* **20**: 380–383.

Greenberg, D.A. and Delgado Escueta, A.V. (1993) The chromosome 6p epilepsy locus – exploring mode of inheritance and heterogeneity through linkage analysis. *Epilepsia* **34**: S12–S18.

Greenberg, D.A., Delgado Escueta, A.V., Widelitz, H., Sparkes, R.S., Treiman, L., Maldonado, H.M., Park, M.S. and Terasaki, P.I. (1988) Juvenile myoclonic epilepsy (JME) may be linked to the BF and HLA loci on human chromosome 6. *Am. J. Med. Genet.* **31**: 185–192.

Greenberg, D.A., Durner, M. and Delgado Escueta, A.V. (1992) Evidence for multiple gene loci in the expression of the common generalized epilepsies. *Neurology* **42**: 56–62.

Greenberg, D.A., Durner, M., Shinnar, S., ***et al.*** (1996) Association of HLA class II alleles in patients with juvenile myoclonic epilepsy compared with patients with other forms of adolescent-onset generalized epilepsy. *Neurology* **47**: 750–755.

Guerrini, R., Bonanni, P., Nardocci, N., Parmeggiani, L., Piccirilli, M., DeFusco, M., Aridon, P., Ballabio, A., Carrozzo, R. and Casari, G. (1999) Autosomal recessive rolandic epilepsy with paroxysmal exercise-induced dystonia and writer's cramp: delineation of the syndrome and gene mapping to chromosome 16p12–11.2. *Ann. Neurol.* **45**: 344–352.

Guipponi, M., Rivier, F., Vigevano, F., Beck, C., Crespel, A., Echenne, B., Lucchini, P., Sebastianelli, R., Baldy Moulinier, M. and Malafosse, A. (1997) Linkage mapping of benign familial infantile convulsions (BFIC) to chromosome 19q. *Hum. Mol. Genet.* **6**: 473–477.

Han, J., Lu, C.M., Brown, G.B. and Rado, T.A. (1991) Direct amplification of a single dissected chromosomal segment by polymerase chain-reaction – a human brain sodium-channel gene is on chromosome-2q22-q23. *Proc. Natl Acad. Sci. USA* **88**: 335–339.

Hauser, W.A., Annegers, J.F., Anderson, V.E. and Kurland, L.T. (1985) The risk of seizure disorders among relatives of children with febrile convulsions. *Neurology* **35**: 1268–1273.

Hosford, D.A., Clark, S., Cao, Z., Wilson, W.A., Lin, F.H., Morrisett, R.A. and Huin, A. (1992) The role of GABA(B) receptor activation in absence seizures of lethargic (lh/lh) mice. *Science* **257**: 398–401.

Jain, S., Dixit, S.N., Andrews, P.I., Radtke, R., Maheshwari, M.C. and McNamara, J.O. (1996) Disease expression among probands with juvenile myoclonic epilepsy and their family members in two population groups. *J. Epilepsy* **9**: 259–267.

Jan, L.Y. and Jan, Y.N. (1997) Cloned potassium channels from eukaryotes and prokaryotes. *Ann. Rev. Neurosci.* **20**: 91–123.

Janz, D. (1997) The idiopathic generalized epilepsies of adolescence with childhood and juvenile age of onset. *Epilepsia* **38**: 4–11.

Janz, D., Beck Mannagetta, G. and Sander, T. (1992) Do idiopathic generalized epilepsies share a common susceptibility gene? *Neurology* **42**: 48–55.

Jarolimek, W., Baurle, J. and Misgeld, U. (1998) Pore mutation in a G-protein-gated inwardly rectifying K+ channel subunit causes loss of K+-dependent inhibition in weaver hippocampus. *J. Neurosci.* **18**: 4001–4007.

Jelacic, T.M., Sims, S.M. and Clapham, D.E. (1999) Functional expression and characterization of G-protein-gated inwardly rectifying K+ channels containing GIRK3. *J. Membr. Biol.* **169**: 123–129.

Jodice, C., Mantuano, E., Veneziano, L., *et al.* (1997) Episodic ataxia type 2 (EA2) and spinocerebellar ataxia type 6 (SCA6) due to cag repeat expansion in the CACNA1A gene on chromosome 19p. *Hum. Mol. Genet.* **6**: 1973–1978.

Johnson, E.W., Dubovsky, J., Rich, S.S., *et al.* (1998) Evidence for a novel gene for familial febrile convulsions, FEB2, linked to chromosome 19p in an extended family from the midwest. *Hum. Mol. Genet.* **7**: 63–67.

Kapfhamer, D., Sweet, H.O., Sufalko, D., Warren, S., Johnson, K.R. and Burmeister, M. (1996) The neurological mouse mutations jittery and hesitant are allelic and map to the region of mouse chromosome 10 homologous to 19p13.3. *Genomics* **35**: 533–538.

Kinton, L., Smith, S.J., Farrell, F., Stevens, J.M., Davis, M.B., Wood, N.W. and Sander, J.W.A.S. (1998) A new autosomal dominant partial epilepsy syndrome: multifocal epilepsy with pericentral spikes. *Epilepsia* **39(suppl. 6)**: 146.

Kofuji, P., Hofer, M., Millen, K.J., Millonig, J.H., Davidson, N., Lester, H.A. and Hatten, M.E. (1996) Functional-analysis of the weaver mutant GIRK2 K+ channel and rescue of weaver granule cells. *Neuron* **16**: 941–952.

Koshy, B.T. and Zoghbi, H.Y. (1997) The CAG/polyglutamine tract diseases: gene products and molecular pathogenesis. *Brain Pathol.* **7**: 927–942.

Kuryatov, A., Gerzanich, V., Nelson, M., Olale, F. and Lindstrom, J. (1997) Mutation causing autosomal dominant nocturnal frontal lobe epilepsy alters Ca2+ permeability, conductance, and gating of human alpha 4 beta 2 nicotinic acetylcholine receptors. *J. Neurosci.* **17**: 9035–9047.

Lee, W.L., Low, P.S. and Rajan, U. (1993) Benign familial infantile epilepsy. *J. Pediatr.* **123**: 588–590.

Lee, W.L., Tay, A., Ong, H.T., Goh, L.M., Monaco, A.P. and Szepetowski, P. (1998) Association of infantile convulsions with paroxysmal dyskinesias (ICCA syndrome): confirmation of linkage to human chromosome 16p12-q12 in a Chinese family. *Hum. Genet.* **103**: 608–612.

Leonard, S., Adams, C., Breese, C.R., *et al.* (1996) Nicotinic receptor function in schizophrenia. *Schiz. Bull.* **22**: 431–445.

Leppert, M., Anderson, V.E., Quattlebaum, T., Stauffer, D., O'Connell, P., Nakamura, Y., Lalouel, J.M. and White, R. (1989) Benign familial neonatal convulsions linked to genetic-markers on chromosome-20. *Nature* **337**: 647–648.

Lerman, P. (1992) Benign partial epilepsy with centro-temporal spikes. In: *Epileptic Syndromes in Infancy, Childhood and Adolescence* (eds J. Roger, M. Bureau, C. Dravet, F.E. Dreifus, A. Perret and P. Wolf). John Libbey, London, Paris, Rome, pp. 201–217.

Letts, V.A., Felix, R., Biddlecome, G.H., Arikkath, J., Mahaffey, C.L., Valenzuela, A., Bartlett, F.S., Mori, Y., Campbell, K.P. and Frankel, W.N. (1998) The mouse stargazer gene encodes a neuronal Ca2+-channel gamma subunit. *Nature Genet.* **19**: 340–347.

Lewis, T.B., Leach, R.J., Ward, K., O'Connell, P. and Ryan, S.G. (1993) Genetic-heterogeneity in benign familial neonatal convulsions – identification of a new locus on chromosome-8q. *Am. J. Hum. Genet.* **53**: 670–675.

Lewis, T.B., Wood, S., Michaelis, E.K., DuPont, B.R. and Leach, R.J. (1996) Localization of a gene for a glutamate binding subunit of a nmda receptor (GRINA) to 8q24. *Genomics* **32**: 131–133.

Liao, Y.J., Jan, Y.N. and Jan, L.Y. (1996) Heteromultimerization of G-protein-gated inwardly rectifying K+ channel proteins GIRK1 and GIRK2 and their altered expression in weaver brain. *J. Neurosci.* **16**: 7137–7150.

Litt, M., Luty, J., Kwak, M., Allen, L., Magenis, R.E. and Mandel, G. (1989) Localization of a human-brain sodium-channel gene (SCN2A) to chromosome-2. *Genomics* 5: 204–208.

Liu, A.W., Delgado Escueta, A.V., Serratosa, J.M., *et al.* (1995) Juvenile myoclonic epilepsy locus in chromosome 6p21.2-p11 – linkage to convulsions and electroencephalography trait. *Am. J. Hum. Genet.* **57**: 368–381.

Liu, A.W., Delgado Escueta, A.V., Gee, M.N., *et al.* (1996) Juvenile myoclonic epilepsy in chromosome 6p12-p11 – locus heterogeneity and recombinations. *Am. J. Med. Genet.* **63**: 438–446.

Loiseau, P. (1992) Childhood absence epilepsy. In: *Epileptic Syndromes in Infancy, Childhood and Adolescence* (eds J. Roger, M. Bureau, C. Dravet, F.E. Dreifus, A. Perret and P. Wolf). John Libbey, London, Paris, Rome, pp. 135–150.

Lorenzon, N.M., Lutz, C.M., Frankel, W.N. and Beam, K.G. (1998) Altered calcium channel currents in Purkinje cells of the neurological mutant mouse leaner. *J. Neurosci.* **18**: 4482–4489.

Lu, C.M., Han, J., Rado, T.A. and Brown, G.B. (1992) Differential expression of 2 sodium-channel subtypes in human brain. *FEBS Lett.* **303**: 53–58.

MacDermott, A.B., Role, L.W. and Siegelbaum, S.A. (1999) Presynaptic ionotropic receptors and the control of transmitter release. *Ann. Rev. Neurosci.* **22**: 443–485.

Mackinnon, R. (1991) Determination of the subunit stoichiometry of a voltage-activated potassium channel. *Nature* **350**: 232–235.

Maihara, T., Tsuji, M., Higuchi, Y. and Hattori, H. (1999) Benign familial neonatal convulsions followed by benign epilepsy with centrotemporal spikes in two siblings. *Epilepsia* **40**: 110–113.

Malo, M.S., Blanchard, B.J., Andresen, J.M., Srivastava, K., Chen, X.N., Li, X., Jabs, E.W., Korenberg, J.R. and Ingram, V.M. (1994a) Localization of a putative human brain sodium-channel gene (SCN1A) to chromosome band 2q24. *Cytogenet. Cell Genet.* **67**: 178–186.

Malo, M.S., Srivastava, K., Andresen, J.M., Chen, X.N., Korenberg, J.R. and Ingram, V.M. (1994b) Targeted gene walking by low stringency polymerase chain-reaction – assignment of a putative human brain sodium-channel gene (SCN3A) to chromosome-2q24–31 *Proc. Natl Acad. Sci. USA* **91**: 2975–2979.

Matsuura, T., Achari, M., Khajavi, M., Bachinski, L.L., Zoghbi, H.Y. and Ashizawa, T. (1999) Mapping of the gene for a novel spinocerebellar ataxia with pure cerebellar signs and epilepsy. *Ann. Neurol.* **45**: 407–411.

McEnery, M.W., Copeland, T.D. and Vance, C.L. (1998) Altered expression and assembly of N-type calcium channel alpha(1b) and beta subunits in epileptic lethargic (lh/lh) mouse. *J. Biol. Chem.* **273**: 21435–21438.

McGehee, D.S. and Role, L.W. (1995) Physiological diversity of nicotinic acetylcholine-receptors expressed by vertebrate neurons. *Ann. Rev. Physiol.* **57**: 521–546.

McGehee, D.S. and Role, L.W. (1996) Presynaptic ionotropic receptors. *Curr. Opin. Neurobiol.* **6**: 342–349.

Miller, L.L., Pellock, J.M., DeLorenzo, R.J., Meyer, J.M. and Corey, L.A. (1998) Univariate genetic analyses of epilepsy and seizures in a population-based twin study: the Virginia twin registry. *Genet. Epidemiol.* **15**: 33–49.

Morita, R., Miyazaki, E., Fong, C.Y.G., Chen, X.N., Korenberg, J.R., Delgado Escueta, A.V. and Yamakawa, K. (1998) JH8, a gene highly homologous to the mouse jerky gene, maps to the region for childhood absence epilepsy on 8q24. *Biochem. Biophys. Res. Commun.* **248**: 307–314.

Nakken, K.O., Magnusson, A. and Steinlein, O.K. (1999) Autosomal dominant nocturnal frontal lobe epilepsy: an electroclinical study of a Norwegian family with ten affected members. *Epilepsia* **40**: 88–92.

Neubauer, B.A., Fiedler, B., Himmelein, B., *et al.* (1998) Centrotemporal spikes in families with rolandic epilepsy-linkage to chromosome 15q14. *Neurology* **51**: 1608–1612.

Nystuen, A., Benke, P.J., Merren, J., Stone, E.M. and Sheffield, V.C. (1996) A cerebellar ataxia locus identified by dna pooling to search for linkage disequilibrium in an isolated population from the Cayman Islands. *Hum. Mol. Genet.* 5: 525–531.

O'Donohoe, N.V. (1992) Febrile convulsions. In: *Epileptic Syndromes in Infancy, Childhood and*

Adolescence (eds J. Roger, M. Bureau, C. Dravet, F.E. Dreifus, A. Perret and P. Wolf). John Libbey, London, Paris, Rome, pp. 45–52.

Odermatt, A., Taschner, P.E.M., Khanna, V.K., Busch, H.F.M., Karpati, G., Jablecki, C.K., Breuning, M.H. and Maclennan, D.H. (1996) Mutations in the gene-encoding serca1, the fast-twitch skeletal muscle sarcoplasmic reticulum Ca2+ ATPase, are associated with Brody disease. *Nature Genet.* **14**: 191–194.

Ohira, M., Seki, N., Nagase, T., *et al.* (1997) Gene identification in 1.6-Mb region of the Down syndrome region on chromosome 21. *Genome Res.* **7**: 47–58.

Ophoff, R.A., Terwindt, G.M., Vergouwe, M.N., *et al.* (1996) Familial hemiplegic migraine and episodic ataxia type-2 are caused by mutations in the Ca2+ channel gene CACNL1A4. *Cell* **87**: 543–552.

Ophoff, R.A., Terwindt, G.M., Frants, R.R. and Ferrari, M.D. (1998) P/Q-type Ca2+ channel defects in migraine, ataxia and epilepsy. *Trends Pharmacol. Sci.* **19**: 121–127.

Orr Urtreger, A., Xu, W., Saeki, M., *et al.* (1997) Animal models with mutations in the alpha 7 and beta 2 subunits and in the beta 4, alpha 3, alpha 5 cluster of neuronal nicotinic acetylcholine receptors. *Am. J. Hum. Genet.* **61**: 39.

Ottman, R., Risch, N., Hauser, W.A., *et al.* (1995) Localization of a gene for partial epilepsy to chromosome 10q. *Nature Genet.* **10**: 56–60.

Panayiotopoulos, C.P., Obeid, T. and Waheed, G. (1989) Absences in juvenile myoclonic epilepsy – a clinical and video-electroencephalographic study. *Ann. Neurol.* **25**: 391–397.

Patil, N., Cox, D.R., Bhat, D., Faham, M., Myers, R.M. and Peterson, A.S. (1995) A potassium channel mutation in weaver mice implicates membrane excitability in granule cell-differentiation. *Nature Genet.* **11**: 126–129.

Peters, H.C., Kammer, G., Volz, A., Kaupmann, K., Ziegler, A., Bettler, B., Epplen, J.T., Sander, T. and Riess, O. (1998) Mapping, genomic structure, and polymorphisms of the human gaba(b)r1 receptor gene: evaluation of its involvement in idiopathic generalized epilepsy. *Neurogenetics* **2**: 47–54.

Phillips, H.A., Scheffer, I.E., Berkovic, S.F., Hollway, G.E., Sutherland, G.R. and Mulley, J.C. (1995) Localization of a gene for autosomal dominant nocturnal frontal lobe epilepsy to chromosome 20q 13.2 *Nature Genet.* **10**: 117–118.

Phillips, H.A., Scheffer, I.E., Crossland, K.M., *et al.* (1998) Autosomal dominant nocturnal frontal-lobe epilepsy: genetic heterogeneity and evidence for a second locus at 15q24. *Am. J. Hum. Genet.* **63**: 1108–1116.

Pieles, U., Zurcher, W., Schar, M. and Moser, H.E. (1993) Matrix-assisted laser-desorption ionization time-of-flight mass-spectrometry – a powerful tool for the mass and sequence-analysis of natural and modified oligonucleotides. *Nucl. Acids Res.* **21**: 3191–3196.

Risch, N. and Merikangas, K. (1996) The future of genetic-studies of complex human-diseases. *Science* **273**: 1516–1517.

Sander, T., Hildmann, T., Janz, D., *et al.* (1995) The phenotypic spectrum related to the human epilepsy susceptibility gene "EJM1". *Ann. Neurol.* **38**: 210–217.

Sander, T., Bockenkamp, B., Hildmann, T., *et al.* (1997a) Refined mapping of the epilepsy susceptibility locus EJM1 on chromosome 6. *Neurology* **49**: 842–847.

Sander, T., Hildmann, T., Kretz, R., Furst, R., *et al.* (1997b) Allelic association of juvenile absence epilepsy with a Glur5 kainate receptor gene (grik1) polymorphism. *Am. J. Med. Genet.* **74**: 416–421.

Sander, T., Kretz, R., Schulz, H., Sailer, U., Bauer, G., Scaramelli, A., Epplen, J.T., Riess, O. and Janz, D. (1998a) Replication analysis of a putative susceptibility locus (EGI) for idiopathic generalized epilepsy on chromosome 8q24. *Epilepsia* **39**: 715–720.

Sander, T., Peters, C., Janz, D., Bianchi, A., Bauer, G., Wienker, T.F., Hildmann, T., Epplen, J.T. and Riess, O. (1998b) The gene encoding the alpha(1a)-voltage-dependent calcium channel (CACN1A4) is not a candidate for causing common subtypes of idiopathic generalized epilepsy. *Epilepsy Res.* **29**: 115–122.

Sander, T., Peters, C., Kammer, G., *et al.* (1999a) Association analysis of exonic variants of the gene encoding the GABA(B) receptor and idiopathic generalized epilepsy. *Am. J. Med. Genet.* **88**: 305–310.

Sander, T., Schulz, H., VieiraSaeker, A.M.M., *et al.* (1999b) Evaluation of a putative major susceptibility locus for juvenile myoclonic epilepsy on chromosome 15q14. *Am. J. Med. Genet.* **88**: 182–187.

Scheffer, I.E. and Berkovic, S.F. (1997) Generalized epilepsy with febrile seizures plus – a genetic disorder with heterogeneous clinical phenotypes. *Brain* **120**: 479–490.

Scheffer, I.E., Bhatia, K.P., LopesCendes, I., *et al.* (1995) Autosomal-dominant nocturnal frontal-lobe epilepsy – a distinctive clinical disorder. *Brain* **118**: 61–73.

Scheffer, I.E., Phillips, H.A., O'Brien, C.E., Saling, M.M., Wrennall, J.A., Wallace, R.H. and Mulley, J.C. (1998) Familial partial epilepsy with variable foci: a new partial epilepsy syndrome with suggestion of linkage to chromosome 2. *Ann. Neurol.* **44**: 890–899.

Schroeder, B.C., Kubisch, C., Stein, V. and Jentsch, T.J. (1998) Moderate loss of function of cyclic-amp-modulated KCNQ2/KCNQ3 K+ channels causes epilepsy. *Nature* **396**: 687–690.

Shinnar, S. and Moshe, S.L. (1991) Age specificity of seizure expression in genetic epilepsies. *Epilepsy Res.* **4**: 69–85.

Shorvon, S. (1991) Epidemiologic studies of epilepsy and their contribution to genetics. *Epilepsy Res.* **4**(Suppl.): 53–68.

Signorini, S., Liao, Y.J., Duncan, S.A., Jan, L.Y. and Stoffel, M. (1997) Normal cerebellar development but susceptibility to seizures in mice lacking G protein-coupled, inwardly rectifying K+ channel GIRK2. *Proc. Natl Acad Sci. USA* **94**: 923–927.

Silverman, S.K., Kofuji, P., Dougherty, D.A., Davidson, N. and Lester, H.A. (1996) A regenerative link in the ionic fluxes through the weaver potassium channel underlies the pathophysiology of the mutation. *Proc. Natl Acad. Sci. USA* **93**: 15429–15434.

Singh, N.A., Charlier, C., Stauffer, D., *et al.* (1998) A novel potassium channel gene, kcnq2, is mutated in an inherited epilepsy of newborns. *Nature Genet.* **18**: 25–29.

Singh, R., Scheffer, I.E., Crossland, K. and Berkovic, S.F. (1999) Generalized epilepsy with febrile seizures plus: a common childhood-onset genetic epilepsy syndrome. *Ann. Neurol.* **45**: 75–81.

Slesinger, P.A., Patil, N., Liao, Y.J., Jan, Y.N., Jan, L.Y. and Cox, D.R. (1996) Functional-effects of the mouse weaver mutation on G-protein-gated inwardly rectifying K+ channels. *Neuron* **16**: 321–331.

Steinlein, O.K., Mulley, J.C., Propping, P., Wallace, R.H., Phillips, H.A., Sutherland, G.R., Scheffer, I.E. and Berkovic, S.F. (1995) A missense mutation in the neuronal nicotinic acetylcholine receptor alpha 4 subunit is associated with autosomal dominant nocturnal frontal lobe epilepsy. *Nature Genet.* **11**: 201–203.

Steinlein, O.K., Magnusson, A., Stoodt, J., Bertrand, S., Weiland, S., Berkovic, SF., Nakken, K.O., Propping, P. and Bertrand, D. (1997) An insertion mutation of the CHRNA4 gene in a family with autosomal dominant nocturnal frontal lobe epilepsy. *Hum. Mol. Genet.* **6**: 943–947.

Surmeier, D.J., Mermelstein, P.G. and Goldowitz, D. (1996) The weaver mutation of girk2 results in a loss of inwardly rectifying K+ current in cerebellar granule cells. *Proc. Natl Acad. Sci. USA* **93**: 11191–11195.

Szepetowski, P., Rochette, J., Berquin, P., Piussan, C., Lathrop, G.M. and Monaco, A.P. (1997) Familial infantile convulsions and paroxysmal choreoathetosis: a new neurological syndrome linked to the pericentromeric region of human chromosome 16. *Am. J. Hum. Genet.* **61**: 889–898.

Terwindt, G.M., Ophoff, R.A., Lindhout, D., Haan, J., Halley, D.J.J., Sandkuijl, L.A., Brouwer, O.F., Frants, R.R. and Ferrari, M.D. (1997) Partial cosegregation of familial hemiplegic migraine and a benign familial infantile epileptic syndrome. *Epilepsia* **38**: 915–921.

Vigevano, F., Fusco, L., Dicapua, M., Ricci, S., Sebastianelli, R. and Lucchini, P. (1992) Benign infantile familial convulsions. *Eur. J. Pediatr.* **151**: 608–612.

Voilley, N., Bassilana, F., Mignon, C., Merscher, S., Mattei, M.G., Carle, G.F., Lazdunski, M. and Barbry, P. (1995) Cloning, chromosomal localization, and physical linkage of the beta-subunit and gamma-subunit (SCNN1B and SCNN1G) of the human epithelial amiloride-sensitive sodium-channel. *Genomics* **28**: 560–565.

Wakamori, M., Yamazaki, K., Matsunodaira, H., *et al.* (1998) Single tottering mutations responsible for the neuropathic phenotype of the P-type calcium channel. *J. Biol. Chem.* **273**: 34857–34867.

Wallace, R.H., Berkovic, S.F., Howell, R.A., Sutherland, G.R. and Mulley, J.C. (1996) Suggestion of a major gene for familial febrile convulsions mapping to 8q13–21. *J. Med. Genet.* **33**: 308–312.

Wallace, R.H., Wang, D.W., Singh, R., *et al.* (1998) Febrile seizures and generalized epilepsy associated with a mutation in the Na+-channel beta 1 subunit gene SCN1B. *Nature Genet.* **19**: 366–370.

Waltz, S., Christen, H.J. and Doose, H. (1992) The different patterns of the photoparoxysmal response – a genetic- study. *Electroencephalog. Clin. Neurophysiol.* **83**: 138–145.

Wang, D.G., Fan, J.B., Siao, C.J., *et al.* (1998) Large-scale identification, mapping, and genotyping of single-nucleotide polymorphisms in the human genome. *Science* **280**: 1077–1082.

Wang, H.S., Pan, Z.M., Shi, W.M., Brown, B.S., Wymore, R.S., Cohen, I.S., Dixon, J.E. and McKinnon, D. (1998) KCNQ2 and KCNQ3 potassium channel subunits: molecular correlates of the M-channel. *Science* **282**: 1890–1893.

Warren, S.T. (1996) The expanding world of trinucleotide repeats. *Science* **271**: 1374–1375.

Weiland, S., Witzemann, V., Villarroel, A., Propping, P. and Steinlein, O. (1996) An amino acid exchange in the second transmembrane segment of a neuronal nicotinic receptor causes partial epilepsy by altering its desensitization kinetics. *FEBS Lett.* **398**: 91–96.

Weissbecker, K.A., Durner, M., Janz, D., Scaramelli, A., Sparkes, R.S. and Spence, M.A. (1991) Confirmation of linkage between juvenile myoclonic epilepsy locus and the HLA region of chromosome 6. *Am. J. Med. Genet.* **38**: 32–36.

Wells, R.G., Mohandas, T.K. and Hediger, M.A. (1993) Localization of the Na+/glucose cotransporter gene-SGLT2 to human chromosome-16 close to the centromere. *Genomics* **17**: 787–789.

Whitehouse, W.P., Rees, M., Curtis, D., Sundquist, A., Parker, K., Chung, E., Baralle, D. and Gardiner, R.M. (1993) Linkage analysis of idiopathic generalized epilepsy (IGE) and marker loci or chromosome 6p in families of patients with juvenile myoclonic epilepsyy: no evidence for an epilepsy locus in the HLA region. *Am. J. Hum. Genet.* **53**: 652–662.

Wischmeyer, E., Doring, F., Spauschus, A., Thomzig, A., Veh, R. and Karschin, A. (1997) Subunit interactions in the assembly of neuronal Kir3.0 inwardly rectifying K+ channels. *Molec. Cell. Neurosci.* **9**: 194–206.

Wolf, P. (1992) Juvenile myoclonic epilepsy. In: *Epileptic Syndromes in Infancy, Childhood and Adolescence* (eds J. Roger, M. Bureau, C. Dravet, F.E. Dreifus, A. Perret and P. Wolf). John Libbey, London, Paris, Rome, pp. 313–327.

Womack, M., Thompson, K., Fanselow, E., Augustine, G.J. and Peterson, A. (1998) Elevated intracellular calcium levels in cerebellar granule neurons of weaver mice. *Neuroreport* **9**: 3391–3395.

Wu, K.D., Lee, W.S., Wey, J., Bungard, D. and Lytton, J. (1995) Localization and quantification of endoplasmic-reticulum Ca^{2+}-ATPase isoform transcripts. *Am. J. Physiol. Cell Physiol.* **38**: C775-C784.

Yang, W.P., Levesque, P.C., Little, W.A., Conder, M.L., Ramakrishnan, P., Neubauer, M.G. and Blanar, M.A. (1998) Functional expression of two KVLQT1-related potassium channels responsible for an inherited idiopathic epilepsy. *J. Biol. Chem.* **273**: 19419–19423.

Yip, T.T. and Hutchens, T.W. (1992) Mapping and sequence-specific identification of phosphopeptides in unfractionated protein digest mixtures by matrix-assisted laser desorption ionization time-of-flight mass-spectrometry. *FEBS Lett.* **308**: 149–153.

Yue, Q., Jen, J.C., Nelson, S.F. and Baloh, R.W. (1997) Progressive ataxia due to a missense mutation in a calcium-channel gene. *Am. J. Hum. Genet.* **61**: 1078–1087.

Zara, F., Bianchi, A., Avanzini, G., Didonato, S., Castellotti, B., Patel, P.I. and Pandolfo, M. (1995) Mapping of genes predisposing to idiopathic generalized epilepsy. *Hum. Mol. Genet.* **4**: 1201–1207.

Zara, F., Labuda, M., Garofalo, P.G.A., Durisotti, C., Bianchi, A., Castellotti, B., Patel, P.I., Avanzini, G. and Pandolfo, M. (1998) Unusual EEG pattern linked to chromosome 3p in a family with idiopathic generalized epilepsy. *Neurology* **51**: 493–498.

Zhuchenko, O., Bailey, J., Bonnen, P., Ashizawa, T., Stockton, D.W., Amos, C., Dobyns, W.B., Subramony, S.H., Zoghbi, H.Y. and Lee, C.C. (1997) Autosomal dominant cerebellar ataxia (SCA6) associated with small polyglutamine expansions in the alpha(1a)-voltage-dependent calcium channel. *Nature Genet.* **15**: 62–69.

Zuberi, S.M., Eunson, L.H., Spauschus, A., DeSilva, R., Tolmie, J., Wood, N.W., McWilliam, R.C., Stephenson, J.P.B., Kullmann, D.M. and Hanna, M.G. (1999) A novel mutation in the human voltage-gated potassium channel gene (Kv1.1) associates with episodic ataxia type 1 and sometimes with partial epilepsy. *Brain* **122**: 817–825.

7

Type 2 diabetes mellitus and maturity-onset diabetes of the young

Robert L. Hanson and William C. Knowler

1. Introduction

Diabetes mellitus is a disorder of glucose homeostasis in which levels of blood glucose are characteristically high. Severe hyperglycemia, if not controlled by therapeutic intervention, can lead to the classic symptoms of the disease, such as increased thirst and increased frequency of urination. Chronic hyperglycemia causes the characteristic complications of diabetes, such as retinopathy, nephropathy and neuropathy, and probably leads to an increased risk of cardiovascular disease (Barzilay *et al.*, 1999; DCCT, 1993; Kannel and McGee, 1979; UKPDS, 1998). Diabetes mellitus is classified into several types on the basis of its physiologic features (ADA, 1997); the most common manifestations of the disease are classified as type 1 and type 2 diabetes. Type 1 diabetes, the subject of another chapter, is characterized by autoimmune destruction of the pancreatic β-cell, and the affected individual, consequently, typically has a severe insulin secretory defect. Type 2 diabetes, in contrast, is characterized by insulin resistance and relative insulin secretory dysfunction and the absence of evidence for autoimmunity directed toward the pancreatic islets. The incidence of type 2 diabetes generally increases with age and is strongly associated with obesity (Chan *et al.*, 1995; Colditz *et al.*, 1990; Knowler *et al.*, 1981; Lipton *et al.*, 1993). Under the current classification system, type 2 diabetes is the most common form of diabetes mellitus. Maturity-onset diabetes of the young (MODY), which is characterized by young age at onset of diabetes and by autosomal dominant inheritance (Tattersall and Fajans, 1975), is generally considered to be distinct from type 2 diabetes. This chapter reviews the genetics of type 2 diabetes; as the genetic studies of MODY are instructive for elucidating the pathophysiology of diabetes mellitus, they will also be briefly reviewed.

2. MODY – a monogenic form of diabetes

In the 1960s an unusual familial form of diabetes with young age of onset, usually before 25 years of age, was recognized (Fajans and Conn, 1965) and, in the 1970s it was recognized that the familial transmission of this form of diabetes followed an autosomal dominant pattern of inheritance (Tattersall, 1974; Tattersall and Fajans, 1975). This form of diabetes was called maturity-onset diabetes of the young (MODY) and it appeared to be clinically distinct from both type 1 diabetes and type 2 diabetes in that there was no evidence of autoimmune destruction of the β-cell and little association with obesity (Fajans *et al.*, 1978; Tattersall, 1982; Tattersall and Fajans, 1975; Winter *et al.*, 1987). Genetic linkage studies identified loci on chromosome 20q in a portion of a single large American pedigree (Bell *et al.*, 1991), and, in French families, on chromosome 7p near the glucokinase locus (Froguel *et al.*, 1992) and on chromosome 12q (Vaxillaire *et al.*, 1995). Glucokinase acts as a glucose sensor for the pancreatic β-cell and mutations in this candidate gene were identified as being responsible for the linkage on chromosome 7p (Vionnet *et al.*, 1992). Yamagata *et al.* (1996a) conducted fine-mapping studies on chromosome 12 by physically mapping the linked region to determine the position of genes in the region, by identifying recombination events in key individuals (which was possible because of the autosomal dominant inheritance of MODY) and by analyzing polymorphisms within genes for association with the disease. With this approach the genetic lesions responsible for chromosome 12-linked MODY were identified as nonsense and missense mutations in the gene coding for hepatocyte nuclear factor-1α (HNF-1α), although different mutations co-segregated with MODY in different families. HNF-1α is a transcription factor that regulates tissue-specific expression of several genes, and it was not considered a candidate gene for MODY prior to its identification by linkage analysis. The knowledge that HNF-1α mutations cause some cases of MODY prompted Yamagata *et al.* (1996b) to examine mutations in another transcription factor located in the region linked to chromosome 20, hepatocyte nuclear factor-4α, and a nonsense mutation which cosegregated with MODY in the linked pedigree was identified; subsequently additional MODY families with mutations in HNF-4α have been described (Bulman *et al.*, 1997). Nonsense mutations in genes encoding for additional transcription factors, namely insulin-promoter factor-1 (IPF1) and hepatocyte nuclear factor-1β (HNF-1β), have also been shown to cause MODY (Horikawa *et al.*, 1997; Stoffers *et al.*, 1997). *Table 1* summarizes the MODY genes identified to date.

Table 1. Genes identified as causes of maturity-onset diabetes of the young (MODY).

Locus	Chromosome	Gene	Reference
MODY1	20q	Hepatocyte nuclear factor-4α	Yamagata *et al.*, 1996b
MODY2	7p	Glucokinase	Vionnet *et al.*, 1992
MODY3	12q	Hepatocyte nuclear factor-1α	Yamagata *et al.*, 1996a
MODY4	13q	Insulin promoter factor-1	Stoffers *et al.*, 1997
MODY5	17q	Hepatocyte nuclear factor-1β	Horikawa *et al.*, 1997

The mechanisms by which mutations in HNF-1α and related transcription factors cause MODY are not well delineated, but they appear to cause diminished insulin secretion (Byrne *et al.*, 1995, 1996). In general, it appears that mutations in the known MODY associated genes are rare causes of type 2 diabetes. Although it is estimated that mutations in HNF-1α are responsible for approximately 70%, 25% and 20% of cases of MODY in the UK, France and Japan, respectively (Frayling *et al.*, 1997; Iwasaki *et al.*, 1997; Vaxillaire *et al.*, 1995), they have been found in less than 1% of individuals with typical type 2 diabetes in European and Japanese populations (Iwasaki *et al.*, 1997; Urhammer *et al.*, 1997). HNF-4α and glucokinase mutations have similarly been found to be rare in individuals with type 2 diabetes (Froguel *et al.*, 1993; Laakso *et al.*, 1995; Moller *et al.*, 1997; Nakajima *et al.*, 1996; Nishi *et al.*, 1994). In an English study missense mutations in IPF-1 were identified in 6% of individuals with familial type 2 diabetes and in 1% of the general population (Macfarlane *et al.*, 1999). In the Oji-Cree, an American Indian population, a missense mutation in codon 319 of HNF-1α which results in a serine for glycine substitution in the protein is common in persons with type 2 diabetes. This mutation is present in 37% of diabetic Oji-Cree and in only 17% of nondiabetic persons, a difference which is highly statistically significant ($P=10^{-6}$) (Hegele *et al.*, 1999a). This mutation appears to be rare in other populations. Thus, some individuals with clinical characteristics of type 2 diabetes may have diabetes due to mutations in genes responsible for MODY. However, except in the Oji-Cree, MODY-associated mutations have not been found to account for a substantial proportion of cases of type 2 diabetes.

3. Familial aggregation of type 2 diabetes

The prevalence of type 2 diabetes is higher in individuals who have a first-degree relative with the disease than in individuals who do not have an affected first-degree relative. This familial aggregation has been demonstrated in a variety of populations, including African-, European- and Mexican-Americans (Harris, 1991; Mitchell *et al.*, 1995), American Indians (Knowler *et al.*, 1981; Lee *et al.*, 1985), and populations from Europe (Charles *et al.*, 1991; Mykkänen *et al.*, 1990), Asia (Pan *et al.*, 1997; Viswanathan *et al.*, 1985), Africa (Mengesha and Abdulkadir, 1997), South America (Oliveria *et al.*, 1996) and Oceania (Zimmet *et al.*, 1977). Diabetes with a young age of onset appears to be more familial than diabetes that occurs at older ages (Knowler *et al.*, 1988a; Köbberling, 1971; Simpson, 1964; Thomas *et al*, 1994). This phenomenon is clearly demonstrated among the Pima Indians, among whom the prevalence of type 2 diabetes is extraordinarily high, even at relatively young ages (Knowler *et al.*, 1988a). As shown in *Figure 1*, the prevalence of diabetes in Pimas is highest among those for whom both parents developed diabetes before the age of 45 years, while it is lowest for individuals with nondiabetic parents.

There is a greater prevalence of diabetes in the offspring of diabetic mothers than in the offspring of diabetic fathers, although the excess maternal transmission is modest in most populations (Alcolado and Alcolado, 1991; Karter *et al.*,

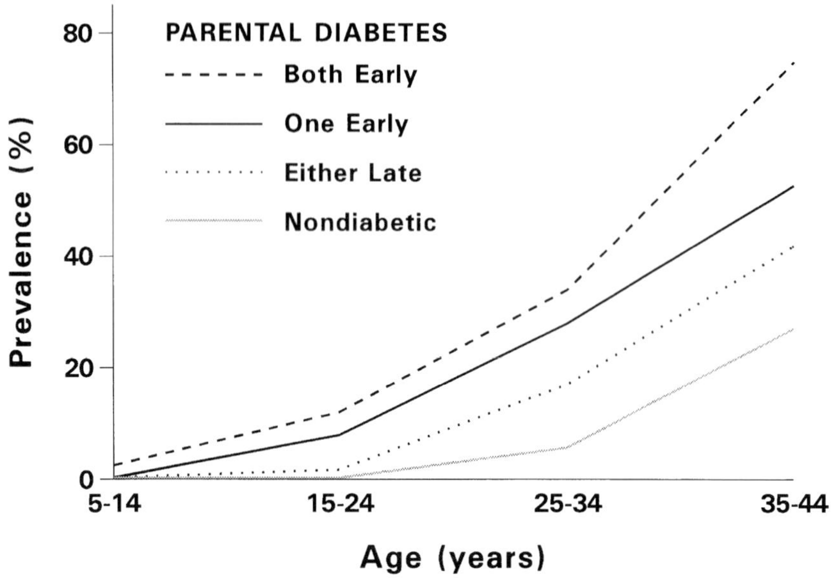

Figure 1. Prevalence of type 2 diabetes by age and by parental diabetes in Pima Indians. If both parents developed diabetes before the age of 45 years individuals are classified as 'both early'; if one parent developed diabetes before the age of 45 years they are classified as 'one early'; if either parent developed diabetes after the age of 45 years (but neither parent developed 'early' diabetes) they are classified as 'either late'; and if both parents were nondiabetic when last examined they are classified as 'nondiabetic'. Adapted from Knowler *et al.*, 1988a.

1999; Mitchell *et al.*, 1995; Thomas *et al.*, 1994). The extent to which this excess maternal transmission is due to genetic factors, such as genomic imprinting and mitochondrial inheritance, or to a greater similarity of diabetogenic environmental factors between mother and child than between father and child is unknown. Mitochondrial mutations are associated with rare syndromic cases of diabetes, often in conjunction with hearing loss (Kadowaki *et al.*, 1994; van den Ouweland *et al*, 1992). On the other hand, exposure to a diabetic environment *in utero* increases an individual's risk for developing diabetes. Among Pima Indians, for example, the prevalence of diabetes is higher in individuals whose mothers had diabetes during the pregnancy than in individuals whose mothers developed diabetes after the pregnancy (Pettitt *et al.*, 1988). Within Pima sibships, individuals born after the onset of maternal diabetes have a significantly higher risk of diabetes than their siblings who were born before their mother developed diabetes; since a similar phenomenon does not occur among the offspring of diabetic fathers, this suggests that the effect of exposure to diabetes *in utero* is at least partially independent of the effects of genes which predispose to young-onset diabetes (Dabelea *et al.*, 1999). Exposure to a diabetic intrauterine environment, therefore, represents a nongenetic mechanism for maternal transmission of type 2 diabetes.

Since obesity, which is also strongly familial, is a strong risk factor for type 2 diabetes, the extent to which the familial, and presumably the genetic, determinants of obesity overlap with those of diabetes has been of considerable interest. As shown in *Figure* 2, the prevalence of diabetes in relatives of diabetic probands is higher for relatives of less obese probands than for relatives of more obese probands (Hanson *et al.*, 1995a; Köbberling, 1971; Kuzuya and Matsuda, 1982). This implies that individuals who develop diabetes despite being relatively lean have a greater 'load' of diabetes-susceptibility genes than more obese persons who develop diabetes, and it suggests that the genetic determinants of diabetes and obesity are at least partially distinct. On the other hand, individuals who have a strong family history of diabetes tend to be more obese than individuals without a family history of diabetes (Mitchell *et al.*, 1999; Stewart *et al.*, 1995), and this suggests that the genetic determinants of diabetes and obesity are partially shared. These findings suggest that, while there is some overlap between the genetic determinants of diabetes and obesity, this overlap is not complete.

4. Twin studies

The extent to which familial aggregation of a disease is due to genetic or environmental factors can be assessed by comparing the extent to which monozygotic twins are concordant for the disease with the concordance in dizygotic twins. Since monozygotic twins are genetically identical while dizygotic twins share on

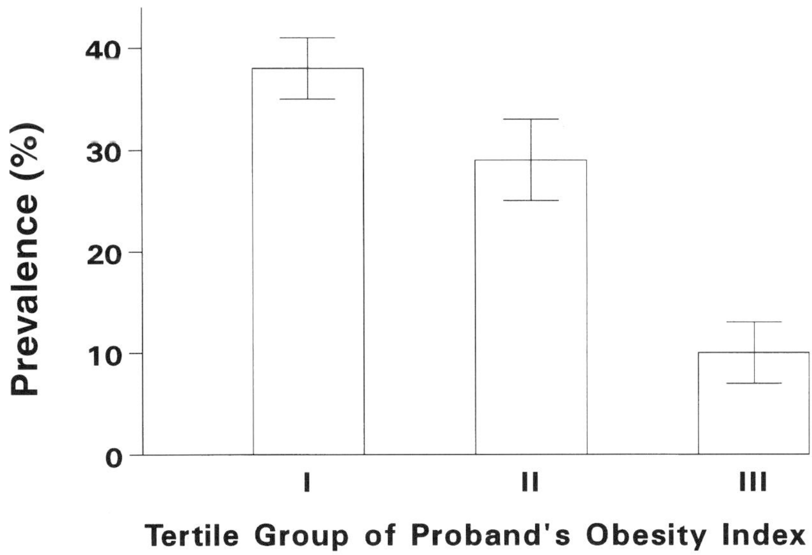

Figure 2. Prevalence of diabetes in siblings of German probands with diabetes, classified by the degree of obesity in the proband. Siblings of the most obese diabetic probands (tertile III) have the lowest prevalence of diabetes, while siblings of the leanest probands (tertile I) have the highest prevalence of diabetes. Bars represent standard errors. Adapted from Köbberling, 1971.

average half of their genes, the difference between the concordance rate in monozygotic and dizygotic twin pairs is related to the contribution of genetic factors to the etiology of the disease. Results of various twin studies of type 2 diabetes are shown in *Table 2*. The concordance rates are higher for monozygotic than for dizygotic twins, and the difference is statistically significant for all studies, except for the Danish study, which was too small to allow for precise estimation of genetic liability. In fact, a quantitative genetic analysis was undertaken to estimate, among these Danish subjects, the proportion of the variance in liability to diabetes which was explained by genetic factors, and this estimate was 0.26 with a 95% confidence interval of 0.0–0.85. The wide confidence interval suggests that this study is too small to be conclusive. (There was, in fact, a significant difference in the concordance rates for monozygotic and dizygotic twins for abnormal glucose tolerance, defined as a 2-hour post-load plasma glucose ≥7.8 mM in these Danish subjects.)

Although the concordance rate for type 2 diabetes is generally higher in monozygotic than in dizygotic twins, the magnitude of the difference is variable among studies. The interpretation of twin studies is limited by the potential for increased sharing of nongenetic factors among monozygotic compared with dizygotic twins, by potentially biased ascertainment of twins and by variable age of onset of diabetes. The concordance rate for monozygotic twins tends to be higher in studies which ascertain twin pairs through a diabetic proband than for those which use population-based sampling to ascertain twin pairs. The high concordance rate in studies which use a proband may be partially due to a tendency for diabetic monozygotic co-twins of probands to participate in the study at a higher rate than nondiabetic co-twins. Most of these studies have not taken into account the variable age of onset of diabetes, which may result in pairs which are originally discordant becoming concordant. In fact, Medici *et al.* (1999) demonstrated a high incidence of diabetes in the initially nondiabetic co-twins of monozygotic twin pairs ascertained as being discordant for diabetes, such that after 15 years of follow-up 76% of pairs became concordant. Despite the difficulties in interpretation, the twin studies generally support a significant genetic component in the pathogenesis of type 2 diabetes, although the magnitude of the genetic effect is difficult to quantify.

Table 2. Pairwise concordance rates for monozygotic (MZ) and dizygotic (DZ) twins in various twin studies.

Reference	Nation	Ascertainment	MZ Concordance	DZ Concordance	*P*-value[a]
Barnett *et al.*, 1981	UK	Proband	0.91	–	–
Kaprio *et al.*, 1992	Finland	Population-based	0.20	0.09	< 0.01
Matsuda and Kuzuya, 1994	Japan	Proband	0.87	0.43	< 0.01
Newman *et al.*, 1987[b]	USA	Population-based	0.41	0.10	< 0.01
Poulsen *et al.*, 1999	Denmark	Population-based	0.33	0.23	0.40

[a] *P*-values are for the null hypothesis that MZ and DZ concordance rates are equal and were calculated using Fisher's exact test.
[b] Concordance rates for the second examination are reported.

5. Segregation analyses

Complex segregation analyses of type 2 diabetes have been conducted for a number of populations, and some of these analyses support a potential role for a single gene with a major influence on the occurrence of type 2 diabetes. A major genetic effect is generally inferred if a model containing a major gene fits better than one with multifactorial inheritance alone, if a model allowing for transmission of this major effect from parent to offspring fits better than one which does not allow transmission, and if a model of Mendelian transmission fits as well as one which allows for non-Mendelian transmission (Lalouel *et al.*, 1983). Among the Pima Indians, for example, segregation analysis supports a major genetic effect on age of onset of diabetes (Hanson *et al.*, 1995b). A similar major genetic effect on age of onset of diabetes was also shown in Mexican-Americans, another population with a relatively high prevalence of type 2 diabetes (Stern *et al.*, 1996). The best-fitting Mendelian models for Pimas and Mexican-Americans are shown in *Figure 3*. The models predict that individuals homozygous for a putative diabetes-susceptibility allele, 'A', have a very high risk of developing diabetes at a young age, whereas at older ages, the incidence of diabetes begins to increase in those who do not carry the hypothetical susceptibility allele.

Segregation analyses in an English population and in a South Asian population do not support major genetic effects (Cook *et al.*, 1994; McCarthy *et al.*, 1994). In both of these studies, the most parsimonious model was that of multifactorial inheritance alone. Given the distributional assumptions which segregation analysis requires, these results do not necessarily imply that genetic factors are of no importance in type 2 diabetes in these populations, they may simply represent a

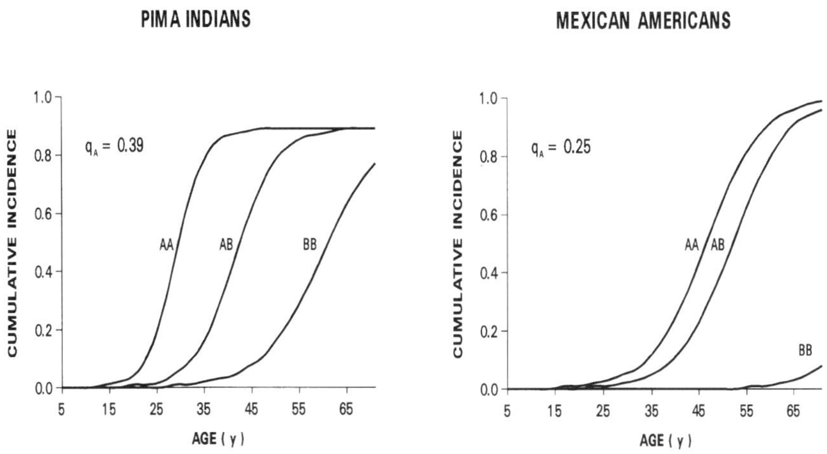

Figure 3. Cumulative incidence of diabetes by age, as predicted by segregation models of a major gene influencing age of onset of diabetes in Pima Indians and Mexican-Americans. The models show the three genotypes at a putative diabetes-susceptibility locus: those homozygous for the 'high risk' allele (AA), heterozygotes (AB) and those homozygous for the 'low risk' allele (BB). The frequency of the 'A' allele is denoted by 'q_A'. Adapted from Hanson *et al.*, 1995b and Stern *et al.*, 1996.

failure of the models as parameterized to capture the underlying genetic architecture. Similarly, although the models are constructed to minimize the possibility of spurious inference of a major genetic effect (Jarvik, 1998), it is possible that the 'major genes' detected in the Pima and Mexican-American populations are due to familial clustering of nongenetic determinants of diabetes. Moreover, these models typically do not distinguish a single gene with a very strong effect from a small number of genes with substantial effects. Nevertheless, these analyses show that, despite the high prevalence of type 2 diabetes in Pima Indians and Mexican-Americans, a strong genetic influence on the occurrence of diabetes, particularly when it develops at a young age, can be detected. These findings have been used in the design and analysis of linkage studies in these populations.

6. Linkage studies

In contrast to studies which examine the overall pattern of familial transmission of a disease, linkage studies examine whether particular alleles at specific genetic markers are co-transmitted with the disease within families. Linkage studies for loci influencing susceptibility to type 2 diabetes have been conducted in a number of populations. Given the difficulty in determining mode of inheritance of type 2 diabetes and the fact that the functions of most human genes are unknown, genome-wide linkage studies using nonparametric methods have been widely employed. As shown in *Table 3*, a number of regions have demonstrated at least suggestive evidence for linkage (lod>2, as defined by Morton (1998)), but only a few regions have shown significant evidence for linkage (lod>3). Hanis *et al.* (1996) detected significant evidence for linkage with a locus, which they called *NIDDM1*, near the q-terminal end of chromosome 2 in a Mexican-American population in Starr County, Texas. Suggestive evidence for linkage (lod=2.1) was also found in this region in European-Americans (Elbein *et al.*, 1999). Multilocus linkage analysis suggests that this locus interacts with a locus on chromosome 15 to influence susceptibility to type 2 diabetes in Mexican-Americans (Cox *et al.*, 1999). Duggirala *et al.* (1999) also demonstrated significant evidence for linkage to age of onset of diabetes on chromosome 10q in Mexican-Americans, and Ehm *et al.* (1999) found significant evidence for linkage with abnormal glucose tolerance on chromosome 3 in Mexican-Americans. In Finnish sib-pairs, significant evidence for linkage with type 2 diabetes was observed on the short arm of chromosome 20, while suggestive evidence was observed on the long arm (Ghosh *et al.*, 1999). The chromosome 20q locus has also shown modest evidence for linkage with type 2 diabetes in several other populations of European origin, as demonstrated by Ji *et al.* (1997; lod=2.4), Bowden *et al.* (1997; lod=1.5) and Zouali *et al.* (1997; lod=1.8). Elbein *et al.* (1999) found significant linkage with type 2 diabetes on chromosome 1q near the *APOA2* locus in European Americans; this same chromosomal location shows linkage with young-onset diabetes in Pima Indians (Hanson *et al.*, 1998). Linkage analyses in European-Americans demonstrated suggestive linkage with abnormal glucose tolerance in a region on chromosome 12q, near the *MODY3* locus (Ehm *et al.*, 1999); the same region was linked to diabetes in a subset of Finnish families participating in a genome-wide linkage study,

Table 3. Chromosomal regions demonstrating at least suggestive (multipoint lod > 2.0) evidence for linkage in genome-wide linkage analysis in various populations

Chrom.	Reference	Population	Trait analyzed	Analysis method	lod
1p	St. Jean *et al.*, 1999	European Americans	Diabetes	Affected relatives	2.3
1q	Elbein *et al.*, 1999	European Americans	Abnormal glucose homeostasis[a]	Parametric	4.3
1q	Hanson *et al.*, 1998	Pima Indians	Young-onset diabetes[b]	Haseman-Elston	2.5
2q	Hanis *et al.*, 1996	Mexican Americans	Diabetes	Affected siblings	4.3
2q	Elbein *et al.*, 1999	European Americans	Abnormal glucose homeostasis[a]	Affected siblings	2.1
3p	Ehm *et al.*, 1999	Mexican Americans	Abnormal glucose homeostasis[c]	Affected relatives	4.2
3p	Duggirala *et al.*, 1999	Mexican Americans	Age of onset of diabetes	Variance components	2.5
5q	Ehm *et al.*, 1999	European Americans	Abnormal glucose homeostasis[c]	Affected relatives	2.7
6q	Hegele *et al.*, 1999	Oji-Cree Indians	Diabetes	Affected siblings	3.3[d]
9p	Duggirala *et al.*, 1999	Mexican Americans	Age of onset of diabetes	Variance components	2.1
10q	Duggirala *et al.*, 1999	Mexican Americans	Age of onset of diabetes	Variance components	3.7
11q	Hanson *et al.*, 1998	Pima Indians	Obesity and diabetes[e]	Variance components	4.9
12q	Ehm *et al.*, 1999	European Americans	Abnormal glucose homeostasis[c]	Affected relatives	2.7
14q	St. Jean *et al.*, 1999	European Americans	Diabetes	Affected relatives	2.1
16	Hegele *et al.*, 1999	Oji-Cree Indians	Diabetes	Affected siblings	3.2[d]
18q	Elbein *et al.*, 1999	European Americans	Abnormal glucose homeostasis[a]	Affected relatives	2.4
20p	Ghosh *et al.*, 1999	Finns	Diabetes	Affected siblings	3.1[f]
20q	Ghosh *et al.*, 1999	Finns	Diabetes	Affected siblings	2.7[f]

[a] Includes persons with diabetes and, depending on age, those with 2-hour post-load plasma glucose ≥ 7.8 mM.
[b] Diabetes with onset before the age of 45 years.
[c] Includes persons with diabetes and with milder degrees of hyperglycemia (fasting or post-load).
[d] Only two-point analyses were reported.
[e] Bivariate variance components analysis assessing linkage to a locus influencing both diabetes-susceptibility and obesity, as measured by body mass index.
[f] Results of analyses in which sibling pairs are treated as independent are reported (i.e., without weighting pairs to adjust for multiple affected pairs in a sibship).

defined by low insulin secretion (Mahtani *et al.*, 1996). Linkage analyses in the Pima Indians demonstrated significant evidence for linkage to obesity, as defined by body mass index, to a location on chromosome 11q (lod=3.6) where there was also tentative evidence for linkage with diabetes (lod=1.8; Hanson *et al.*, 1998). Bivariate analyses of diabetes and obesity demonstrated strong evidence for a locus influencing both traits in this region of chromosome 11 (lod=4.9). Two regions, on chromosomes 6 and 16, showed evidence for linkage with diabetes in a small study of Oji-Cree Indians involving 49 affected sibling pairs (Hegele *et al.*, 1999b).

The genetic linkage studies of type 2 diabetes are, thus, somewhat inconsistent. While some of the putative diabetes-susceptibility loci derive at least modest support from several different populations, no locus shows strong evidence for linkage in multiple populations. This suggests that no type 2 diabetes-susceptibility locus has a strong effect in most populations – unlike type 1 diabetes where the HLA locus nearly uniformly shows strong evidence for linkage (Concannon *et al.*, 1998; Mein *et al.*, 1998). Some of the putative loci may be spurious, but variation in linkage results is to be expected for multifactorial diseases; the ability to detect the initial linkage may result from a fortuitous combination of sampling variation, gene–environment interaction and gene–gene interaction, so much larger sample sizes may be required to replicate the initial finding (Suarez *et al.*, 1994). Thus, some of the loci detected by linkage studies are likely to represent true diabetes-susceptibility genes. Linkage studies of data pooled from several genome-wide scans may help to identify some of the more important loci (Boehnke *et al.*, 1998). Fine-mapping studies to identify polymorphisms in particular genetic elements that account for the linkage results are also currently under way.

7. Association studies

In association studies, the frequencies of particular alleles in a genetic marker are compared between individuals with the disease (cases) and those without the disease (controls). A number of case-control studies, typically using markers within candidate genes, have been conducted for type 2 diabetes. In most of these studies controls who were not biologically related to the cases were chosen. Alleles at the major histocompatibility (HLA) locus on chromosome 6p are associated with type 2 diabetes in a variety of populations. However in different populations different HLA alleles occur with increased frequency in diabetic individuals: HLA-A2 in Xhosa and Pima Indians (Briggs *et al.*, 1980; Williams *et al.*, 1981), HLA-B61 in Asian Indians (Omar *et al.*, 1988; Serjeantson *et al.*, 1981), HLA-B45 in Chinese (Zhao *et al.*, 1982) and HLA-B62 in Papuans (Bhatia *et al.*, 1984). Among Finns, a number of HLA haplotypes, most of which are associated with type 1 diabetes, are also associated with type 2 diabetes (Tuomilehto-Wolf *et al.*, 1993). A variety of polymorphisms in other genes have been associated with type 2 diabetes in some populations including alleles of the glycogen synthase gene (Groop *et al.*, 1993), alleles of the insulin receptor substrate-1 gene (Zhang *et al.*, 1996), the glycogen targeting subunit of type 1 protein phosphatase gene (Xia *et al.*, 1998), a Trp-Arg missense mutation in the beta-3 adrenergic receptor

(Fujisawa *et al.,* 1996) and the frataxin gene, which causes Friedrich's ataxia (Ristow *et al.,* 1998). In most of these cases, however, there are also populations in which there is no association of diabetes with these polymorphisms (e.g. Dalgaard *et al.,* 1999; Rissanen *et al.,* 1997; 't Hart *et al.,*1999; Zhang *et al.,* 1996).

The rationale for association studies is that the associated alleles may directly increase susceptibility to the disease or, more likely, be in linkage disequilibrium with a disease-causing polymorphism. However, when controls that are unrelated to the cases are analyzed, these studies are potentially confounded by population stratification. This is illustrated by a study of haplotypes of the Gm system of immunoglobulin G in Pima Indians (Knowler *et al.,* 1988b). The haplotype $Gm^{3;5,13,14}$ has a strong inverse association with the prevalence of diabetes, being 0.5 (95% confidence interval 0.4–0.7) times less common in persons with diabetes than among nondiabetic individuals. However, the $Gm^{3;5,13,14}$ haplotype is also more common in individuals with European admixture, who are less likely to have diabetes than full-heritage American Indians. As shown in *Figure 4*, within strata defined by the degree of Indian heritage, there is no association between the $Gm^{3;5,13,14}$ haplotype and diabetes (heritage adjusted prevalence ratio=0.8, 95% confidence interval 0.6–1.2). Thus, while case-control studies can be quite powerful for detection of genetic effects, they are prone to produce spurious results due to confounding by population stratification (which can be difficult to detect). To control for population stratification, it has been proposed that unaffected siblings be used as controls (Curtis, 1997; Horvath and Laird, 1998; Spielman and Ewens, 1998) or that control genotypes be constructed from parental alleles that were not transmitted to an affected offspring, the transmission/disequilibrium test (Spielman *et al.,* 1993). These family-based analyses will only produce positive

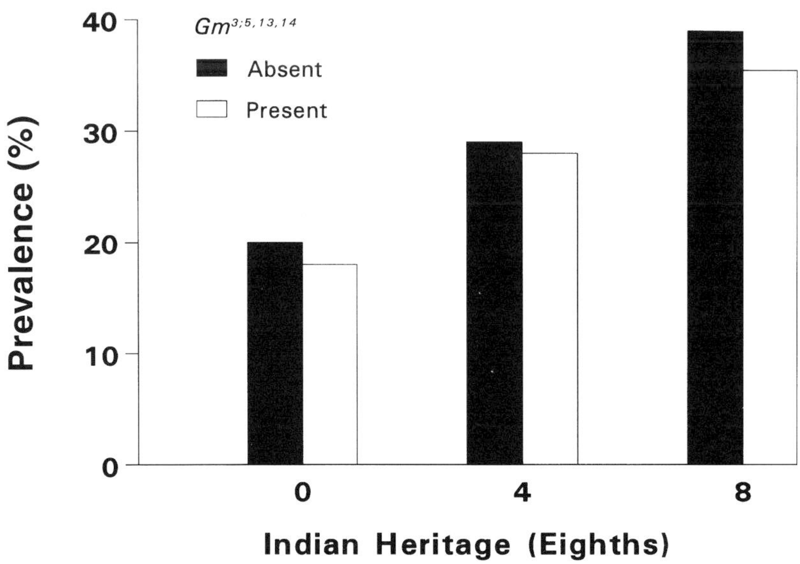

Figure 4. Prevalence of diabetes according to presence of the $Gm^{3;5,13,14}$ haplotype in strata defined by the degree of American Indian heritage in the Gila River Indian Community in Arizona. Adapted from Knowler *et al.,* 1988b.

results if the marker is both linked and associated with the disease and, thus, they do not detect association that is due to causes other than linkage disequilibrium. This robustness, however, comes at the cost of diminished power compared with studies that use unrelated controls. The family-based association methods have not been widely applied to type 2 diabetes, but in a Finnish population a sibling case-control study showed no linkage disequilibrium between type 2 diabetes and glycogen synthase alleles that were associated with diabetes in a study using unrelated cases and controls (Orho-Melander *et al.*, 1999). Regardless of which analysis method is used, the same considerations that require a very low P-value to declare linkage significant also apply to association studies (Kidd, 1993). Since few association studies have achieved this threshold and since the results of these studies have been inconsistent, the pathogenetic role of most polymorphisms detected by case-control studies remains uncertain. In the near future, the ability to type a large number of very densely spaced genetic markers may make feasible genome-wide scans for linkage disequilibrium (Kruglyak, 1999; Risch and Merikangas, 1996); the utility of this approach for detecting diabetes-susceptibility genes remains to be determined.

8. Conclusions

Despite considerable progress in studies of the genetic and non-genetic causes of type 2 diabetes, these causes remain complex and our understanding of them incomplete. There is clear evidence for important environmental and socio-economic factors in the development of type 2 diabetes which complicate elucidation of its genetic causes. Yet the strong familial aggregation provides evidence for genes having strong effects. None of the family studies, however, support a simple Mendelian inheritance of the disease (i.e. solely due to a single gene with high penetrance). The major unsolved problems are whether there are a few genes having large effects, or many genes having small effects, and the degree of heterogeneity in susceptibility genes among different ethnic groups or even among different families in the same ethnic group.

In certain types of familial diabetes, called MODY, linkage studies performed in the last several years have led to identification of responsible genes. These genes do not appear to play a large role in most cases of type 2 diabetes, although the HNF-1α gene causing one form of MODY is associated with what had been considered typical type 2 diabetes in a small, isolated community of Oji-Cree Indians in Canada. While these studies of MODY have demonstrated the power of linkage studies to lead to gene identification in some forms of diabetes, the genetics of typical type 2 diabetes remain elusive.

Linkage and association studies of type 2 diabetes can be interpreted as indicating either marked heterogeneity or some degree of consistency. Many such studies have not been replicated, suggesting that the findings are spurious or apply to only one limited population. Some, such as the HLA associations, have been reported in many populations, but generally with different HLA types being associated with type 2 diabetes in different populations, and most of the associations being relatively weak. Certain recent linkage studies, however, provide some degree of replication and hence there is hope that they will lead to gene identification. For example, type 2 diabetes has been linked to chromosome 1q in Pima Indians (Hanson *et al.*, 1998) and in Americans of northern European descent (Elbein *et al.*, 1999), and it has been linked to chromosome 20q in Finns and in three studies of persons of European descent (Bowden *et al.*, 1997; Ghosh *et al.*, 1999; Ji *et al.*, 1997; Zouali *et al.*, 1997).

We anticipate that present and future efforts in gene mapping and identification will lead to the discovery of genes influencing risk of type 2 diabetes, but that the effects of these genes on disease risk will be modified by many other factors. The likelihood of discovery of a single gene accounting for most cases of type 2 diabetes in most populations appears low, but further genetic research promises to help elucidate the complex set of causes of this disease.

References

Alcolado, J. and Alcolado, R. (1991) Importance of maternal history of non-insulin-dependent diabetes mellitus. *Brit. Med. J.* **302:** 1178–1180.

American Diabetes Association (ADA) Expert Committee on the Diagnosis and Classification of Diabetes Mellitus. (1997) Report of the expert committe on the diagnosis and classification of diabetes mellitus. *Diabetes Care* **20:** 1183–1197.

Barnett, A.H., Eff, C., Leslie, R.D.G., *et al.* (1981) Diabetes in identical twins: a study of 200 pairs. *Diabetologia* **20:** 87–93.

Barzilay, J.I., Spickerman, C.F., Wahl, P.W., *ct al.* (1999) Cardiovascular disease in older adults with glucose disorders: comparison of American Diabetes Association criteria for diabetes with WHO criteria. *Lancet* **354:** 622–625.

Bell, G.I., Xiang, K.S., Newman, M.V., *et al.* (1991) Gene for non-insulin-dependent diabetes mellitus (maturity-onset diabetes of the young subtype) is linked to DNA polymorphism on human chromosome 20q. *Proc. Natl Acad. Sci. USA.* **88:** 1484–1488.

Bhatia, K., Patel, M. and Gorogo, M. (1984) Type 2 (non-insulin-dependent) diabetes mellitus and HLA antigens in Papua, New Guinea. *Diabetologia* **27:** 370–373.

Boehnke, M. and the International Type 2 Diabetes Linkage Analysis Consortium. (1998) Lessons learned in a combined linkage analysis (24 datasets, >2000 families) of type 2 diabetes on chromosome 20. *Am. J. Hum. Genet.* **63 (Suppl):** A282.

Bowden, D.W., Sale, H., Howard, T.D., *et al.* (1997) Linkage of genetic markers on human chromosomes 20 and 12 to NIDDM in Caucasian sib pairs with a history of diabetic nephropathy. *Diabetes* **46:** 882–886.

Briggs, B.R, Botha, M.C, Jackson, W.P.U., *et al.* (1980) The histocompatibility (HLA) antigen distribution in South African blacks (Xhosa). *Diabetes* **29:** 68–70.

Bulman, M.P., Dronsfield, M.J., Frayling, T., *et al.* (1997) A missense mutation in the hepatocyte nuclear factor 4 alpha gene in a UK pedigree with maturity-onset diabetes of the young. *Diabetologia* **40:** 859–862.

Byrne, M.M, Sturis, J., Fajans, S.S., *et al.* (1995) Altered insulin secretory responses to glucose in subjects with a mutation in the MODY1 gene on chromosome 20. *Diabetes* **44:** 699–704.

Byrne, M.M., Sturis, J., Menzel, S., *et al.* (1996) Altered insulin secretory responses to glucose in subjects with mutations in the diabetes susceptibility gene MODY on chromsome 12. *Diabetes* **45:** 1503–1510.

Chan, J.M., Rimm, E.B., Colditz, G.A., *et al.* (1994) Obesity, fat distribution, and weight gain as risk factors for clinical diabetes in men. *Diabetes Care* **17:** 961–969.

Charles, M.A., Fontbonne, A., Thibult, N., *et al.* (1991) Risk factors for NIDDM in white population: Paris prospective study. *Diabetes* **40:** 796–799.

Colditz, G.A., Willett, W.C., Stampfer, M.J., *et al.* (1990) Weight as a risk factor for clinical diabetes in women. *Am. J. Epidemiol.* **132:** 501–513.

Concannon, P., Gogolin-Ewens, K.J., Hinds, D.A., *et al.* (1998) A second-generation screen of the human genome for susceptibility to insulin-dependent diabetes mellitus. *Nature Genet.* **19:** 292–296.

Cook, J.T.E., Shields, D.C., Page, R.C.L., *et al.* (1994) Segregation analysis of NIDDM in Caucasian families. *Diabetologia* **37:** 1231–1240.

Cox, N.J., Frigge, M., Nicolae, D.L., *et al.* (1999) Loci on chromosomes 2 (*NIDDM1*) and 15 interact to increase susceptibility to diabetes in Mexican Americans. *Nature Genet.* **21:** 213–215.

Curtis, D. (1997) Use of siblings as controls in case-control association studies. *Ann. Hum. Genet.* **61:** 319–333.

Dabelea, D., Hanson, R.L., Bennett, P.H., *et al.* (1999) Intrauterine exposure to diabetes conveys risk for diabetes and obesity in offspring above that attributable to genetics. *Diabetes* **48(Suppl 1):** A52.

Dalgaard, L.T., Hansen, T., Urhammer, S.A., *et al.* (1999) Intermediate expansions of a GAA repeat in the frataxin gene are not associated with type 2 diabetes or altered glucose-induced beta-cell function in Danish Caucasians. *Diabetes* **48:** 914–917.

Diabetes Control and Complications Trial (DCCT) Research Group. (1993) The effect of intensive treatment of diabetes on the development and progression of long-term complications in insulin-dependent diabetes mellitus. *N. Engl. J. Med.* **329:** 977–986.

Duggirala, R., Blangero, J., Almasy, L., *et al.* (1999) Linkage of type 2 diabetes mellitus and of age at onset to a genetic location on chromosome 10q in Mexican Americans. *Am. J. Hum. Genet.* **64:** 1127–1140.

Ehm, M.G., Karnoub, M.C., Sakul, H., *et al.* (1999) Genome-wide search for type 2 diabetes susceptibility genes in Caucasians, Mexican Americans, African Americans and Japanese Americans. *Diabetes* **48(Suppl 1):** A46–A47.

Elbein, S.C., Hoffman, M.D., Teng, K. *et al.* (1999) A genome-wide search for type 2 diabetes susceptibility genes in Utah Caucasians. *Diabetes* **48:** 1175–1182.

Elston, R.C. (1975) The prior probability of autosomal linkage. *Ann. Hum. Genet.* **38:** 341–350.

Fajans, S.S. and Conn, J.W. (1965) Prediabetes, subclinical diabetes and latent clinical diabetes: interpretation, diagnosis and treatment. In: *On the Nature and Treatment of Diabetes* (eds. Leibel, D.S. and Wrenshall, G.S.) Excerpta Medica, Amsterdam, pp. 641–656.

Fajans, S.S., Cloutier, M.C. and Crowther, R.L. (1978) Clinical and etiologic heterogeneity of idiopathic diabetes mellitus. *Diabetes* **27:** 1112–1125.

Frayling, T.M., Bulman, M.P., Ellard, S., *et al.* (1997) Mutations in the hepatocyte nuclear factor-1α gene are a common cause of maturity-onset diabetes of the young in the U.K. *Diabetes* **46:** 720–725.

Froguel, P., Vaxillaire, M. and Sun, F. (1992) Close linkage of glucokinase locus on chromosome 7p to early-onset non-insulin-dependent diabetes mellitus. *Nature* **356:** 162–164.

Froguel, P., Zouali, H., Vionnet, N., *et al.* (1993) Familial hyperglycemia due to mutations in glucokinase: definition of a subtype of diabetes mellitus. *N. Engl. J. Med.* **328:** 697–702.

Fujisawa, T., Ikegami, H., Yamato, E., *et al.* (1996) Association of Trp64Arg mutation of the beta3-adrenergic receptor with NIDDM and body weight gain. *Diabetologia* **39:** 349–352.

Ghosh S., Watanabe, R.M., Hauser, E.B., *et al.* (1999) Type 2 diabetes: evidence for linkage on chromosome 20 in 716 Finnish affected sib pairs. *Proc. Natl Acad. Sci. USA.* **96:** 2198–2203.

Groop, L.C., Kankuri, M., Schalin-Jantti, C., *et al.* (1993) Association between polymorphism of the glycogen synthase gene and non-insulin-dependent diabetes mellitus. *N. Engl. J. Med.* **328:** 10–14.

Hanis, CL, Boerwinkle, E., Chakraborty, R., *et al.* (1996) A genome-wide search for human non-insulin-dependent (type 2) diabetes genes reveals a major susceptibility locus on chromosome 2. *Nature Genet.* **13:** 161–166.

Hanson, R.L., Pettitt, D.J., Bennett, P.H., *et al.* (1995a) Familial relationships between obesity and NIDDM. *Diabetes* **44:** 418–422.

Hanson, R.L., Elston, R.C., Pettitt, D.J., *et al.* (1995b) Segregation analysis of non-insulin-dependent diabetes mellitus in Pima Indians: evidence for a major-gene effect. *Am. J. Hum. Genet.* **57:** 160–170.

Hanson, R.L., Ehm, M.G., Pettitt, D.J., *et al.* (1998) An autosomal genomic scan for loci linked to type II diabetes mellitus and body-mass index in Pima Indians. *Am. J. Hum. Genet.* **63:** 1130–1138.

Harris, M.I. (1991) Epidemiological correlates of NIDDM in Hispanics, whites and blacks in the U.S. population. *Diabetes Care* **14:** 639–648.

Hegele, R.A., Cao, H., Harris, S.B., *et al.* (1999a) The hepatocyte nuclear factor-1α G319S variant is associated with early-onset type 2 diabetes in Canadian Oji-Cree. *J. Clin. Endocrinol. Metab.* **84:** 1077–1082.

Hegele, R.A., Sun, F., Harris, S.B., *et al.* (1999b) Genome-wide scanning for type 2 diabetes susceptibility genes in Canadian Oji-Cree, using 190 microsatellite markers. *J. Hum. Genet.* **44:** 10–14.

Horikawa, Y., Iwasaki, N., Hara, M., *et al.* (1997) Mutation in hepatocyte nuclear factor-1α gene (*TCF2*) associated with MODY. *Nature Genet.* **17:** 384–385.

Horvath, S. and Laird, N.M. (1998) A discordant-sibship test for disequilibrium and linkage: no need for parental data. *Am. J. Hum. Genet* **63:** 1886–1897.

Iwasaki, N., Oda, N., Hara, M., *et al.* (1997) Mutations in the hepatocyte nuclear factor-1α/MODY3 gene in Japanese subjects with early- and late-onset NIDDM. *Diabetes* **46:** 1504–1508.

Jarvik, G.P. (1998) Complex segregation analyses: uses and limitations. *Am. J. Hum. Genet.* **63:** 942–946.

Ji, L., Malecki, M., Warram, J.H., *et al.* (1997) New susceptibility locus for NIDDM is localized to human chromosome 20q. *Diabetes* **46:** 876–881.

Kadowaki, T., Kadowaki, H., Mori, H., *et al.* (1994) A subtype of diabetes mellitus associated with a mutation of mitochondrial DNA. *N. Engl. J. Med.* **330:** 962–968.

Kannel, W.B. and McGee, D.L. (1979) Diabetes and cardiovascular disease: the Framingham study. *JAMA* **241:** 2035–2038.

Kaprio, J., Tuomilehto, J., Koskenkvuo, M., *et al.* (1992) Concordance for type 1 (insulin-dependent) and type 2 (non-insulin-dependent) diabetes mellitus in a population-based cohort of twins in Finland. *Diabetologia* **35:** 1060–1067.

Karter, A.J., Rowell, S.E., Ackerson, L.M., *et al.* (1999) Excess maternal transmission of type 2 diabetes: the Northern California Kaiser Permante diabetes registry. *Diabetes Care* **22:** 938–943.

Kidd, K.K. (1993) Associations of disease with genetic markers: déjà vu all over again. *Am. J. Med. Genet.* **48:** 71–73.

Knowler, W.C., Pettitt, D.J., Savage, P.J., *et al.* (1981) Diabetes incidence in Pima Indians: contributions of obesity and parental diabetes. *Am. J. Epidemiol.* **113:** 144–156.

Knowler, W.C., Pettitt, D.J., Lillioja, S., *et al.* (1988a) Genetic and environmental factors in the development of diabetes mellitus in Pima Indians. In: *Genetic Susceptibility to Environmental Factors – a Challenge for Public Intervention*. (eds. Smith, U., Eriksson, S. and Lindgärde, F.). Almqvist and Wiksell International, Stockholm, pp. 67–76.

Knowler, W.C., Williams, R.C., Pettitt, D.J., *et al.* (1988b) $Gm^{3;5,13,14}$ and type 2 diabetes mellitus: an association in American Indians with genetic admixture. *Am J. Hum. Genet.* **43:** 520–526.

Köbberling, J. (1971) Studies on the genetic heterogeneity of diabetes mellitus. *Diabetologia* **7:** 46–49.

Kruglyak, L. (1999) Prospects for whole-genome linkage disequilibrium mapping of common disease genes. *Nature Genet.* **22:** 139–144.

Kuzuya, T. and Matsuda, A. (1982) Family histories of diabetes among Japanese patients with type 1 (insulin-dependent) and type 2 (non-insulin-dependent) diabetes. *Diabetologia* **22:** 372–374.

Laakso, M., Malkki, M., Kekalainen, P., *et al.* (1995) Glucokinase gene variants in subjects with late-onset NIDDM and impaired glucose tolerance. *Diabetes Care* **18:** 398–400.

Lalouel, J.M., Rao, D.C., Morton, N.E., *et al.* (1983) A unified model for complex segregation analysis. *Am J. Hum. Genet.* **35:** 816–826.

Lander, E. and Kruglyak, L. (1995) Genetic dissection of complex traits: guidelines for interpreting and reporting linkage results. *Nature Genet.* **11:** 241–247.

Lee, E.T., Anderson, P.S., Bryan, J., *et al.* (1985) Diabetes, parental diabetes and obesity in Oklahoma Indians. *Diabetes Care* **8:** 107–113.

Lipton, R.B., Liao, Y., Cao, G., *et al.* (1993) Determinants of incident non-insulin-dependent diabetes mellitus among blacks and whites in national sample: the NHANES I epidemiologic follow-up study. *Am. J. Epidemiol.* **138:** 826–839.

Macfarlane, W.M., Frayling, T.M., Ellard, S., *et al.* (1999) Missense mutations in the insulin promoter factor-1 gene predispose to type 2 diabetes. *J. Clin. Invest.* **104:** R33–R39.

Mahtani, M.M., Widén, E., Lehto, M. *et al.* (1996) Mapping of a gene for type 2 diabetes associated with an insulin secretion defect by a genome scan in Finnish families. *Nature Genet.* **14:** 90–94.

Matsuda, A. and Kuzuya, T. (1994) Diabetic twins in Japan. *Diabetes Res Clin Pract.* **24 (Suppl):** S63–S67.

McCarthy, M.I., Hitman, G.A., Shields, D.C., *et al.* (1994) Family studies of non-insulin-dependent diabetes mellitus in South Indians. *Diabetologia* **37:** 1221–1230.

Medici, F., Hawa, M., Ianari, A., *et al.* (1999) Concordance rate for type II diabetes mellitus in monozygotic twins: actuarial analysis. *Diabetologia* **42:** 146–150.

Mein, C.A., Esposito, L, Dunn, M.G., *et al.* (1998) A search for type 1 diabetes susceptibility genes in families from the United Kingdom. *Nature Genet* **19:** 297–300.

Mengesha, B. and Abdulkadir, J. (1997) Heritability of diabetes mellitus in Ethiopian diabetics. *East. Afr. Med. J.* **74:** 37–40.

Mitchell, B.D, Kammerer, C.M., Reinhart, L.J. *et al.* (1995) Is there an excess in maternal transmission of NIDDM? *Diabetologia* **38:** 314–317.

Mitchell, B.D., Almasy, L.A., Rainwater, D.L., *et al.* (1999) Diabetes and hypertension in Mexican American families: relation to cardiovascular risk. *Am. J. Epidemiol.* **149:** 1047–1056.

Moller, A.M., Urhammer, S.A., Dalgaard, L.T., *et al.* (1997) Studies of the genetic variability of the coding region for the hepatocyte nuclear factor-4alpha in Caucasians with maturity onset NIDDM. *Diabetologia* **40:** 980–983.

Morton, N.E. (1998) Significance levels in complex inheritance. *Am. J. Hum. Genet.* **62:** 690–697.

Mykkänen, L., Laakso, M., Uusitupa, M., *et al.* (1990) Prevalence of diabetes and impaired glucose tolerance in elderly subjects and their association with obesity and family history of diabetes. *Diabetes Care* **13:** 1099–1105.

Nakajima, H., Yoshiuchi, I., Hamaguchi, T., *et al.* (1996) Hepatocyte nuclear factor-4 alpha gene mutations in Japanese non-insulin-dependent diabetes mellitus (NIDDM) patients. *Res. Commun. Mol. Pathol. Pharmacol.* **94:** 327–330.

Nishi, S., Hinata, S., Matsukage, T., *et al.* (1994) Mutations in the glucokinase gene are not a major cause of late onset type 2 (non-insulin-dependent) diabetes mellitus in Japanese subjects. *Diabet. Med.* **11:** 193–197.

Newman, B., Selby, J.V., King, M-C., *et al.* (1987) Concordance for type 2 (non-insulin-dependent) diabetes mellitus in male twins. *Diabetologia* **30:** 763–768.

Oliveira, J.E., Milech, A., Franco, L.J., *et al.* (1996) The prevalence of diabetes in Rio de Janeiro, Brazil. *Diabetes Care* **19:** 663–666.

Omar, M.A.K., Hammand, M.G., Motala, A.A., *et al.* (1988) HLA class I and II antigens in South African Indians with non-insulin-dependent diabetes mellitus. *Diabetes* **37:** 769–799.

Orho-Melander, M., Almgren, P., Kanninen, T., *et al.* (1999) A paired-sibling analysis of the XbaI polymorphism in the muscle glycogen synthase gene. *Diabetologia* **42:** 1138–1145.

Pan, X.R., Yang, W.Y., Li, G.W., *et al.* (1997) Prevalence of diabetes and its risk factors in China, 1994. *Diabetes Care* **20:** 1664–1669.

Pettitt, D.J., Aleck, K.A., Baird, H.R., *et al.* (1988) Congenital susceptibility to NIDDM: role of intrauterine environment. *Diabetes* **37:** 622–628.

Poulsen, P., Ohm Kyvik, K., Vaag, A., *et al.* (1999) Heritability of type II (non-insulin-dependent) diabetes mellitus and abnormal glucose tolerance: a population-based twin study. *Diabetologia* **42:** 139–145.

Risch, N. and Merikangas, K. (1996) The future of genetic studies of complex human diseases. *Science* **273:** 1516–1517.

Rissanen, J., Kuopusjarvi, J., Pihlajamaki, J., *et al.* (1997) The Trp64Arg polymorphism of the beta3-adrenergic receptor gene: lack of association with NIDDM and features of insulin resistance syndrome. *Diabetes Care* **20:** 1319–1323

Ristow, M., Giannakidou, E., Hebinck, J., *et al.* (1998) An association between NIDDM and a GAA trinucleotide repeat polymorphism in the X25/frataxin (Friedrich's ataxia) gene. *Diabetes* **47:** 851–854.

St. Jean, P.L., Mitchell, B.D., Hsueh, W.C., *et al.* (1999) Type 2 diabetes in the Old Order Amish. *Diabetes* **48(Suppl. 1):** A46.

Serjeantson, S.W., Ryan, D.P., Ram, P., *et al.* (1981) HLA and non-insulin-dependent diabetes mellitus in Fiji Indians. *Med. J. Aust.* **i:** 462–463.

Simpson, N. (1964) Multifactorial inheritance: a possible hypothesis for diabetes. *Diabetes* **13:** 462–471.

Spielman, R.S. and Ewens, W.J. (1998) A sibship test for linkage in the presence of association: the sib transmission/disequilibrium test. *Am. J. Hum. Genet.* **62:** 450–458.

Spielman, R.S., McGinnis, R.E. and Ewens, W.J. (1993) Transmission test for linkage disequilibrium: the insulin gene region and insulin-dependent diabetes mellitus (IDDM). *Am. J. Hum. Genet.* **52:** 506–516.

Stern, M.P., Mitchell, B.D., Blangero, J., *et al.* (1996) Evidence for a major gene for type II diabetes and linkage analyses with selected candidate genes in Mexican-Americans. *Diabetes* **45:** 563–568.

Stewart, M.W., Humphriss, D.B., Berrish, T.S., *et al.* (1995) Features of syndrome X in first-degree relatives of NIDDM patients. *Diabetes Care* **18:** 1020–1022.

Stoffers, D.A., Ferrer, J., Clarke, W.L., *et al.* (1997) Early-onset type-II diabetes mellitus (MODY4) linked to *IPF1*. *Nature Genet.* **17:** 138–139.

Suarez, B.K., Hampe, C.L. and van Eerdewegh, P. (1994) Problems of replicating linkage claims in psychiatry. In: *Genetic Approaches to Mental Disorders*. (eds., Gershon, E.S. and Cloninger, C.R.) American Psychiatric Press, Washington, pp. 23–46.

't Hart, L.M., Stolk, R.P., Dekker, J.M., *et al.* (1999) Prevalence of variants in candidate genes for type 2 diabetes mellitus in The Netherlands: the Rotterdam study and the Hoorn study. *J. Clin. Endocrinol. Metab.* **84:** 1002–1006.

Tattersall, R.B. (1974) Mild familial diabetes with dominant inheritance. *Q. J. Med.* **43:** 339–357.

Tattersall, R.B. (1982) The present status of maturity-onset type diabetes of young people (MODY). In: *The Genetics of Diabetes Mellitus: Proceedings of the Serono Symposia, Volume 47*. (eds. Köbberling, J. and Tattersall, R.) Academic Press, London, pp. 261–270.

Tattersall, R.B. and Fajans, S.S. (1975) A difference between the inheritance of classic juvenile-onset and maturity-onset type diabetes of young people. *Diabetes* **24:** 44–53.

Thomas, F., Balkau, B., Vauzelle-Kervroedan, V., *et al.* (1994) Maternal effect and familial aggregation in NIDDM: the CODIAB study. *Diabetes* **43:** 63–67.

Tuomilehto-Wolf, E., Tuomilehto, J., Hitman, G.A., *et al.* (1993) Genetic susceptibility to non-insulin-dependent diabetes mellitus and glucose intolerance are located in HLA region. *Brit. Med. J.* **307:** 155–159.

U.K. Prospective Diabetes Study (UKPDS) Group. (1998) Intensive blood-glucose control with sulphonylureas or insulin compared with conventional treatment and risk of complications in patients with type 2 diabetes. *Lancet* **352:** 837–853.

Urhammer, S.A., Rasmussen, S.K., Kaisaki, P.J., *et al.* (1997) Genetic variation in the hepatocyte nuclear factor-1β gene in Danish Caucasians with late-onset NIDDM. *Diabetologia* **40:** 473–475.

Van den Ouweland, J.M.W., Lemkes, H.H.P.J., Ruitenbeek, W., *et al.* (1992) Mutation in mitochondrial tRNA$^{(Leu,UUR)}$ gene in a large pedigree with maternally transmitted type II diabetes mellitus and deafness. *Nature Genet* **1:** 368–371.

Vaxillaire, M., Boccio, V., Philippi, A., *et al.* (1995) A gene for maturity onset diabetes of the young (MODY) maps to chromosome 12q. *Nature Genet* **9:** 418–423.

Vionnet, N., Stoffel, M., Takeda, J., *et al.* (1992) Nonsense mutation in the glucokinase gene causes early-onset non-insulin-dependent diabetes mellitus. *Nature* **356**. 721–722.

Viswanathan, M., Mohan, V., Snehelatha, C., *et al.* (1985) High prevalence of type 2 (non-insulin-dependent) diabetes among the offspring of conjugal type 2 diabetic parents in India. *Diabetologia* **28:** 907 910.

Williams, R.C., Knowler, W.C., Butler, W.J., *et al.* (1981) HLA-A2 and type 2 (non-insulin-dependent) diabetes mellitus in Pima Indians: an association of allele frequency with age. *Diabetologia* **21:** 460–463.

Winter, W.E., Maclaren, N.K., Riley, W.J., *et al.* (1987) Maturity-onset diabetes of youth in black Americans. *N. Engl. J. Med.* **316:** 285–291.

Xia, J., Scherer, S.W., Cohen, P.T.W., *et al.* (1998) A common variant in *PPP1R3* associated with insulin resistance and type 2 diabetes. *Diabetes* **47:** 1519–1524.

Yamagata, K., Oda, N., Kaisaki, P., *et al.* (1996a) Mutations in the hepatocyte nuclear factor-1α gene in maturity-onset diabetes of the young (MODY3). *Nature* **384:** 455–458.

Yamagata, K., Furuta, H., Oda, N., *et al.* (1996b) Mutations in the hepatocyte nuclear factor-4α gene in maturity-onset diabetes of the young (MODY1). *Nature* **384:** 458–460.

Zhang, Y., Wat, N., Stratton, I.M., *et al.* (1996) UKPDS19: hetereogeneity in NIDDM: separate contributions of IRS-1 and beta3-adrenergic-receptor mutations to insulin resistance and obesity respectively with no evidence for glycogen synthase gene mutations. UK Prospecitive Diabetes Study. *Diabetologia* **39**: 1505–1511.

Zhao, T., Chi, Z., Wang, H., *et al.* (1982) HLA and diabetes mellitus in China. *Chin. Med. J.* **95:** 609–612.

Zimmet, P., Taft, P., Guinea, A., *et al.* (1977) The high prevalence of diabetes mellitus on a central Pacific island. *Diabetologia* **13:** 111–115.

Zouali, H., Hani, E.H., Philippi, A., *et al.* (1997) A susceptibility locus for early-onset non-insulin-dependent (type 2) diabetes mellitus maps to chromosome 20q, proximal to the phosphoenolpyruvate carboxykinase gene. *Hum. Mol. Genet.* **6:** 1401–1408.

8

Type 1 Diabetes

L. Leigh Field

1. Introduction

Of all the genetically-complex diseases, type 1 diabetes probably has been subjected to the most intensive and extensive investigation in order to locate the multiple loci capable of conferring genetic predisposition. Thus, there may be insights to be garnered from this collective experience that are relevant to analysis of other multifactorial disorders. This chapter will attempt to focus on issues which have arisen, especially during the last half-decade 'gene search', presented in the context of what is currently known about the genetics of type 1 diabetes. The primary issues can be briefly summarized as follows:

(i) Which genes detected by linkage/association analysis are 'real'? Does non-replication in independent studies indicate type 1 error, or sampling and/or ethnic variation? Can alternate definitions of the disease phenotype (e.g. presence of autoantibodies in nondiabetic siblings of diabetics) assist in confirming gene localizations?

(ii) Once a predisposing gene has been located to a chromosomal region, how can that region be narrowed and the culprit gene isolated? Is association (linkage disequilibrium) analysis the most powerful approach? Are genetic isolates (founder populations) useful? How will functional candidates be tested for involvement in susceptibility?

(iii) When multiple susceptibility genes have been identified, how can the interactions between these loci be analyzed and interpreted? Is there sharing of genes with disorders having similar pathogenesis (e.g. other autoimmune disorders)? Finally, how can gene interactions with specific environmental factors be investigated in order to have a comprehensive understanding of disease etiology?

In the ensuing pages, we will first discuss the genetic epidemiology of type 1 diabetes, then the results from genome screen and candidate gene linkage and association studies searching for the susceptibility loci.

Analysis of Multifactorial Disease, edited by T. Bishop and P. Sham.

2. Genetic epidemiology

2.1 Pathogenesis of type 1 diabetes

Type 1 diabetes (formerly called insulin-dependent diabetes mellitus, or IDDM) is the result of T-cell-mediated autoimmune destruction of the insulin-producing β cells of the pancreas. This form of diabetes has also been called 'juvenile onset' diabetes, since the peak age of onset is around puberty, in contrast to type 2 non-insulin-dependent diabetes which usually develops during adulthood. It appears that type 1 diabetes has a long preclinical phase during which T cells infiltrate the islets of the pancreas and autoantibodies to a variety of islet cell components are formed, presumably because the immune system has not been tolerized to these newly exposed proteins. Antibodies towards islet cells (ICA), insulin (IAA), glutamic acid decarboxylase (GAD), and islet cell antigen 512 (IA-2α) are common at time of diagnosis of type 1 diabetes. Recent studies have shown that these autoantibodies occur at increased frequencies in blood taken from infants and very young children who *subsequently* develop diabetes, suggesting that the autoimmune process may begin very early, perhaps even *in utero* (Lindberg *et al.*, 1999; Roll *et al.*, 1996; Ziegler *et al.*, 1999). The long preclinical phase of type 1 diabetes is followed by accelerated destruction of pancreatic β cells, resulting in clinically overt disease when about 90% of β cells are gone and requiring daily injection of exogenous insulin for survival.

The greatest unsolved mystery of type 1 diabetes is – what antigen initiates this autoimmune process? Is there a specific β cell antigen with primary involvement in the autoimmune response (with other autoantigens being secondary by-products of β cell destruction)? It has been suggested that viral infection is involved in initiating the disease process. For example, it is well established that children with congenital rubella infection have an elevated risk of type 1 diabetes (e.g. Ginsberg-Fellner *et al.*, 1985 reported that 30/242 = 12% of children with congenital rubella syndrome developed type 1 diabetes). One mechanism proposed to explain this association is structural similarity ('molecular mimicry') between an immunogenic epitope on the rubella virus and a pancreatic β cell protein (Karounos *et al.*, 1993), such that exposure to rubella virus elicits an immune response directed towards both antigens. There is also considerable circumstantial evidence for involvement of Coxsackie B enteroviruses in the etiology of type 1 diabetes. It has proven difficult to detect consistent differences between diabetic children and controls in enterovirus exposure at time of *onset* of diabetes, as some studies produce negative results (e.g. Tuvemo *et al.*, 1989) while others produce positive results (e.g. Nairn *et al.*, 1999). However, recent studies have shown increased frequencies of enterovirus antibodies during the preclinical phase in children who *subsequently* became diabetic (Roivainen *et al.*, 1998) and in mothers at time of delivery of children who subsequently became diabetic (Dahlquist *et al.*, 1995). In addition, Hiltunen *et al.* (1997) showed temporal association between appearance of anti-viral antibodies and islet cell autoantibodies in siblings of type 1 diabetics. Similarly, Juhela *et al.* (1999) reported that non-diabetic children with ICA or GAD autoantibodies demonstrated stronger T cell responses to enterovirus proteins than did children without those autoantibodies. Thus, there is now convincing evidence for a relationship between enterovirus infection and autoimmune diabetes in humans, although a causal role remains to be demonstrated (Hyoty *et al.*, 1998). Molecular mimicry with GAD antigen has been invoked as the mechanism underlying association of enterovirus infection with diabetes (Maclaren and Alkinson, 1997; Vreugdenhil *et al.*, 1998), but this has been disputed (Horwitz *et al.*, 1998).

2.2 Familial aggregation and genetic etiology

The prevalence of type 1 diabetes is highest in individuals of European ancestry, and unless otherwise specified, this is the reference population assumed in the discussions which follow. Family and twin studies clearly show that type 1 diabetes has a genetic basis, and studies of specific predisposing HLA genes (see below) suggest that genetic susceptibility to the disease occurs in most (if not all) affected individuals. However, additional (non-genetic) factors presumably are required to precipitate the disease in these genetically-susceptible persons, since studies of monozygote twin pairs in which one twin has type 1 diabetes have shown that less than 100% of the co-twins develop diabetes (i.e. the 'concordance rate' is less than 100%). It is unknown whether there are any 'purely genetic' cases of type 1 diabetes, in which non-genetic factors are unnecessary. There are no documented families with Mendelian subforms of type 1 diabetes (as exist, for example, in non-insulin-dependent diabetes).

The genetic basis of type 1 diabetes is demonstrated by many studies showing an increased frequency of the disorder in close relatives of diabetics compared with the frequency in the general population. For example, the frequency in siblings of diabetics is about 6% by age 30 (reviewed in Thomson *et al.*, 1988) compared with the prevalence in the general population of about 0.4% by age 30 (Spielman *et al.*, 1980). Thus, type 1 diabetes is about 6/0.4 = 15 times more common in siblings of diabetics than in the general population (this ratio between risk to siblings versus prevalence in the general population is referred to as λ_{sib} [Risch, 1987]). When siblings are followed to an older age, they have an even higher risk of developing diabetes (Lorenzen *et al.*, 1994 cited 10% by age 60). The frequency of diabetes is also elevated in children of type 1 diabetics, and there is an intriguing difference in risk depending on the sex of the affected parent: the risk to the child is higher if the diabetic parent is the father compared to if the diabetic parent is the mother (e.g. 6% compared to 1% by age 20, for children of diabetic fathers versus mothers, respectively, in the study by Warram *et al.*, 1984). More recent studies have consistently shown a 2–3 times increased risk to children of diabetic fathers compared with diabetic mothers (9% versus 3% by age 20, Bleich *et al.*, 1993; 8% versus 4% by age 20, Tuomilehto *et al.*, 1995). Interestingly, this difference has also been observed with respect to the frequency of diabetes-related autoantibodies in early childhood: children of diabetic fathers were positive for at least one autoantibody more often than children of diabetic mothers (at 9 months of age, 9% versus 4%, and at 2 years, 17% versus 8%; Roll *et al.*, 1996). Several mechanisms might explain this sex-of-parent effect, including: (i) selective pregnancy loss of diabetes-prone fetuses of diabetic mothers, (ii) fetal protection against diabetes if mother diabetic, and (iii) imprinting, such that diabetogenic genes are expressed differently when derived from fathers versus mothers. Warram *et al.* (1988) presented evidence against mechanism (i) by demonstrating no relation between rates of abortion or perinatal death and frequency of diabetes in children of diabetic mothers. Bleich *et al.* (1993) found that a reduced diabetes risk was observed in offspring of diabetic mothers only for mothers who developed their own diabetes at an older age (>8 years old). If there are genetic differences between early- and late-onset diabetics, some of the genes predisposing to later-onset disease may be imprinted (mechanism (iii)). Further studies aimed at understanding the differences in risk between children of diabetic mothers and fathers may lead to important clues about the etiology of type 1 diabetes.

Close relatives of diabetics (such as siblings and children) also share common environmental factors, so it could be argued that environment, not heredity, accounts for the higher frequency of type 1 diabetes among relatives of diabetics. Studies comparing diabetes frequencies in monozygotic twins (MZ, 100% shared genes) versus dizygotic twins (DZ, 50% shared genes) are more definitive, since these two types of twins probably experience a comparable degree of similarity in their pre- and post-natal environments. These studies have routinely shown that MZ twins have a higher 'concordance rate' (frequency of diabetes in co-twin of diabetic individual) than DZ twins (Barnett *et al.*, 1981; Kyvik *et al.*, 1995; Olmos *et al.*, 1988), clearly demonstrating that genetic factors are involved in type 1 diabetes. The frequency of diabetes in MZ co-twins of diabetics also gives some indication of the degree of genetic versus nongenetic determination of a disorder. Olmos *et al.* (1988) found that only 34% of MZ co-twins of individuals with type 1 diabetes were also diabetic by age 30. The average penetrance (probability of an individual being affected given that he/she has a susceptible genotype) can be calculated as follows: penetrance = MZ concordance rate/(MZ concordance rate + 0.5[MZ discordance rate]) (Spielman *et al.* 1979). Thus, an MZ concordance rate of 34% corresponds to an average penetrance of 50%, suggesting that nongenetic factors must also strongly influence the etiology of type 1 diabetes. These nongenetic factors could include environmental agents (viruses or other infectious organisms; cow's milk proteins [Levy-Marchal *et al.*, 1995]) and/or stochastic processes operating during the development of the immune system or growth of the pancreatic β cell mass. However, a recent Danish population-based study estimated a much higher concordance rate of 70% in MZ twins and 13% in DZ twins by age 35 (Kyvik *et al.*, 1995). An MZ concordance rate of 70% corresponds to an average penetrance of about 82%, suggesting that genetic factors may be more important than previously suspected.

Since autoantibodies to various islet cell components arise early in the long preclinical phase of type 1 diabetes, it is of interest to know whether these autoantibodies show evidence for familial aggregation and genetic determination. In the general population, the frequencies of ICA, IAA, and anti-GAD are quite low – for example, 3–4% in a recent study of Swedish children (Hagopian *et al.*, 1995). The frequencies among siblings of type 1 diabetics are higher but not strikingly so – for example, 6% of siblings were ICA+ in a large study by Lipton *et al.* (1992). Again, the higher frequency of diabetes-related autoantibodies in siblings of individuals with type 1 diabetes could be due to shared environmental factors, so twin studies are crucial. Petersen *et al.* (1997) found that among nondiabetic co-twins of type 1 diabetics, the frequencies of ICA, IAA and anti-GAD were *not* higher in MZ co-twins (20%, 50%, 40% for the three antibody types) than in DZ co-twins (26%, 49%, 40%), suggesting that islet cell autoimmunity is environmentally determined. In contrast, Redondo *et al.* (1999) detected higher frequencies of islet autoantibodies among MZ non-diabetic siblings (42%) than among DZ non-diabetic siblings (20%), nontwin nondiabetic siblings (11%), or unrelated controls (6%). While the latter results do suggest genetic contribution to islet cell autoimmunity, the difference in autoantibody frequencies between nondiabetic DZ twin siblings (20%) and nondiabetic nontwin siblings (11%), both sharing an average of 50% of genes with the diabetic, also points to a possible role for shared environment. However, caution must be used in interpreting differences in frequencies of detected autoantibodies, since they may fluctuate over time. For example, Wilkin *et al.* (1985) took two blood samples one or more years apart from 11 discordant (nondiabetic) MZ twins of diabetics and found IAA in all 11 nondiabetics (in only one sample in 10/11 and in both samples in the remaining nondiabetic twin). Similarly, Verge *et al.* (1995) found that in a sample of 12 discordant MZ twins of diabetics who had remained nondiabetic for 8–39 years, 66% (8 of 12) nevertheless had autoantibodies and/or evidence of β cell damage. Studies such as these suggest that the MZ twin concordance rate for islet cell autoimmunity may be higher than for overt diabetes. Among siblings, presence of antibodies to islet cell components (especially multiple autoantibodies at high titers) is strongly associated with risk of progressing to clinical diabetes (e.g. Bonifacio *et al.*, 1990). Some of the genes predisposing to type 1 diabetes may also contribute genetic susceptibility to initial islet cell autoimmunity. Thus, it may be worthwhile to genetically analyze the 'islet cell autoimmune' phenotype, rather than just the 'diabetic' phenotype, in order to understand the genetics of the earliest stages of type 1 diabetes. This will be discussed in greater detail below.

The age at diagnosis of overt disease may define clinical subtypes of type 1 diabetes influenced by different predisposing genes. A recent study reported significant correlations for age at diagnosis in type 1 diabetic MZ twins ($r = 0.94$) and diabetic non-twin siblings ($r = 0.53$), suggesting that most of the variability of age at diagnosis is genetically determined (Fava *et al.*, 1998). It is now recognized that adults who present after age 30 with apparent non-insulin-dependent diabetes, which later converts to insulin-dependency, may have 'latent autoimmune diabetes in adults' (LADA). This slow-onset form of diabetes is characterized by presence of GAD autoantibodies, and HLA diabetes-predisposing gene frequencies that are lower than found in classical childhood-onset diabetics but higher than found in the general population (Tuomi *et al.*, 1999; Zimmet *et al.*, 1994). It will be of interest to determine which genes identified as susceptibility genes for childhood-onset type 1 diabetes are also relevant to LADA, and conversely, whether there are LADA susceptibility genes that do not affect susceptibility to type 1 diabetes.

2.3 Environmental factors and gene-environment interaction

As indicated above, there is strong circumstantial evidence that infection with certain viruses increases risk of developing diabetes. However, paradoxically, there is also evidence that an environment providing early and frequent exposure to pathogens may decrease the risk of type 1 diabetes. Thus, epidemiologic studies over small geographic areas in Britain have shown significantly lower incidences of type 1 diabetes in regions with high population density and household overcrowding (Staines *et al.*, 1997). In some countries, the incidence of type 1 diabetes is increasing over time (e.g. in Denmark: Green *et al.*, 1996), which could conceivably be related to general improvement in public health conditions. If the pathogenesis of diabetes begins in very early life, then the immune status (susceptibility to infection) of the mother during pregnancy may be as relevant as the immune status of her diabetes-at-risk offspring. It follows that in order to understand the genetics of type 1 diabetes, researchers will have to include analysis of the effects of maternal genotype. Very few studies of this nature have been conducted.

In addition to viral infections, exposure to cow's milk has been implicated in the etiology of type 1 diabetes. This has been shown by positive correlation between cow's milk consumption and incidence of autoimmune diabetes. For example, Finland has the highest incidence of type 1 diabetes and the highest consumption of cow's milk in the world, and even within single countries these two variables have been significantly correlated (e.g. Italy: Fava *et al.*, 1994). A relationship between diabetes and cow's milk has also been shown by studies demonstrating higher levels of antibodies to cow's milk proteins (such as bovine serum albumin) in children with type 1 diabetes compared with control children (e.g. Levy-Marchal *et al.*, 1995; Savilahti *et al.*, 1993). Finnish investigators have also found that high milk consumption is associated with emergence of pancreatic islet cell autoantibodies and progression to clinical diabetes in siblings of type 1 diabetics (Virtanen *et al.*, 1998). A recent Finnish study suggested that bovine insulin might be the culprit protein in cow's milk, since children exposed to cow's milk before 3 months of age had significantly higher levels of bovine insulin antibodies (which cross-reacted with human insulin) than children who were exclusively breast-fed (Vaarala *et al.*, 1999). However, not all studies have been able to show a relationship between cow's milk and type 1 diabetes, and it is unclear how consumption of cow's milk could be causally involved in diabetes pathogenesis. Kolb and Pozzilli (1999) in their review of the topic suggested that the relationship is indirect, with early exposures to immunogenic dietary factors, including cow's milk proteins, modulating the maturation of the gastrointestinal (GI) immune system. The status of the GI immune system could then influence tolerance to self-antigens and/or susceptibility to infection with potentially-diabetogenic enteroviruses. Pococco *et al.* (1997) reported that antibodies to *Helicobacter pylori* were found significantly more often in diabetic children (20%) than in control children (5%; $P < 0.001$), and that among diabetics, antibodies to *H. pylori* were significantly associated with antibodies to cow's milk protein. The authors suggested (i) that *H. pylori* infection may be increased in diabetics due to possession of HLA genes that predispose both to *H. pylori* infection and to type 1 diabetes, and (ii) that *H. pylori* infection leads to abnormal permeability of gastric mucosa, allowing absorption of large immunogenic milk proteins.

In summary, the nature of the relationship between type 1 diabetes and putative environmental etiologic factors, such as enteroviruses and cow's milk proteins, is still very unclear. Nevertheless, it is anticipated that studies of interaction between genetic factors and these environmental agents will intensify in the near future. These studies have already begun. For example, Finnish researchers recently showed that among diabetic-nondiabetic sibling pairs who were HLA-identical, the diabetic siblings still had significantly higher levels of cow's milk antibodies than the nondiabetic siblings, suggesting that HLA and cow's milk antibodies are independent risk markers for diabetes (Saukkonen *et al.*, 1998).

3. Linkage and association studies

3.1 Analytical methods

A variety of methods have been used to attempt to locate the genes which family and twin studies (such as those discussed above) have concluded must be involved in type 1 diabetes. The primary methods are linkage analysis and association (linkage disequilibrium) analysis. This topic is dealt with in greater detail elsewhere in this volume, but a brief summary will be presented here. Highly polymorphic 'marker' genes (not expected to be the true disease-susceptibility genes) are employed to test for linkage in 'genome screen linkage studies'. However, 'candidate gene studies' use marker variants near or within genes having a possible causal role in disease to test for both linkage and association. In the future, very dense maps of markers will be used to conduct 'genome screen association studies'.

Linkage analysis is a method to determine if there is significant evidence for co-segregation of alleles at a marker locus and alleles at a hypothetical disease-susceptibility locus. In *lod score linkage analysis*, the probability of co-segregation (linkage) between alleles at marker and disease loci is compared to the probability of no linkage, and the log of this odds ratio is taken to derive the 'lod' (log of odds) score. By convention, when analyzing a simple single-locus disorder, a lod score $\geq$ 3.0 is considered significant evidence for linkage (as will be seen below, this criterion is rarely met when analyzing complex multigenic traits such as type 1 diabetes). Lod score linkage analysis requires the user to specify parameters for the hypothetical disease locus (e.g. mode of inheritance, penetrance, disease allele frequency), so this method is often called 'model-dependent'. Since the mode of inheritance is unknown for multigenic disorders, model-independent methods have been advocated for linkage analysis of such disorders. One 'model-independent' method is *affected sibpair linkage analysis*, in which the mean proportion of marker alleles shared by pairs of affected siblings is compared with the 50% shared proportion expected by random Mendelian segregation. Significantly increased marker sharing in affected sibpairs suggests that the region containing the marker locus also contains a disease susceptibility locus. The observed proportion of marker sharing in affected sibpairs can also be converted to a maximum lod score equivalent (Davies *et al.*, 1994; Field *et al.*, 1996; Risch, 1990), referred to as an MLS.

Association analysis is a method to determine whether *particular* marker alleles are more prevalent in disease patients compared with control subjects. Linkage only requires that particular marker alleles co-occur (co-segregate) with disease *within* each family (different families can have different marker alleles segregating with disease). Association requires that particular marker alleles tend to co-occur with disease across the entire population (different families have the same marker alleles co-occurring with disease). In traditional 'case-control' (*population-based*) tests of association, marker allele frequencies in unrelated patients and unrelated controls are compared. Identification of a significant difference between cases and controls suggests that the marker locus is close to (generally within 1–2 cM of) a disease susceptibility locus, with the association between marker alleles and disease being caused by *linkage with linkage disequilibrium* between the marker and disease loci in the population. However, this traditional method can produce spurious association if the cases and controls are not accurately matched for ethnic background (not drawn from the same ethnic population), since different ethnic groups may have differing frequencies of both disease and marker alleles. This problem is especially relevant to studies of North Americans and other mixed populations, even when restricted to persons of European ancestry ('Europeans' are not genetically homogeneous). To avoid the possibility of spurious association resulting from mismatched controls, association analysis methods have been developed which are *family-based* rather than population-based. AFBAC (for Affected-Family-BAsed Controls; Thomson, 1988) is one such family-based association method. Suggested by Field *et al.* (1986) and Field (1989), and more fully developed by Thomson (1995), this method involves comparing the frequency distributions of alleles transmitted versus not transmitted from parents to the affected child. Statistical significance is assessed by a $(2 \times n)\ \chi^2$ test, where n is the number of different marker alleles (rare alleles can be lumped to avoid small expected cell values). The nontransmitted ('control') alleles form an unbiased sample of genes from the general population (Thomson, 1995). If association is detected, differences in maternal versus paternal contributions to susceptibility can be tested by comparing alleles transmitted from mothers versus from fathers. In families with two affected children, the AFBAC computer program allows one to either: (i) weight the transmissions to affected children such that alleles transmitted to two affected are given twice the weight of alleles transmitted to only one affected (this 'multiplex' technique also produces unbiased control allele frequencies) or, (ii) consider each diabetic child independently as if it were in its own single-affected-child family (this 'double simplex' technique is more powerful for detecting association, but produces [conservatively] biased control allele frequencies). Another family-based association test is the 'TDT' (transmission disequilibrium test) of Spielman *et al.* (1993). Unlike AFBAC, the TDT analyzes only marker alleles transmitted from heterozygous parents, testing the observed frequency of transmission of a particular allele with the 50% expected by random Mendelian segregation. Very recent strong admixture effects may adversely influence the AFBAC method (Thomson, 1995). However, we prefer AFBAC to screen for disease-marker association, since: (i) by using all four parental alleles, it is a more powerful test than TDT; (ii) nontransmitted alleles can be used to estimate population gene frequencies (e.g. for use in linkage analyses); (iii) it simultaneously considers all marker alleles in one $(2 \times n)$ statistical test, avoiding the problem of multiple comparisons created when each allele is tested separately; (iv) if recent strong admixture is suspected on historical grounds, another test can be employed.

There is one potential *disadvantage* of performing family-based, rather than case-control, association testing. This is: if disease association is stronger with specific marker *genotypes* rather than specific marker alleles, the effect may be more difficult to detect examining only transmitted and non-transmitted alleles. Therefore, any disease association detected with family-based methods should be followed up with case-control genotype studies from ethnically homogeneous populations.

3.2 HLA region predisposing genes (IDDM1)

For more than 25 years, it has been known that certain genotypes at the human leukocyte antigen (HLA) loci render an individual at increased risk of type 1 diabetes (Cudworth and Woodrow, 1974; Nerup *et al.*, 1974; Singal and Blajchman, 1973). HLA molecules participate in the binding and presentation of foreign (or self) antigens to T cells, and therefore have a critical role in the regulation of the immune response system. The large amounts of data supporting both association and linkage of type 1 diabetes to genes in the HLA region cannot be discussed in detail in this chapter, only briefly summarized (for review, see Cucca and Todd, 1996). Many studies have shown that diabetics of European ancestry have higher frequencies of HLA-DR3 and DR4 (variants at the class II HLA-DRB1 locus on chromosome 6p21.3) compared to the general population – for example, 96% of IDDM patients from Alberta, Canada had at least one of these alleles (45% expected), and, even more dramatically, 38% had *both* alleles (3% expected) (Field and McArthur, 1987). This excess of DR3/DR4 heterozygotes, relative to either homozygote, led to the hypothesis that there were at least two *different* HLA region susceptibility alleles with synergistic actions, either DR3 and DR4 themselves or genes in disequilibrium with them. Nepom *et al.* (1986) showed that among DR4 haplotypes found in diabetics, there was an increased frequency of DQB1*0302 at the nearby HLA-DQB1 locus, which suggested that DQB1 rather than DRB1 alleles might be exerting the primary effects on IDDM predisposition. Similarly, Todd *et al.* (1987) showed that several HLA haplotypes positively associated with IDDM (including DR4-DQB1*0302) encode an amino acid *other than aspartate* at position 57 of the DQB1 chain, again implying that DQB1 was the primary susceptibility locus. However, Sheehy *et al.* (1989) showed that DR4 haplotypes encoding *both* the DRB*0401 subtype of DR4 *and* DQB*0302 were more diabetogenic than DR4 haplotypes encoding only one of these – thus, DRB1 and DQB1 together may confer susceptibility. Many studies have shown that DR2 haplotypes carrying DRB1*1501 and DQB1*0602 confer (apparently dominant) protection against diabetes. HLA-DQA1 alleles have also been implicated in susceptibility (Khalil *et al.*, 1990; Todd *et al.*, 1989), suggesting that the whole DQ molecule (dimers of DQA1 and DQB1) determines predisposition. Huang *et al.* (1995) demonstrated that type 1 diabetes in Chinese is determined by combinations of predisposing and protecting DRB1, DQA1 and DQB1. The mechanism by which HLA molecules produce predisposition to or protection from autoimmunity is not yet understood. One recent model is that protective HLA-DR and DQ molecules bind diabetogenic antigens with high affinity and allow thymic negative selection of self-reactive T cells, while susceptible HLA molecules bind with low affinity and allow escape into the periphery of autoreactive T cells (Ridgway and Fathman, 1998). This model would account for the dominance of protective alleles. In addition to genes at HLA loci, there is also evidence in both humans and animal models that additional (? non-HLA) genes within the HLA region contribute independently to type 1 diabetes susceptibility (Hattori *et al.*, 1999; Lie *et al.*, 1999; Yan *et al.*, 1997). Some studies have found associations between diabetes and genes within the HLA class II region which are involved in antigen processing – e.g. TAP2 (Caillat-Zucman *et al.*, 1993), and LMP1 and LMP7 (Deng *et al.*, 1995). However, subsequent studies demonstrated that the associations may be secondary to linkage disequilibrium of these genes with known HLA susceptibility alleles (Caillat-Zucman *et al.*, 1995; Cucca *et al.*, 1994; Ronningen *et al.*, 1993; Undlien *et al.*, 1997). In summary, the precise identities of the HLA region predisposing genes are not yet fully delineated. It appears that different combinations of HLA-DRB, DQA, DQB alleles may confer varying degrees of susceptibility/resistance and that as-yet-unidentified nearby non-HLA genes may also be involved.

All of the HLA investigations discussed above were *population-based* association studies. It has also been conclusively shown by numerous *linkage studies* that a gene or genes in the HLA region influence IDDM susceptibility – for example, Payami *et al.* (1985) presented a compilation of literature data showing that among 538 diabetic sibling pairs, 54% shared two HLA haplotypes and 7.3% shared zero haplotypes, compared with the 25% expected in each category (highly significant differences). From this data, it can be calculated that the increased risk to siblings (λ_{sib}) attributable to HLA region genes is about 3.4 (ratio of expected to observed sharing of zero haplotypes in siblings = 0.25/0.073 = 3.4; Risch, 1987). Since the total λ_{sib} is about 15 (see above), the HLA contribution to total familial clustering of type 1 diabetes is about 44% (assuming that λ_{sib} values are multiplicative, 15/3.4 = 4.4 and 3.4/[3.4 + 4.4] = 44%). Thus, even prior to the initiation of the genome screen and candidate gene studies described below, there was theoretical evidence for the existence of loci predisposing to type 1 diabetes located *outside* the HLA region.

Insights from HLA (IDDM1) studies. Even when a region containing strongly-predisposing genes has been clearly identified, isolating the one or more genetic variants that produce disease susceptibility can be difficult and require many years of research.

3.3 Insulin gene region predisposition (IDDM2)

Significant association between type 1 diabetes and a variable number tandem repeat (VNTR) marker in the 5′ flanking region of the insulin gene (*INS*) on chromosome 11p15 was first detected about 15 years ago: the short repeat 'class I' alleles were increased in diabetics compared with controls (e.g., 88% versus 67%, Bell *et al.*, 1984). Paradoxically, researchers could *not* find significant evidence for increased gene sharing (linkage) of *INS* region markers in pairs of diabetic siblings (Cox and Spielman, 1989; Field, 1989). Even the largest linkage study of type 1 diabetes to date (616 families; Concannon *et al.*, 1998) could not detect significant evidence for linkage to this region (maximum lod score 0.6). However, the *INS* region association was verified using the AFBAC family-based method of association analysis (Field, 1991; Thomson *et al.*, 1989), demonstrating that the population-based finding of association was not spuriously created by use of ethnically-mismatched controls. The frequencies of class I alleles transmitted and not transmitted to diabetic children using family-based analysis (e.g. 89% versus 71%, Field, 1991) were very similar to those seen in random diabetics and unrelated controls. One of the most interesting aspects of the *INS* region studies is that they clearly demonstrate that significant association (linkage disequilibrium) can be detected in the absence of significant evidence for linkage (Field and Nagatomi, 1994). In other words, providing that the marker being tested is at or near the predisposing genetic locus (as would be done, for example, in candidate gene studies), association analysis is more powerful than linkage analysis for detecting the effects of this susceptibility locus. The same conclusion has more recently been clearly stated by Risch and Merikangas (1996) in a theoretical study. In summary, it is very unlikely that the contribution of the insulin gene region to diabetes susceptibility would have been detected by a random genome linkage screen. It was detected because a specific candidate locus (insulin) was examined by association analysis using genetic variants within or immediately adjacent to the locus. As argued by Risch and Merikangas (1996), most genes predisposing to genetically-complex traits such as type 1 diabetes will likely be located using this more powerful association approach.

The reason for the association between type 1 diabetes and the insulin gene region is still unknown, despite intensive effort. Molecular studies suggested that the 5′ VNTR itself might directly influence insulin gene expression (Catasti *et al.*, 1996; Hammond-Kosack *et al.*, 1992). However, studies of the relationship between insulin expression and VNTR alleles were conflicting. Some studies showed increased insulin gene transcription in the pancreas to be associated with high risk class I VNTR alleles (Bennett *et al.*, 1995; Lucassen *et al.*, 1995), while another study showed increased pancreatic insulin expression to be associated with protective class III alleles (Kennedy *et al.*, 1995). Two subsequent studies have shown increased insulin expression in the thymus also to be associated with protective class III alleles (Pugliese *et al.*, 1997; Vafiadis *et al.*, 1997). These latter investigators suggested that higher levels of (pro)insulin in the developing thymus could promote induction of immune tolerance towards (pro)insulin and thereby reduce predisposition to diabetes-related autoimmunity. However, this suggestion has been contested by Sarugeri *et al.* (1998), who found no relationship between T cell or antibody responses to insulin and VNTR genotype in type 1 diabetics. Insulin autoantibodies, found in about 40% of newly diagnosed diabetics (e.g. Schoenfeld *et al.*, 1988), were also found *not* to be associated with the INS region VNTR in a study of 105 new diabetics (L. Field and J. Palmer, unpublished). The insulin-like growth factor 2 (*IGF2*) locus is tightly linked to *INS*, downstream of the VNTR, but Vafiadis *et al.* (1998) reported no association between the VNTR and lymphocytic *IGF2* gene expression. In summary, it appears that the 5′ INS VNTR does influence insulin transcription, but the mechanism by which this modifies susceptibility to type 1 diabetes is still unclear.

Insights from INS (IDDM2) studies. (i) Predisposing genes that are readily detectable by association analysis may not show significant evidence for linkage and would be overlooked in a standard linkage genome screen; (ii) Even when a very small region has been identified as containing a diabetes-predisposing gene and excellent candidate genes exist, it may be very difficult to determine which genetic sequence is responsible for predisposition, and the mechanism of its effect; (iii) Despite the high prevalence of autoantibodies to insulin at time of onset of diabetes, there does not appear to be a relationship between presence of such antibodies and genotype at the *INS* region.

3.4 Predisposing genes outside the HLA and insulin gene regions

Within the past 5 years, a number of centers throughout the world have performed genome-wide linkage screens using families containing multiple siblings affected with type 1 diabetes (multiplex families). These studies have resulted in the identification of more than 15 regions containing putative type 1 diabetes-predisposing genes. Most of these localizations have been assigned unique 'IDDM' gene symbols by the Human Gene Mapping Nomenclature Committee, including *IDDM3* through *IDDM17*, but several other regions demonstrating linkage have not been assigned *IDDM* gene symbol designations (these are also discussed below).

IDDM3 was localized by linkage to marker D15S107 on chromosome 15q26 (maximum lod score of 2.5; affected sibpair P value = 0.001) by Field *et al.* (1994) using 250 families from Canada, the US, and Britain with two or more diabetic children. This localization was confirmed independently by Luo *et al.* (1995) in 31 American families (affected sib pair P = 0.020), although they later reported failure to detect *IDDM3* in a sample of Italian families (Luo *et al.*, 1996). Confirmation was also reported by Zamani *et al.* (1996) in 81 Danish families (affected sibpair P value = 0.010). Interestingly, in the data of Field *et al.* (1994), most of the evidence for linkage to D15S107 was contained in families showing no linkage to HLA region genes, suggesting that the *IDDM3* susceptibility locus and HLA region predisposition act in a biologically independent manner.

Linkage has also been reported to the *IDDM3* region on chromosome 15q26 for celiac disease (Houlston *et al.*, 1997; Zhong *et al.*, 1996), a disorder associated with HLA-DR3 (DRB1*03-DQB1*02), as is type 1 diabetes, and with the CTLA-4 locus (*IDDM12*) (Djilali-Saiah *et al.*, 1998). Thus, in addition to HLA region genes, it is possible that some non-HLA diabetes-predisposing genes will be shared with other autoimmune/inflammatory disorders, reflecting common pathogenic pathways (Becker 1999).

To date, the largest genome-wide linkage screens for type 1 diabetes genes are: (i) that of Concannon *et al.* (1998) who studied two sets of families (Set 1: 101 British, 65 American, 22 Philadelphian, 188 total; Set 2: 149 British, 279 American, 428 total) for a grand total of 616 families, and (ii) that of Mein *et al.* (1998) who studied 356 British families (250 of which were apparently also included in Concannon *et al.*). There was significant evidence for linkage to the *IDDM3* region in family set 1 of Concannon *et al.* (as expected, since there was a large overlap with the families analyzed by Field *et al.*, 1994), but there was no evidence for linkage in their family set 2, or in the families of Mein *et al.* What does it mean when larger datasets fail to detect linkages which more than one smaller dataset has shown? In fact, considering linkages detected with MLS equivalents of 1.0 or greater, the *only overlap* between the studies of Concannon *et al.* and Mein *et al.* was linkage to the HLA region (*IDDM1*) and to the chromosome 6q25-q27 region (*IDDM5-IDDM8*, see below). The question is now rephrased to: What does it mean when two large datasets show minimal overlap in regions showing linkage? It is now uncomfortably clear that there is linkage heterogeneity between datasets for *most* of the detected non-HLA diabetes predisposing loci. This may be due to:

(i) lack of true linkage (no diabetes predisposing gene in the region);
(ii) sampling variation (successful replication of weak linkage is more difficult than initial detection of the linkage);
(iii) population genetic variation (differing frequencies of susceptibility alleles in different ethnic groups);
(iv) population environmental variation (differing penetrances of susceptible genotypes dependent on variable interacting environmental factors); or
(v) combinations of (ii), (iii) and (iv).

If heterogeneity were only due to sampling variation, then pooling families to create larger sample sizes would ameliorate the problem, while if heterogeneity is also due to genetic and gene-environment variation, then pooling families from a variety of ethnic backgrounds and geographic locations (e.g. as in Concannon *et al.*, 1998) may be a poor approach for detecting the effects of any other than the strongest predisposing genes common to most populations. What is needed are reliable methods for confirming the existence of minor predisposing genes; large studies from mixed sources may not be the best approach. Field and colleagues propose that special populations (e.g. young populations which may show greater linkage disequilibrium between marker and disease loci) may be uniquely valuable for verification of putative predisposing loci detected by linkage genome screens, and are currently in the process of analyzing French Canadian families with type 1 diabetes for that purpose.

Insights from IDDM3 studies. (i) Weak linkages to putative IDDM genes are difficult to verify and may require more sensitive methods (e.g. association analysis in special populations) to confirm their existence; (ii) linkage and/or association analysis with other autoimmune disorders (possibly showing stronger effects than for diabetes) may assist in fine mapping the responsible genes.

IDDM4 was reported simultaneously by three research groups (Davies *et al.*, 1994; Field *et al.*, 1994; Hashimoto *et al.*, 1994) by linkage to the fibroblast growth factor 3 (*FGF3*) locus on chromosome 11q13. Davies *et al.* and Hashimoto *et al.* found linkage to *FGF3* during random genome screens. In contrast, Field *et al.* found evidence for linkage to *FGF3* (maximum lod score 1.1, affected sibpair

P = 0.004) while examining 11q13 as a candidate region for diabetes predisposition, since a gene for atopy had been reported between *FGF3* and *FCER1B* (8 cM proximal to *FGF3*). Subsequently, Luo *et al.* (1995, 1996) reported that maximum linkage was obtained with D11S1337, a marker about 3 cM proximal to *FGF3* (MLS 3.9). Similarly, Nakagawa *et al.* (1998), analyzing 707 ethnically-heterogeneous families, obtained a maximal MLS of 2.7 at D11S1337. However, MLS values of at least 1.7 extended over a 15 cM region (Nakagawa *et al.*, 1998). This type of low but broad curve is characteristic of linkage signals for genes predisposing to genetically-complex traits. Field and Tobias (1997), using the AFBAC program, reported significant linkage disequilibrium (*P* = 0.012) with D11S480, located about 6 cM proximal to *FGF3*. Recently, Nakagawa *et al.*, (1998) analyzed markers across the 15 cM linked region for evidence of disequilibrium and found significant association with D11S1917 (*P* = 0.01), very close to D11S1337. A second marker which they then isolated near D11S1917 was also significantly associated (*P* = 0.003), as were some haplotypes formed using the two markers. An interesting feature of the association analyses of Nakagawa *et al.* (1998) is that they also examined transmissions and non-transmissions from parents to *unaffected* siblings, presumably to detect protective rather than predisposing effects, and not as a check for segregation distortion independent of disease. While this approach has not proved very powerful for diabetes linkage studies (i.e. analysis of non-sharing of genes by discordant sibpairs), it may be more useful in association studies.

Insights from IDDM4 studies. (i) Even markers showing only moderate evidence for linkage by conventional standards (lod scores only about 3) may be able to detect significant association (linkage disequilibrium) in ethnically heterogeneous datasets; (ii) however, until a susceptibility locus is actually isolated, it is unclear how powerful and accurate association effects will have been in assisting fine-mapping of that locus.

IDDM5 near the estrogen receptor (*ESR*) locus on chromosome 6q25 was reported by Davies *et al.* (1994) and confirmed in an independent dataset by Luo *et al.* (1996; MLS 4.5). In the study by Concannon *et al.* (1998), their first set of 188 families showed significant evidence for linkage (this dataset overlapped with that of Davies *et al.*, 1994), but the second larger set of 428 families showed no evidence for linkage. Their total dataset of 616 families produced an MLS of only 1.46 at *ESR*. This illustrates once again the dilemma of nonreplication, even in a large dataset, although more than one independent report has previously demonstrated linkage. Some evidence for association has been reported for one marker near *ESR* (Davies *et al.*, 1996), but it is unclear if these results are significant after correction for multiple tests performed on 15 marker loci.

IDDM6 on chromosome 18q21 has an interesting history. Back in the 'good old days' prior to the advent of abundant DNA-based markers, evidence for linkage of type 1 diabetes to the Kidd (Jk) blood group locus (lod score 2.8) was detected in a random marker study utilizing a limited (then large) set of 27 serum protein and blood group loci (Hodge *et al.*, 1981). This result generated much excitement among the investigators, but (the now routine) follow-up by typing additional markers could not be performed since they were nonexistent (the Jk locus was not even accurately mapped; it was then thought to be on chromosome 2, but is now known to be on 18q). An independent set of diabetes families (Dunsworth *et al.*, 1982) showed no linkage to Jk. It was not until 1994, in the 'DNA era', that linkage to 18q was again detected (Davies *et al.*, 1994; MLS 1.1). Subsequently, Merriman *et al.* (1997) demonstrated evidence for both linkage and association with markers near D18S487 (about 13 cM from Jk), with an MLS of 1.6 in 722 multiplex families and an association *P* value of 0.003 in 1457 families from various ethnic groups. However, it appeared that the evidence for association was highly heterogeneous, since some groups showed no association and others strong association. As with linkage studies, then, we are encountered with the question: What does it mean when differing datasets show different results for association testing?

Insights from IDDM6 studies. Even markers showing weak evidence for linkage by conventional standards (lod scores about 1.5) may be able to detect significant association.

IDDM7 on chromosome 2q31 was reported independently by two groups, Owerbach and Gabbay (1995) by linkage to the HOXD8 locus (MLS 2.6) and Copeman *et al.* (1995) by association and linkage with marker D2S152 (MLS 1.3). Antibodies to GAD are present in the majority of diabetics at time of diagnosis, arise early in the preclinical phase of IDDM and can even be detected in cord blood of some infants who subsequently become diabetic (see above), leading some investigators to suggest that GAD could be the initiating target of autoimmunity (Kaufman *et al.*, 1992). *GAD1*, which encodes GAD_{67}, has been mapped 2q31 (Bu *et al.*, 1992) and thus is a candidate for *IDDM7*. However, Rambrand *et al.* (1997) reported no evidence for linkage to markers spanning *GAD1* in 58 Danish diabetic sibpairs, and no evidence for association between *GAD1* genotype and presence of anti-GAD antibodies.

IDDM8 was initially localized to 6q25-q27, more than 28 cM distal to *IDDM5*, by linkage to D6S264 (MLS 2.0) and to D6S446 (MLS 2.8) by Luo *et al.* (1995). Davies *et al.* (1994) also reported some evidence for linkage to marker D6S264. More recently, Luo *et al.* (1996) reported strongest linkage to D6S281, near D6S446, in their total dataset of 265 American and Italian families (MLS 3.6), and Davies *et al.* (1996) reported strongest linkage to a marker closer to D6S264 in their total 299 British families (MLS 1.5). It therefore appears that there are two distinct susceptibility loci on the distal long arm of chromosome 6, *IDDM5* and *IDDM8*. It is unclear which locus confers larger effects on diabetes susceptibility (in Luo *et al.*, 1996, it was *IDDM5*, while in Davies *et al.*, 1996 it was *IDDM8*). Ironically, in the largest dataset yet studied (616 British and American families in Concannon *et al.*, 1998), evidence for linkage to both loci was not strong (MLS 1.8 for *IDDM5* and MLS 1.3 for *IDDM8*). However, as noted above in the discussion of *IDDM3*, the genome screen of Concannon *et al.* only revealed substantial overlap (lod > 1.0) with the second largest genome screen (Mein *et al.*, 1998) in three regions – HLA, *IDDM5* and *IDDM8*.

In an effort to identify genes that might be involved in the earliest stages of the pathogenesis of type 1 diabetes, we have been analyzing gene sharing in sibpairs consisting of a diabetic child and a sibling who has ICAs but has not progressed to overt diabetes (Field *et al.*, unpublished data). To date, 57 such pairs have been genotyped at most putative diabetes susceptibility loci. Even though the sample size was small, these sibpairs have shown significantly increased sharing in the HLA and *IDDM8* regions ($P = 0.03$ and 0.04, respectively), suggesting that *IDDM8* may contribute susceptibility to an early event in the disease process. The identity of the HLA region genes predisposing to ICA formation remains undetermined. Studies of the genetics of β cell autoimmunity may be especially valuable, since the early pathogenic events may be those most amenable to successful intervention to prevent further disease progression.

Insights from IDDM8 studies. (i) Large pooled datasets are not necessarily the best approach for mapping susceptibility genes; a better approach might be to analyze ethnically and/or geographically distinct populations separately; (ii) redefining the disorder of interest to include a 'pre-diabetes' phenotype may be a powerful approach to verifying putative susceptibility loci and/or to identifying loci important early in disease pathogenesis.

IDDM9 has not been published in detail, but appears in various publications (e.g. Concannon *et al.*, 1998; Mein *et al.*, 1998) to refer to the linkage on chromosome 3q21-q25 first reported by Davies *et al.* (1994) using D3S1303. While Mein *et al.* (1998) continue to see evidence for linkage to this region in British families (MLS 1.1), very little evidence for linkage was found by Concannon *et al.* (1998) in mixed British and American families (MLS 0.2).

IDDM10 in the centromeric region of chromosome 10 (10p11-q11) was reported by Reed *et al.* (1997) following results of Davies *et al.* (1994). In 236 British families, Reed *et al.* found an MLS of 2.96 at D10S208/D10S193. These results were strengthened by addition of more families, with an MLS of 4.7 reported in 356 British families (Mein *et al.*, 1998). In fact, this was the largest non-HLA lod score observed in the genome screen of Mein *et al.* (1998). However, the genome screen of Concannon *et al.* (1998), using many of the same British families as well as a large number of American families, found only weak linkage to this region of chromosome 10 (MLS 0.4). In view of the high prevalence of anti-GAD autoantibodies in type 1 diabetics, it is of great interest that the *GAD2* locus, encoding GAD_{65}, has been mapped to 10p, about 6 cM distal to D10S193 (Bu *et al.*, 1992; Wapelhorst *et al.*, 1995). However, no association between diabetes and a marker at *GAD2* was seen by Wapelhorst *et al.* (1995), and no association between *GAD2* genotype and presence of anti-GAD antibodies was seen by Rambrand *et al.* (1997).

Insights from IDDM10 studies. There is no evidence that GAD autoantibodies are associated with *GAD2* genotypes (similarly, GAD autoantibodies show no association with *GAD1* region genotypes, and insulin autoantibodies show no association with insulin region genotypes). However, one might ask whether it is a coincidence that three regions containing diabetes-predisposing genes (*IDDM2, IDDM7, IDDM10*) also contain loci encoding three antigens (insulin, glutamic acid decarboxylase 67 and 65) that elicit autoantibodies in type 1 diabetics.

IDDM11 was localized by Field *et al.* (1996) to chromosome 14q24.3-q31 by strong evidence for linkage to D14S67 (maximum lod score 4.0, affected sibpair $P = 0.00001$) in 254 Canadian, American and British families. An intriguing feature of this linkage was the significant heterogeneity ($P = 0.009$) in strength of linkage to D14S67 between affected sibpairs with and without increased HLA gene sharing. The evidence for linkage to D14S67 was strong (lodscore 4.6) in 99 families whose affected sibpairs *did not have* increased sharing of HLA genes (HLA sharing $\leqslant$50%), while evidence for linkage to the region was weak (maximum lodscore 0.4) in 147 families whose affected sibpairs *did have* increased HLA sharing (HLA sharing >50%). In the Field *et al.* dataset, *IDDM11* demonstrated stronger evidence for linkage than any susceptibility locus other than HLA. However, Mein *et al.* (1998) reported no evidence for linkage to the *IDDM11* region (although they did not genotype all the British families included in Field *et al.*, 1996). Similarly, though Concannon *et al.* (1998) found significant linkage to D14S67 in their first set of families (which overlapped with those in Field *et al.*, 1996), overall evidence for linkage to this region in their total dataset was weak. Pociot *et al.* (1998) also reported nonsignificantly increased gene sharing at D14S67 in Danish families.

Field *et al.* (1996) found no association between diabetes and D14S67. To enrich for families in which *IDDM11* was more likely to be involved, association analyses were subsequently performed only on the subset of families showing linkage to the region (AFBAC simplex analysis, employing only one diabetic child to test for association). These families showed no evidence for association to D14S67, but significant association with D14S256, located <1 cM distal to D14S67 (P=0.019) (Field *et al.*, unpublished data).

The heterogeneity of *IDDM11* linkage between families with and without increased HLA sharing reported by Field *et al.* (1996) suggests that *IDDM11* and HLA region genes act in a *biologically* independent manner on susceptibility. In other words, families with strong HLA region predisposition are less likely to also carry *IDDM11* predisposition (although they may), while families without strong HLA predisposition may require additional contributions from other susceptibility genes such as *IDDM11*. In contrast, some diabetes predisposing genes may interact in producing susceptibility. For example, there is evidence that *IDDM11* and *IDDM3* interact (Field and Tobias, 1997).

Insights from IDDM11 studies. (i) Even strong evidence for linkage in an initial dataset may be difficult to replicate in subsequent samples; (ii) it may be more powerful to search for evidence of association only using families which demonstrate increased gene sharing (linkage) to the region of interest – i.e. analyzing allele transmissions and non-transmissions to one diabetic child in families whose affected sibpair shares two marker genes.

IDDM12 and *IDDM13* have both been localized to chromosome 2q, which with *IDDM7* at 2q31 brings to three the number of susceptibility loci reported on 2q. Nistico *et al.* (1996) identified *IDDM12* by linkage (MLS 3.2) and association with the *CTLA4* (cytotoxic T lymphocyte associated-4) locus at 2q33, about 10 cM distal to *IDDM7*, in Italian and Spanish datasets but not in American, British, or Sardinian datasets. They suggested that *IDDM7* and *IDDM12* are distinct loci, since association was detected at both regions in the Italian data. Marron *et al.* (1997) confirmed *IDDM12* by demonstrating highly significant evidence for association between diabetes and an A/G (Thr/Ala) exon 1 polymorphism of *CTLA4* in 669 multiplex and 357 simplex families. There was strong population heterogeneity for this association: Italian, Spanish, French, Mexican, and Korean samples demonstrated significantly increased transmission of allele G to diabetics, while British, Sardinian, Chinese, and European-Americans demonstrated little or no increased transmission of this allele. Case-controls studies of German (Donner *et al.*, 1997) and Polish (Krokowski *et al.*, 1998) diabetics have also found significant increases of allele G. Since *CTLA4* is critically involved in down-regulation of T cell activation (Tivol *et al.*, 1995) and is therefore a plausible candidate gene for predisposition to type 1 diabetes, the accumulated association evidence strongly supports the hypothesis that *CTLA4* itself is *IDDM12*.

Insights from IDDM12 studies. The IDDM12 susceptibility locus was identified by positional-candidate mapping – that is, detection of linkage followed by intensive analysis of association for a candidate gene within the linked region (*CTLA4*). IDDM12 is only the third diabetes-predisposing locus to be fine mapped (HLA and *INS* are the other two), but demonstrates that the positional-candidate approach holds promise for other diabetes susceptibility genes.

IDDM13 was identified by Morahan *et al.* (1996) in Australian families by linkage to D2S164 at 2q35 (MLS 3.3). Interestingly, they reported that most of the *IDDM13* linkage evidence was contained in affected sibpairs *not* showing linkage to the HLA region, a phenomenon also observed for *IDDM3* (Field *et al.*, 1994) and *IDDM11* (Field *et al.*, 1996). Perhaps loci demonstrating this phenomenon form interacting (epistatic) networks with strong predisposing effects independent of HLA. *IDDM13* was also detected by Fu *et al.* (1998), who reported evidence for association with D2S137, near D2S164, in Japanese type 1 diabetic patients with low risk HLA genotypes.

To date, no information concerning *IDDM14* has been published (the symbol has been requested by an investigator and reserved by the Human Gene Mapping Nomenclature Committee).

IDDM15 was reported by Delepine *et al.* (1997) to lie at 6q21, proximal to *IDDM5* and *IDDM8*, following analysis of 376 French and American families. Identification of this linkage was problematic because of its proximity to the HLA region which shows very strong linkage to diabetes. After adjusting to account for linkage to HLA, Delephine *et al.* detected highly significant residual linkage to 6q21 ($P = 0.0002$ at D6S283). Concannon *et al.* (1998) also identified an independent locus at 6q21 (MLS 1.7–2.3, after adjustment for linkage with HLA).

IDDM16 was localized to the immunoglobulin heavy chain (*IGH*) region on chromosome 14q32.3 by linkage to D14S292 (affected sibpair $P = 0.005$), D14S293 ($P = 0.004$) and D14S542 ($P = 0.003$) in 284 Canadian, American, and British families (Field *et al.*, 1997; Field *et al.*, unpublished). These three markers lie in a 1 cM region, with D14S542 at the IGH gene cluster. Significant linkage was not detected at the 14q terminal marker D14S1007. Field and colleagues, analyzing IgG (Gm) serological allotypes, have long contended that genes at or near *IGH* influence susceptibility to type 1 diabetes (reviewed in Field, 1991). The large genome screen of Concannon *et al.* also detected evidence for linkage to the *IDDM16* region, with an MLS approaching 1.0 (see Figure 3 in Concannon *et al.*, 1998). Veijola *et al.* (1996) analyzed four different immunoglobulin heavy-chain variable gene markers in 133 diabetic sibpairs, and in 101 diabetics and 114 unrelated controls from Finland. They found no evidence for linkage in the affected sibpairs, but significant association in 2 of the 4 markers in the case-control series ($P = 0.02$ and 0.004). Similarly, a set of Danish families typed for D14S542 showed no linkage but significant association ($P = 0.019$) (Field, Pociot and Nerup, unpublished). Interestingly, marker D14S292 has also demonstrated linkage to another autoimmune disorder, multiple sclerosis, in 227 British families (MLS 1.4, Sawcer *et al.*, 1996).

Insights from IDDM16 studies. Historically, this was one of the first non-HLA regions suggested to be involved in susceptibility to type 1 diabetes (using serological markers not readily available) (Field, 1991). Use of more polymorphic and readily available DNA markers at the immunoglobulin loci may now enable their contribution to diabetes predisposition to be clarified.

IDDM17 has been localized to chromosome 10q25 by detection of strong linkage ($P = 0.00004$) in a large Arab family with 19 affected members (Verge *et al.*, 1998). Linkage to this region was also detected by Concannon *et al.* (1998) with MLS 1.1.

3.5 Predisposing genes without 'IDDM' designations.

In addition to the 16 '*IDDM*' designated regions (*IDDM1-17* minus *IDDM14* not yet published), there are six regions of the genome to which linkage has been detected that should be mentioned: *GCK* (on 7p), 1q, 16q, 19p, 19q, and Xp.

Mutations of the glucokinase (*GCK*) gene on chromosome 7p13 are responsible for a proportion of autosomal dominant families with 'maturity-onset diabetes of the young' (MODY) (reviewed in Hattersley and Turner, 1993). *GCK* has been designated *MODY2 (MODY1* and *MODY3* are on chromosomes 20 and 12, respectively). However, significant linkage ($P = 0.003$) and association ($P = 0.001$) between type 1 diabetes and a marker at *GCK* has been detected by Rowe *et al.* (1995) in 339 American and British multiplex families. It is unclear why this gene, encoding an enzyme which may control glucose homeostasis by determining the insulin response to glucose (Sturis *et al.*, 1994), should be involved in susceptibility to type 1 diabetes (similar comments pertain to *IDDM2* at the insulin gene). Field *et al.* (unpublished) found no significant evidence for increased *GCK* gene sharing in 298 affected sibpairs (average sharing 0.51, $P = 0.25$), but these same families produced another form of evidence for linkage: significantly *decreased* GCK gene sharing in 228 *affected-unaffected* (discordant) sibpairs (average sharing 0.47, $P = 0.032$) as well as increased sharing in 21 unaffected sibpairs (average sharing 0.55, $P = 0.15$). It is possible that decreased gene sharing in discordant sibpairs may reflect the presence of *resistance genes*. Perhaps there are uncommon *GCK* mutations associated with resistance to development of diabetes, which may be especially relevant to individuals who also carry some susceptibility genes (such as siblings of diabetics).

Insights from GCK studies. Although analysis of gene sharing in discordant (affected-unaffected) sibpairs is generally not a powerful approach to identification of susceptibility loci in a multigene disorder, it may assist in identifying resistance genes on a background of predisposing genes.

Linkage to chromosome 1q42 (MLS 3.3 at D1S1617) was detected in the large genome screen of Concannon *et al.* (1998). In fact, in their study of 616 American and British families, this region produced the *largest non-HLA lod score*. In contrast, Mein *et al.* (1998) did not report any evidence for linkage to this region in their 356 British families (250 of which appear to overlap with those in Concannon *et al.*), suggesting that most of the evidence for linkage in Concannon *et al.* was obtained from the American families. This is (yet another) striking example of linkage heterogeneity, the possible reasons for which were discussed above with respect to *IDDM3*.

Linkage to chromosomal regions 16q22-q24 (MLS 3.4), 19p13 (MLS 1.7), and 19q13 (MLS 1.6) was reported by Mein *et al.* (1998), but, conversely, none of these regions produced an MLS of 1.0 in the study by Concannon *et al.* (1998). The MLS for 16q22-q24 was the third largest lod score (after those for the HLA and *IDDM10* regions) in the study of Mein *et al.* It is noteworthy that 16q22-q24 and 19q13 have each been reported to be involved in at least two other autoimmune disorders (Becker *et al.*, 1998), thereby strengthening the evidence that these regions contain genes predisposing to autoimmune diabetes.

Linkage to the short arm of the X chromosome (Xp13-p11) was initially reported by Davies *et al.* (1994) and extended by Cucca *et al.* (1998). The latter reported that most of the evidence for linkage was from families with DR3/X (where X was not DR4) affected sibpairs, consistent with their finding that DR3/X diabetics were more often male than female. This region of the X chromosome has also been reported to show linkage to other autoimmune disorders (Becker *et al.*, 1998).

Insights from 1q, 16q, 19p, 19q, and Xp linkage studies. In the two large genome screens of Concannon *et al.* (1998) and Mein *et al.* (1998), some of the largest reported lod scores were obtained in regions not previously producing positive signals (1q and 16q), but ironically neither were found in both screens. This suggests that additional regions showing linkage remain to be discovered, but again will be difficult to verify.

4. Concluding remarks

More than 20 regions have been identified, primarily through linkage studies, as containing putative genes predisposing to type 1 diabetes. The positive signals in many of these regions need to be confirmed using additional linkage datasets and/or using alternate approaches to detecting susceptibility genes (such as detailed linkage disequilibrium mapping, analysis of special founder populations, or redefinition of the phenotype of interest). Since the two recent large genome screens have uncovered previously unreported linkages, it is possible that additional susceptibility genes remain to be discovered even by routine affected sib-pair linkage studies. However, it is likely that the most significant advances in the future will be made using the above-mentioned alternate approaches. In particular, a dense disequilibrium (association) genome screen will need to be performed to detect true susceptibility genes that do not show strong linkage. Rather than using strictly random markers, this association screen may consist of analyzing all remotely reasonable candidate genes (e.g. genes involved in the immune system broadly defined, genes involved in pancreatic growth and development, genes implicated in other autoimmune disorders, etc.). While the λ_{sib} value for type 1 diabetes is about 15 and the HLA λ_{sib} is about 3.4 (see above), the λ_{sib} values estimated for other diabetes susceptibility genes have generally been below 1.5. In summary, we now know that the genetic basis of type 1 diabetes consists of one or more 'major' gene contributions from the HLA region, acting independently or in concert with multiple contributions from other 'minor' loci distributed across the genome. The challenge will be to identify and isolate all susceptibility loci and determine how they produce the autoimmune destruction of pancreatic β cells which leads to type 1 diabetes.

Acknowledgments

Dr Field's research is funded by the Medical Research Council of Canada (grant MT-7910), by the Juvenile Diabetes Foundation International, and by the Canadian Genetic Diseases Network of the Network of Centres of Excellence program of the Canadian and Alberta governments. Dr Field is a Scientist of the Alberta Heritage Foundation for Medical Research. The following contributions are gratefully acknowledged: Drs E. Colle, H. Dean, J. Dupre, M. Gonthier, J. Mahon, R. McArthur, E. Ryan, D. Stephure, L. Stewart, and R. Trussell for collection of Canadian families; M. Engels, E. Swiergala, R. Tobias, and J. Nagatomi for laboratory expertise; T. Magnus, M. Nelson, G. Thomson, and Z. Zhang for computing assistance; R. Hodorek for secretarial work; and all the families who generously participated in Dr Field's research.

References

Barnett, A.H., Eff, C., Leslie, R.D.G. and Pyke, D.A. (1981) Diabetes in identical twins. *Diabetologia* **20**:87–93.

Becker, K.G. (1999) Comparative genetics of type 1 diabetes and autoimmune disease: common loci, common pathways? *Diabetes* **48**: 1353–1358.

Becker, K.G., Simon, R.M., Bailey-Wilson, J.E., *et al.* (1998) Clustering of non-major histocompatibility complex susceptibility candidate loci in human autoimmune diseases. *Proc. Natl. Acad. Sci. USA* **95**: 9979–9984.

Bell, G.I., Horita, S. and Karam, J.H. (1984) A polymorphic locus near the human insulin gene is associated with insulin-dependent diabetes mellitus. *Diabetes* **33**: 176–183.

Bennett, S.T., Lucassen, A.M., Gough, S.C.L., *et al.* (1995) Susceptibility to human type 1 diabetes at IDDM2 is determined by tandem repeat variation at the insulin gene minisatellite locus. *Nature Genet.* **9**: 284–292.

Bleich, D., Polak, M., Eisenbarth, G.S. and Jackson, R.A. (1993) Decreased risk of type I diabetes in offspring of mothers who acquire diabetes during adrenarchy. *Diabetes* **42**: 1433–1439.

Bonifacio, E., Bingley, P.J. and Shattock, M., *et al.* (1990) Quantification of islet-cell antibodies and prediction of insulin-dependent diabetes. *Lancet* **335**: 147–149.

Bu D-F., Erlander, M.G. and Hitz, B.C., (1992) Two human glutamate decarboxylases, 65-kDa GAD and 67-kDa GAD, are each encoded by a single gene. *Proc. Natl. Acad. Sci. USA* **89**: 2115–2119.

Caillat-Zucman, S., Bertin, E., Timsit, J., Boitard, C., Assan, R. and Bach, J.F., *et al.* (1993) Protection from insulin-dependent diabetes mellitus is linked to a peptide transporter gene. *Eur. J. Immunol.* **23**: 1784–1788.

Caillat-Zucman, S., Daniel, S., Djilali-Saiah, I., *et al.* (1995) Family study of linkage disequilibrium between TAP2 transporter and HLA class II genes. Absence of TAP2 contribution to association with insulin-dependent diabetes mellitus. *Hum. Immunol.* **44**: 80–87.

Catasti, P., Chen, X., Moyzis, R.K., Bradbury, E.M. and Gupta, G. (1996) Structure-function correlations of the insulin-linked polymorphic region. *J. Mol. Biol.* **264**: 534–545.

Concannon, P., Gogolin-Ewens, K.J., Hinds, D.A., *et al.* (1998) A second-generation screen of the human genome for susceptibility to insulin-dependent diabetes mellitus. *Nat. Genet.* **19**: 292–296.

Copeman, J.B., Cucca, F., Hearne, C.M., *et al.* (1995) Linkage disequilibrium mapping of a type 1 diabetes susceptibility gene (IDDM7) to chromosome 2q31-q33. *Nature Genet.* **9**: 80–85.

Cox, N.J. and Spielman, R.S. (1989) The insulin gene and susceptibility to IDDM. *Genet. Epidemiol.* **6**: 65–69.

Cucca, F. and Todd, J.A. (1996) HLA susceptibility to type 1 diabetes: methods and mechanisms. In: *HLA and MHC: Genes, Molecules and Function*. (eds M.J. Browning and A.J. McMichael). BIOS Scientific Publishers Ltd, Oxford, pp. 383–406.

Cucca, F., Congia, M., Trowsdale, J. and Powis, S.H. (1994) Insulin-dependent diabetes mellitus and the major histocompatibility complex peptide transporters TAP1 and TAP2: no association in a population with a high disease incidence. *Tissue Antigens* **44**: 234–240.

Cucca, F., Goy, J.V., Kawaguchi, Y. *et al.* (1998) A male-female bias in type 1 diabetes and linkage to chromosome Xp in MHC HLA-DR3-positive patients. *Nat. Genet.* **19**: 301–302.

Cudworth, A.G. and Woodrow, J.C. (1974) Letter: HL-A antigens and diabetes mellitus. *Lancet* **2**: 1153.

Dahlquist, G.G., Ivarsson, S., Lindberg, B. and Forsgren, M. (1995) Maternal enteroviral infection during pregnancy as a risk factor for childhood IDDM. A population-based case-control study. *Diabetes* **44**: 408–413.

Davies, J.L., Kawaguchi, Y., Bennett, S.T., *et al.* (1994) A genome-wide search for human type 1 diabetes susceptibility genes. *Nature* **371**: 130–136.

Davies, J.L., Cucca, F., Goy, J.V., *et al.* (1996) Saturation multipoint linkage mapping of chromosome 6q in type 1 diabetes. *Hum. Mol. Genet.* **5**: 1071–1074.

Delepine, M., Pociot, F., Habita, C., *et al.* (1997) Evidence of a non-MHC susceptibility locus in type I diabetes linked to HLA on chromosome 6. *Am. J. Hum. Genet.* **60**: 174–187.

Deng, G.Y., Muir, A., Maclaren, N.K. and She, J.X. (1995) Association of LMP2 and LMP7 genes within the major histocompatibility complex with insulin-dependent diabetes mellitus: population and family studies. *Am. J. Hum. Genet.* **56**: 528–534.

Djilali-Saiah, I., Schmitz, J., Harfouch-Hammoud, E., Mougenot, J.F., Bach J.F. and Caillat-Zucman, S. (1998) CTLA-4 gene polymorphism is associated with predisposition to coeliac disease. *Gut* **43**: 187–189.

Donner, H., Rau, H., Walfish, P.G., *et al.* (1997) CTLA4 alanine-17 confers genetic susceptibility to Graves' disease and to type 1 diabetes mellitus. *J. Clin. Endocrinol. Metab.* **82**: 143–146.

Dunsworth, T.S., Rich, S.S., Swanson, J. and Barbosa, J. (1982) No evidence for linkage between diabetes and the Kidd marker. *Diabetes* **31**: 991–993.

Fava, D., Leslie, R.D. and Pozzilli, P. (1994) Relationship between dairy product consumption and incidence of IDDM in childhood in Italy. *Diabetes Care* **17**: 1488–1490.

Fava, D., Gardner, S., Pyke, D. and Leslie, R.D. (1998) Evidence that the age at diagnosis of IDDM is genetically determined. *Diabetes Care* **21**: 925–929.
Field, L.L. (1989) Genes predisposing to IDDM in multiplex families. *Genet. Epidemiol.* **6**: 101–106.
Field, L.L. (1991) Non-HLA region genes in insulin dependent diabetes mellitus. *Baillière's Clin. Endocrinol. Met.* 5: 413–438.
Field, L.L. and McArthur, R.G. (1987) The genetics of susceptibility to insulin-dependent diabetes: possible new markers. *Clin. Invest. Med.* **10**: 437–443.
Field, L.L. and Nagatomi, J. (1994) Linkage analysis: inadequate for detecting susceptibility loci in complex disorders? *Am. J. Hum. Genet.* 55: A347.
Field, L.L. and Tobias, R. (1997) Unravelling a complex trait: the genetics of insulin-dependent diabetes mellitus. *Clin. Invest. Med.* **20**: 41–49.
Field, L.L., Tobias, R. and Magnus, T. (1994) A locus on chromosome 15q26 (*IDDM3*) produces susceptibility to insulin-dependent diabetes mellitus. *Nature. Genet.* **8**: 189–194.
Field, L.L., Tobias, R. and Thomson, G. (1996) Susceptibility to insulin-dependent diabetes mellitus maps to a locus (*IDDM11*) on chromosome 14q24.3-q31. *Genomics* **33**: 1–8.
Field, L.L., Tobias, R. and Zhang, Z. (1997) Do genes in the immunoglobulin heavy chain (IGH) region on 14q32.3 influence IDDM susceptibility? Abstracts of the Genetics of Diabetes Meeting, Ystad, Sweden.
Field, L.L., Fothergill-Payne, C., Bertrams, J. and Baur, M.P. (1986) HLA-DR effects in a large German IDDM dataset. *Genet. Epi. Suppl.* **1**: 323–328.
Fu, J., Ikegami, H., Kawaguchi, Y., *et al.* (1998) Association of distal chromosome 2q with IDDM in Japanese subjects. *Diabetologia* **41**: 228–232.
Ginsberg-Fellner, F., Witt, M.E., Fedun, B., *et al.* (1985) Diabetes mellitus and autoimmunity in patients with the congenital rubella syndrome. *Rev. Infect. Dis.* **7** (Suppl 1): S170–S176.
Green, A., Sjolie, A.K. and Eshoj, O. (1996) Trends in the epidemiology of IDDM during 1970–2020 in Fyn County, Denmark. *Diabetes Care* **19**: 801–806.
Hagopian, W.A., Sanjeevi, C.B. and Kockum, I. (1995) Glutamate decarboxylase-, insulin-, and islet cell-antibodies and HLA typing to detect diabetes in a general population-based study of Swedish children. *J. Clin. Invest.* **95**: 1505–1511.
Hammond-Kosack, M.C.U., Dobrinski, B., Lurz, R., Docherty, K. and Kilpatrick, M.W. (1992) The human insulin gene linked polymorphic region exhibits an altered DNA structure. *Nucleic Acids Res.* **20**: 231–236.
Hashimoto, L., Habita, C., Beressl, J.P., *et al.* (1994) Genetic mapping of a susceptibility locus for insulin-dependent diabetes mellitus on chromosome 11q. *Nature* **371**: 161–164.
Hattersley, A.T. and Turner, R.C. (1993) Mutations of the glucokinase gene and type 2 diabetes. *Q. J. Med.* **86**: 227–232.
Hattori, M., Yamato, E., Itoh, N., *et al.* (1999) Cutting edge: homologous recombination of the MHC class I K region defines new MHC-linked diabetogenic susceptibility gene(s) in nonobese diabetic mice. *J. Immunol.* **163**: 1721–1724.
Hiltunen, M., Hyoty, H., Knip, M., *et al.* (1997) Islet cell antibody seroconversion in children is temporally associated with enterovirus infections. Childhood Diabetes in Finland (DiMe) Study Group. *J. Infect. Dis.* **175**: 554–560.
Hodge, S.E., Anderson, C.E., Neiswanger, K., *et al.* (1981) Close genetic linkage between diabetes mellitus and kidd blood group. *Lancet* **2**: 893–895.
Horwitz, M.S., Bradley, L.M., Harbertson, J., Krahl, T., Lee, J. and Sarvetnick, N. (1998) Diabetes induced by Coxsackie virus: initiation by bystander damage and not molecular mimicry. *Nature Med.* **4**: 781–785.
Houlston, R.S., Tomlinson, I.P., Ford, D. *et al.* (1997) Linkage analysis of candidate regions for coeliac disease genes. *Hum. Mol. Genet.* **6**: 1335–1339.
Huang, H.S., Peng, J.T., She, J.Y., *et al.* (1995) HLA-encoded susceptibility to insulin-dependent diabetes mellitus is determined by DR and DQ genes as well as their linkage disequilibria in a Chinese population. *Hum. Immunol.* **44**: 210–219.
Hyoty, H., Hiltunen, M. and Lonnrot, M. (1998) Enterovirus infections and insulin dependent diabetes mellitus–evidence for causality. *Clin. Diagn. Virol.* **9**: 77–84.
Juhela, S., Hyoty, H., Hinkkanen, A., *et al.* (1999) T cell responses to enterovirus antigens and to beta-cell autoantigens in unaffected children positive for IDDM-associated autoantibodies. *J. Autoimmun.* **12**: 269–278.

Karounos, D.G., Wolinsky, J.S. and Thomas, J.W. (1993) Monoclonal antibody to rubella virus capsid protein recognizes a beta-cell antigen. *J. Immunol.* **150**: 3080–3085.

Kaufman, D.L., Erlander, M.G., Clare-Salzler, M., Atkinson, M.A., Maclaren, N.K. and Tobin, A.J. (1992) Autoimmunity to two forms of glutamate decarboxylase in insulin-dependent diabetes mellitus. *J. Clin. Invest.* **89**: 283–292.

Kennedy, G.C., German, M.S. and Rutter, W.J. (1995) The minisatellite in the diabetes susceptibility locus IDDM2 regulates insulin transcription. *Nat. Genet.* **9**: 293–298.

Khalil, I., d'Auriol, L., Gobet, M., *et al.* (1990) A combination of HLA-DQ beta Asp57-negative and HLA DQ alpha Arg52 confers susceptibility to insulin-dependent diabetes mellitus. *J. Clin. Invest.* **85**: 1315–1319.

Kolb, H. and Pozzilli, P. (1999) Cow's milk and type I diabetes: the gut immune system deserves attention. *Immunol. Today* **20**: 108–110.

Krokowski, M., Bodalski, J., Bratek, A., Machejko, P. and Caillat-Zucman, S. (1998) CTLA-4 gene polymorphism is associated with predisposition to IDDM in a population from central Poland. *Diabetes Metab.* **24**: 241–243.

Kyvik, K.O., Green, A. and Beck-Nielsen, H. (1995) Concordance rates of insulin dependent diabetes mellitus: a population based study of young Danish twins. *BMJ* **311**: 913–917.

Levy-Marchal, C., Karjalainen, J., Dubois, F., Karges, W., Czernichow, P. and Dosch, H.M. (1995) Antibodies against bovine albumin and other diabetes markers in French children. *Diabetes Care* **18**: 1089–1094.

Lie, B.A., Todd, J.A., Pociot, F., *et al.* (1999) The predisposition to type 1 diabetes linked to the human leukocyte antigen complex includes at least one non-class II gene. *Am. J. Hum. Genet.* **64**: 793–800.

Lindberg, B., Ivarsson, S.A., Landin-Olsson, M., Sundkvist, G., Svanberg, L. and Lernmark, A. (1999) Islet autoantibodies in cord blood from children who developed type I (insulin-dependent) diabetes mellitus before 15 years of age. *Diabetologia* **42**: 181–187.

Lipton, R.B., Atchison, J., Dorman, J.S., *et al.* (1992) Genetic, immunological, and metabolic determinants of risk for type 1 diabetes mellitus in families. *Diabet. Med.* **9**: 224–232.

Lorenzen, T., Pociot, F., Hougaard, P. and Nerup, J. (1994) Long-term risk of IDDM in first-degree relatives of patients with IDDM. *Diabetologia* **37**: 321–327.

Lucassen, A.M., Screaton, G.R., Julier, C., Elliott, T.J., Lathrop, M. and Bell, J.I. (1995) Regulation of insulin gene expression by the IDDM associated, insulin locus haplotype. *Hum. Mol. Genet.* **4**: 501–506.

Luo, D-F., Bui, M.M., Muir, A., Maclaren, N.K., Thomson, G. and She, J-X. (1995) Affected sibpair mapping of a novel susceptibility gene to insulin-dependent diabetes mellitus (IDDM8) on chromosome 6q25-q27. *Am. J. Hum. Genet.* **57**: 911–919.

Luo, D-F., Buzzetti, R., Rotter, J.I., *et al.* (1996) Confirmation of three susceptibility genes to insulin-dependent diabetes mellitus: *IDDM4*, *IDDM5*, and *IDDM8*. *Hum. Mol. Genet.* **5**: 693–698.

Maclaren, N.K. and Alkinson, M.A. (1997) Insulin-dependent diabetes mellitus: the hypothesis of molecular mimicry between islet cell antigens and microorganisms. *Mol. Med. Today* **3**: 76–83.

Marron, M.P., Raffel, L.J., Garchon, H.J., *et al.* (1997) Insulin-dependent diabetes mellitus (IDDM) is associated with CTLA4 polymorphisms in multiple ethnic groups. *Hum. Mol. Genet.* **6**: 1275–1282.

Mein, C.A., Esposito, L., Dunn, M.G., *et al.* (1998) A search for type 1 diabetes susceptibility genes in families from the United Kingdom. *Nat. Genet.* **19**: 297–300.

Merriman, T., Twells, R., Merriman, M., *et al.* (1997) Evidence by allelic association-dependent methods for a type 1 diabetes polygene (IDDM6) on chromosome 18q21. *Hum. Mol. Genet.* **6**: 1003–1010.

Morahan, G., Huang, D., Tait, B.D., Colman, P.G. and Harrison L.C. (1996) Markers on distal chromosome 2q linked to insulin-dependent diabetes mellitus. *Science* **272**: 1811–1813.

Nairn, C., Galbraith, D.N., Taylor, K.W. and Clements, G.B. (1999) Enterovirus variants in the serum of children at the onset of type 1 diabetes mellitus. *Diabet. Med.* **16**: 509–513.

Nakagawa, Y., Kawaguchi, Y., Twells, R.C., *et al.* (1998) Fine mapping of the diabetes-susceptibility locus, IDDM4, on chromosome 11q13. *Am. J. Hum. Genet.* **63**: 547–556.

Nepom, B.S., Palmer, J., Kim, S-J., Hansen, J.A., Holbeck, S.L. and Nepom, G.T. (1986) Specific genomic markers for the HLA-DQ subregion discriminate between DR4+ insulin-dependent diabetes and DR4+ seropositive juvenile rheumatoid arthritis. *J. Exp. Jed.* **164**: 345–350.

Nerup, J., Platz, P., Andersen, O.O., *et al.* (1974) HL-A antigens and diabetes mellitus. *Lancet* **2**: 864–866.

Nistico, L., Buzzetti, R., Pritchard, L.E., *et al.* (1996) The *CTLA-4* gene region of chromosome 2q33 is linked to, and associated with, type 1 diabetes. *Hum. Mol. Genet.* **5**: 1075–1080.

Olmos, P., A'Hern, R., Heaton, D.A., *et al.* (1988) The significance of the concordance rate for type 1 (insulin-dependent) diabetes in identical twins. *Diabetologia* **31**: 747–750.

Owerbach, D. and Gabbay, K.H. (1995) The HOXD8 locus (2q31) is linked to type I diabetes: Interaction with chromosome 6 and 11 disease susceptibility genes. *Diabetes* **44**: 132–136.

Payami, H., Thomson, G., Motro, U., Louis, E.J. and Hudes, E. (1985) The affected sib method. IV. Sib trios. *Ann. Hum. Genet.* **49**: 303–314.

Petersen, J.S., Kyvik, K.O., Bingley, P.J., *et al.* (1997) Population based study of prevalence of islet cell autoantibodies in monozygotic and dizygotic Danish twin pairs with insulin dependent diabetes mellitus. *BMJ* **314**: 1575–1579.

Pocecco, M., Buratti, E., Tommasini, A., Torre, G. and Not, T. (1997) High risk of *Helicobacter pylori* infection associated with cow's milk antibodies in young diabetics. *Acta Paediatr.* **86**: 700–703.

Pociot, F., Larsen, Z.M., Johannesen, J., Kristiansen, O.P. and Nerup, J. (1998) Linkage analysis of *IDDM11* in type 1 diabetes Danish multiplex families. *Diabetologia* **41** (Suppl. 1): A12.

Pugliese, A., Zeller, M., Fernandez, A. Jr, *et al.* (1997) The insulin gene is transcribed in the human thymus and transcription levels correlated with allelic variation at the INS VNTR-IDDM2 susceptibility locus for type 1 diabetes. *Nature Genet.* **15**: 293–297.

Rambrand, T., Pociot, F., Ronningen, K., Nerup, J. and Michelsen, B.K. (1997) Genetic markers for glutamic acid decarboxylase do not predict insulin-dependent diabetes mellitus in pairs of affected siblings. The Danish Study Group of Diabetes in Childhood. *Hum. Genet.* **99**: 177–185.

Redondo, M.J., Rewers, M., Yu, L., *et al.* (1999) Genetic determination of islet cell autoimmunity in monozygotic twin, dizygotic twin, and non-twin siblings of patients with type 1 diabetes: prospective twin study. *BMJ* **318**: 698–702.

Reed, P., Cucca, F., Jenkins, S., *et al.* (1997) Evidence for a type 1 diabetes susceptibility locus (IDDM10) on human chromosome 10p11-q11. *Hum. Mol. Genet.* **6**: 1011–1016.

Ridgway, W.M. and Fathman, C.G. (1998) The association of MHC with autoimmune diseases: understanding the pathogenesis of autoimmune diabetes. *Clin. Immunol. Immunopathol.* **86**: 3–10.

Risch, N. (1987) Assessing the role of HLA-linked and unlinked determinants of disease. *Am. J. Hum. Genet.* **40**: 1–14.

Risch, N. (1990) Linkage strategies for genetically complex traits. II. The power of affected relative pairs. *Am. J. Hum. Genet.* **46**: 229–241.

Risch, N. and Merikangas, K. (1996) The future of genetic studies of complex human diseases. *Science* **273**: 1516–1517.

Roivainen, M., Knip, M., Hyoty, H., *et al.* (1998) Several different enterovirus serotypes can be associated with prediabetic autoimmune episodes and onset of overt IDDM. Childhood Diabetes in Finland (DiMe) Study Group. *Med. Virol.* **56**: 74–78.

Roll, U., Christie, M.R., Fuchtenbusch, M., Payton, M.A., Hawkes, C.J. and Ziegler, A.G. (1996) Perinatal autoimmunity in offspring of diabetic parents. The German Multicenter BABY-DIAB study: detection of humoral immune responses to islet antigens in early childhood. *Diabetes* **45**: 967–973.

Ronningen, K.S., Undlien, D.E., Ploski, R., *et al.* (1993) Linkage disequilibrium between TAP2 variants and HLA class II alleles; no primary association between TAP2 variants and insulin-dependent diabetes mellitus. *Eur. J. Immunol.* **23**: 1050–1056.

Rowe, R.E., Wapelhorst, B., Bell, G.I., Risch, N., Spielman, R.S. and Concannon, P. (1995) Linkage and association between insulin-dependent diabetes mellitus (IDDM) susceptibility and markers near the glucokinase gene on chromosome 7. *Nature Genet.* **10**: 240–242.

Sarugeri, E., Dozio, N., Belloni, C., Meschi, F., Pastore, M.R. and Bonifacio, E. (1998) Autoimmune responses to the beta cell autoantigen, insulin, and the INS VNTR-IDDM2 locus. *Clin. Exp. Immunol.* **114**: 370–376.

Saukkonen, T., Virtanen, S.M., Karppinen, M., *et al.* (1998) Significance of cow's milk protein antibodies as risk factor for childhood IDDM: interactions with dietary cow's milk intake and HLA-DQB1 genotype. Childhood Diabetes in Finland Study Group. *Diabetologia* **41**: 72–78.

Savilahti, E., Saukkonen, T.T., Virtala, E.T., Tuomilehto, J. and Akerblom, H.K. (1993) Increased levels of cow's milk and beta-lactoglobulin antibodies in young children with newly diagnosed IDDM. The Childhood Diabetes in Finland Study Group. *Diabetes Care* **16**: 984–989.

Sawcer, S., Jones, H.B., Feakes, R., *et al.* (1996) A genome screen in multiple sclerosis reveals susceptibility loci on chromosome 6p21 and 17q22. *Nature Genet.* **13**: 464–468.

Schoenfeld, S.L., Asplin, C.M., Field, L.L. and Palmer, J.P. (1988) Insulin autoantibodies are associated with HLA-DR4 and Km1+ in IDDM. *Diabetes Res. Clin. Practice.* 5: S261.

Sheehy, M.J., Scarf, S.J., Rowe, J.R., *et al.* (1989) A diabetes susceptibility HLA haplotype is best defined by a combination of HLA-DR and -DQ alleles. *J. Clin. Invest.* **83**: 830–835.

Singal, D.P. and Blajchman, M.A. (1973) Histocompatibility (HL-A) antigens, lymphocytotoxic antibodies and tissue antibodies in patients with diabetes mellitus. *Diabetes* **22**: 429–432.

Spielman, R.S., Baker, L. and Zmijewski, C.M. (1979) Inheritance of susceptibility to juvenile onset diabetes. In: *Genetic Analysis of Common Disease: Applications to Predictive Factors in Coronary Disease.* (eds C.F. Sing and M. Skolnick), Alan Liss, New York, pp. 567–585.

Spielman, R.S., Baker, L. and Zmijewski, C.M. (1980) Gene dosage and susceptibility to insulin-dependent diabetes. *Ann. Hum. Genet.* **44**: 135–150.

Spielman, R.S., McGinnis, R.E. and Ewens, W.J. (1993) Transmission distortion test for linkage: the insulin gene and susceptibility in insulin-dependent diabetes mellitus. *Am. J. Hum. Genet.* **52**: 506–516.

Staines, A., Bodansky, H.J., McKinney, P.A., *et al.* (1997) Small area variation in the incidence of childhood insulin-dependent diabetes mellitus in Yorkshire, UK: links with overcrowding and population density. *Int. J. Epidemiol.* **26**: 1307–1313.

Sturis, J., Kurland, I.J., Byrne, M.M., *et al.* (1994) Compensation in pancreatic beta-cell function in subjects with glucokinase mutations. *Diabetes* **43**: 718–723.

Thomson, G. (1988) HLA disease associations: a model for the study of complex human genetic disorders. *Ann. Rev. Genet.* **22**: 31–50.

Thomson, G. (1995) Mapping disease genes: family based association studies. *Am. J. Hum. Genet.* 57: 487–498.

Thomson, G., Robinson, W.P., Kuhner, M.K., *et al.* (1988) Genetic heterogeneity, modes of inheritance and risk estimates for a joint study of Caucasians with insulin-dependent diabetes mellitus. *Am. J. Hum. Genet.* **43**: 799–816.

Thomson, G., Robinson, W.P., Kuhner, M.K., Joe, S. and Klitz, W. (1989) HLA and Insulin gene associations with IDDM. *Genet. Epi.* **6**: 155–160.

Tivol, E.A., Borriello, F., Schweitzer, A.N., Lynch, W.P., Bluestone, J.A. and Sharpe, A.H. (1995) Loss of CTLA-4 leads to massive lymphoproliferation and fatal multiorgan tissue destruction, revealing a critical negative regulatory role of CTLA-4. *Immunity* **3**: 541–547.

Todd, J.A., Bell, J.I. and McDevitt H.O. (1987) HLA-DQB gene contributes to susceptibility and resistance to insulin-dependent diabetes mellitus. *Nature* **329**: 599–604.

Todd, J.A., Mijovic, C., Fletcher, J., Jenkins, D., Bradwell, A.R. and Barnett, A.H. (1989) Identification of susceptibility loci for insulin-dependent diabetes mellitus by transracial gene mapping. *Nature* **338**: 587–589.

Tuomi, T., Carlsson, A., Li, H., *et al.* (1999) Clinical and genetic characteristics of type 2 diabetes with and without GAD antibodies. *Diabetes* **48**: 150–157.

Tuomilehto, J., Podar, T., Tuomilehto-Wolf, E. and Virtala, E. (1995) Evidence for importance of gender and birth cohort for risk of IDDM in offspring of IDDM parents. *Diabetologia* **38**: 975–982.

Tuvemo, T., Dahlquist, G., Frisk, G. *et al.* (1989) The Swedish childhood diabetes study III: IgM against coxsackie B viruses in newly diagnosed type 1 (insulin-dependent) diabetic children – no evidence of increased antibody frequency. *Diabetologia* **32**: 745–747.

Undlien, D.E., Akselsen, H.E., Joner, G., *et al.* (1997) No independent associations of LMP2 and LMP7 polymorphisms with susceptibility to develop IDDM. *Diabetes* **46**: 307–312.

Vaarala, O., Knip, M., Paronen, J., *et al.* (1999) Cow's milk formula feeding induces primary immunization to insulin in infants at genetic risk for type 1 diabetes. *Diabetes* **48**: 1389–1394.

Vafiadis, P., Bennett, S.T. and Todd, J.A., *et al.* (1997) Insulin expression in human thymus is modulated by INS VNTR alleles at the IDDM2 locus. *Nature Genet.* **15**: 289–292.

Vafiadis, P., Bennett, S.T., Todd, J.A., Grabs, R. and Polychronakos, C. (1998) Divergence between genetic determinants of IGF2 transcription levels in leukocytes and of IDDM2-encoded susceptibility to type 1 diabetes. *Clin. Endocrinol. Metab.* **83**: 2933–2939.

Veijola, R., Knip, M., Puukka, R., Reijonen, H., Cox, D.W. and Ilonen, J. (1996) The immunoglobulin heavy-chain variable region in insulin-dependent diabetes mellitus: affected-sib-pair analysis and association studies. *Am. J. Hum. Genet.* **59**: 462–470.

Verge, C.F., Gianani, R., Yu, L., *et al.* (1995) Late progression to diabetes and evidence for chronic beta-cell autoimmunity in identical twins of patients with type I diabetes. *Diabetes* **44**: 1176–1179.

Verge, C.F., Vardi, P., Babu, S., *et al.* (1998) Evidence for oligogenic inheritance of type 1 diabetes in a large Bedouin Arab family. *J. Clin. Invest.* **102**: 1569–1575.

Virtanen, S.M., Hypponen, E., Laara, E., *et al.* (1998) Cow's milk consumption, disease-associated autoantibodies and type 1 diabetes mellitus: a follow-up study in siblings of diabetic children. Childhood Diabetes in Finland Study Group. *Diabet. Med.* **15**: 730–738.

Vreugdenhil, G.R., Geluk, A., Ottenhoff, T.H., Melchers, W.J., Roep, B.O. and Galama, J.M. (1998) Molecular mimicry in diabetes mellitus: the homologous domain in coxsackie B virus protein 2C and islet autoantigen GAD65 is highly conserved in the coxsackie B-like enteroviruses and binds to the diabetes associated HLA-DR3 molecule. *Diabetologia* **41**: 40–46.

Wapelhorst, B., Bell, G.I., Risch, N., Spielman, R.S. and Concannon, P. (1995) Linkage and association studies in insulin-dependent diabetes with a new dinucleotide repeat polymorphism at the GAD65 locus. *Autoimmunity* **21**: 127–130.

Warram, J.H., Krolewski, A.S., Gottlieb, M.S. and Kahn, C.R. (1984) Differences in risk of insulin-dependent diabetes in offspring of diabetic mothers and diabetic fathers. *N. Engl. J. Med.* **311**: 149–152.

Warram, J.H., Krolewski, A.S. and Kahn, C.R. (1988) Determinants of IDDM and perinatal mortality in children of diabetic mothers. *Diabetes* **37**: 1328–1334.

Wilkin, T., Hoskins, P.J., Armitage, M., *et al.* (1985) Value of insulin autoantibodies as serum markers for insulin-dependent diabetes mellitus. *Lancet* **1**: 480–481.

Yan, G., Fu, Y. and Faustman, D.L. (1997) Reduced expression of Tap1 and Lmp2 antigen-processing genes in the nonobese diabetic (NOD) mouse due to a mutation in their shared bidirectional promoter. *J. Immunol.* **159**: 3068–3080.

Zamani, M., Pociot, F., Raeymaekers, P., Nerup, J. and Cassiman, J.J. (1996) Linkage of type I diabetes to 15q26 (IDDM3) in the Danish population. *Hum. Genet.* **98**: 491–496.

Zhong, F., McCombs, C.C., Olson, J.M., *et al.* (1996) An autosomal screen for genes that predispose to celiac disease in the western counties of Ireland. *Nature Genet.* **14**: 329–333.

Ziegler, A.G., Hummel, M., Schenker, M. and Bonifacio, E. (1999) Autoantibody appearance and risk for development of childhood diabetes in offspring of parents with type 1 diabetes: the 2-year analysis of the German BABYDIAB Study. *Diabetes* **48**: 460–468.

Zimmet, P.Z., Tuomi, T., Mackay, I.R., *et al.* (1994) Latent autoimmune diabetes mellitus in adults (LADA): the role of antibodies to glutamic acid decarboxylase in diagnosis and prediction of insulin dependency. *Diabet. Med.* **11**: 299–303.

9

Bipolar disorder

Nick Craddock and Ian Jones

1. Introduction

Bipolar disorder (manic depressive illness) has the highest heritability of the common forms of mental illness that afflict human populations. In this chapter we aim to illustrate the range of molecular genetic approaches being used in bipolar disorder, provide an overview of the current state of the field and likely future directions. Before doing this we will set the scene by briefly describing the pheno-type, outlining the evidence accumulated from classical genetic studies, and flagging up the important methodological issues. We have assumed throughout that the reader has some familiarity with the theoretical material covered in Chapters 1–3 of this volume.

2. Bipolar disorder: definition and properties of the phenotype

The core feature of bipolar disorder is pathological disturbance in mood (affect) ranging from extreme elation (mania) to severe depression usually accompanied by disturbances in thinking and behavior, which may include psychotic symptoms, such as delusions and hallucinations. For a diagnosis of bipolar disorder, all modern classification systems e.g. DSMIV (APA, 1994) or ICD10 (WHO, 1993) require that an individual has suffered at least one episode of mania (see *Table 1*) with or without episodes of depression at other times. The occurrence of an episode of mania during the course of illness distinguishes bipolar disorder from unipolar disorder (also called unipolar major depression or unipolar depression) the more common form of mood disorder in the population, in which subjects suffer episodes only of depression. (Genetic approaches to unipolar depression are discussed in detail in Chapter 14 of this volume). DSMIV divides bipolar illness into bipolar I disorder in which subjects experience at least one full-blown manic episode at some time during the course of illness, and bipolar II disorder in which the subject never experiences a full manic episode but has one or more milder manic-like episodes (known as 'hypomania'). This subclassification requires a judgement of the degree of functional

Analysis of Multifactorial Disease, edited by T. Bishop and P. Sham.

Table 1. Operational diagnostic criteria for a manic episode

(a)	Elevated, irritable or expansive mood for at least 1 week (or any duration if hospitalized)
(b)	During mood disturbance 3+ of the following symptoms (4+ if mood only irritable): (i) Inflated self esteem or grandiosity (ii) Decreased need for sleep (iii) Increased speech (iv) Racing thoughts (v) Distractibility (vi) Increased activity (vii) Reckless behavior
(c)	Mood disturbance causes marked impairment in functioning

Simplified from DSMIV. All of a, b and c are required for diagnosis of a full manic episode.

impairment caused by the episode (making subjects near the threshold difficult to classify reliably). It awaits robust validation, although some researchers view bipolar II as the most common bipolar phenotype (Simpson *et al.*, 1993). Others regard it as a relatively distinct nosological entity (Coryell *et al.*, 1989).

Narrowly defined bipolar disorder has a lifetime prevalence of approximately 1% with no evidence that rates vary widely among different populations. Rates of illness are similar in males and females and the mean age of onset is around the age of 21 years (Smith and Weissman, 1992). Bipolar illness is associated with high morbidity and service utilization and approximately 15% of patients eventually die by suicide (Guze and Robins, 1970).

3. Methodological issues

We briefly consider some of the major methodological issues and challenges confronting the researcher investigating bipolar disorder and indicate the implication for study design and conduct. Although some of these are specific to psychiatric disorders in general or bipolar illness in particular, the reader will recognize others as common to genetic investigation of all complex diseases.

3.1 Lifetime psychiatric diagnoses are required

The issue. A 'lifetime diagnosis' is necessary in order to classify subjects on the basis of a presumed diathesis for illness. A person's abnormal behavior and experience over his or her lifetime must be reduced to a small number of diagnostic categories, usually just one. This used to be a major methodological obstacle, but a robust methodology has been developed that produces diagnoses with a reliability comparable to, or better than, those achieved for clinical diagnosis for nonpsychiatric diseases (Farmer *et al.*, 1994; Leckman *et al.*, 1982).

The implication. Studies require detailed clinical work-up of subjects including both structured psychiatric interview and review of hospital case notes. This is both costly and time-consuming.

3.2 Age of onset of illness is variable

The issue. The peak age of onset of bipolar disorder is late teens and early-to-mid 20s. Approximately 50% of cases have onset below age 30 and nearly 90% before age 50 (Goodwin and Jamison, 1990). However, subjects can develop the first episode of bipolar disorder at any time during life.

The implication. Unaffected subjects (particularly those below age 30) are much less useful than are affected subjects because they provide less information about genetic risk. This favours 'affected only' forms of analysis.

3.3 Secular changes have been reported

The issue. A change in the rate of mood disorder in successive birth cohorts has been described in some studies although the effect has been less clearly demonstrated in bipolar disorder than in unipolar illness (Klerman, 1988). For example, in a study of first degree relatives of bipolar probands, Gershon *et al.* (1987) reported approximately twice the lifetime risk of affective disorders in relatives born after 1940 compared with those born before this date. However, this is by no means a universal finding (e.g. Pauls *et al.*, 1992) and may simply be an artefact of methodology (Giuffra and Risch, 1994) – important potential contributors include an increased willingness of younger subjects to report episodes of mood disturbance and increased forgetting of past episodes in older subjects.

The implication. This effect (whether real or artefactual) complicates prevalence-dependent analyses that include differing birth cohorts, further reducing the usefulness of unaffected individuals.

3.4 Pathogenesis is poorly understood

The issue. Like most complex traits, bipolar disorder results from the interplay of genetic and environmental factors. The illness has been subjected to study at many levels of analysis, ranging from molecular through to social, but there are few robust findings and pathogenesis remains poorly understood. Two consistent themes that have emerged are (i) the very persuasive body of evidence that highlights the important role of genes (see Section 9.3 below) and (ii) many studies that point to the importance of *stress* in predisposition to mood disorders and to involvement of the hypothalamic pituitary adrenal (HPA) endocrine system which mediates stress responses (Goodwin and Jamison, 1990; Hendrick *et al.*, 1998). To date, most theories about neurochemical pathology have been overly simplistic and based on the presumed mode of action of effective medications used in treatment (e.g. neuroleptics, antidepressants) and prophylaxis (e.g. lithium and anticonvulsants) of bipolar disorder, many of which have important actions at one or more of the major neurotransmitter systems (e.g. dopaminergic, noradrenergic and serotonergic) (Goodwin and Jamison, 1990). More complex (and plausible) hypotheses are being developed that postulate mechanisms involving modulation of a range of neurotransmitter systems through action at secondary messenger systems (e.g. Manji *et al.*, 1995) but the field lacks coherent theories of pathogenesis.

The implication. In view of the above observations, positional genetic approaches are relatively attractive because, in their pure form, they are independent of prior knowledge of pathophysiology. The corollary is that for most candidate gene studies there is a disappointingly low prior probability that the gene is involved.

3.5 Unknown validity of bipolar spectrum

The issue. Although clinical, outcome and genetic studies suggest that bipolar disorder is a relatively distinct nosological entity, a spectrum of phenotypes is associated with bipolar susceptibility genes (Farmer and McGuffin, 1989; Kendell, 1987). These include schizoaffective disorder in which individuals have features both of schizophrenia and bipolar disorder, and also some cases of recurrent unipolar major depression. However, in the absence of a clear understanding of the biology of illness the most appropriate boundaries and hierarchies between these categories remains unclear.

The implication. Most researchers employ a set of diagnostic models (usually between 2 and 5 in number) from the most narrowly defined illness (bipolar I disorder) to a broadly defined phenotype (typically including (i) bipolar I disorder, (ii) bipolar II disorder, (iii) bipolar disorder, not otherwise specified, (iv) schizoaffective disorder, bipolar type, and (v) recurrent major depression). Although these different diagnostic models are not independent, such an approach introduces issues of multiple testing that complicate interpretation of the results.

3.6 Phenotypic subtypes

The issue. Focusing on phenotypic subtypes thought to represent more genetically homogeneous forms of the disorder can increase the power to detect genes involved in a complex disease (Lander and Schork, 1994). Researchers are beginning to take an interest in some potentially useful subtypes in bipolar illness. It should be noted that some subtypes may reflect the action of genes (and/or evironmental factors) that *modify* the disease phenotype rather than influencing disease *susceptibility*, although for complex genetic diseases it may prove difficult to classify unambiguously all pathogenically relevant genes using this simple dichotomy.

The implication. Several phenotypic subtypes are currently under study including:

Puerperal psychosis. Puerperal psychosis describes severe (usually psychotic) psychiatric disorder occurring within days of childbirth (Brockington, 1996). The great majority of individuals who suffer episodes of puerperal psychosis have a bipolar disorder diathesis together with a susceptibility to puerperal triggering of episodes (usually manic). There is compelling evidence that the puerperal trigger is familial (Jones and Craddock, unpublished data) and plausible biological systems include steroid hormone pathways (Jones *et al.*, 1998). The existence of this clearly defined subset of bipolar female patients together with a clear biological hypothesis regarding the pathogenesis of triggering facilitates focused candidate gene studies on a relatively homogenous set of patients and, in principle should offer higher prior probabilities for the genes examined.

Rapid cycling. Rapid cycling describes the occurrence of four or more episodes of mood disorder within a 12 month period. It occurs in a substantial proportion of bipolar patients, is more common in females and can be induced by a variety of medications including antidepressants. Family studies have produced inconsistent results regarding the familiality of rapid cycling (Coryell *et al.*, 1992; Nurnberger *et al.*, 1988; Wehr *et al.*, 1988) but there are interesting findings emerging from candidate gene association studies (see Section 9.7 below).

Seasonal affective disorder. Seasonal affective disorder (SAD) describes mood disorder that has a characteristic seasonal variation, typically with winter depressions. Many subjects with SAD are bipolar. Genetic factors influence seasonal variation in mood (Madden *et al.*, 1996) and SAD aggregates in families (Allen *et al.*, 1993; Sher *et al.*, 1999). Biological systems implicated in the pathogenesis of SAD include systems involved in serotonergic neurotransmission and circadian and circannual clocks (Thompson *et al.*, 1997), again providing the opportunity for focused hypotheses in a clearly defined subset of bipolar patients.

Lithium responsiveness. Lithium is widely used as a prophylaxis against recurrence of both manic and depressive episodes. Despite its chemical simplicity it has a complex pharmacology and although its therapeutic mode of action remains poorly understood, it is believed that its effect on the phosphatidyl inositol signal transduction pathway is important (Berridge *et al.*, 1989). Up to 70% of patients benefit from lithium prophylaxis, some remaining free from all episodes for prolonged periods. Lithium responsiveness appears to be familial (Alda *et al.*, 1994; Grof *et al.*, 1994) and samples are being used for molecular genetic studies (e.g. Turecki *et al.*, 1999). Such samples are, however, difficult to recruit because accurate measurement of responsiveness to lithium (or any other prophylactic treatment) requires extended periods of observation of the course of illness both before and after initiation of the treatment. This can usually only be achieved satisfactorily in prospective designs.

4. Classical genetic studies

Classical genetic investigations provide a compelling and consistent body of evidence for the importance of genes influencing susceptibility to bipolar disorder (Craddock and Jones, 1999; McGuffin and Katz, 1989; Tsuang and Faraone, 1990). Using a modern definition of bipolar disorder 21 family studies, six twin studies and two adoption studies have been undertaken and all support the involvement of genes (Craddock and Jones, 1999). Approximate lifetime rates of

Table 2. Approximate lifetime rates of mood disorder in various classes of relative of bipolar probands

Degree of relationship to bipolar proband	Risk of bipolar disorder	(Additional) risk of unipolar depression
Monozygotic co-twin	40–70%	15–25%
First degree relative	5–10%	10–20%
General population (i.e. unrelated)	0.5–1.5%	5–10%

The lifetime risk of major mood disorder in a relative is obtained by *adding* the risk of bipolar disorder and the risk of unipolar depression. General population lifetime risk of unipolar depression is notoriously difficult to quantify but the figures in the table are based on a definition of clinically significant depression comparable to that used in the genetic studies. This table is reproduced with permission from Craddock and Jones (1999).

mood disorder in various classes of relative of a bipolar proband are shown in *Table 2*. Although there are many methodological impediments to producing a consistent quantification of risk, family studies suggest $\lambda_S \approx 7$ (95% confidence interval 5–10) and $\lambda_{MZ} \approx 50$ (95% confidence interval 40 – 60). Broadening the definition of phenotype increases the prevalence within relatives of a proband and in the general population, resulting in an overall reduction in familial recurrence risk. The implication for genetic studies is that concentrating on the narrowest definitions of phenotype is likely to provide optimal power to detect susceptibility genes. It is important to note that the twin studies provide persuasive evidence for the role of nongenetic factors in influencing susceptibility to illness, with an identical co-twin of a bipolar proband having 40–50% chance of escaping bipolar disorder (although a substantial proportion of these individuals will not escape recurrent depressive episodes). Unfortunately, the understanding of nongenetic risk factors for bipolar disorder is poor, making it difficult to follow designs in which specific and relevant nongenetic factors can be investigated alongside genetic factors.

5. Genetic model

A small number of very large multiply affected bipolar pedigrees has been described (e.g. Barden *et al.*, 1996; Blackwood *et al.*, 1996; Egeland *et al.*, 1987; McInnes *et al.*, 1996) and it seems likely that in at least some of these a single gene may play a major role in determining disease susceptibility. However, it should not be forgotten that polygenic inheritance can itself produce occasional large densely affected pedigrees consistent with a single gene model (McGue and Gottesman, 1989).

As any clinical psychiatrist will readily confirm, apparent 'single gene' pedigrees are very much the exception and the vast majority of bipolar disorder in the population shows a much more modest degree of familial clustering with absence of a compelling simple model. Some complex segregation analyses on systematically ascertained pedigree sets have produced results consistent with single gene models (Crowe and Smouse, 1977; Pauls *et al.*, 1992; Rice *et al.*, 1987; Spence *et al.*, 1995) whilst others have been unable to demonstrate major locus transmission (Bucher *et al.*, 1981; Goldin *et al.*, 1983; Sham *et al.*, 1991). However, caution is needed in interpreting results of these complex segregation analyses because of (i) limited power to distinguish between single gene and oligogenic models, and (ii) failure to take account of an important, known parameter, λ_{MZ} (Craddock *et al.*, 1997). Consideration of the rapid decrease in recurrence risk from an identical co-twin to first degree relative and back to the general population (*Table 1*) is *not* consistent with single gene modes of inheritance or simple locus heterogeneity models (i.e. collections of single genes) (Craddock *et al.*, 1995). The recurrence risk data are consistent with epistatic interaction of multiple genes (Craddock *et al.*, 1995) or with more complex genetic mechanisms such as dynamic mutation (McInnis *et al.*, 1993), imprinting or mitochondrial inheritance (McMahon *et al.*, 1995).

X-linkage. X-linkage was first suggested 65 years ago (Rosanoff *et al.*, 1935) and received vigorous debate during the 1970s and 1980s because of several reports of families showing co-segregation of bipolar disorder and X-linked markers (e.g. Baron *et al.*, 1987; Mendlewicz *et al.*, 1972; Reich *et al.*, 1969). However, these reports have been criticised on methodological grounds (Hebebrand, 1992) particular issues being ascertainment bias and nonblinding to marker status. If X-linkage does occur, it is likely to account for only a minority of cases (Bucher *et al.*, 1981; Risch *et al.*, 1986).

6. Positional studies

Linkage and association studies using DNA markers are the workhorse of modern approaches to complex genetic diseases and, conceptually, these studies may be divided into *positional* and *candidate gene* approaches. In the classical positional approaches that have been used with success for single gene disorders, the aim is to determine the chromosomal locations of disease-related genes using purely genetic methods. Conventionally this is approached initially by linkage studies in multiply-effected families, followed by finer localization using linkage disequilibrium (allelic association) studies and then physical mapping of the region. Such approaches are attractive for bipolar disorder because they are not dependent upon knowledge of disease pathophysiology. However, it is recognized that in practice the genetic dissection of most diseases requires a combination of positional *and* candidate approaches (the *positional-candidate* paradigm) (Collins 1995) and most research groups are pursuing both approaches in parallel.

For convenience in structuring our discussion we will divide linkage approaches into (i) traditional parametric lod score studies that use large densely affected pedigrees usually in the hope and expectation of identifying a single major gene in that pedigree or collection of pedigrees, and (ii) nonparametric linkage (or allele sharing) methods that use smaller, less densely affected family units usually with the expectation of identifying genes that make more modest influences on susceptibility to disease. It is, however, important to stress the arbitrary nature of this classification. Both parametric and nonparametric approaches *can* be used in pedigrees of differing size, structure and density of affection and many research groups have applied, and continue to apply, both forms of analysis to their differing pedigree sets.

6.1 Large family parametric linkage approach: early set-backs

Early linkage studies of bipolar disorder used traditional parametric lod score approaches in large densely affected families and were based on the implausible assumption that *all* illness was caused by one or a very small number of major 'Bipolar genes'. Two high profile claims of linkage were published in the journal *Nature* in the same year and have since entered the folklore of psychiatric genetics. Baron *et al.* (1987) reported linkage to classical genetic markers (color blindness and glucose-6-phosphate dehydrogenase deficiency) at Xq28 in several Israeli pedigrees and Egeland and colleagues (1987) reported linkage to DNA markers on chromosome 11p15 in a single very large pedigree of the Old Order Amish community in Pennsylvania. Despite impressive maximum lod scores in the original reports (>6 and >4, respectively) independent researchers were unable to replicate the findings and eventually in both cases the original groups published updated and extended analyses in their own data showing that the linkage support had all but vanished (Baron *et al.*, 1993; Kelsoe *et al.*, 1989). The reasons for such major changes in findings included:

(i) individuals initially believed unaffected developed illness for the first time during follow-up;
(ii) new branches of pedigrees were investigated which did not show evidence for linkage; and
(iii) extracting additional linkage information by typing new DNA markers resulted in a reduction of evidence for linkage.

Although very disappointing at the time, these reversals served to focus attention on a number of methodological issues in psychiatric molecular genetics, including the hazards of analyses that depend crucially upon a subject being *unaffected* and the need to think of the possibility of complex modes of inheritance, even within large apparently homogeneous pedigrees from a relatively genetically isolated population.

6.2 Ongoing large family parametric lod score studies

Several large densely affected pedigrees are being investigated in current positional studies. Although the main statistical approach continues to be the parametric lod score method, it is now usual to set relaxed parameters in order to allow for low penetrance, high phenocopy rate and locus heterogeneity. In addition, most researchers also apply nonparametric methods of analysis although, perhaps not surprisingly, available nonparametric methods generally fail to make optimal use of the rich genetic information contained in such large pedigrees. Regions of interest, defined by a haplotype segregating with affection status, have been identified in some such pedigrees (see *Table 3*) e.g. 12q23-q24 in a large French Canadian genetic isolate in Quebec (Barden *et al.*, 1996) and 4p16 in a large Scottish pedigree (Blackwood *et al.*, 1996). Such findings keep alive the hope that classical positional cloning can be used to identify the responsible gene. However, the mode of inheritance does not seem as simple as might be hoped even in these rare pedigrees. For example, the large pedigree from Quebec (Barden *et al.*, 1996) has a branch, indistinguishable phenotypically from the rest of the pedigree, that does not show 12q23-q24 linkage but that shows modest evidence for linkage with markers on 21q22. A further sobering observation is that, although persuasive, the lod score of 4.9 in this pedigree is much lower than the maximum lod score that this pedigree could generate if the mode of inheritance were both known and simple. Such observations make it difficult for researchers to have confidence that any apparent recombination provides robust positional information about the disease gene.

6.3 Nonparametric linkage studies in large samples of smaller families

The application of nonparametric methods of linkage analysis to larger samples of smaller less densely affected families has become increasingly popular. Potential advantages include:

(i) Nonparametric methods of analysis are independent of knowledge of mode of inheritance of disease and are relatively robust to nonpenetrance, phenocopies etc;
(ii) Such pedigrees are more representative of illness in the population;
(iii) Although having less power for linkage studies, such families are more common in the population and therefore more straightforward to recruit;
(iv) Small family designs facilitate focusing on the core phenotype which has the highest λ_S and therefore provides maximum power to detect linkage. In contrast, in large family designs many of the 'affected' family members exhibit spectrum phenotype with consequently less clearly defined relationship with the core phenotype.

Several groups around the world are undertaking large scale molecular genetic linkage studies using nonparametric approaches in collections of families of small to moderate size (e.g. Aita *et al.*, 1999; Berrettini *et al.*, 1997; Gill and Craddock, 1998; NIMH Genetics Initiative Bipolar Group, 1997). Because of the relative ease of collection and ready availability of analytic methods, sibpair designs have been popular although many samples contain other affected nonsibling members necessitating analysis using both sibpair (e.g. ASPEX, MAPMAKER/SIBS) and generic nonparametric (e.g. GENEHUNTER, MFLINK) methods. As with large family studies, interesting results are emerging, as described in the section below.

6.4 Summary of linkage findings to date

The pattern of results that has been emerging from linkage studies provides compelling evidence that *no single major gene exists that explains the majority of cases of bipolar disorder.* However, it is important for the reader to be aware that a number of methodological differences between studies complicates the task of assessing consistency between different findings (Craddock and Lendon, 1999):

(i) *Clinical factors vary between samples.* These include operational classification system used and spectrum of phenotype included in the sample;
(ii) *Genetic factors vary between samples.* This includes size and configuration of pedigrees, density of affected individuals, and the genetic background of the sample;

(iii) *Marker maps vary between studies*. Variation includes the particular map of markers chosen, the informativeness, and the inter-marker spacing used;
(iv) *Statistical analyses vary greatly between studies*. This includes use of a large variety of parametric and nonparametric methods, variously quoting results as *P* values, lod scores, NPL scores etc., and undertaking multiple testing using different diagnostic and genetic models;
(v) *Power of studies varies greatly*. This is of major importance for interpreting 'negative' results;
(vi) *Many data sets reported in the literature overlap*. Different reports may include different subsets of a data set thus making it difficult to come to a clear judgement of the overall weight of evidence for a particular finding.

With these limitations in mind *Table 3* provides information about four regions currently attracting interest in linkage studies of bipolar disorder. Of these 12q23-q24 and 21q22 are arguably the most promising (and, in passing we note that the apparent reduced incidence of mania in trisomy 21 is consistent with a chromosome 21 locus (Craddock and Owen, 1994a,b)). Despite, or perhaps because of, being the focus of a great deal of attention since the report by Berrettini and colleagues (1994), chromosome 18 is the most confusing 'region' with positive reports scattered throughout the chromosome in a pattern that is not consistent with a single susceptibility gene (Van Broeckhoven and Verheyen, 1999). Other regions of interest include 16p13 (Ewald *et al*., 1995); Xq24-q26 (Pekkarinen *et al*., 1995) and 15q11-q13 (Craddock and Lendon, 1999). New data are published regularly. Detailed and updated reviews of linkage data for each chromosome can be found in the chromosome workshop reports of the World Congress of Psychiatric Genetics, held annually. Reports from the Sixth World Congress held in Bonn, Germany, in 1998 can be found in *American Journal of Medical Genetics* (Neuropsychiatric Genetics), 1999; 88 (3).

For the reasons described, it is difficult to make a truly meaningful assessment of the weight of statistical evidence for each of these regions of interest. Lander and Kruglyak (1995) have proposed apparently simple criteria by which linkage findings can be judged. These apply for whole genome screens conducted and reported without bias. Unfortunately, many of the findings in the literature do not represent unbiased, systematic genome screens, making the 'L&K criteria' difficult to apply with confidence. This having been said, the findings on 12q and 21q can probably be regarded as satisfying 'L&K' significant or highly significant linkage together with independent replication.

The common features that are emerging from current studies include:

(i) No finding replicates in all data sets;
(ii) Levels of statistical significance and estimated effect sizes are usually modest;
(iii) Chromosomal regions of interest are typically broad (often >20–30 cM);
(iv) The same region may be implicated as (a) containing a susceptibility gene in large samples comprised of small families, and (b) harboring a putative major locus within one or more large densely affected kindreds;
(v) Some chromosome regions have been implicated in linkage studies of different psychiatric phenotypes. For example, 15q11-q13 is implicated in studies both of bipolar illness and schizophrenia (Craddock and Lendon, 1999).

Table 3. Some regions of interest from molecular genetic linkage studies of bipolar disorder

Chromosomal location	Study reference	Comment
4p16	Blackwood *et al.*, 1996	Single large UK pedigree with max. lod = 4.8
	Asherson *et al.*, 1998	Single moderate size UK pedigree (schizoaffective disorder) with max. lod =1.9
12q23-q24	Craddock *et al.*, 1994	Single UK pedigree in which Darier's disease (Craddock *et al.*, 1993; Sakhuntabhai *et al.*, 1999) and bipolar disorder co-segregate. Max. lod = 2.1
	Dawson *et al.*, 1995	29 small-moderate UK pedigrees. Max. lod = 2.0. No evidence in 16 German families
	Barden *et al.*, 1996	Single very large pedigree from an isolated French Canadian community. Max. lod = 4.9
	Ewald *et al.*, 1998	2 moderate size Danish pedigrees. Max. lod = 3.4
18 centromeric	Berrettini *et al.*, 1994	22 moderate size US pedigrees. Max. nonparametric evidence $P < 0.0001$
	Stine *et al.*, 1995	28 moderate size US pedigrees particularly those with paternal transmission (particularly those showing paternal transmission). Max. nonparametric evidence $P < 0.001$
18q22	Stine *et al.*, 1995	28 moderate size US pedigrees (particularly those showing paternal transmission). Max. sib pair evidence $P < 0.001$
18q22-q23	Freimer *et al.*, 1996	2 large Costa Rican pedigrees. Max. lod for combined linkage-association = 4.06
21q22	Straub *et al.*, 1994	Single large US pedigree. max. lod = 3.4
	Detera-Wadleigh *et al.*, 1996	22 moderate size US pedigrees (particularly those showing maternal transmission). Max. sib pair evidence $P < 0.001$
	Detera-Wadleigh *et al.*, 1997	97 moderate size US pedigrees. Max. sib pair evidence $P < 0.001$
	Smyth *et al.*, 1997	23 UK and Icelandic pedigrees. Max. lod = 2.2

This table illustrates some current regions of interest. Readers seeking a systematic overview of findings including both negative and positive reports should consult Chromosome Workshop reports (see text). This table is reproduced with permission from Craddock and Jones (1999).

It is worth noting that the first three of these are to be expected in the search for genes involved in any complex genetic disease (Kruglyak and Lander, 1995; Lander and Schork, 1994; Suarez *et al.*, 1994) and will be a feature of the findings described in other chapters in this book.

7. Attempts to identify dynamic mutations involved in susceptibility to bipolar disorder

Several neurological/neuropsychiatric diseases, such as Huntington's disease, myotonic dystrophy, and fragile X syndrome, are caused by DNA sequences of trinucleotide repeats that expand as they are passed from parent to offspring (Richards and Sutherland, 1997). Such illnesses often display the clinical phenomenon of anticipation (progressively more severe disease/early age at onset as the disease is transmitted through successive generations) and repeat expansion provides the molecular mechanism by which this anticipation is explained. Anticipation has been reported in some recent studies of bipolar disorder (e.g. McInnis *et al.*, 1993) leading to the hypothesis that trinucleotide repeat expansion may play a role in pathogenesis (reviewed by O'Donovan and Owen, 1996). Although the reported anticipation may be an artefact of ascertainment (O'Donovan and Owen, 1996) the molecular hypothesis has received support from four independent association studies using the repeat expansion detection (RED) method (Schalling *et al.*, 1993) that show trinucleotide repeat sequences to be significantly larger in bipolar subjects compared with controls (Lindblad *et al.*, 1995; O'Donovan *et al.*, 1995; O'Donovan *et al.*, 1996; Oruc *et al.*, 1997) although some studies have failed to find this association (Vincent *et al.*, 1996). The RED method is a biological assay that determines the size of the single largest trinucleotide repeat in an individual's genome but does not determine the chromosomal location of this largest repeat which, because all individuals in the population have a range of loci with such repeats, varies from individual to individual. Direct approaches have been used to examine known trinucleotide repeat loci in bipolar patients and controls in an attempt to identify the putative locus or loci associated with susceptibility to illness, so far without success (Franks *et al.*, 1999; Guy *et al.*, 1997). One of the groups with a positive RED finding has reported that most of their large RED products may be explained by two specific loci, CTG18.1 and ERDA7 (Lindblad *et al.*, 1998), and that CTG18.1 shows a modest association with bipolar disorder. However, the finding awaits independent replication.

8. Association studies in the candidate gene approach

The association paradigm provides an important experimental approach both to follow-up regions of interest detected in linkage studies (using both systematic linkage disequilibrium mapping and positional candidate studies) as well as for pure functional candidate gene studies. It is recognized that the approach provides relative robustness to genetic heterogeneity together with the ability to detect much smaller effect sizes than are achievable using feasible sample sizes in linkage studies (Camp, 1997; Risch and Merikangas, 1996). Linkage disequilibrium approaches are being used to follow-up regions of interest identified in linkage studies (e.g. Escamilla *et al.*, 1999) but robust positive findings have yet to emerge.

Most association studies have used the candidate gene approach. The main difficulty with this design is that poor understanding of the pathophysiology of illness compromises choice of candidate. To date, most candidate gene studies have focused on neurotransmitter systems that are influenced by medications used in clinical management of the disorder. Thus, as with other psychiatric phenotypes such as schizophrenia, studies of bipolar disorder have typically examined polymorphisms within genes encoding receptors or proteins and enzymes involved in metabolism or re-uptake of dopamine, serotonin (5HT) and noradrenalin. Early studies often used anonymous markers in the hope of detecting linkage disequilibrium but recently direct examination of polymorphisms of known or presumed functional relevance has become more usual. Most studies in the literature have been of the unrelated case-control design with samples rarely exceeding 200–300 subjects. No clear positive findings have emerged but we will illustrate the approach with three examples.

Tyrosine hydroxylase. A plausible functional candidate that has been subjected to considerable scrutiny over the last decade is tyrosine hydroxylase (TH), the rate limiting enzyme that catalyzes the first step in the synthesis of catecholamines. TH has been considered as an important candidate because catecholamine dysfunction has long been proposed as a possible biochemical mechanism in bipolar disorder. Moreover, the TH gene maps to 11p15, in the region implicated by the original Old Order Amish positive linkage report (Egeland *et al.*, 1987). Leboyer and colleagues (1990) aroused interest when they reported significant association with alleles at 2 RFLPs at TH in a small French case-control sample ($P < 0.01$). However, several groups have examined these and other polymorphisms at TH and a recent meta-analysis of 547 patients and 522 comparison individuals revealed no evidence that variation at this gene influences susceptibility to bipolar disorder (Turecki *et al.*, 1997). However, it should be recognized that even in this meta-analysis the sample sizes have modest power to detect plausible effect sizes for susceptibility genes for complex diseases. Further, although a useful technique to enhance sample size, meta-analysis has limited utility in situations where there is substantial genetic heterogeneity between samples and is compromised in situations in which publication bias exists.

Serotonin transporter. An extremely plausible gene that has attracted recent interest is the human serotonin transporter (hSERT). This is undoubtedly an excellent functional candidate because it is the site of action of the selective serotonin re-uptake inhibitors, a major class of antidepressants which are affective in treatment of bipolar depression and which can induce mania in bipolar individuals (Prozac™ is a well known example of this class of drug). Several studies have produced modest evidence implicating this gene using both case-control (Collier *et al.*, 1996; Kunugi *et al.*, 1997; Rees *et al.*, 1997) and family-based (Kirov *et al.*, 1999) approaches. However, the picture is far from clear with other groups reporting negative findings with the same polymorphisms (Battersby *et al.*, 1996; Bellivier *et al.*, 1997; Furlong *et al.*, 1998; Hoehe *et al.*, 1998; Stöber *et al.*, 1996). To make the situation more confusing, some groups have found association with one allele of a VNTR in intron 2, whereas others have found association with the short allele of a functional deletion/insertion polymorphism in the promoter and it does not appear that the different observations simply reflect linkage disequilibrium. hSERT is certainly not a gene of major effect. For example, in the studies reporting a statistically significant result, effect sizes, as measured by genotype odds ratios are typically in the region of 1.5–2.5. The findings may result from a modest influence on disease susceptibility, perhaps in a specific subset of bipolar disorder, or may simply reflect sampling variance. Further studies are required and undoubtedly genes of modest effect size will require an accumulation of large independent studies to provide robust statistical evidence to support their involvement in disease. However, no matter how compelling the statistical significance, biological evidence from a cellular, tissue and ultimately animal model will be necessary to confirm the pathogenic relevance of any putative susceptibility allele.

Catechol-O-methyl transferase in rapid cycling. Candidate gene association studies suggest that the low activity allele at a common polymorphism within the catechol-*O*-methyl transferase (COMT) gene may be a *course modifier* associated with increased susceptibility to rapid cycling within bipolar patients (Kirov *et al.*, 1998; Lachman *et al.*, 1996; Papolos *et al.*, 1998). This possibility has pleasing biological consistency with the observed tendency for antidepressants to induce rapid cycling in that both increase the availability of catecholamines at neuronal synapses. The data suggest that COMT makes only a modest contribution to rapid cycling observed in the bipolar population and confirmation is, of course, required in large independent samples.

Candidate gene association approaches are potentially very powerful, particularly when used within the context of a VAPSE (Variation Affecting Protein Sequence or Expression) paradigm (Sobell *et al.*, 1992) and using family-based association samples to minimize effects of population stratification. Studies to date have only scratched at the surface of the problem. A deeper understanding of the pathogenesis of bipolar disorder will almost certainly extend to systems involved in signal transduction and modulation of gene expression. As the human genome project unfolds and genes involved in these systems are cloned, candidate gene studies will offer increasingly powerful approaches to explore their role in pathogenesis and modification of the course of bipolar disorder.

In anticipation of this several research groups are assembling large samples of unrelated probands and appropriate comparison subjects for use in systematic linkage disequilibrium and candidate gene studies that are powered to detect genes of modest effect size. Although family-based paradigms are currently popular, such samples can be surprisingly difficult to collect for bipolar disorder, with unavailability of parents resulting frequently from death, family break up, migration and refusal. This practical issue combined with power considerations (Risch and Teng, 1998) suggests that both case-control and family-based samples will continue to have a place in association studies of bipolar disorder.

9. Summary and conclusion

Bipolar disorder (manic depressive illness) is a complex genetic disorder in which the core feature is pathological disturbance in mood, ranging from extreme elation (mania) to severe depression, usually accompanied by disturbances in thinking and behavior. The lifetime prevalence of 1% is similar in males and females and does not appear to vary widely between populations. Important methodological issues that impact genetic investigation of bipolar disorder include (i) the necessity to reduce lifetime experience of illness to a single diagnosis, (ii) variable age of onset, (iii) possible secular changes in lifetime risk, (iv) poor understanding of pathogenesis, and (v) unknown validity of the bipolar phenotypic spectrum. Classical genetic studies provide robust evidence for a major genetic contribution to susceptibility. The approximate lifetime risk of bipolar disorder in relatives of a bipolar proband are: monozygotic co-twin 40–70%; first degree relative 5–10%; unrelated person 0.5–1.5%. Occasional families exist in which a single gene appears to play the major role in determining susceptibility, but the majority of bipolar disorder involves interaction of multiple genes (epistasis) or more complex genetic mechanisms (such as dynamic mutation or imprinting). Molecular genetic positional and candidate gene approaches are being used for the genetic dissection of bipolar disorder. No gene has yet been identified but promising findings are emerging. Regions of interest identified in linkage studies include 4p16, 12q23-q24, 16p13, 21q22, and Xq24-q26. Chromosome 18 has received a great deal of attention but the findings are confusing with up to three possible regions implicated. Most candidate gene studies to date have focused on neurotransmitter systems influenced by medication used in clinical management of disorder but no robust positive findings have emerged. Research groups are recruiting larger samples that are appropriately powered to detect genes of modest effect size and this, together with the knowledge that will flow from the human genome project should facilitate the identification of bipolar susceptibility genes over the opening years of the new millennium. This will have a major impact on our understanding of disease pathophysiology and will provide important opportunities to investigate the interaction between genetic and environmental risk factors. This can be expected to lead to major improvements in patient care but will also raise important ethical issues that will need to be addressed with the same vigor applied to finding the genes.

Ackowledgments

Ian Jones is a Wellcome Trust Research Fellow in Mental Health. Nick Craddock is a Wellcome Trust Senior Research Fellow in Clinical Sciences.

References

Aita, V.M., Liu, J, Knowles, J.A., *et al.* (1999) A comprehensive linkage analysis of chromosome 21q22 supports prior evidence for a putative bipolar affective disorder locus. *Am. J. Hum. Genet.* **64**: 217.

Alda, M., Grof, P., Grof, E., Zvolsky, P. and Walsh, M. (1994) Mode of inheritance in families of patients with lithium-responsive affective disorders. *Acta Psychiatr. Scand.* **90**: 304–310.

Allen, J.M., Lam, R.W., Remick, R.A. and Sadovnick, A.D. (1993) Depressive symptoms and family history in seasonal and non seasonal mood disorders. *Am. J. Psychiat.* **150**: 443–448.

APA. (1994) *Diagnostic and Statistical Manual for Mental Disorders*, 4th Edn. American Psychiatric Association: Washington, DC.

Asherson, P., Mant, R., Williams, N., *et al.* (1998) A study of chromosome 4p markers and dopamine D5 receptor gene in schizophrenia and bipolar disorder. *Mol. Psychiat.* **3**: 310–320.

Barden, N., Plante, M., Rochette, D., *et al.* (1996) Genome-wide microsatellite marker linkage study of bipolar affective disorders in a very large pedigree derived from a homogeneous population in Quebec points to susceptibility locus on chromosome 12. *Psychiat. Genet.* **6**: 145–146.

Baron, M., Risch, N., Hamburger, R., *et al.* (1987) Genetic linkage between X-chromosome markers and bipolar affective illness. *Nature* **326**: 289–292.

Baron, M., Freimer, N.F., Risch, N., *et al.* (1993) Diminished support for linkage between manic-depressive illness and X-chromosome markers in three Israeli pedigrees. *Nature Genet.* **3**: 49.

Battersby, S., Ogilvie, A.D., Smith, C.A.D., *et al.* (1996) Structure of a variable number tandem repeat of the serotonin transporter gene and association with affective disorder. *Psychiat. Genet.* **6**: 177–181.

Bellivier, F., Laplanche, J.L., Leboyer, M., *et al.* (1997) Serotonin transporter gene and manic depressive illness: An association study. *Biol. Psychiat.* **41**: 750–752.

Berrettini, W.H., Ferraro, T.N., Goldin, L.R., *et al.* (1994) Chromosome 18 DNA markers and manic-depressive illness: evidence for a susceptibility gene. *Proc. Natl Acad. Sci. USA* **91**: 5918–5921.

Berrettini, W.H., Ferraro, T.N., Goldin, L.R., *et al.* (1997) A linkage study of bipolar illness. *Arch. Gen. Psychiat.* **54**: 27.

Berridge, M.J., Downes, C.P. and Hanley, M.R. (1989) Neural and developmental actions of lithium: A unifying hypothesis. *Cell* **59**: 411–419.

Blackwood, D.H.R., He, L., Morris, S.W., *et al.* (1996) A locus for bipolar affective disorder on chromosome 4p. *Nature Genet.* **12**: 427–430.

Brockington, I.F. (1996) *Motherhood and Mental Health.* Oxford University Press: Oxford.

Bucher, K.D., Elston, R.C., Green, R., *et al.* (1981) The transmission of manic depressive illness – II Segregation analysis of three sets of family data. *J. Psychiat. Res.* **16**: 65–78.

Camp, N.J. (1997) Genomewide transmission/disequilibrium – consideration of the genotypic relative risks at disease loci. *Am. J. Hum. Genet.* **61**: 1424.

Collier, D.A., Stober, G., Li, T., *et al.* (1996) A novel functional polymorphism within the promoter of the serotonin transporter gene: possible role in susceptibility to affective disorder. *Mol. Psychiat.* **1**: 453–460.

Collins, F.S. (1995) Positional cloning moves from periditional to traditional. *Nature Genet.* **9**: 347–350.

Coryell, W., Keller, M., Endicott, J., Andreasen, N., Clayton, P. and Hirschfeld, R. (1989) Bipolar II illness: course and outcome over a five year period. *Psychol. Med.* **19**: 129.

Coryell, W., Endicott, J. and Keller, M. (1992) Rapidly cycling affective disorder. Demographics, diagnosis, family history, and course. *Arch. Gen. Psychiat.* **49**: 126–131.

Craddock, N. and Jones, I. (1999) Genetics of bipolar disorder. *J. Med. Genet.* 36: 585–594.

Craddock, N. and Lendon, C. (1999) Chromosome workshop: Chromosomes 11, 14, and 15. *Am. J. Med. Genet. (Neuropsychiat. Genet.)* **88**: 244–254.

Craddock, N. and Owen, M. (1994a) Chromosomal aberrations and bipolar affective disorder. *Br. J. Psychiat.* **164**: 507–512.

Craddock, N. and Owen, M. (1994b) Is there an inverse relationship between Down's syndrome and bipolar affective disorder? Literature review and genetic implications. *J. Intell. Disabil. Res.* **38**: 613–620.

Craddock, N., Dawson, E., Burge, S., *et al.* (1993) The gene for Darier's disease maps to chromosome 12q23-q24.1. *Hum. Mol. Genet.* **2**, 1941–1943.

Craddock, N., Owen, M., Burge, S., Kurian, B., Thomas, P. and McGuffin, P. (1994) Familial cosegregation of major affective disorder and Darier's disease (keratosis follicularis). *Br. J. Psychiat.* **164**: 355–358.

Craddock, N., Khodel, V., Van Eerdewegh, P. and Reich, T. (1995) Mathematical limits of multilocus models: the genetic transmission of bipolar disorder. *Am. J. Hum. Genet.* **57**: 690–702.

Craddock, N., Van Eerdewegh, P. and Reich, T. (1997) Single major locus models for bipolar disorder are implausible. *Am. J. Med. Genet.* **74**: 18.

Crowe, R.R. and Smouse, P.E. (1977) The genetic implications of age-dependent penetrance in manic-depressive illness. *J. Psychiat. Res* **13**: 273.

Dawson, E., Parfitt, E., Roberts, Q., *et al.* (1995) Linkage studies of bipolar disorder in the region of the Darier's disease gene on chromosome 12q23–24.1. *Am. J. Med. Genet. (Neuropsychiat. Genet.)* **60**: 94–102.

Detera-Wadleigh, S.D., Badner, J.A., Goldin, L.R., *et al.* (1996) Affected sib-pair analyses reveal support of a prior evidence for a susceptibility locus for bipolar disorder, on 21q. *Am. J. Hum. Genet.* **58**: 1279–1285.

Detera-Wadleigh, S.D., Badner, J.A., Yoshikawa, T., *et al.* (1997) Initial genome scan of the NIMH genetics initiative bipolar pedigrees: Chromosomes 4, 7, 9, 18, 19, 20, and 21q. *Am. J. Med. Genet.* **74**: 254–262

Egeland, J.A., Gerhard, D.S., Pauls, D.L., *et al.* (1987) Bipolar affective disorders linked to DNA markers on chromosome 11. *Nature* **325**: 783–787.

Escamilla, M.A., McInnes, L.A., Spesny, M., *et al.* (1999) Assessing the feasibility of linkage disequilibrium methods for mapping complex traits: an initial screen for bipolar disorder loci on chromosome 18. *Am. J. Hum. Genet.* **64**: 1670–1678.

Ewald, H., Mors, O., Flint, T., Koed, K., Eiberg, H. and Kruse, T.A. (1995) A possible locus for manic depressive illness on chromosome 16p13. *Psychiat. Genet.* **5**: 71–81.

Ewald, H., Degn, B., Mors, O. and Kruse, T.A. (1998) Significant linkage between bipolar affective disorder and chromosome 12q24. *Psychiat. Genet.* **8**: 131–140.

Farmer, A. and McGuffin, P. (1989) The classification of depression. Contemporary confusion revisited. *Br. J. Psychiat.* **155**: 437.

Farmer, A.E., Williams, J. and Jones, I. (1994) Phenotypic definitions of psychotic illness for molecular genetic research. *Am. J. Med. Genet. (Neuropsychiat. Genet.)* **54**: 365–371.

Franks, E., Guy, C., Jacobsen, N., *et al.* (1999) Eleven trinucleotide repeat loci that map to chromosome 12 excluded from involvement in the pathogenesis of bipolar disorder. *Am. J. Med. Genet. (Neuropsychiat. Genet.).* **88**: 67–70.

Freimer, N.B., Reus, V.I., Escamilla, M.A., *et al.* (1996) Genetic mapping using haplotype, association and linkage methods suggests a locus for severe bipolar disorder (BPI) at 18q22–123. *Nature Genet.* **12**: 436–441.

Furlong, R.A, Ho, L., Walsh, C., *et al.* (1998) Analysis and meta-analysis of two serotonin transporter gene polymorphisms in bipolar and unipolar affective disorders. *Am. J. Med. Genet. (Neuropschiat. Genet.)* **81**: 58–63.

Gershon, E.S., Hamovit, J.H., Guroff, J.J. and Nurnberger, J.I. (1987) Birth-cohort changes in manic and depressive disorders in relatives of bipolar and schizoaffective patients. *Arch. Gen. Psychiat.* **44**: 314–319.

Gill, M. and Craddock, N. (1998) The Wellcome Trust UK-Irish bipolar sib pair study: Sample and study design. *Am. J. Med. Genet.* **81**: 461.

Giuffra, L.A. and Risch, N. (1994) Diminished recall and the cohort effect of major depression: a simulation study. *Psychol. Med* **24**: 375–383.

Goldin, L.R., Gershon, E.S., Targum, S.D., Sparkes, R.S. and McGinnis, M. (1983) Segregation and linkage analyses in families of patients with bipolar, unipolar, and schizoaffective mood disorders. *Am. J. Hum. Genet.* **45**: 274–287.

Goodwin, F.K. and Jamison, K.R. (1990) *Manic-Depressive Illness*. New York: Oxford University Press.

Grof, P., Alda, M., Grof, E., Zvolsky, P. and Walsh, M. (1994) Lithium response and genetics of affective disorders. *J. Affect. Disord.* **32**: 85–95.

Guy, C., Bowen, T., Daniels, J.K., *et al.* (1997) Exclusion of expansion of 50 CAG/CTG trinucleotide repeats in bipolar disorder. *Am. J. Psychiat.* **154**: 1146–1147.

Guze, S.B. and Robins, E. (1970) Suicide and primary affective disorders. *Br. J. Psychiat.* **117**: 437–438.

Hebebrand, J. (1992) A critical appraisal of X-linked bipolar illness. Evidence for the assumed mode of inheritance is lacking. *Br. J. Psychiat.* **160**: 7–11.

Hendrick, V., Altshuler, L. and Whybrown, P. (1998) Psychoneuroendocrinology of mood disorders: The hypothalamic-pituitary-thyroid axis. *Psychoneuroendocrinology (Psychiat. Clinics North Am.)* **21**, 277–292.

Hoehe, M.R., Wendel, B., Grunewald, I., *et al.* (1998) Serotonin transporter (5-HTT) gene polymorphisms are not associated with susceptibility to mood disorders. *Am. J. Med. Genet. (Neuropsychiat. Genet.)* **81**: 1–3.

Jones, I.R., McCandless, F., Morey, J., Benjamin, J., Brockington, I. and Craddock, N. (1998) Molecular genetic approaches to puerperal psychosis. *Am. J. Med. Genet.* **81**: 544.

Kelsoe, J.R., Ginns, E.I., Egeland, J.A, *et al.* (1989) Re-evaluation of the linkage relationship between chromosome 11p loci and the gene for bipolar affective disorder in the Old Order Amish. *Nature* **342:** 238–243.

Kendell, R. E. (1987) Diagnosis and classification of functional psychoses. *Br. Med. Bull.* **43**(3): 499.

Klerman, G.L. (1988) The current age of youthful melancholia: Evidence for increase in depression among adolescents and young adults. *Br. J. Psychiat.* **152**: 4–14.

Kirov, G., Murphy, K.C., Arranz, M.J., *et al.* (1998) Low activity allele of catechol-O-methyltransferase gene associated with rapid cycling bipolar disorder. *Mol. Psychiat.* **3**: 342–345.

Kirov, G., Rees, M., Jones, I., McCandless, F., Owen, M.J. and Craddock, N. (1999) Bipolar disorder and the serotonin transporter gene: A family-based association study. *Psychol. Med.* **29**: 1249–1254.

Kruglyak, L. and Lander, E.S. (1995) High-resolution genetic mapping of complex traits. *Am. J. Hum. Genet.* **56**: 1212–1223.

Kunugi, H., Hattori, M., Kato, T., *et al.* (1997) Serotonin transporter gene polymorphisms; ethnic difference and possible association with bipolar affective disorder. *Mol. Psychiat.* **2**: 457–462.

Lachman, H.M., Morrow, B., Shprintzen, R., *et al.* (1996) Association of codon 108/158 catechol-O-methyltransferase gene polymorphism with the psychiatric manifestations of velo-cardio-facial syndrome. *Am. J. Med. Genet. (Neuropsychiat. Genet.)* **67**: 468–472.

Lander, E. and Kruglyak, L. (1995) Genetic dissection of complex traits: guidelines for interpreting reporting linkage results. *Nature Genet.* **11**: 241.

Lander, E.S. and Schork, N.J. (1994) Genetic dissection of complex traits. *Science* **265**: 2037–2048.

Leboyer, M., Malafosse, A., Boularand, S., *et al.* (1990) Tyrosine hydroxylase polymorphisms associated with manic-depressive illness. *Lancet* **335**: 1219.

Leckman, J.F., Sholomskas, D., Thompson, D., Belanger, A. and Weissman, M.M. (1982) Best estimate of lifetime psychiatric diagnosis. *Arch. Gen. Psychiat.* **39**: 879–883.

Lindblad, K., Nylander, P.O., De bruyn, A., *et al.* (1995) Detection of expanded CAG repeats in bipolar affective disorder using the repeat expansion detection (RED) method. *Neurobiol. Dis.* **2**: 55–62.

Lindblad, K., Nylander, P.O., Zander, C., *et al.* (1998) Two commonly expanded CAG/CTG repeat loci: involvement in affective disorders? *Mol. Psychiat.* **3**: 405–410.

Madden, P.A.F., Heath, A.C., Rosenthal, N.E. and Martin, N.G. (1996) Seasonal changes in mood and behavior – The role of genetic factors. *Arch. Gen. Psychiat.* **53**: 47–55.

Manji, H.K., Potter, W.Z. and Lenox, R.H. (1995) Signal transduction pathways. Molecular targets for lithium's actions. *Arch. Gen. Psychiat.* **52**: 531–543.

McGue, M. and Gottesman, I.I. (1989) Genetic linkage in schizophrenia: perspectives from genetic epidemiology. *Schizophr. Bull.* **15**: 453.

McGuffin, P. and Katz, R. (1989) The genetics of depression and manic-depressive disorder. *Br. J. Psychiat.* **155**: 294–304.

McInnes, L.A., Escamilla, M.A. and Service, S.K. (1996) A complete genome screen for genes predisposing to severe bipolar disorder in two Costa Rican pedigrees. *Proc. Natl. Acad. Sci. USA* **93**: 13060.

McInnis, M.G., McMahon, F.J., Chase, G.A., Simpson, S.G., Ross, C.A. and DePaulo, J.R. Jr. (1993) Anticipation in bipolar affective disorder. *Am. J. Hum. Genet.* **53**: 385–390.

McMahon, F.J., Stine, O.C., Meyers, D.A., Simpson, S.G. and DePaulo, J.R. (1995) Patterns of maternal transmission in bipolar affective disorder. *Am. J. Hum. Genet.* **56**: 1277–1286.

Mendlewicz, J., Fleiss, J.L. and Fieve, R.R. (1972) Evidence for X-linkage in the transmission of manic-depressive illness. *J. Am. Med. Assoc.* **222**: 1624.

NIMH Genetics Initiative Bipolar Group: Nurnberger, J.I., DePaulo, J.R., Gershon, E.S., Reich, T., Blehar, M.C. *et al.* (1997) Genomic survey of bipolar illness in the NIMH Genetics Initiative Pedigrees: A preliminary report. *Am. J. Med. Genet. (Neuropsychiatr. Genet.)* **74**: 227–237.

Nurnberger, J., Guroff, J.J., Hamovit, J., Berrettini, W. and Gershon, E. (1988) A family study of rapid-cycling bipolar illness. *J. Affect. Disord.* **15**: 87–91.

O'Donovan, M.C. and Owen, M.J. (1996) Dynamic mutations and psychiatric genetics. *Psychol. Med.* **26**: 1–6.

O'Donovan, M.C., Guy, C., Craddock, N., *et al.* (1995) Expanded CAG repeats in schizophrenia and bipolar disorder. *Nature Genet.* **10**: 380–381.

O'Donovan, M.C., Guy, C., Craddock, N., *et al.* (1996) Confirmation of association between expanded CAG/CTG repeats and both schizophrenia and bipolar disorder. *Psychol. Med.* **26**: 1145–1153.

Oruc, L., Lindblad, K., Verheyen, G.R., *et al.* (1997) CAG repeat expansions in bipolar and unipolar disorders. *Am. J. Hum. Genet.* **60**: 730–732.

Papolos, D.F., Veit, S., Faedda, G.L., Saito, T. and Lachman, H.M. (1998) Ultra-ultra rapid cycling bipolar disorder is associated with the low activity catecholamine-O-methyl transferase allele. *Mol. Psychiat.* **3**: 346–349.

Pauls, D.L., Morton, L.A. and Egeland, J.A. (1992) Risks of affective illness among first-degree relatives of bipolar I Old-Order Amish probands. *Arch. Gen. Psychiat.* **49**: 703–708.

Pekkarinen, P.J., Bredbacka, P.-E., Loonqvist, J. and Peltonen, L. (1995) Evidence of a predisposing locus to bipolar disorder on Xq24-q27.1 in an extended Finnish pedigree. *Genome Res.* **5**: 105–115.

Rees, M., Norton, N., Jones, I., *et al.* (1997) Association studies of bipolar disorder at the human serotonin transporter gene (hSERT; 5HTT) *Mol. Psychiat.* **2**: 398–402.

Reich, T., Clayton, P. and Winokur, G. (1969) Family history studies: V. The genetics of mania. *Am. J. Psychiat.* **125**: 1358–1368.

Rice, J., Reich, T., Andreasen, N.C., *et al.* (1987) The familial transmission of bipolar illness. *Arch. Gen. Psychiat.* **44**: 441–447.

Richards, R.I. and Sutherland, G.R. (1997) Dynamic mutation: possible mechanisms and significance in human disease. *TIBS* **22**: 432–436

Risch, N. and Merikangas, K. (1996) The future of genetic studies of complex human diseases. *Science* **273**: 1516.

Risch, N. and Teng, J. (1998) The relative power of family-based and case-control designs for linkage disequilibrium studies of complex human diseases. I. DNA pooling. *Genome Res.* **8**: 1273–1288.

Risch, N., Baron, M. and Mendlewicz, J. (1986) Assessing the role of X-linked inheritance in bipolar-related major affective disorder. *J. Psychiat. Res.* **20**: 275–288.

Rosanoff, A.J., Handy, L.M. and Plesset, I.R. (1935) The etiology of manic-depressive syndromes with special reference to their occurrence in twins. *Am. J. Psychiat.* **91**: 725–762.

Sakuntabhai, A., Ruiz-Perez, V., Carter, S., *et al.* (1999) Mutations in ATP2A2, encoding a Ca^{2+} pump, cause Darier disease. *Nature Genet.* **21**: 271–277.

Schalling, M., Hudson, T.J, Buetow, K.H. and Houseman, D.E. (1993) Direct detection of novel expanded trinucleotide repeats in the human genome. *Nature Genet.* **4**: 135–139.

Sham, P.C., Morton, N.E. and Rice, J.P. (1991) Segregation analysis of the NIMH collaborative study. Family data on bipolar disorder. *Psychiat. Genet.* **2**: 175–184.

Sher L., Goldman D., Ozaki N. and Rosenthal N.E. (1999) The role of genetic factors in the etiology of seasonal affective disorder and seasonality. *Affect. Disord.* **53**: 203–210.

Simpson, S.G., Folstein, S.E., Meyers, D.A., McMahon, F.J., Brusco, D.M. and DePaulo, J.R. (1993) Bipolar II: The most common bipolar phenotype? *Am. J. Psychiat.* **150**: 901.

Smith, A.L. and Weissman, M.M. (1992) Epidemiology. In: *Handbook of Affective Disorders* (ed. E.S. Paykel), Churchill Livingstone, Edinburgh, pp. 111–129.

Smyth, C., Kalsi, G., Curtis, D., *et al.* (1997) Two-locus admixture linkage analysis of bipolar and unipolar affective disorder supports the presence of susceptibility loci on chromosomes 11p15 and 21q22. *Genomics* **39**: 271–278.

Sobell, J.L., Heston, L.L. and Sommer, S.S. (1992) Delineation of genetic predisposition to multifactorial disease: a general approach on the threshold of feasibility. *Genomics* **12**: 1–6.

Spence, M.A., Flodman, P.L., Sadovnik, A.D., Bailey-Wilson, J.E., Ameli, H. and Remick, R.A. (1995) Bipolar disorder: evidence for a major locus. *Am. J. Med. Genet. (Neuropsychiat. Genet.)* **60**: 370–376.

Stine, O.C., Xu, J., Koskela, R., *et al.* (1995) Evidence for linkage of bipolar disorder to chromosome 18 with a parent-of-origin effect. *Am. J. Hum. Genet.* **57**: 1384–1394.

Stöber, G., Heils, A. and Lesch, K.P. (1996) Serotonin transporter gene polymorphism and affective disorder. *Lancet* **347**: 1340–1341.

Straub, R.E., Lehner, T., Luo, Y., *et al.* (1994) A possible vulnerability locus for bipolar affective disorder on chromosome 21q22.3. *Nature Genet.* **8**: 291–296.

Suarez, B.K., Hample, C.L. and Van Eerdewegh, P. (1994) Problems of replicating linkage claims in psychiatry. In: *Genetic Approaches to Mental Disorders*, (eds Gershon, E.S. and Cloninger, C.R.). American Psychiatric Press, Inc., Washington DC, pp. 23–46.

Thompson, C., Childs, P.A., Martin, N.J., Rodin, I. and Smythe, P.J. (1997) Effects of monitoring phototherapy on circadian markers in seasonal affective disorder. *Br. J. Psychiat.* **170**: 431–435.

Tsuang, M.T. and Faraone, S.V. (1990) *The Genetics of Mood Disorders.* The Johns Hopkins University Press, Baltimore.

Turecki, G., Rouleau, G.A., Mari, J., Joober, R. and Morgan, K. (1997) Lack of association between bipolar disorder and tyrosine hydroxylase: A meta-analysis. *Am. J. Med. Genet. (Neuropsychiat. Genet.)* **74**: 348–352.

Turecki, G., Grof, P., Cavazzoni, P., *et al.* (1999) MAOA: association and linkage studies with lithium responsive bipolar disorder. *Psych. Genet.* **9**: 13–16.

Van Broeckhoven, C. and Verheyen, G. (1999) Report of the chromosome 18 workshop. *Am. J. Med. Genet. (Neuropsychiat. Genet.)* **88**: 263–270.

Vincent, J.B., Klempan, T., Parikh, S.S., *et al.* (1996) Frequency analysis of large CAG/CTG trinucleotide repeats in schizophrenia and bipolar affective disorder. *Mol. Psychiat.* **1**: 141–148.

Wehr, T.A., Sack, D.A., Rosenthal, N.E. and Cowdry, R.W. (1988) Rapid cycling affective disorder: contributing factors and treatment responses in 51 cases. *Am. J. Psychiat.* **145**: 179–184.

WHO (1993) *The ICD10 Classification of Mental and Behavioural Disorders.*

10

Schizophrenia

M.J. Owen, N.M. Williams and M.C. O'Donovan

1. Introduction

Schizophrenia is a severe and debilitating mental disorder which, despite the move towards community care, still accounts for more hospital bed occupancy than any other disease (Gottesman, 1991). It is a disorder with a relatively low population incidence (0.17 – 0.57 per 1000), a relatively high point prevalence (2.4 – 6.7 per 1000) and a lifetime morbid risk of slightly less than 1% (Gottesman, 1991; Jablensky, 1999). The incidence of schizophrenia is surprisingly uniform across a variety of different populations with different cultures, and the evidence for pockets of particularly high and low incidence and prevalence is not strong (Jablensky, 1999). Most individuals with schizophrenia have an onset in early adulthood, have long periods of illness, are unable to work, and have difficulty in sustaining family relationships because of their illness. Schizophrenia also has an enormous impact on families, including the necessity of providing informal care. Epidemiological studies have identified a number of putative associations with schizophrenia. These include: certain pregnancy and delivery complications (Hultman *et al.*, 1999; Jones *et al.*, 1998), delayed developmental milestones (Jones *et al.*, 1994), low IQ score (David *et al.*, 1997), personality characteristics concerned with social relations (Malmberg *et al.*, 1998), urban upbringing (Lewis *et al.*, 1992), immigration (Hutchinson *et al.*, 1996) and the use of illegal drugs especially cannabis (Andreasson *et al.*, 1987). However, in each case the relative risks are small compared with those associated with having an affected first degree relative.

Pathogenic theories have for many years been dominated by ideas based on neuropharmacological studies suggesting that abnormalities in monoamine neurotransmission, in particular of the dopaminergic and serotonergic systems, underlie schizophrenia. The predominant hypothesis for more than 3 decades was that schizophrenia is caused by excessive dopaminergic neurotransmission. This view was based on two observations. First, most conventional antipsychotic drugs block dopamine receptors. Second, dopaminergic drugs such as amphetamine are psychotomimetic. More recently it has become apparent that many of the newer drugs that are used to treat schizophrenia may also owe their efficacy, in part, to effects on the serotonergic transmitter system, specifically the 5HT2a receptor. However, evidence for primary neurotransmitter abnormalities has not been

Analysis of Multifactorial Disease, edited by T. Bishop and P. Sham.

forthcoming. Moreover the effects of antipsychotic drugs are not specific to schizophrenia, and these compounds are also effective in affective and organic psychoses. It seems likely therefore that while primary disturbances in monoamine neurotransmission may play a role in the etiology of psychosis, other pathogenic mechanisms must also be involved.

There is a growing body of empirical evidence in support of abnormal glutamate function in schizophrenia from neuroimaging and postmortem studies (Olney and Farber, 1995; Deakin *et al.*, 1997). Unfortunately, one of the problems that has plagued studies of the biology of schizophrenia is distinguishing primary changes in brain function from those secondary to the disorder or its treatment. Therefore, it is of particular importance that pharmacological interventions at NMDA glutamatergic receptors can either precipitate a psychosis similar to schizophrenia by inhibition of NMDA function (Javitt and Zuchen, 1991), or possibly ameliorate the course of schizophrenia by potentiation of NMDA function (Heresco-Levy *et al.*, 1996). Furthermore, mice expressing low levels of NMDAR1 receptor subunits display behaviors that are frequently used to model schizophrenia in pharmaceutical research, and these can be attenuated by treatment with anti-psychotic drugs (Mohn *et al.*, 1999).

In the last 13 years the view that schizophrenia is, at least in part, a neurodevelopmental disorder has become widely accepted. This is based on some of the epidemiological findings cited above and on the observation that cerebral anatomical abnormalities revealed by neuroimaging studies are present at the first episode of psychosis (Murray *et al.*, 1988). However, attempts to identify the nature of the neurodevelopmental processes that are disturbed have not yet succeeded and may have to await further advances in developmental neurobiology. This means that at the present time it is difficult to base rational selection of candidate genes on the 'neurodevelopmental hypothesis'.

Ignorance of pathogenesis has meant that, despite considerable effort and ingenuity, it has not been possible to delineate etiologically distinct subgroups and we remain largely ignorant of pathogenic mechanisms. The diagnosis of schizophrenia therefore still embraces conditions with a variety of symptoms, courses, outcomes, responses to treatment and probably etiologies. This means that the disorder is effectively syndromic and diagnosis has to rely on a combination of heterophenomenology and clinical observation. In spite of the fact that the diagnosis of schizophrenia is a clinical one that cannot be validated by laboratory testing, the use of structured and semi-structured interviews together with explicit operational diagnostic criteria means that it is possible to achieve high degrees of diagnostic reliability. Moreover, the syndrome so defined has a high heritability. However, we also have to accept that the disorder, as defined by current diagnostic criteria, may well include a number of heterogeneous disease processes and that this will hinder any attempts at identifying etiological factors.

Given the difficulties above, it should not come as a surprise that to date, disappointments abound in schizophrenia genetic research, particularly in the field of molecular genetics. Nevertheless, progress has been made. In this article, we shall review these advances, consider some of the difficulties that lie ahead, and suggest possible solutions and the trajectory future research is likely to follow.

2. Genetic epidemiology

2.1 Evidence for genetic influences

Molecular genetic approaches to schizophrenia have been stimulated by a rich literature in genetic epidemiology. Family studies over many years have confirmed an increased risk of schizophrenia in relatives of probands with the disorder (McGuffin *et al.*, 1995). Whereas the lifetime risk for schizophrenia in the general population is generally reported to be around 1%, the risk in siblings and offspring is ten times this. The risk in a particular class of relative compared to that in the general population is often denoted as λ, the λ for siblings (λs) in schizophrenia is thus approximately ten (Risch, 1990).

Schizophrenia has also been the focus of a large number of twin and adoption studies, which have allowed the relative contributions of nature and nurture to familal aggregation to be estimated. The evidence here is clear-cut; shared genes rather than shared environments underlie the increased risk of illness in relatives of those with the disorder (McGuffin *et al.*, 1994). Undoubtedly, as yet unknown nonshared environmental factors play an important role in determining liability to schizophrenia. However, twin and adoption studies have revealed a substantial genetic contribution, with estimates of heritability greater than 60% (McGuffin *et al.*, 1994) which are almost certainly conservative (Cannon *et al.*, 1998; Cardno *et al.*, 1999).

2.2 What is inherited and how?

Whilst it is clear from family, twin and adoption studies that there is a genetic contribution to schizophrenia, the existence of monozygotic twins that are discordant for the clinical diagnosis strongly suggests that what is inherited is not the certainty of disease accompanying a particular genotype but rather a predisposition or liability to develop the disorder. Moreover twin and adoption studies have also shown that schizophrenia shares familial, and probably genetic, liability with a range of other psychotic illnesses (Kendler *et al.*, 1993b) and personality disorders, such as schizotypal personality disorder (Kendler *et al.*, 1993a), collectively known as schizophrenia spectrum disorders.

Studies on the recurrence risk in various classes of relative allow us to exclude the possibility that schizophrenia is a single-gene disorder or collection of single gene disorders even when incomplete penetrance is taken into account. Rather, the mode of transmission like that of other complex disorders is complex and nonMendelian (McGue *et al.*, 1989). The commonest mode of transmission is probably oligogenic (a small number of genes of moderate effect) or polygenic (many genes of small effect) or a mixture of the two (McGuffin *et al.*, 1995). However, the number of susceptibility loci, the disease risk conferred by each locus, and the degree of interaction between loci all remain unknown. Just as the risk in a particular class of relative compared to that in the general population is often denoted as λ, so the contribution of an individual allelic variant to the familiality of a disorder can be expressed in these terms. Thus, for example for a given risk allele, λ_S is the relative risk to siblings resulting from possession of the disease allele (Risch, 1990). Risch (1990) has calculated that the data for recurrence risks in the relatives of probands with schizophrenia are incompatible with the existence of a single locus of $\lambda_S > 3$ and, unless extreme epistasis exists, models with two or three loci of λ_S less than or equal to 2 are more plausible. It should be emphasized that these calculations are based upon a homogeneous population, and under a model of heterogeneity, it is quite possible that genes of larger effect are operating in some subpopulations of patients.

In spite of the problems and uncertainties we have described, and the difficulties that ensue, to many of us, schizophrenia seems a compelling candidate for molecular genetic approaches. This is because genes make such a strong and fairly specific contribution to its etiology, and the brain is still difficult to study directly despite advances in neuroimaging. In particular it is hoped that one or other variation of the nonhypothesis driven positional cloning paradigm will enable us to identify susceptibility genes in spite of our incomplete understanding of pathophysiology.

3. Molecular genetic approaches

3.1 Linkage studies

The first generation of systematic molecular genetic studies of schizophrenia effectively ignored the evidence for genetic complexity and targeted large multiply affected pedigrees for analysis. This was done with the hope that etiological heterogeneity exists and that such families, or at least a proportion of them, are segregating genes of major effect. This approach has been successful in other complex disorders, for example Alzheimer's disease (AD), where mutations in three genes, *APP*, *PS1* and *PS2* are now known to cause rare forms of the disorder. In such cases, the disease is of unusually early onset and is transmitted through multiplex pedigrees as an autosomal dominant fashion (Goate *et al.*, 1991; Levy-Lahad *et al.*, 1995, Sherrington *et al.*, 1995).

Unfortunately, it has not proven possible to identify a phenotypic trait analogous to age at onset in AD by which to classify multiplex families segregating schizophrenia. Nevertheless, studies of such large families also initially produced positive findings in schizophrenia (Sherrington *et al.*, 1988) but unfortunately these could not be replicated. The reasons for this have become clear as data from systematic genome scans have accumulated; highly penetrant mutations causing schizophrenia are at best extremely rare and quite possibly nonexistent (Craddock *et al.*, 1995; McGuffin *et al.*, 1996). The 'false-positives' were due largely to a combination of multiple testing and the use of statistical methodology and significance levels derived from work on single-gene disorders.

In spite of the failure to identify regions of unambiguous linkage in multiply affected families, modest evidence for several regions has been reported in more than one data set. Areas implicated for which supportive data have also been obtained from international collaborative studies include chromosome 6p24–22, 8p22–21 and 22q11–12 (Gill *et al.*, 1996; Levinson *et al.*, 1996, reviewed by McGuffin *et al.*, 1996 and Moldin *et al.*, 1997). There are also a number of other promising areas of putative linkage currently under investigation by international consortia. These include 13q14.1-q32 (Blouin *et al.*, 1998; Brzustowicz *et al.*, 1999; Lin *et al.*, 1995), 5q21-q31, 18p 22–21 and 10p15-p11 (Faraone *et al.*, 1998; Schwab *et al.*, 1997, 1998a,b; Straub *et al.*, 1997, 1998;). However, in each case there are negative as well as positive findings, and in only one case, that of chromosome 13q, did any single study achieve genome-wide significance at $\alpha=0.05$.

Many groups are now moving away from the analysis of multiplex families in favor of smaller families with two or more affecteds in a single sibship in the belief that such families may be more suited for the analysis of complex traits. We have recently reported the largest systematic search for linkage yet published using such a sample of 196 affected sibling pairs (ASP; Williams *et al.*, 1999). The results of simulation studies suggested that the power of this study was >0.95 to detect a susceptibility locus of $\lambda_S = 3$ with a genome wide significance of 0.05 but only 0.70 to detect a locus of $\lambda_S = 2$ with the conservative assumption that a locus lies midway between two adjacent markers. This study yielded evidence at the level of Lander and Kruglyak's definition of 'suggestive' linkage to chromosomes 4p, 18p and Xcen (Lander *et al.*, 1995). However, none of our findings approached a genome-wide significance of 0.05 corresponding to Lander and Kruglyak's definition of 'significant' linkage.

The findings from linkage studies of schizophrenia to date demonstrates several features that are to be expected in the search for genes for complex traits (Lander et al., 1994, 1995; Roberts *et al.*, 1999; Suarez *et al.*, 1994). First, no finding replicates in all data sets. Second, levels of statistical significance are unconvincing and estimated effect sizes are usually modest. Third, chromosomal regions of interest are typically broad (often >20–30 cM).

At the present time therefore, the linkage literature supports the predictions made by Risch (1990): it is highly unlikely that a commonly occurring locus of effect size $\lambda_S > 3$ exists, but there is suggestive evidence implicating a number of regions which is consistent with the existence of some susceptibility alleles of moderate effect (λ_S 1.5–3).

Linkage methods in sample sizes that are realistically achievable can detect smaller genetic effects than the studies to date. For example it is possible to detect alleles with λ_S 1.5–3 in a sample of 600–800 ASPs (Hauser *et al.*, 1996; Scott *et al.*, 1997). We therefore suggest that priority should now be given to collecting such samples using robust clinical methodology that is comparable across all interested research groups. However, if liability to schizophrenia is entirely due to the operation of many genes of small effect, then even these large-scale studies will be unsuccessful.

3.2 Candidate gene association studies

Once genes of smaller effect than $\lambda_S = 1.5$ are sought the number of affected family members required becomes prohibitively large (Hauser *et al.*, 1996; Risch *et al.*, 1996; Scott *et al.*, 1997). For this reason many researchers have sought to take advantage of the potential of candidate gene association studies to identify such loci (Owen and McGuffin, 1993; Risch *et al.*, 1996). Although a potentially powerful means of identifying genes of small effect, association studies are not without their problems. First, for enigmatic disorders such as schizophrenia, the choice of candidate genes is limited largely by the imagination and resources of the researcher. This places a stringent burden of statistical proof on positive results because of low prior probability and multiple testing (Owen *et al.*, 1997). Second, case-control association studies have the potential to generate false positives due to population stratification. This problem can be addressed by using family-based association methods (Schaid *et al.*, 1994), but because of stigma, adult age at onset and the disruptive effects of mental illness on family relationships, family-based samples may be unrepresentative as well as limited in size. Consequently, family-based studies may introduce more spurious results than do case–control studies (Risch *et al.*, 1998). It would seem unwise therefore to discard the case–control study design, which has served epidemiology so well over the years. A third problem is common to all molecular genetic studies in complex diseases is that they are prone to type 2 errors simply because they are often underpowered, and therefore in order to draw satisfactory conclusions from negative studies, larger sample sizes are required than have been used typically to date in psychiatric genetics (Owen *et al.*, 1997). Fourth, even with larger samples, it is by no means certain that a given replication study will be sufficiently powered to replicate a particular effect. This is because there may be differences in the contribution of a given susceptibility allele in different patient populations as a result of different allele frequencies at the locus of interest or at interacting loci. Further potential for heterogeneity occurs if the association with the marker is a result of tight linkage with the true susceptibility allele, or if different subtypes of the disease exist. Given that all the above may influence power, and that none of the above are known in advance, it is difficult to obtain an accurate measure of the power of a replication study. Since we cannot specify accurately the prior probability of a candidate gene, nor can we know the true power for replication, it is difficult to draw definitive conclusions from conflicting findings. However, the purpose of experiment is to reject a null hypothesis and that in the face of uncertainty, the burden of proof remains with the proponents of a particular candidate gene.

Most candidate gene studies have derived their functional authority from neuropharmacological studies suggesting that abnormalities in monoamine neurotransmission, in particular dopaminergic and serotonergic systems, play a role in the etiology of schizophrenia. Overall, the results in this extensive literature are disappointing, but it should be noted that the sample sizes in many of the older studies would now generally be regarded as inadequate, particularly in view of the fact that the polymorphic markers in question did not in themselves represent functional variants and few genes have been systematically screened even for common variants. However, recently, there have been more promising reports of candidate gene associations, three of which are considered here.

Serotonin 5-HT2a receptor. Many of the newer drugs that are used to treat schizophrenia were selected because of their effects on the serotonergic system. The first genetic evidence that serotonergic receptors might play a role in schizophrenia came from a Japanese group who reported association between a T to C polymorphism at nucleotide 102 in the *5HT2a* receptor gene in a small sample

(Inayama *et al.*, 1996). A large European consortium consisting of seven centers and involving 571 patients and 639 controls then replicated this finding (Williams *et al.*, 1996), a result that was further replicated by using a family-based design (Spurlock *et al.*,1998). While many other studies have followed with mixed results, a recent meta-analysis of all available data including over 3000 subjects support the original finding ($P = 0.0009$) and this does not appear to have been due to publication bias (Williams *et al.*, 1997).

Since this meta-analysis was undertaken, a few further negative reports have followed, but none has approached the sample sizes required. If we assume homogeneity of the genetic architecture of schizophrenia and if the association is true, the putative odds ratio (OR) for the C allele can be expected to be around 1.2 in any replication sample. Sample sizes of 1000 subjects are then required for 80% power to detect an effect of this size even at a relaxed criterion of $\alpha = 0.05$. Thus, the negative studies are effectively meaningless, but it is also true that the evidence for association even in the meta-analysis is not definitive ($P= 0.0009$). At present, all we can conclude is that the evidence favors association between the T102C 5HT2a polymorphism and schizophrenia, but the burden of proof has not yet been met.

If the association is real, it is unlikely that the T102C is the susceptibility variant because this nucleotide change does not alter the predicted amino acid sequence of the receptor protein, nor is it in a region of obvious significance for regulating gene expression. It is therefore of some interest that we have shown association between schizophrenia and a new polymorphism in the promoter region of this gene. Unfortunately, there is no evidence that this has a functional effect either (Spurlock *et al.*, 1998), but recent evidence of polymorphic monoallelic expression of the *5HT2a* gene points to the possible existence of functional sequence variants that have not yet been identified (Bunzel *et al.*, 1998), and one or more of these may be the true susceptibility variants.

Dopamine D3 receptor gene. For more than 30 years, thinking about the biology of schizophrenia has been dominated by the idea that the illness reflects a disorder of dopaminergic neurotransmission, with disordered transmission at the D2 receptor widely favored. However, with few exceptions, candidate gene studies of D2 dopamine receptor gene (*DRD2*) and schizophrenia have been negative. In contrast, association has been widely reported between schizophrenia and homozygosity for a Ser9Gly polymorphism in exon 1 of the D3 dopamine receptor gene (*DRD3*; Crocq *et al.*, 1992). This receptor is functionally similar to *DRD2* and is expressed in regions of the CNS which are believed to be involved in the pathogenesis of schizophrenia. As with the *5HT2a* association, the results have now been confirmed in several independent samples, including one family based study (Williams *et al.*, 1998), but several negative studies were also reported. Fortunately, as before, interpretation of these findings is simplified by a meta-analysis of data from over 5000 individuals, which revealed a small (O.R.=1.23) but significant ($P=0.0002$) association between homozygosity at Ser9Gly and schizophrenia (Williams *et al.*, 1998). Again this could not easily be ascribed to selective publication (Williams *et al.*, 1998). Because it is a fairly uncommon genotype, we cannot be certain that homozygosity for the 9Gly allele alone is not associated with increased risk of schizophrenia and therefore it is not certain that the findings at D3 are an example of heterosis. However, a plausible biological explanation for D3 heterosis has been put forward in that possession of two different molecular forms of the receptor might allow the dopaminergic neuron to respond more flexibly (Crocq *et al.*, 1992).

At present then, the status of the *DRD3* finding is similar to that of *5HT2a*, that is the balance of evidence at present favors association, but the null hypothesis still cannot be confidently rejected. Those wishing to replicate or reject these findings should bear in mind that in order to obtain power of >0.80 to detect an effect of this size at a criterion of $\alpha=0.05$, a sample of 1500 cases and 1500 controls will be required. So far, no other polymorphisms have been found that might explain the putative D3 association, but our group has recently identified and sequenced several new polymorphisms in previously unknown exons 5′ to the exon referred to above as exon 1 (unpublished). We are currently testing these in our patient samples to establish whether variants in this region in LD with the Ser9Gly polymorphism provide a more functionally plausible explanation of our findings.

Anticipation and trinucleotide repeats. Anticipation was first described in connection with severe mental disorder (Morel, 1857), and a series of recent studies that have applied modern diagnostic criteria have now confirmed that the inheritance of schizophrenia is at least consistent with this phenomenon, although ascertainment biases offer an alternative explanation (O'Donovan and Owen, 1996). Since pathogenic expanded trinucleotide repeats are the only known genetic mechanisms for anticipation, these findings have been taken as suggesting that such mutations might account for at least some of the complexity in the pattern of inheritance in this disorder (Petronis *et al.*, 1996). This hypothesis was supported by two groups who observed that the maximum length of the most common known pathogenic trinucleotide repeat, CAG/CTG, was greater in patients with schizophrenia than in unaffected controls (O'Donovan *et al.*, 1995; Morris *et al.*, 1995), findings that were later replicated in a European multicentre study (O'Donovan *et al.*, 1996). Unfortunately, the early RED studies were followed by a series of unsuccessful attempts to identify the relevant repeat-containing loci by a variety of methods, by several failures to replicate the RED findings (Laurent *et al.*, 1998; Li *et al.*, 1998b; Vincent *et al.*, 1999) thus casting doubt over the CAG/CTG repeat hypothesis.

The trinucleotide repeat hypothesis was however rejuvenated with the report of association between schizophrenia and alleles of a member of the family of calcium-activated potassium channels KCa3 (hKCa3/KCCN3; Chandy *et al.*, 1998). For several reasons, *KCa3* seemed a remarkable candidate gene for schizophrenia. First, the gene contained two CAG repeats, one of which is highly polymorphic. Second, the family of genes to which it belongs is thought to play an important role regulating neuronal activity and it was therefore considered a functional candidate gene. Third, the gene was thought also to be a positional candidate, as it was (erroneously) believed to map to chromosome 22q11. Two further case–control studies have subsequently supported the findings of Chandy and colleagues (Bowen *et al.*, 1998; Dror *et al.*, 1999).

However, while evidence from three case–control studies lends support to the hypothesis of *KCa3* as a susceptibility gene for schizophrenia, in other respects the case for this gene as a candidate weakened considerably. First *KCa3* maps not to 22q11, but to 1q21 (Dror *et al.*, 1999). Second, the RED data cited above lend no support to *KCa3* as a candidate because the polymorphic trinucleotide repeat in this gene is far too short to account for the RED associations (Bowen *et al.*, 1998). Third a series of case–control and family studies (Austin *et al.*, 1999; BonnetBrilhault *et al.*, 1999; Joober *et al.*, 1999; Li *et al.*, 1998a; Stöber *et al.*, 1998; Tsai *et al.*, 1998; Wittekindt *et al.*, 1999) now failed to replicate the findings. Thus, we are back to our familiar position in candidate gene analysis while there are insufficient data to draw firm conclusions, we believe at present, the case for *KCa3* remains firmly with the null hypothesis.

What then is the status of the original associations between large CAG/CTG repeats and schizophrenia? It has been reported that large CAG/CTG RED products (repeat size > 40) are explained by repeat size at two autosomal loci, one at 18q21.1 and the other at 17q21.3 (Lindblad *et al.*, 1998; Sidransky *et al.*, 1998). If correct, then it follows that one or both of these loci should be associated with schizophrenia. Unfortunately, data from Vincent and colleagues (1999) and data from our group (unpublished) unequivocally show that expansions at these loci are not responsible for the RED associations. However, in both samples only around 50% of large CAG/CTG repeats detected by RED could be explained by polymorphisms at these two loci suggesting that at least one further locus is responsible for the RED data, a possibility supported by two recent studies on protein extracts from schizophrenic tissues (Ross, 1999).

4. High rates of schizophrenia in adults with velo-cardio-facial syndrome

Velo-cardio-facial syndrome (VCFS), also known as DiGeorge or Shprintzen syndrome, is associated with small interstitial deletions of chromosome 22q11 in 80% to 85% of individuals (Driscoll *et al.*, 1992). It was first described by Shprintzen and colleagues (1978) and has an estimated prevalence of 1 in 4000 births (Papolos *et al.*, 1996). Distinctive dysmorphology, congenital heart disease and learning disabilities characterize the syndrome, although considerable phenotypic variability occurs. As the first recognized cohort of children with VCFS was followed up into adolescence and early adulthood, evidence began to accumulate for a high prevalence of major mental illness. Early reports suggested that more than 10% of the cohort had developed psychiatric disorders which mostly resembled chronic schizophrenia with paranoid delusions, although operational criteria were not used (Shprintzen *et al.*, 1992). In a follow-up study of adults (aged ≥17 years) using DSM-III-R criteria, Pulver *et al.* (1994b) reported that 11 (79%) of their sample of 14 patients were given a psychiatric diagnosis: 29% had schizophrenia (22%) or schizoaffective disorder (7%), 29% had simple or social phobia, 21% had depression, and 14% had obsessive-compulsive disorder. More recently, Papolos *et al.* (1996) reported that 4 (16%) of their sample of 15 children and 10 adults had psychotic symptoms and 16 (64%) met DSM-III-R criteria for a spectrum of bipolar affective disorders. While none had schizophrenia, the two oldest members of their patient cohort (aged 29 and 34 years) both had schizoaffective disorder.

In order to try and gain a more precise determination of the prevalence and nature of psychopathology in adults with VCFS, rather than rely upon clinical diagnosis, we recently evaluated 50 cases using a structured clinical interview to establish a DSM-IV diagnosis (Murphy *et al.*, 1999). Fifteen individuals with VCFS (30%) had a psychotic disorder, with 24% ($n=12$) fulfilling DSM-IV criteria for schizophrenia. In addition, 6 (12%) had major depression without psychotic features. We were unable to replicate the findings of Papolos and colleagues (1996) of a high prevalence of bipolar spectrum disorders in VCFS. However, these workers studied a small sample that included few adults, and in view of the fact that their oldest cases satisfied criteria for schizoaffective disorder it is possible that the psychotic phenotype in VCFS varies with age. Prospective studies are now required to test this hypothesis.

The current balance of evidence thus favors the view that the high prevalence of psychosis results from hemizygosity for a gene or genes at chromosome 22q11 rather than ascertainment bias or a non-specific association with low IQ (Murphy *et al.*, 1999). If this is true, where is the connection between VCFS and schizophrenia? Several lines of evidence now suggest the view that schizophrenia is a neurodevelopmental disorder (Andreasen, 1984). In VCFS, defective development and migration of mesencephalic and cardiac neural crest cells are believed to play a significant role in the pathogenesis of midfacial and cardiac abnormalities (Scambler *et al.*, 1992). Consequently, it has been postulated that a gene or genes causing disruption of neural cell migration may be a common neurodevelopmental mechanism for both VCFS and schizophrenia (Chow *et al.*, 1994).

What then is the importance of 22q11 and VCFS in the etiology of schizophrenia as a whole? Karayiorgou *et al.* (1995) reported that two of 100 randomly ascertained individuals with schizophrenia were found to have a 22q11 deletion. In contrast, when subjects with schizophrenia were selected for the presence of clinical features consistent with VCFS, 22q11 deletions were identified in 20–59% of cases (Bassett *et al.*, 1998; Gothelf *et al.*, 1997). These findings suggest that a small proportion of cases of schizophrenia may result from deletions of 22q11, and that clinicians should be vigilant especially when psychosis occurs in the presence of dysmorphology, mild learning disability or a history of cleft palate or congenital heart disease (Murphy *et al.*, 1997). The question of whether genetic variation in 22q11 confers susceptibility to schizophrenia in cases without a deletion is more difficult to answer. As we have seen the results of some linkage studies suggest the presence of a schizophrenia susceptibility locus on 22q. However, the linkage findings tend to point telomeric to the VCFS region (Gill *et al.*, 1996; Pulver *et al.*, 1994a). Nevertheless, modest evidence for linkage to the VCFS region has also been claimed (Blouin *et al.*, 1998; Lasseter *et al.*, 1995; Shaw *et al.*, 1998), and as we have noted above, linkage mapping in complex diseases is somewhat imprecise.

Another important question concerns the factors that determine whether an individual with VCFS develops schizophrenia. When adult VCFS cases were tested with a quantitative measure of schizotypy, individuals with psychosis had the highest scores. However, perhaps of greater interest, those without psychosis had intermediate scores compared with controls (Murphy *et al.*, 1999). If schizotypy is a trait marker for increased liability to psychosis, this suggests that the majority if not all, of those with 22q11 deletions are at increased risk of psychosis, but that other genetic or environmental factors are required for this risk to be expressed. The genetic loci involved may reside elsewhere in the genome, and include those involved more widely in psychosis, or lie within 22q11. The occurrence of psychosis does not appear to be related to the size of the deletion (Karayiorgou *et al.*, 1995). However, it is possible that susceptibility to psychosis reflects allelic variation of a hemizygous gene or genes within the deletion. The gene encoding Catechol-o-methyl transferase (COMT), an enzyme involved in the catabolism of catecholamine neurotransmitters, maps to that VCFS region and is therefore an obvious candidate for influencing the expression of psychosis in VCFS probands. This gene exists in two allelic forms encoding high and low activity isoforms of the enzyme, and it has been suggested that possession of the low-activity COMT allele might be associated with the occurrence of schizophrenia in VCFS (Dunham *et al.*, 1992). However, we have found no evidence for an association between the low-activity COMT allele and either schizophrenia or schizotypy individuals with VCFS (Murphy *et al.*, 1999).

5. Future directions

Research to date has largely excluded the possibility that genes of major effect exist even in a sub-set of families. Evidence has been obtained for the location of some genes of moderate effect. However, none of these findings can be regarded as conclusive and proof in each case will probably have to await identification of the susceptibility locus itself. Finally, there are data suggesting that allelic association in at least two genes, those encoding the 5HT2a and DRD3 receptors, confer a small degree of susceptibility. However, further work is required to determine the specificity of these associations as well as whether they reflect the existence of true susceptibility or modifying alleles (Owen *et al.*, 2000).

As in other common diseases, it is hoped that advances will come through the use of a new generation of genetic markers and new methods of laboratory and statistical analysis (Owen *et al.*, 2000). The raw data from the Human Genome Project combined with novel automated technologies for mutation detection and genotyping will allow researchers to replace opportunistic analyses of the occasional candidate gene with more systematic analyses of large complex candidate gene systems. For example, we are currently engaged in a systematic molecular genetic analysis of the glutamatergic and GABAergic neurotransmission systems with others planned. This requires over 50 genes to be subjected to mutation screening and around 200 different polymorphisms to be genotyped. Even with current technology, this is far from a trivial undertaking, but it is certainly now feasible and will allow us to determine if these systems are involved in the primary pathogenesis of schizophrenia. However, successful application of these methods also requires access to large, well-characterized patient samples and collection of these is a priority at the present time (Owen *et al.*, 2000). This has practical implications for sampling strategies and study design. For example, while in some cases, particularly very common childhood disorders, family based association samples might be almost as easy to collect as case–control samples, in schizophrenia, large samples will be difficult to collect. The place for family based studies is therefore probably as part of a first or even second replication tier. This will allow the lower power both intrinsic and due to smaller sample-size (due to difficulty of collection) to be offset by higher prior odds, the ability to predict the specific allele associated and the direction of effect, and the fact that few markers will reach this stage of the sequential process. In addition, samples for genetic studies are often ascertained in widely differing and unsystematic ways. Reported associations might therefore be with confounding variables such as comorbidity or with modifying factors such as those influencing disease severity or age at onset. The design of studies using only incident cases may be required before explanations of this sort can be excluded.

Another important challenge will be the development of suitable animal models to allow functional studies of putative disease loci (Owen *et al.*, 2000). Disorders that predominantly involve higher cognitive function such as schizophrenia are likely to prove difficult to model in animals. However, there are features of the human phenotype such as subtle abnormalities of cell migration, enlarged cerebral ventricles and information processing abnormalities including defects in pre-pulse inhibition that can be detected in animals (Wood *et al.*, 1998). In fact a possible approach to producing a mouse model for at least some of the schizophrenia phenotype is suggested by the finding of increased prevalence of schizophrenia in VCFS. Currently attempts are underway to produce transgenic mice deleted for the syntenic region of mouse chromosome 16 (Skoultchi *et al.*, 1997). These animals should be investigated closely for neuroanatomical and behavioral phenotypes of possible relevance to schizophrenia, and such studies are already yielding encouraging findings (Kimber *et al.*, 1999).

The effectiveness of molecular genetic studies depends upon the genetic validity of the phenotypes studied. Perhaps if we were better at defining phenotypes we would be better at finding genes. It is worth reiterating at this point that the commonly used diagnostic criteria define phenotypes with high heritability. In principle, therefore, it should be possible to identify the genes predisposing to schizophrenia if sufficiently large samples are studied. However, perhaps genetic validity could be improved by focusing upon aspects of clinical variation such as age of onset or symptom profiles, or by identifying biological markers that predict degree of genetic risk or define more homogeneous subgroups. Unfortunately, despite much work, it has not been possible to identify genetically distinct subtypes of schizophrenia. Instead, clinical variation is likely to reflect in part at least a combination of quantitative variation in genetic risk for the disorder and the effect of modifying genes which influence illness expression rather than the risk of illness per se. Examples of this phenomenon relating to age at onset and symptom pattern in schizophrenia have already been described (Kendler *et al.*, 1987, 1997).

The search for trait markers aims to move genetic studies beyond the clinical syndrome by identifying indices of genetic risk that can be measured in asymptomatic individuals and/or by identifying markers of pathophysiological processes that are closer to the primary effects of susceptibility genes than clinical symptoms. Work in this area is developing fast; for example, candidate trait markers for schizophrenia include schizotypal personality traits, measures of cognitive processing, brain evoked potentials and abnormalities in eye movements (De Lisi, 1999). It is also hoped that advances in brain imaging will lead to the identification of genetically valid trait markers. However, it seems unlikely that these phenotypes will provide a rapid solution to the problem. First, we will need measures that can be practicably applied to a sufficient number of families or unrelated patients and second, we will need to ensure that the traits identified are highly heritable, which will itself require a return to classic genetic epidemiology and model fitting.

Efforts to improve the selection of phenotypes are also concerned with enhancing the traditional categorical approach to defining psychiatric disorders, by identifying genetically valid phenotypes that can be measured quantitatively. These can be used in quantitative trait locus (QTL) approaches to gene mapping. Work in this area has begun but still faces problems particularly relating to the confounding influence of state related effects. Perhaps the best hope of taking account of the complexity and heterogeneity of the schizophrenia phenotype comes from new methods of analysis in which aspects of the phenotype can be entered as covariates in linkage analyses (Almasy and Blangero, 1999).

In conclusion, attempts to identify the genes that predispose to schizophrenia face formidable challenges arising from both genetic and phenotypic complexity. Technical and methodological advances are required, but success will also depend as much upon the traditional medical disciplines of clinical description and epidemiology and our ability to integrate these with genetic approaches.

Ackowledgments

We are supported by a Cooperative Group Grant from the Medical Research Council.

References

Almasy, L. and Blangero, J. (1999) Linkage strategies for mapping genes for complex traits in man. In: *Handbook of Molecular-Genetic Techniques for Brain and Behaviour Research*. (eds Cruisio, W.E. and Gerlai, R.T.). Elsevier, Amsterdam, pp. 100–112.

Andreasen, N.C. (1984) The scale for the assessment of positive symptoms (SAPS). University of Iowa, Iowa city.

Andreasson, S., Allebeck, A., Engstrom, A. and Rydberg, U. (1987) Cannabis and schizophrenia: a longitudinal study of Swedish conscripts. *Lancet* **2**: 1483–1486.

Austin, C.P., Holder, D.J., Ma, L., Mixson, L.A. and Caskey, C.T. (1999) Mapping of hKCa3 to chromosome 1q21 and investigation of the linkage of CAG repeat polymorphism to schizophrenia. *Mol. Psychiatry* **4**: 261–266.

Bassett, A., Hodgkinson, K., Chow, E.W.C., Correia, S., Scutt, L.E. and Welsberg, R. (1998) 22q11 deletion syndrome in adults with schizophrenia. *Am. J. Med. Gene.* **81**: 328–337.

Blouin, J.L., Dombroski, B.A., Nath, S.K., *et al.* (1998) Schizophrenia susceptibility loci on chromosomes 13q32 and 8p21. *Nature Genet.* **20**: 70–73.

BonnetBrilhault, F., Laurent, C., Campion, D., *et al.* (1999) No evidence for involvement of

KCNN3 (hSKCa3) potassium channel gene in familial and isolated cases of schizophrenia. *Eur. J. Hum. Genet.* **7**: 247–250.

Bowen, T., Guy, C.A., Craddock, N., *et al.* (1998) Further support for an association between a polymorphic CAG repeat in the hKCa3 gene and schizophrenia. *Mol. Psychiatry* **3**: 266–269.

Brzustowicz, L.M., Honer, W.G., Chow, E.W.C., Little, D., Hogan, J., Hodgkinson, K. and Bassett A.S. (1999) Linkage of familial schizophrenia to chromosome 13q32. *Am. J. Hum. Genet.* **65**: 1096–1103.

Bunzel, R., Blumcke, I., Cichon, S., Normann, S., Schramm, J., Propping, P. and Nothen, M.M. (1998) Polymorphic imprinting of the serotonin-2A (5-HT2A) receptor gene in human adult brain. *Molec. Brain Res.* **59**: 90–92.

Cannon, T.D., Kaprio, J., Lonnqvist, J., Huttunen, M. and Koskenvuo, M. (1998) The genetic epidemiology of schizophrenia in a Finnish twin cohort – a population-based modeling study. *Arch. Gen. Psychiatry* 55: 67–74.

Cardno, A.G., Marshall, E.J., Coid, B., *et al.* (1999) Heritability estimates for psychotic disorders – the Maudsley twin psychosis series. *Arch. Gen. Psychiatry* **56**: 162–168.

Chandy, K.G., Fantino, E., Wittekindt O., *et al.* (1998) Isolation of a novel potassium channel gene hSKCa3 containing a polymorphic **CAG** repeat: a candidate for schizophrenia and bipolar disorder? *Mol. Psychiatry* **3**: 32–37.

Chow, E.W.C., Bassett, A.S. and Weksberg, R. (1994) Velo-cardio-facial syndrome and psychotic disorders: implications for psychiatric genetics. *Am. J. Med. Genet.* **54**: 107–112.

Craddock, N., Khodel, V., Van Eerdewegh, P. and Reich, T. (1995) Mathematical limits of multilocus models: the genetic transmission of bipolar disorder. *Am. J. Hum. Genet.* **57**: 690–702.

Crocq, M.A., Mant, R., Asherson. P., *et al.* (1992) Association between schizophrenia and homozygosity at the dopamine-D3 receptor gene. *J. Med. Genet.* **29**: 858–860.

David, A.S., Malmberg, A., Brandt, L., Allebeck, A. and Lewis, G. (1997) IQ and risk for schizophrenia: a population-based cohort study. *Psychol. Med.* **27**: 1311–1323.

Deakin, J.F.W. and Simpson, M.D.C. (1997). A two-process theory of schizophrenia: Evidence from studies in post-mortem brain. *J. Psychiatr. Res.* **31**: 277–295.

DeLisi, L.E. (1999) A critical overview of recent investigations into the genetics of schizophrenia. *Curr. Opin. Psychiatry.* **12**: 29–39.

Driscoll, D.A., Spinner, N.B., Budarf, M.L., *et al.* (1992) Deletions and microdeletions of 22q11.2 in velo-cardio-facial syndrome. *Am. J. Med. Genet.* **44**: 261–268.

Dror, V., Shamir, E., Ghanshani, S., *et al.* (1999) hKCa3/KCNN3 potassium channel gene: association of longer CAG repeats with schizophrenia in Israeli Ashkenazi Jews, expression in human tissues and localization to chromosome 1q21. *Mol. Psychiatry* **4**: 254–260.

Dunham, I., Collins, J., Wadey, R. and Scambler, P. (1992) Possible role for COMT in psychosis associated with velo-cardio-facial syndrome. *Lancet* **340**: 1361–1362.

Faraone, S.V., Matise, T., Svrakic, D., *et al.* (1998) Genome scan of European-American schizophrenia pedigrees: results of the NIMH genetics initiative and millennium consortium. *Am. J. Med. Genet* **81**: 290–295.

Gill, M., Vallada, H., Collier D., *et al.* (1996) A combined analysis of D22s278 marker alleles in affected sib-pairs – support for a susceptibility locus for schizophrenia at chromosome 22Q12. *Am. J. Med. Genet.* **67**: 40–45.

Goate, A., Chartierharlin, M.C., Mullan, M., *et al.* (1991) Segregation of a missense mutation in the amyloid precursor protein gene with familial Alzheimers-disease. *Nature* **349**: 704–706.

Gothelf, D., Frisch, A., Munitz, H., Rockah, R., Aviram, A. Mozes, T., Birger, M., Weizman, A., and Frydman, M. (1997) Velocardiofacial manifestations and microdeletions in schizophrenic patients. *Am. J. Med. Genet.* 72: 455–461.

Gottesman, I.I. (1991) *Schizophrenia Genesis: The Origins of Madness*. Freeman, New York.

Hauser, E.R., Boehnke, M., Guo, S.W. and Risch, N.J. (1996) Affected sib pair interval mapping and exclusion for complex traits: Sampling considerations. *Gene. Epidemiol.* **13**: 117–137.

Heresco-Levy, U., Silipo, G. and Javitt, D.C. (1996) Glycinergic augmentation of NMDA receptor-mediated neurotransmission in the treatment of schizophrenia. *Psychopharmacology Bull.* **32**: 731–740.

Hultman, C.M., Sparen, P., Takei, N., Murray, R.M. and Cnattingius, S. (1999) Prenatal and perinatal risk factors for schizophrenia, affective psychosis and reactive psychosis: case–control study. *BMJ* **318**: 421–426.

Hutchinson, G., Takei, N., Fahy, T., *et al.* (1996) Morbid risk of psychotic illness in first degree relatives of white and African-Caribbean patients with psychosis. *Brit. J. Psychiatry* **169**: 776–780.

Inayama, Y., Yoneda, H., Sakai, T., *et al.* (1996) Positive association between a DNA sequence variant in the serotonin 2A receptor gene and schizophrenia. *Am. J. Med. Genet.* **67**: 103–105.

Jablensky, A. (1999) The 100-year epidemiology of schizophrenia. *Schizophrenia Res.* **28** : 111–125.

Javitt, D.C. and Zukin, S.R. (1991) Recent advances in the phencyclidine model of schizophrenia. *Am. J. Psychiatry* 148: 1301–1308.

Joober, R., Benkelfat, C., Brisebois, K., *et al.* (1999) Lack of association between the hSKCa3 channel gene CAG polymorphism and schizophrenia. *Am. J. Med. Genet.* **88**: 154–157.

Jones, P., Rantakallio, P., Hartikainen, A.L., Isohanni, M. and Sipilä, P. (1998) Schizophrenia as long-term outcome of pregnancy, delivery and perinatal complications: A 28-year follow-up of the 1966 North Finland general population birth cohort. *Am. J. Psychiatry* **155**: 355–364.

Jones, P., Rodgers, B., Murray, R. and Marmot, M. (1994) Child development risk factors for adult schizophrenia in the British 1946 birth cohort. *Lancet* **344**: 1398–1402.

Karayiorgou, M., Morris, M.A., Morrow, B., *et al.* (1995) Schizophrenia susceptibility associated with interstitial deletions of chromosome 22q11. *Proc. Natl Acad. Sci. USA* **92**: 7612–7616.

Kendler, K.S., Tsuang, M.T. and Hays, P. (1987) Age at onset in schizophrenia: a familial perspective. *Arch. Gen. Psychiatry* **44**: 881–890.

Kendler, K.S., Mcguire, M., Gruenberg, A.M., O'Hare, A., Spellman, M. and Walsh, D. (1993a) The Roscommon Family Study III. Schizophrenia-related personality disorders in relatives. *Arch. Gen. Psychiatry* **50**: 781–788.

Kendler, K.S., Mcguire, M., Gruenberg, A.M., Spellman, M., O'Hare. A. and Walsh, D. (1993b) The Roscommon Family Study II. The risk of nonschizophrenic nonaffective psychoses in relatives. *Arch. Gen. Psychiatry* **50**: 645–652.

Kendler, K.S., Karkowski-Shuman, L., O'Neill, F.A., Straub, R.E., MacLean, C.J. and Walsh, D. (1997) Resemblance of psychotic symptoms and syndromes in affected sibling pairs from the Irish study of high-density schizophrenia families: evidence for possible etiologic heterogeneity. *Am. J. Psychiatry* **154**: 191–198.

Kimber, W.L., Hsieh, P., Hirotsune, S., *et al.* (1999) Deletion of 150 kb in the minimal DiGeorge/velocardiofacial syndrome critical region in mouse. *Hum. Mol. Genet* **8**: 2229–2237.

Lander, E. and Kruglyak, L. (1995) Genetic dissection of complex traits – guidelines for interpreting and reporting linkage results. *Nature Genet.* **11**: 241–247.

Lander, E.S. and Schork, N.J. (1994) Genetic dissection of complex traits. *Science* **265**: 2037–2048.

Lasseter, V.K., Pulver, A.E., Wolyniec, P.S., *et al.* (1995) Follow-up report of potential linkage for schizophrenia on chromosome 22q.3. *Am. J. Med. Genet.* **60**: 172–173.

Laurent, C., Zander, C., Thibaut, F., Brilhault, F., Chavand, O., Jay, M. and Samolyk, D. (1998) Anticipation in schizophrenia: no evidence of expanded CAG/CTG repeat sequences in French families and sporadic cases. *Am. J. Med. Genet.* **81**: 342–346.

Levinson, D.F., Wildenauer, D.B., Schwab, S.G., *et al.* (1996) Additional support for schizophrenia linkage on chromosome-6 and chromosome-8 – a multicenter study. *Am. J. Med. Genet* **67**: 580–594.

Levylahad, E., Wasco, W., Poorkaj, P., *et al.* (1995) Candidate gene for the chromosome-1 familial Alzheimers-disease locus. *Science* **269**: 973–977.

Lewis, G., David, A., Andreasson, S. and Allebeck, P. (1992) Schizophrenia and city life. *Lancet* **340**: 137–140.

Li, T., Hu, X., Chandy, K.G., *et al.* (1998a) Transmission disequilibrium analysis of a triplet repeat within the hKCa3 gene using family trios with schizophrenia. *Biochem. Biophys. Res. Comm.* **251**: 662–665.

Li, T., Vallada, H.P., Liu, X.H., *et al.* (1998b) Analysis of CAG/CTG repeat size in Chinese subjects with schizophrenia and bipolar affective disorder using the repeat expansion detection method. *Biol. Psychiatry* **44**: 1160–1165.

Lin, M.W., Curtis, D., Williams, N., *et al.* (1995) Suggestive evidence for linkage of schizophrenia to markers on chromosome 13q14.1-q32. *Psychiatric Genet.* **5**: 117–126.

Lindblad, K., Nylander, P.O., Zander, C., *et al.* (1998) Two commonly expanded CAG/CTG repeat loci: involvement in affective disorders? *Mol. Psychiatry* **3**: 405–410.

Malmberg, A., Lewis, G., David, A.S. and Allebeck, A. (1998) Premorbid adjustment and personality in schizophrenia. *British J. Psychiatry* **172**: 308–313.

McGue, M. and Gottesman, I.I. (1989) A single dominant gene still cannot account for the transmission of schizophrenia. *Arch. Gen. Psychiatry* **46**: 478–479.

McGuffin, P. and Owen, M.J. (1996) Molecular genetic studies of schizophrenia. *Cold Spring Harbor Symposium on Quantitative Biology* **61**: 815–822.

McGuffin, P., Owen, M.J., O'Donovan, M.C., Thapar, A. and Gottesman, I.I. (1994) *Seminars in Psychiatric Genetics*. Gaskell Press, London.

McGuffin, P., Owen, M.J. and Farmer, A.E. (1995) Genetic-basis of schizophrenia. *Lancet* **346**: 678–682.

Moldin, S.O. and Gottesman, I.I. (1997) At issue: genes, experience, and chance in schizophrenia-positioning for the 21st century. *Schizophrenia Bull.* **23**: 547–561.

Morel, B.A. (1857) *Traite des degenerescences*. J.B. Bailliere, Paris.

Morris, A.G., Gaitonde, E., Mckenna, P.J., Mollon, J.D. and Hunt, D.M. (1995) CAG repeat expansions and schizophrenia: association with disease in females and with early age at onset. *Hum. Mol. Genet.* **4**: 1957–1961.

Mohn, A.R., Gainetdinov, R.R., Caron, M.G. and Koller, B.H. (1999). Mice with reduced NMDA receptor expression display behaviors related to schizophrenia. *Cell* **98**: 427–436.

Murphy, K.C. and Owen, M.J. (1997) The behavioural phenotype in velo-cardio-facial syndrome. *Am. J. Hum. Genet.* **61: (4.SS)**: 15.

Murphy, C., Jones, L.A. and Owen, M.J. (1999) High rates of schizophrenia in adults with velo-cardio-facial syndrome. *Arch. Gen. Psychiatry* **56**: 940–945.

Murray, R.M., Lewis, S.L., Owen, M.J. and Foerster, A. (1988) The neurodevelopmental origins of dementia praecox. In: *Schizophrenia: the Major Issues*, (Eds. P. McGuffin and P. Bebbington). Heinemann, pp. 90–106.

O'Donovan, M.C. and Owen, M.J. (1996) Dynamic mutations and psychiatric genetics. *Psychol. Med.* **28**: 1–6.

O'Donovan, M.C., Guy, C., Craddock, N., Murphy, K.C., Cardno, A.G., Jones, L.A., Owen, M.J. and McGuffin, P. (1995) Expanded CAG repeats in schizophrenia and bipolar disorder. *Nature Genet.* **10**: 380–381.

O'Donovan, M.C., Guy, C., Craddock, N., *et al.* (1996) Confirmation of association between expanded cag/ctg repeats and both schizophrenia and bipolar disorder. *Psychol. Med.* **26**: 1145–1153.

Olney, J.W. and Farber, N.B. (1995). Glutamate receptor dysfunction and schizophrenia. *Arch. Gen. Psychiatry* **52**: 998–1007.

Owen, M.J. and McGuffin, P. (1993) Association and linkage: complementary strategies for complex disorders. *J. Med. Genet.* **30**: 638–639.

Owen, M.J., Holmans, P. and McGuffin, P. (1997) Association studies in psychiatric genetics. *Mol. Psychiatry* **2**: 270–273.

Owen, M.J., Cardno, A.G. and O'Donovan, M.C. (2000) Psychiatric genetics: back to the future. *Mol. Psychiatry* **5**: 22–31.

Papolos, D.F., Faedda, G.I., Veit, S., Goldberg, R.B., Morrow, B., Kucherlapati, R. and Shprintzen, R. (1996) Bipolar spectrum disorders in patients diagnosed with velo-cardio-facial syndrome: does a hemizygous deletion of chromosome 22q11 result in bipolar affective disorder? *Am. J. Psychiatry* **153**: 1541–1547.

Petronis, A. and Kennedy, J.L. (1996) Unstable genes – unstable mind? *Am. J. Hum. Genet.* **152**: 164–172.

Pulver, A.E., Karayiorgou, M., Wolyniec, P.S. *et al.* (1994a) Sequential strategy to identify a susceptibility gene for schizophrenia – report of potential linkage on chromosome 22Q12-Q13.1. 1. *Am. J. Med. Genet.* **54**: 36–43.

Pulver, A.E., Nestadt, G., Goldberg, R. *et al.* (1994b) Psychotic illness in patients diagnosed with velo-cardio-facial syndrome and their relatives. *J. Nervous Mental Dis.* **182**: 476–478.

Risch, N. (1990) Linkage strategies for genetically complex traits. 2. The power of affected relative pairs. *Am. J. Hum. Genet.* **46**: 229–241.

Risch, N. and Merikangas, K. (1996) The future of genetic studies of complex human diseases. *Science* **273**: 1516–1517.

Risch, N.J. and Teng, J. (1998) The relative power of family based and case control designs for linkage disequilibrium studies of complex human diseases 1. DNA pooling. *Genome Res.* **8**: 1273–1288.

Roberts, S.B., Maclean, C.J., Neale, M.C., Eaves, L.J. and Kendler, K.S. (1999) Replication of linkage studies of complex traits: An examination of variation in location estimates. *Am. J. Hum. Genet.* **65**: 876–884.

Ross, C.A. (1999) Schizophrenia genetics: expansion of knowledge? *Mol. Psychiatry* **4**: 4–5.

Scambler, P., Kelly, D., Lindsay, E. *et al.* (1992) Velo-cardio-facial syndrome associated with chromosome 22 deletions encompassing the DiGeorge locus. *Lancet* **339**: 1138–1139.

Schaid, D.J. and Sommer, S.S. (1994) Comparison of statistics for candidate gene association studies using cases and parents. *Am. J. Hum. Genet.* **55**: 402–409.

Schwab, S.G., Eckstein, G.N., Hallmayer, J. *et al.* (1997) Evidence suggestive of a locus on chromosome 5q31 contributing to susceptibility for schizophrenia in German and Israeli families by multipoint affected sib-pair linkage analysis. *Mol. Psychiatry* **2**: 156–160.

Schwab, S.G., Hallmayer, J., Albus, M. *et al.* (1998a) Further evidence for a susceptibility locus on chromosome 10p14-p11 in 72 families with schizophrenia by nonparametric linkage analysis. *Am. J. Medical Genet.* **81**: 302–307.

Schwab, S.G., Hallmayer, J., Lerer, B. *et al.* (1998b) Support for a chromosome 18p locus conferring susceptibility to functional psychoses in families with schizophrenia, by association and linkage analysis. *Am. J. Hum. Genet.* **63**: 1139–1152.

Scott, W., Pericak-Vance, M. and Haines, J. (1997) Genetic analysis of complex diseases. *Science* **275**: 1327.

Shaw, S.H., Kelly, M., Smith, A.B., *et al.* (1998) A genome wide search for schizophrenia susceptibility genes. *Am. J. Med. Genet.* **81**: 364–376.

Sherrington, R., Brynjolfsson, J., Petursson, H., Potter, M., Dudleston, K., Barraclough, B., Wasmuth, J., Dobbs, M. and Gurling, H. (1988) Localization of a susceptibility locus for schizophrenia on chromosome-5. *Nature* **336**: 164–167.

Sherrington, R., Rogaev, E.I., Liang, Y., *et al.* (1995) Cloning of a gene bearing missense mutations in early-onset familial Alzheimers-disease. *Nature* **375**: 754–760.

Shprintzen, R., Goldberg, R.B., Lewin, M.L., Sidoti, E.J., Berkman, M.D., Argamaso, R.V. and Young, D. (1978) A new syndrome involving cleft palate, cardiac anomalies, typical facies and learning disabilities: velo-cardio-facial syndrome. *Cleft Palate J.* **15**: 56–62.

Shprintzen, R.J., Goldberg, R.B. and Golding-Kushner, K.J. (1992) Late-onset psychosis in the velo-cardio-facial syndrome. *Am. J. Med. Genet.* **42**: 141–142.

Sidransky, E., Burgess, C., Ikeuchi, T., *et al.* (1998) A triplet repeat on 17q accounts for most expansions detected by the repeat-expansion-detection technique. *Am. J. Human Genet.* **62**: 1548–1551.

Skoultchi, A.I., Puech, A., Saint-Jore, B. *et al.* (1997) Comparative mapping of the human and mouse VCFS/DGS syntenic region discloses the presence of a large internal rearrangement. *Am. J. Hum. Genet.* **61: A296.**

Spurlock, G., Heils, A., Holmans, P., *et al.* (1998) A family based association study of T102C polymorphism in 5HT2A and schizophrenia plus identification of new polymorphisms in the promotor. *Mol. Psychiatry* **3**: 42–49.

Straub, R.E., Maclean, C.J., Oneill, F.A., Walsh, D. and Kendler, K.S. (1997) Support for a possible schizophrenia vulnerability locus in region 5q22–31 in Irish families. *Mol. Psychiatry* **2**: 148–155.

Straub, R.E., Maclean, C.J., Martin, R.B., *et al.* (1998) A schizophrenia locus may be located in region 10p15-p11. *Am. J Med. Genet.* **81**: 296–301.

Stöber, G., Jatzke, S., Meyer, J., Okladnova, O., Knapp, M., Beckmann, H. and Lesch, K.P. (1998) Short CAG repeats within the hSKCa3 gene associated with schizophrenia: results of a family-based study. *Neuroreport* **9**: 3595–3599.

Suarez, B.K., Hampe, C.L. and Van Eerdewegh, P. (1994) Problems of replicating linkage claims in psychiatry. In: *Genetic Approaches to Mental Disorders*. (Eds. E.S. Gershon and C.R. Cloninger). American Psychiatric Press, Washington, DC, pp. 23–46.

Tsai, M.T., Shaw, C.K., Hsiao, K.J. and Chen, C.H. (1998) Genetic association study of a polymorphic CAG repeat array of calcium activated potassium channel (KCNN3) gene and schizophrenia among the Chinese population fron Taiwan. *Mol. Psychiatry* **4**: 271–273.

Vincent, J.B., Petronis, A., Strong, E., Parikh, S.V., Meltzer, H.Y., Lieberman, J. and Kennedy,

J.L. (1999) Analysis of genome-wide CAG/CTG repeats, and at SEF2–1B and ERDA1 in schizophrenia and bipolar affective disorder. *Mol. Psychiatry* **4**: 229–234.

Williams, J., Spurlock, G., McGuffin, P., *et al.* (1996) Association between schizophrenia and T102C polymorphism of the 5-hydroxytryptamine type 2A-receptor gene. *Lancet* **347**: 1294–1296.

Williams, J., McGuffin, P., Nothen, M., Owen, M.J., and the EMASS Collaborative Group (1997) Meta-analysis of association between the 5-HT2a receptor T102C polymorphism and schizophrenia. *Lancet* **349**: 1221.

Williams, J., Spurlock, G., Holmans, P., *et al.* (1998) A meta-analysis and transmission disequilibrium study of association between the dopamine D3 receptor gene and schizophrenia. *Mol. Psychiatry* **3**: 141–149.

Williams, N.M., Rees, M.I., Holmans, P., *et al.* (1999) A two-stage genome scan for schizophrenia susceptibility genes in 196 affected sibling pairs. *Hum. Mol. Genet.* **8**: 1729–1739.

Wittekindt, O., Schwab, S.G., Burgert, E., *et al.* (1999) Association between hSKCa3 and schizophrenia not confirmed by transmission disequilibrium test in 193 offspring parents trios. *Mol. Psychiatry* **4**: 267–270.

Wood, G.K., Tomasiewicz, H., Rutishauser, U., Magnuson, T., Quirion, R., Rochford, J. and Srivastava, L.K. (1998) NCAM-180 knockout mice display increased lateral ventricle size and reduced prepulse inhibition of startle. *Neuroreport* **9**: 461–466.

11

Atopy and asthma

Lyle J. Palmer and William O.C.M. Cookson

1. Introduction

Asthma is characterized by variable symptoms such as wheeze, shortness of breath and coughing. The symptoms of asthma are primarily due to excessive airway narrowing, which leads to an increased resistance to airflow. Most asthma is thought to be the result of airway inflammation (Beasley *et al.*, 1993; Holgate, 1996). Airway inflammation in asthma is commonly initiated by an enhanced immunoglobin E (IgE)-mediated response to inhaled allergens such as house dust mite and grass pollen proteins (Godfrey, 1993; Howarth, 1995). This enhanced state of IgE responsiveness is known as atopy. The atopic syndrome includes hay fever and eczema as well as asthma.

Atopic diseases are very common in the developed world. Most surveys have found a population prevalence of around 20–30% with a peak of 40–50% in young adolescents (Barbee *et al.*, 1981; Roorda *et al.*, 1994). Asthma is the most serious of these diseases, and is the most common chronic childhood disease in developed nations (Asher *et al.*, 1995). Asthma carries a very substantial direct and indirect economic cost worldwide (Toelle *et al.*, 1995; Weiss *et al.*, 1992).

It has been widely observed that the prevalence of asthma and other allergic diseases have risen over the past 2 decades in developed nations (McNally *et al.*, 1998; Sly, 1999; Woolcock, 1991). During the same period, the genetic etiology of asthma has been increasingly emphasized as a means of better understanding its pathogenesis, with the eventual goal of improving preventive strategies, diagnostic tools and therapies (Sandford *et al.*, 1996). Considerable effort and expense is thus currently being expended in attempts to detect genetic loci contributing to asthma susceptibility (CSGA, 1997; Daniels *et al.*, 1996; Ober *et al.*, 1998; Wjst *et al.*, 1999). However, many of the most fundamental questions relating to the genetic epidemiology of asthma and associated factors remain poorly defined.

This chapter will review what is currently known about the genetics of asthma, with a focus on methodological issues. The theoretical basis and experimental evidence for genetic causation in asthma will therefore be examined from both epidemiological and molecular perspectives.

Analysis of Multifactorial Disease, edited by T. Bishop and P. Sham.

2. Phenotype definition

The study of genetics always begins with the study of phenotype. The choice of phenotype is critical for the success of programs of gene discovery. In the case of asthma, however, the choice is far from simple.

2.1 Asthma

Although few clinicians would have any difficulty recognizing a wheezing child as asthmatic, there is no generally accepted definition of clinical asthma. Consensus clinical statements defining asthma (NIH, 1995; Warner *et al.*, 1989) have not proved useful for epidemiological or genetic purposes. Many epidemiological and genetic studies of asthma have been based upon phenotypes derived from questionnaire data. Most questionnaires in general use (MRC, ATS, ISAAC, ECRHS) contain standard questions about the presence of wheeze and shortness of breath, and doctor-diagnosed asthma. Extensive studies have shown that asthma may be effectively recognized by this sort of questionnaire (O'Connor and Weiss, 1994), and that these simple responses seem as effective and reproducible as phenotypes derived from complex algorithms. However, data regarding disease history collected from questionnaires are subject to recall bias and recollection loss (Speight, 1978); this may have acted to bias the reported familial aggregation of disease state in some studies. For instance, asthmatic parents may be more likely to recognize and remember the disease in their offspring; thus inflating estimates of familial aggregation.

The relationship of childhood-onset asthma to adult-onset asthma is unclear. Adult-onset asthma is a poorly defined illness that is not obviously part of the atopic syndrome. The diagnosis of asthma in older subjects is seriously confounded by the presence of cigarette smoking. This may be dealt with most simply by concentrating on asthma in children and young adults, and by excluding smokers from the study population.

2.2 Asthma-associated pathophysiological traits

Asthma and atopy are associated with increased levels of total serum IgE and detectable levels of IgE specific to common aero-allergens such as house dust mite and grass pollens (Burrows *et al.*, 1989). Specific IgE can be detected by ELISA serum assays, or by prick skin tests in which the wheal response to inoculation with allergen is measured. Asthma is also associated with increased nonspecific airway responsiveness (AR) to a variety of spasmogens such as histamine and methacholine (Burrows *et al.*, 1995a; Sears *et al.*, 1993a). AR is quantifiable by a number of techniques which measure airflow limitation as a function of increasing doses of histamine or methacholine. Asthma is commonly accompanied by elevated blood eosinophil counts, and variably reduced spirometric measures of pulmonary function such as the forced expiratory volume in 1 s (FEV_1) and the forced vital capacity (FVC).

Although none of the pathophysiological factors known to be associated with symptoms are specific to or diagnostic of asthma (Barbee *et al.*, 1981; Bousquet *et al.*, 1994; Pattemore *et al.*, 1990; Sears *et al.*, 1993b; Shibasaki *et al.*, 1988; Spry, 1988), they may be used as intermediate phenotypes to dissect the multiple pathways that are likely to interact to produce clinical asthma. The total serum IgE measurement is log-normally distributed, with a well-established normal range. Skin and serum ELISA tests of specific IgE are approximately normally distributed, but with a left-hand tail which is effectively censored, as the majority of subjects do not have recordable responses to specific allergens. Nevertheless, positive skin and ELISA tests have a well-understood relationship to disease (Burrows *et al.*, 1989). AR is commonly expressed as the provocative concentration (PC_{20}) or dose (PD_{20}) of spasmogen required to produce a 20% fall in airflow. This measurement is heavily right-censored, with a positive value recordable in less than 10% of subjects. For this reason, the dose–response slope (DRS) (O'Connor *et al.*, 1987), which is log-normal and only 10% censored, may be a better phenotype for genetic studies. Eosinophil counts are log-normally distributed.

All of these quantitative traits are closely associated with each other in symptomatic and asymptomatic subjects (Barbee *et al.*, 1981; Bousquet *et al.*, 1990; Burrows *et al.*, 1992; Ferguson and Wong, 1989; Halonen *et al.*, 1982; Kartasamita *et al.*, 1994; Mensinga *et al.*, 1992; Paoletti *et al.*, 1995; Peat *et al.*, 1987; Rijcken *et al.*, 1988; Sears *et al.*, 1993a; Taylor and Luksza, 1987; Wardlaw *et al.*, 1988). However, the inter-relationships of these traits are variable. The extent to which the close inter-relationships of asthma-associated traits reflects shared genetic determinants remains uncertain. The few studies that have addressed this issue suggest that total and specific serum IgE levels, AR, blood eosinophil counts and spirometric measures may be genetically distinct traits (Cookson and Palmer, 1998; Postma *et al.*, 1995).

3. Familial aggregation of asthma and atopy

Since the 1916 study of Cooke and Van der Veer (Cooke and Veer, 1916) it has been clear that most asthma and other allergic disease has a familial component. First degree relatives of asthmatic subjects are at increased risk of developing asthma and other allergic disease (Dold *et al.*, 1992; Sibbald *et al.*, 1980; Turner *et al.*, 1974; Wittig *et al.*, 1978) and have a higher prevalence of elevated serum IgE levels and increased AR (Laprise and Boulet, 1996). In childhood, a family history of asthma and other atopic disorders is one of the strongest and best-documented risk factors for asthma, increased AR, lower respiratory symptoms and atopy (Panhuysen *et al.*, 1995; Sandford *et al.*, 1996; Tariq *et al.*, 1998). Family history of asthma is also a risk factor for the persistence of childhood asthma into adulthood (Jenkins *et al.*, 1994). The age-dependent distribution of atopic disease and associated quantitative traits suggests a possible reduced penetrance of genetic susceptibility in early and later life.

Twin studies have generally shown that concordance rates for asthma are significantly higher in monozygotic (MZ) twins than dizygotic (DZ) twins (Duffy *et al.*, 1990; Edfors-Lubs, 1971; Hopp *et al.*, 1984; Nieminen *et al.*, 1991; Sarafino and Goldfedder, 1995), whether reared apart or together (Hanson *et al.*, 1991). Broad sense heritability estimates derived from twin studies range from 36% (Nieminen *et al.*, 1991) to 75% (Duffy *et al.*, 1990).

There have been few studies specifically analyzing the segregation of asthma within families, probably because the phenotype 'asthma' is itself controversial and difficult to define. Those studies which have been published have produced results consistent with a general oligogenic model of inheritance (Holberg *et al.*, 1996; Jenkins *et al.*, 1997; Lawrence *et al.*, 1994).

4. Familial aggregation of asthma-associated traits

Parental history of atopy has been associated with elevated serum total IgE levels in Caucasian infants (Kjellman and Johansson, 1976) and children (Burrows *et al.*, 1995b). Numerous pedigree studies have suggested that total and specific serum IgE levels aggregate in families (Fujii *et al.*, 1995; Kuehr *et al.*, 1993; Turner *et al.*, 1974) and that total serum IgE levels have a narrow-sense heritability of around 50–60% (Grundbacher, 1975; Lawrence *et al.*, 1994). Twin studies have consistently indicated that concordance rates for serum total IgE levels are significantly higher in MZ twins than DZ twins (Bazaral *et al.*, 1974; Hopp *et al.*, 1984), even when reared apart (Hanson *et al.*, 1991), and have suggested a broad-sense heritability of around 50–84% (Bazaral *et al.*, 1974; Blumenthal and Bonini, 1990; Hanson *et al.*, 1991; Hopp *et al.*, 1984). Numerous studies of the segregation of serum total IgE levels within families have been undertaken. Results have been contradictory, and recessive (Gerrard *et al.*, 1978; Marsh *et al.*, 1974; Meyers *et al.*, 1987, 1994) (Dizier *et al.*, 1995), dominant (Blumenthal *et al.*, 1981; Cookson and Hopkin, 1988; Marsh *et al.*, 1994), codominant (Martinez *et al.*, 1994; Meyers *et al.*, 1982), and polygenic (Hasstedt *et al.*, 1983) models have all been proposed. These studies suggest that major gene effects – though not necessarily the same major gene – may be influencing total serum IgE levels in many different populations, and that atopy as reflected by serum IgE levels is characterized by genetic heterogeneity (Blumenthal *et al.*, 1981, 1986; Xu *et al.*, 1995). Some studies (Dizier *et al.*, 1995; Marsh *et al.*, 1974; Meyers *et al.*, 1987) have also suggested that the mechanisms regulating serum total and specific IgE titers are, to some degree, under separate genetic control.

There has been little study of the genetic regulation of blood eosinophil counts in the general population. A study of familial aggregation of blood eosinophil counts in three endemically parasitized Brazilian populations (Moro-Furlani and Krieger, 1992) found evidence of significant familial aggregation accounting for around 58% of the total phenotypic variance, but was unable to resolve the phenotypic variance into genetic and environmental components. It may be of interest that two segregation analyses of human resistance to parasitization have been undertaken in an endemically parasitized tropical population from Brazil. Segregation analysis of both infection intensity (fecal egg counts) by *Schistosoma mansoni* (Abel *et al.*, 1991) and IL-5 levels (Rodrigues *et al.*, 1996) suggested major codominant genes regulating both traits.

The aggregation of increased AR to cholinergic agents within families has been well documented (Hopp *et al.*, 1988; Longo *et al.*, 1987; Postma *et al.*, 1995; Townley *et al.*, 1986). Parental history of asthma has been associated with increased prevalence of AR in Caucasian infants (Hopp *et al.*, 1990; Young *et al.*,

1991). Twin and family studies have suggested that the area under the methacholine or histamine dose–response curve has a significantly higher concordance in MZ twins than DZ twins and that this trait has a heritibility of 27–66% (Hopp *et al.*, 1984; Lawrence *et al.*, 1994). A segregation analysis of the area under the methacholine-ΔFEV_1 dose–response curve demonstrated a bimodal distribution of AR in families ascertained via asthmatic probands and a unimodal distribution in families ascertained via nonatopic probands. This study adjusted for the effects of age, gender, recent respiratory tract infection and ascertainment, and did not suggest determination by a single major gene (Townley *et al.*, 1986). There is weak evidence that AR to carbachol is inherited in humans as an autosomal dominant trait with incomplete penetrance (Longo *et al.*, 1987).

Many of these studies of the descriptive genetic epidemiology of asthma and associated traits have experienced difficulties in adjusting for familial and non-familial factors known to have major effects on the traits investigated, including sex, age, race, body size, active and passive tobacco smoke exposure, infection and recent allergen exposure. These factors are themselves likely to aggregate in families, and need to be adjusted for if rigorous investigation of genetic influences on asthma are to be performed. Other potentially important factors complicating the comparison of studies include the varying ages of subjects studied and differing methods of phenotypic assessment.

Many of the reported twins studies of asthma- and atopy-associated pathophysiological traits also suffer from limitations. These include small sample sizes and associated low power; a focus on adult twins which may be biased by age-dependent penetrance of genes and the cumulative effects of environmental factors; lack of consistency in study design; and the use of different statistical methods to estimate heritability. It is also uncertain to what extent results derived from adults can be applied to children.

The differing results of segregation analyses of asthma- and atopy-associated traits may be partly due to the analysis of families from different and often genetically-isolated populations. Failure to adequately address distributional assumptions may also have contributed to these diverse results. In addition, inter-population heterogeneity may affect variance components and segregation analyses, and could explain some of the reported differences between populations. Some studies have been conducted upon populations selected on the basis of a proband with atopic disease, without correction for ascertainment (Blumenthal *et al.*, 1981; Elston, 1995b; Marsh *et al.*, 1974). Further, it is possible that environment-dependent expression of susceptibility loci and geographical variability in environmental exposure may affect the perceived mode of inheritance of a trait. For instance, allergy to sporadic pollens in Europe and North America has an overall susceptibility-dominant pattern of inheritance (Marsh and Huang, 1991). In contrast, allergy to ubiquitous pollens in Japan has an overall resistance-dominant pattern of inheritance (Sasazuki *et al.*, 1983).

The balance of evidence suggests that it is likely that the commonly measured asthma-associated pathophysiological factors are under some degree of significant genetic control. However, the degree of genetic control and the extent to which these traits are genetically distinct is still unclear.

4.1 Animal models

The genetics of asthma- and atopy-associated traits have also been studied in inbred strains of mice, rats and dogs. Most studies of inbred strains and backcrosses have suggested strong genetic control of serum IgE levels; broad-sense heritability has been estimated at 60–80% (Biozzi *et al.*, 1979; Sapin *et al.*, 1984). The genetics of eosinophilia, particularly in relation to parasitic infection, has been extensively studied in inbred strains of experimental animals (Lammas *et al.*, 1988; Rothwell *et al.*, 1988). These studies have consistently shown marked strain-dependent variation in capacity to mount an eosinophilic response (Dawkins *et al.*, 1989; Lammas *et al.*, 1988; Wakelin, 1985). Previous research in animals indicates that AR to cholinergic agents in inbred models is heritable (Hirshman *et al.*, 1984; Pauwels *et al.*, 1985) and can be transmitted either as an autosomal recessive (Levitt and Mitzner, 1988) or polygenic (De Sanctis *et al.*, 1995) trait. Heritability of AR in inbred murine strains has been estimated to be around 50% (De Sanctis *et al.*, 1995).

It is uncertain to what extent these traits, and their underlying genetic control, correspond to their human counterparts. Nevertheless, it seems likely that animal models hold considerable potential for understanding the genetics of asthma and associated disease.

5. Environmental factors

The important role of environmental factors in the etiology of asthma is demonstrated in the rising incidence and prevalence rates, particularly in children, of asthma and other allergic disease in the developed nations (McNally *et al.*, 1998; Sly, 1999; Woolcock, 1991). These changes have been too rapid to be accounted for by changes in gene frequencies due to admixture, migration, selection or random genetic drift (Shirakawa and Morimoto, 1991). Nor is it likely that they can be fully accounted for by changes in either clinical diagnostic patterns or increased recognition of asthmatic symptoms by the general population (Hopkin, 1995; Hopper *et al.*, 1995; Sears, 1991). These changes in prevalence suggest environmental factors are modulating substantial increases in asthma penetrance from early life.

Environmental factors thus play a key role in the pathogenesis of atopic asthma (Ownby, 1990), and it is likely that asthma reflects exposure to environmental trigger factors in genetically susceptible individuals (Panhuysen *et al.*, 1995; Sandford *et al.*, 1996). Three primary environmental exposures that lead to airway inflammation and allergic symptoms have been identified (Becker and Chan-Yeung, 1998; Holt, 1996; Martinez, 1995; McNally *et al.*, 1998; McNichol *et al.*, 1973; Sporik *et al.*, 1992; Tariq *et al.*, 1998; Weiss, 1994): exposure to common aero-allergens such as house dust mite, environmental tobacco smoke exposure, and exposure to viral respiratory infections. In genetically susceptible individuals, these exposures may result in clinical disease from early childhood.

Studies from the developing nations suggest that 'Westernization' is an important risk factor for the development and increased prevalence of atopic asthma (Leung and Jenkins, 1994; Massicot and Cohen, 1986). Relevant lifestyle factors related to 'Westernization' (Shirakawa and Morimoto, 1991; Shirakawa *et al.*, 1996a) may include changes in housing (Cullinan and Newman-Taylor, 1994; Platts-Mills, 1995; Seaton *et al.*, 1994), and decreasing family size and concomitant reduced exposure to parasitic (helminthic) and viral infections (Cookson and Moffatt, 1997; vonMutius, 1996). Psychosocial factors, diet, air pollution, various socioeconomic factors and birthweight have also been associated with atopy and asthma (Becker and Chan-Yeung, 1998; McNally *et al.*, 1998; Morgan and Martinez, 1992; Tariq *et al.*, 1998).

Environmental factors common to families will inflate intrafamilial covariance and, if interpreted naively, can inflate estimates of the relative importance of genetic factors in determining asthma and associated traits (Hopper, 1983; Lange, 1986). It is hence important to attempt to adjust for known important environmental exposures in epidemiological and genetic studies of families. A major limitation of most descriptive and molecular studies of the genetics of asthma is that, with the exception of tobacco smoke exposure, few of the known environmental exposures affecting asthma and atopy risk, most of which also aggregate within families, have been effectively studied and adjusted for. This may have systematically biased previous studies in unknown ways.

5.1 Maternal effects

Many epidemiological studies have suggested that maternal phenotype influences the inheritance of asthma and atopy (Moffatt and Cookson, 1998), although in general the studies were not designed specifically to test for parent-of-origin effects. The presence of asthma, eczema, elevated serum IgE levels, and skin prick test positivity in children have all been accompanied by an increased prevalence of asthma or atopy in mothers (Moffatt and Cookson, 1998). The differential risk of transmission between parents may be four-fold. Parent-of-origin effects have been noted in other immunological disorders, most notably type I diabetes (Warram *et al.*, 1984), rheumatoid arthritis (Akolkar *et al.*, 1997), inflammatory bowel disease (Koumantaki *et al.*, 1997), and selective IgA deficiency (Vorechovsky *et al.*, 1999), suggesting that parental effects on the developing infant's immune system may be an important common process. The mechanism, or mechanisms for these parent-of-origin effects are unknown, but possibilities include genomic imprinting or maternal modification of the developing infant's immune system by transmission of immune factors across the placenta or through breast milk.

6. Study design for gene discovery

Two fundamental approaches have been used to investigate the molecular genetics of asthma: candidate gene approaches and whole genome screens followed by positional cloning attempts.

6.1 Factors complicating genetic studies of asthma

The mapping of human susceptibility loci for asthma is made difficult by a high population frequency, incomplete penetrance, phenocopies and genetic heterogeneity, and phenotypic pleiotropy (Moffatt *et al.*, 1992; Daniels *et al.*, 1996; Sandford *et al.*, 1996).

The high population frequency of symptoms of asthma and atopy imply a high population frequency of susceptibility alleles and concomitant etiological heterogeneity, complicating linkage studies of this disease (Moffatt *et al.*, 1992). With a very common disease such as asthma, there may be both phenocopies and more than one major gene segregating within a large pedigree.

Incomplete penetrance arises in asthma due to the variability of gene–environmental interactions. The lack of total concordance among MZ twins for asthma is evidence of variable and incomplete penetrance (Duffy *et al.*, 1990; Edfors-Lubs, 1971; Nieminen *et al.*, 1991), as is the widespread variation in the prevalence of asthma in different populations (The International Study of Asthma and Allergies in Childhood (ISAAC) Steering Committee, 1998). Incomplete penetrance may also be associated with a high prevalence of susceptibility alleles in the general population (a likely scenario in asthma and atopy). Systematic quantitation of the penetrance of asthma has not been carried out.

Known phenocopies of asthma may arise as the result of viral respiratory infections (Folkerts and Nijkamp, 1995; Hegele *et al.*, 1995) and occupational exposure to various inhaled environmental pollutants (Bardana, 1995; Chan-Yeung and Malo, 1995). The extent of other nongenetic causes of asthma remains unknown.

The evidence from studies of the descriptive genetic epidemiology of asthma phenotypes strongly suggests substantial genetic heterogeneity (Cookson and Palmer, 1998; Sandford *et al.*, 1996), which is likely to operate at both allelic and loci levels. Strategies to minimize the effects of genetic heterogeneity in studies of asthma genetics have included the use of: large pedigrees, genetically isolated populations likely to exhibit founder effects, and the division of study populations into phenotypically homogenous subgroups such as childhood-onset asthma. Importantly, the issue of age-censoring has seldom been addressed in genetic studies using a dichotomous 'asthma' phenotype.

Asthma is characterized by a great variability of symptoms and signs within the same individual, almost certainly as a response to haphazard environmental exposure to allergens. It is therefore to be anticipated that the same genetic polymorphisms may give rise to a wide variety of asthma-associated traits, different components of which may predominate in different study populations. This phenotypic pleiotropy has already been observed for some loci (Daniels *et al.*, 1996), and further complicates unraveling the genetics of allergic disease.

6.2 Choice of linkage strategy

Linkage strategy essentially involves a choice between parametric and nonparametric approaches. Linkage replication may be inherently difficult for a multigenic trait such as asthma (Suarez *et al.*, 1991) and it is likely that the power to detect linkage to any single locus will be low (Davies *et al.*, 1994; Lander and Schork, 1994).

Lod score statistics depend critically upon assumptions about mode of inheritance, gene frequency and penetrance. A clear, generally accepted model for the inheritance of 'asthma' or 'atopy' has not been defined. Therefore, asthma is not presently amenable to analysis by traditional lod score genetic linkage techniques (Weeks and Lathrop, 1995). Nevertheless, refinement of the genetic models underlying asthma is possible, and the combination of linkage and segregation analyses may add power to the detection of linkage to asthma-associated loci.

While less powerful than correctly-specified parametric lod score techniques of linkage analysis, sibpair methods are more robust to genetic heterogeneity and incomplete penetrance than other linkage strategies as they make fewer assumptions. Sibpair strategies also have the advantage of ease of data collection and are relatively straightforward to analyze statistically. Both affected (Blackwelder and Elston, 1985) and Haseman-Elston IBD sibpair methods (Haseman and Elston, 1972) have been used in studies of asthma phenotypes. As for any linkage test, the power to detect linkage with these techniques depends upon the degree of magnitude of effect, mode of ascertainment, genetic heterogeneity, sample size, and the informativeness of the markers used.

Nonparametric sibpair approaches (Khoury *et al.*, 1993) have become the norm in studies of asthma genetics. Other nonparametric linkage techniques such as extensions of the affected sibpair method using maximum likelihood techniques (Hyer *et al.*, 1991) and the affected-pedigree-member method (Weeks and Lange, 1988) have generally not been applied to asthma.

6.3 Choice of association strategy

Association analyses are intrinsically more powerful than linkage analyses in detecting weak genetic effects (Elston, 1995a). It is likely that many of the genes affecting asthma and atopy susceptibility will be of modest effect. Therefore, while linkage studies may be used to detect chromosomal loci containing 'oligogenes' influencing asthma, association may be preferable to linkage in candidate gene investigations where smaller effects on asthma and atopy susceptibility might be anticipated.

The most common design for association studies of asthma phenotypes has been the case–control study. However, case–control studies can be sensitive to population stratification (Ewens and Spielman, 1995) resulting from recent admixture or from poorly matched cases and controls. This may be a particular problem in genetically diverse populations such as those of the United States. Therefore, family-based association strategies are likely to become increasingly common. However, many of the nongenetic factors modulating asthma- and atopy-associated traits are unknown, and it is therefore likely that there will be strong residual phenotypic correlations between family members. In order to avoid bias and decreased power to detect associations, these correlations must be modelled appropriately (Burton *et al.*, 1999; George and Elston, 1987; Tregouet *et al.*, 1997).

The transmission disequilibrium test (TDT) (Spielman *et al.*, 1994) has not often been used in association studies of asthma (Barnes *et al.*, 1996; Hizawa *et al.*, 1998), possibly because the TDT has been restricted to binary phenotypes. The recent availability of quantitative TDT tests (Abecasis *et al.*, 2000) is likely to increase the use of family-based studies of association.

6.4 Choice of study population

Population-based studies provide important background for genetic studies of disease, and may be used to estimate prevalences, incidences and associations. A current limitation to our understanding of the genetic epidemiology of asthma is that the determinants and interactions of asthma and associated intermediate phenotypes have not been fully investigated in true general population samples. Ultimately, it is likely that large, prospective cohort studies will be necessary to fully investigate genetic causation in asthma and atopy, particularly in regard to the role of early life events in disease pathogenesis.

Nuclear families have most often been utilized for genetic studies of asthma. Extended families and interrelated families from inbred populations have also been used for descriptive genetic epidemiological, linkage and association studies. Problems for family studies of asthma for both epidemiological and molecular investigations include age and sex differences in the expression of clinical asthma, classification of family members as 'affected' or 'nonaffected', appropriate adjustment for potentially confounding environmental factors, and ascertainment bias. A further problem is the change in prevalence of the asthma phenotype from one generation to the next, because the prevalence of childhood asthma and other allergic diseases was low in the 1960s but was much higher in the 1990s (Woolcock, 1991).

The use of genetically isolated populations is likely to have some significant advantages in the study of asthma and associated traits (see Chapter 3). Linkage and association studies of genetic isolates from the United States, such as the Hutterites (Ober *et al.*, 1998) and the Amish (Barnes *et al.*, 1996), or the islanders of Tristan da Cunha (Zamel *et al.*, 1996) for asthma phenotypes are currently underway. However, the applicability of results obtained from isolated or founder populations to the general population is unclear.

6.5 Which phenotype?

Molecular genetic studies of a disease depend upon accurate classification of phenotype; definitions of binary outcomes such as 'asthma' are subjective, often study- and population-specific, and likely to be inaccurate to some degree. Hence, quantitative traits have often been used as proxy 'asthma' or 'atopy' phenotypes in molecular genetic investigations. Continuous phenotypes are inherently more informative, objective and statistically powerful than binary categorizations of 'asthma' or 'atopy', and avoid the problems of arbitrary dichotomization and categorization.

One measure of the familial aggregation of disease status is the recurrence risk ratio (λ_S) of the prevalence of disease amongst siblings of affected individuals relative to the prevalence in the general population (Risch, 1990). The λ_S of past or current physician-diagnosed asthma has a value <2 (Cookson and Palmer, 1998). This suggests that over 1000 affected sibpairs will be required to achieve 90% power to detect linkage to this phenotype using affected sibpairs (Palmer *et al.*, 1998b). Conversely, power estimates based upon heritabilities for asthma-associated quantitative traits suggest that these traits are associated with far greater power to detect linkage using sibpair methods (Cookson and Palmer, 1998; Palmer *et al.*, 1998b). A further advantage of using asthma-associated quantitative traits is that this allows the selection of extremely discordant sibpairs, leading to substantially enhanced statistical power compared to an equivalent sample size of unselected sibships (Risch and Zhang, 1995).

For these reasons, the use of intermediate, quantitative phenotypes such as serum IgE levels and AR has become widespread. Importantly, complex variance components modelling of asthma-associated phenotypic traits has suggested that traits such as serum IgE and AR are genetically distinct, and should therefore be examined separately in genetic studies of asthma (Cookson and Palmer, 1998; Palmer, 1998).

7. Molecular studies

7.1 Candidate gene approaches

An ideal candidate gene for asthma would be a known gene of defined sequence that produced a protein plausibly involved in the pathogenesis of asthma, which has been shown to be altered in asthmatic compared to normal subjects. However, obvious limitations of this approach are the large number of potential candidates (Sandford *et al.*, 1996), and that only known genes can be investigated.

There is now an extensive and growing list of candidate genes investigated for linkage and association with asthma- and atopy-associated traits. A region with extensive evidence of linkage and association is the cytokine cluster within the 5q31 region (Hay *et al.*, 1995; Marsh *et al.*, 1994; Meyers *et al.*, 1994; Palmer *et al.*, 1998a; Rossenwasser *et al.*, 1995; Xu *et al.*, 1995).

Several associations have been noted between measures of atopy and genes of the cluster, including IL4 and CD14 (Baldini *et al.*, 1999; Rosenwasser, 1997). The congregation of cytokine genes in the region may have evolved for their co-regulation, and claims for the importance of particular polymorphisms within the cluster should be interpreted in the context of possible linkage disequilibrium with other known or unknown genes. An effect on atopy and serum IgE levels of IL4-receptor (on chromosome 16) has been recognized (Hershey *et al.*, 1997), and may be stronger than the effects of polymorphism in IL4 itself.

Coding variants within the β-adrenergic receptor have been shown *in vitro* to be functionally important (Green *et al.*, 1994), but associations with asthma are weak and inconsistent (Hall, 1996). It remains to be seen if these polymorphisms are important regulators of the response to the commonly prescribed asthma treatment of β-adrenergic agonists. A variant within the 5-lipoxygenase gene has recently been suggested to predict the response to the anti-leukotriene ABT-761 (Drazen *et al.*, 1999). Confirmation of these findings may mark the beginning of the clinical use of genotyping as an adjunct to pharmacotherapy for asthma and many other disorders.

The human MHC on chromosome 6p, particularly HLA and tumor necrosis factor (TNF) locus polymorphism, has also been extensively investigated (Aron *et al.*, 1996; Blumenthal *et al.*, 1992; Campbell *et al.*, 1995; Freidhoff *et al.*, 1988; Howell and Holgate, 1995; Marsh and Huang, 1991; Moffatt and Cookson, 1997; Young *et al.*, 1994), as has polymorphism in the 12q15–24 region (Barnes *et al.*, 1996; Wilkinson *et al.*, 1996). Other candidate genes investigated include: the α region of the T-cell receptor (TCR) α/δ locus (Moffatt *et al.*, 1994); the α_1-antitrypsin gene (α_1-AT) (Hyde *et al.*, 1979; Katz *et al.*, 1976; Liebermann and Colp, 1990); histo-blood-group genetic systems (Kauffmann *et al.*, 1996); the cystic fibrosis gene (DF508) (Mennie *et al.*, 1995; Schroeder *et al.*, 1995); Gm allotypes of IgG genes (Oxelius, 1990); the Ig heavy chain gamma 4 locus (IGHG4) (Amoroso *et al.*, 1996); and angiotensin-converting enzyme (ACE) gene (Benessiano *et al.*, 1997). A number of these studies have not yet been replicated in independent populations.

The clear association between exuberant IgE and eosinophil responses and helminthic infection (Butterworth, 1977; Lynch *et al.*, 1983; Pritchard *et al.*, 1995) has led to the hypothesis that helminthic infections may have provided the selective pressure for high-IgE-response- and hypereosinophilia-alleles that now predispose to atopic disease in nonparasitized populations (Palmer *et al.*, 1997). Studies of candidate genes previously associated with atopy phenotypes have begun to address these hypotheses (Palmer *et al.*, 1997).

Chromosome 11q13. Chromosome 11q13 was the first region to be intensively studied for effects on asthma and atopy. The history of studies of the 11q13 region and the $Fc_{\epsilon}R1$-β candidate gene are instructive as to the difficulties of genetic investigations in asthma and atopy.

Initial studies suggested linkage between 'atopy' (defined by one or more of positive skin test response, positive ELISA tests or elevated serum IgE levels) and an anonymous minisatellite marker on chromosome 11q13 (Cookson *et al.*, 1989; Young *et al.*, 1992). It was subsequently recognized that the gene for the β chain of the high-affinity receptor for IgE ($Fc_{\epsilon}R1$-β) localized to the same region (Sandford *et al.*, 1993; Shirakawa *et al.*, 1994a). Linkage was predominately through maternally derived alleles (Cookson *et al.*, 1992). Coding and noncoding polymorphisms within the gene were subsequently related to serum total IgE and other measures of atopy (Hill *et al.*, 1995; Shirakawa *et al.*, 1994b; Shirakawa *et al.*, 1996c) in several populations.

The initial linkage studies were criticized for the very broad definition of 'atopy' (Marsh and Meyers, 1992; Morton, 1992). Subsequently, a number of studies (Amelung *et al.*, 1992; Brereton *et al.*, 1994; Hizawa *et al.*, 1992; Hizawa *et al.*, 1995; Lympany *et al.*, 1992; Rich *et al.*, 1992; Watson *et al.*, 1995) were not able to replicate the initial linkage of 11q13 to atopy (Cookson *et al.*, 1989) using either lod score or sibpair linkage methods. These studies were of small data sets, generally less than 15 sibpairs. Contradictory findings were published both refuting (Hizawa *et al.*, 1992) and confirming (Shirakawa *et al.*, 1994a) linkage of atopy to 11q13 markers in Japanese families. However, several later studies did confirm linkage or association of 11q13 markers to asthma and atopy (Boguniewicz and Hayward, 1996; Collée *et al.*, 1993; Cox, H. *et al.*, 1998; Palmer *et al.*, 1997; Palmer *et al.*, 1998a; Shirakawa *et al.*, 1995; Shirakawa *et al.*, 1996b; vanHerwerden *et al.*, 1995).

Polymorphism within the $Fc_{\epsilon}R1$-β gene has been related in different studies to atopy (Hill and Cookson, 1996), asthma (Shirakawa *et al.*, 1996b), bronchial hyper-responsiveness (vanHerwerden *et al.*, 1995) and severe atopic dermatitis (Cox, H. *et al.*, 1998). Polymorphism within the gene has also been associated with levels of IgE in heavily parasitized Australian aborigines, implying a protective role for the gene in helminth infestation (Palmer *et al.*, 1997). Although a few coding changes have been identified within $Fc_{\epsilon}R1$-β (Hill and Cookson, 1996;

Shirakawa *et al.*, 1994b), they are conservative and do not seem to alter gene function. The functional mechanism for the influence of the gene or nearby gene(s) on atopic disorders has yet to be described.

The study of this locus has therefore identified several important general issues and problems for the investigation of candidate asthma and atopy loci (Cookson, 1995): ascertainment of family panels; differing phenotypic definitions; small sample sizes; parent-of-origin effects; genetic heterogeneity; and phenotypic pleiotropy.

7.2 Whole genome approaches

A number of whole genome screens for asthma and atopy phenotypes have now been published (Bleecker *et al.*, 1999; Cox, N. *et al.*, 1998; CSGA, 1997; Daniels *et al.*, 1996; Dizier *et al.*, 1999; Ober *et al.*, 1998; Wjst *et al.*, 1999). All of these studies have been of nuclear families, and have investigated a number of asthma-associated phenotypes. Most have utilized nonparametric sibpair linkage strategies, with the exception of the studies of Hutterite (Ober *et al.*, 1998) and Dutch (Bleecker *et al.*, 1999) populations which used semiparametric and parametric methods respectively. Caucasian populations have generally been studied (Bleecker *et al.*, 1999; Daniels *et al.*, 1996; Dizier *et al.*, 1999; Ober *et al.*, 1998; Wjst *et al.*, 1999), although the US Collaborative Studies on the Genetics of Asthma (CSGA) study has investigated ethnically diverse populations (Caucasian, Hispanic, Afro-American; CSGA, 1997). Three studies have used samples of highly selected affected sibpairs (CSGA, 1997; Dizier *et al.*, 1999; Wjst *et al.*, 1999), one study has been of a sample subselected on the basis of nonconcordant sibships from a population-based sample of families (Daniels *et al.*, 1996), one of a sample selected on the basis of an adult proband with asthma (Bleecker *et al.*, 1999) and one study has been of a genetically isolated founder population (the Hutterites; Ober *et al.*, 1998).

Strategies aimed at minimizing type I (false-positive) and type II (false-negative) errors in these whole genome screens for asthma have included: the adoption of stringent criteria for significance (including calculation of empirical *P* values) (Daniels *et al.*, 1996; Dizier *et al.*, 1999); replication in multiple independent data sets (Daniels *et al.*, 1996; Dizier *et al.*, 1999; Ober *et al.*, 1998); adjustment either analytically or via subject selection for known potential confounders (CSGA, 1997); and the use of stringent selection criteria for affected subjects (CSGA, 1997; Dizier *et al.*, 1999; Ober *et al.*, 1998; Wjst *et al.*, 1999).

While there are some inconsistencies between the locations of the important linkages found in the genome screens undertaken to date (Bleecker *et al.*, 1999; CSGA, 1997; Cox, N. *et al.*, 1998; Daniels *et al.*, 1996; Dizier *et al.*, 1999; Ober *et al.*, 1998; Wjst *et al.*, 1999), there are also 'consensus' areas of linkage to asthma, atopy or a related phenotype such as serum IgE levels. These include chromosome 5q (Bleecker *et al.*, 1999; CSGA, 1997; Dizier *et al.*, 1999; Ober *et al.*, 1998), chromosome 6p (Daniels *et al.*, 1996; Wjst *et al.*, 1999), chromosome 11q (Cox, N. *et al.*, 1998; Daniels *et al.*, 1996) and chromosome 12q (CSGA, 1997; Dizier *et al.*, 1999; Ober *et al.*, 1998; Wjst *et al.*, 1999). These regions all contain biologically plausible candidate genes, and most have been previously linked to asthma phenotypes in candidate gene studies (see section 7.1).

There are a number of limitations and problems with the genome screens for asthma and atopy phenotypes so far published. The precise number of affected sibpairs genotyped has not always been explicitly stated (CSGA, 1997). The relatively small numbers studied would have both tended to limit the power of these genome screens to detect linkage to dichotomous phenotypes ('asthma'; Cookson and Palmer, 1998) and increased the possibility of type I experimental error. The use of an appropriate genome-wide threshold for statistical significance is a contentious area (see Chapter 2). Both the use of widely differing thresholds for 'significance' and the use of differing statistical methods in these studies makes it somewhat difficult to compare them. Although multipoint linkage is generally likely to be more informative than singlepoint analysis (see Chapter 2), only three studies utilized multipoint analysis (Bleecker *et al.*, 1999; CSGA, 1997; Wjst *et al.*, 1999). In three of the studies, no replication of putative novel linkages was attempted in independent population samples of similar ethnicity (CSGA, 1997; Bleecker *et al.*, 1999; Wjst *et al.*, 1999). Finally, the calculation of empirical *P* values by simulation in order to provide some impression of the plausibility of the results does not appear to have been attempted in several of the studies (CSGA, 1997; Ober *et al.*, 1998; Wjst *et al.*, 1999). In one study, only the first 40% of the sample were presented in the initial paper (CSGA, 1997). Unsurprisingly, when further sibpairs were analysed (Cox, N. *et al.*, 1998) substantially different results were found.

A further problem in interpreting all of the whole genome screens thus far published is that the phenotypic traits examined are highly correlated and that many of the markers in each genome screen showed evidence of linkage to several traits (CSGA, 1997; Daniels *et al.*, 1996; Ober *et al.*, 1998; Wjst *et al.*, 1999); it is thus unclear to what extent the different linkages might reflect differing, trait-specific genes. Further, the implicit or explicit assumption underlying the reporting of these screens, that positional cloning of susceptibility genes for asthma and atopy-associated quantitative traits is identical to mapping susceptibility genes for asthma and atopy, may not be valid.

8. Conclusions

Asthma clearly aggregates within families, and many studies have demonstrated evidence of a heritable component to the familial aggregation. In the absence of clear patterns of monogenic inheritance, the recurrence of asthma in families is suggestive of a multifactorial etiology. The physiological factors associated with asthma also exhibit familial aggregation. The evidence suggests that some of these (in particular the serum IgE) are due to an oligogenetic background modified by strong environmental factors.

Despite much progress in defining the genetic epidemiology of asthma and atopy in the last decade, much further research is required. In particular, progress to date in phenotype definition and in identifying specific functional mutations encoding asthma susceptibility has been poor. However, a number of groups are currently active in addressing methodological problems in phenotypic assessment and in positional cloning attempts using fine mapping techniques.

Future studies need to directly address the issue of etiological heterogeneity in asthma and atopy, and should begin to study the genetic determinants of asthma-associated factors other than atopy. Despite their close associations, asthma, IgE levels, eosinophil counts, spirometric parameters and AR are not necessarily proxy measures of each other, and may well comprise distinct phenotypes under some degree of separate genetic control.

Given the multifactorial nature of atopy and asthma, any attempt to resolve genetic factors must do so within an epidemiological framework that includes both genetic and environmental elements. Future studies should also start to address with the heterogeneity characteristic of asthma from both study design and analytical viewpoints, and must begin to include information on the known environmental factors associated with asthma which aggregate within families. Success in defining specific susceptibility genes involved in the pathogenesis of asthma holds the potential for substantial benefits to public health.

References

Abecasis, G.R., Cardon L.R. and Cookson W.O. (2000) A general orthogonal test of association for quantitative traits in nuclear families. *Am. J. Hum. Genet.* **66**: 279–292.

Abel, L., Demenais, F., Prata, A., Souza A. and Dessein, A. (1991) Evidence for the segregation of a major gene in human susceptibility/resistance to infection by *Schistosoma mansoni*. *Am. J. Hum. Genet.* **48**: 959–970.

Akolkar, P.N., Gulwani-Akolkar, B., Heresbach, D., *et al.* (1997) Differences in risk of Crohn's disease in offspring of mothers and fathers with inflammatory bowel disease. *Am. J. Gastroenterol.* **92**(12): 2241–2244.

Amelung, P.J., Panhuysen, C.I.M., Postma, D. S., *et al.* (1992) Atopy and bronchial hyperresponsiveness: exclusion of linkage to markers on chromosomes 11q and 6p. *Clin. Exp. Allergy* **22**: 1077–1084.

Amoroso, A., Berrino, M., Bottaro, A., *et al.* (1996) The genetics of allergy – RFLP analysis of HLA and immunoglobulin heavy chain constant genes in Italian patients. *Fundam. Clin. Immunol.* **4**(1): 35–44.

Aron, Y., Desmazes-Dufeu, N., Matran, R., *et al.* (1996) Evidence of a strong, positive association between atopy and the HLA class II alleles DR4 and DR7. *Clin. Exp. Allergy* **26**: 821–828.

Asher, M.I., Keil, U., Anderson, H.R., *et al.* (1995) International Study of Asthma and Allergies in Childhood (ISAAC): rationale and methods. *Eur. Respir. J.* **8**(3): 483–491.

Baldini, M., Carla Lohman, I., Halonen, M., Erickson, R.P., Holt, P.G. and Martinez, F.D. (1999) A polymorphism in the 5′ flanking region of the CD14 gene is associated with circulating soluble CD14 levels and with total serum immunoglobulin E. *Am. J. Respir. Cell. Mol. Biol.* **20**(5): 976–983.

Barbee, R., Halonen, M., Lebowitz, M. and Burrows, B. (1981) Distribution of IgE in a community population sample: correlations with age, sex, and allergen skin test reactivity. *J. Allergy Clin. Immunol.* **68**(2): 106–111.

Bardana, E.J. Jr. (1995) Occupational asthma and related respiratory disorders. *Disease-A-Month* **41**(3): 143–199.

Barnes, K.C., Neely, J.D., Duffy, D.L., *et al.* (1996) Linkage of asthma and total serum IgE concentration to markers on chromosome 12q: evidence from Afro-Caribbean and Caucasian populations. *Genomics* **37**(1): 41–50.

Bazaral, M., Orgel, A. and Hamburger, R. (1974) Genetics of IgE and allergy: serum IgE levels in twins. *J. Allergy Clin. Immunol.* **54**: 288–304.

Beasley, R., Burgess, C., Crane, J., Pearce, N. and Roche, W. (1993) Pathology of asthma and its clinical implications. *J. Allergy Clin. Immunol.* **92**(1 Pt 2): 148–154.

Becker, A. and Chan-Yeung, M. (1998) Primary prevention of asthma and other allergic disorders. *Semin. Respir. Crit. Care Med.* **19**(6): 563–568.

Benessiano, J., Crestani, B., Mestari, F., *et al.* (1997) High frequency of a deletion polymorphism of the angiotensin-converting enzyme gene in asthma. *J. Allergy Clin. Immunol.* **99**(1): 53–57.

Biozzi, G., Mouton, D., Sant'Anna, O., *et al.* (1979) Genetics of immunoresponsiveness to natural antigens in the mouse. *Curr. Top. Microbiol. Immunol.* **85**: 31–98.

Blackwelder, W. and Elston, R. (1985) A comparison of sibpair linkage tests for disease susceptibility loci. *Genet. Epidemiol.* **2**: 85–97.

Bleecker, E.R., Postma, D.S., Howard, T.D., *et al.* (1999) Genome screen for susceptibility loci for bronchial hyperresponsiveness in a genetically homogeneous Dutch population. *Am. J. Respir. Crit. Care Med.* **159**(3): A645.

Blumenthal, M. and Bonini, S. (1990) Immunogenetics of specific immune responses to allergens in twins and families. In: *Genetic and Environmental Factors in Clinical Allergy.* (eds D. Marsh and M. Blumenthal). University of Mineapolis Press, Minneapolis, pp. 132–142.

Blumenthal, M., Namboordiri, K., Mendell, N., Gleich, P., Elston, R. and Yunis, E. (1981) Genetic transmission of serum IgE levels. *Am. J. Med. Genet.* **10**: 219–228.

Blumenthal, M., Yunis, E., Mendell, N. and Elston, R. (1986) Preventive allergy: Genetics of IgE-mediated diseases. *J. Allergy Clin. Immunol.* **78**(5(2)): 962–968.

Blumenthal, M., Marcus-Bagley, D., Awdeh, Z., Johnson, B., Yunis, E.J. and Alper, C.A. (1992) HLA-DR2, [HLA-B7, SC31, DR2], and [HLA-B8, SCO1, DR3] haplotypes distinguish subjects with asthma from those with rhinitis only in ragweed pollen allergy. *J. Immunol.* **148**(2): 411–416.

Boguniewicz, M. and Hayward, A. (1996) Atopy, airway responsiveness, and genes. *Thorax* **51**(Suppl 2): S55–S59.

Bousquet, J., Chanez, P., Lacoste, J., *et al.* (1990) Eosinophilic inflammation in asthma. *N. Engl. J. Med.* **323**: 1033–1039.

Bousquet, J., Chanez, P., Vignola, A.M., Lacoste, J.Y. and Michel, F.B. (1994) Eosinophil inflammation in asthma. *Am. J. Respir. Crit. Care Med.* **150**(5 Pt 2): S33–38.

Brereton, H., Ruffin, R., Thompson, P. and Turner, D. (1994) Familial atopy in Australian pedigrees: adventitious linkage to chromosome 8 is not confirmed nor is there evidence of linkage to the high affinity IgE receptor. *Clin. Exp. Allergy* **24**: 868–877.

Burrows, B., Martinez, F., Halonen, M., Barbee, R. and Cline, M. (1989) Association of asthma with serum IgE levels and skin-test reactivity to allergens. *N. Engl. J. Med.* **320**: 271–277.

Burrows, B., Sears, M.R., Flannery, E.M., Herbison, G.P. and Holdaway, M.D. (1992) Relationships of bronchial responsiveness assessed by methacholine to serum IgE, lung function, symptoms, and diagnoses in 11-year-old New Zealand children. *J. Allergy Clin. Immunol.* **90**: 376–385.

Burrows, B., Sears, M.R., Flannery, E.M., Herbison, G.P., Holdaway, M.D. and Silva, P.A. (1995a) Relation of the course of bronchial responsiveness from age 9 to age 15 to allergy. *Am. J. Respir. Crit. Care Med.* **152**(4 Pt 1): 1302–1308.

Burrows, B., Martinez, F., Cline, M. and Lebowitz, M. (1995b) The relationship between parental and children's serum IgE and asthma. *Am. J. Respir. Crit. Care Med.* **152**: 1497–1500.

Burton, P., Tiller, K., Gurrin, L., Cookson, W., Musk, A. and Palmer, L. (1999) Genetic variance components analysis for binary phenotypes using generalized linear mixed models (GLMMs) and Gibbs sampling. *Genet. Epidem.* [in press].

Butterworth, A. (1977) The eosinophil and its role in immunity to helminthic infections. *Curr. Top. Microbiol. Immunol.* **77**: 127–168.

Campbell, D., Britton, J., Pavord, I., *et al.* (1995) LTa NcoI polymorphism at the TNF locus correlates with clinical symptoms of asthma. *Eur. J. Resp. Dis.* **8**: 552.

Chan-Yeung, M. and Malo, J.L. (1995) Occupational asthma. *N. Engl. J. Med.* **333**(2): 107–112.

Collée, J.M., ten Kate, L.P., de Vries, H.G., *et al.* (1993) Allele sharing on chromosome 11q13 in sibs with asthma and atopy. *The Lancet* **342**: 936.

Cooke, R. and Veer, A.V. (1916) Human sensitization. *J. Immunol.* **1**: 201–205.

Cookson, W. (1995) 11q and high-affinity IgE receptor in asthma and allergy. *Clin. Exp. Allergy* **25**(S2): 71–73.

Cookson, W. and Hopkin, J. (1988) Dominant inheritance of atopic immunoglobulin-E responsiveness. *Lancet* **i**: 86–88.

Cookson, W. and Moffatt, M. (1997) Asthma: An epidemic in the absence of infection? *Science* **275**: 41–42.

Cookson, W. and Palmer, L. (1998) Investigating the asthma phenotype. *Clin. Exp. Allergy* **28** (Suppl. 1): 88–89.

Cookson, W., Sharp, P., Faux, J. and Hopkin, J. (1989) Linkage between immunoglobulin E responses underlying asthma and rhinitis and chromosome 11q. *Lancet* **i**: 1292–1295.

Cookson, W.O.C.M., Young, R.P., Sandford, A.J., *et al.* (1992) Maternal inheritance of atopic IgE responsiveness on chromosome 11q. *Lancet* **340**: 381–384.

Cox, H., Moffatt, M., Faux, J., *et al.* (1998) Association of atopic dermatitis to the beta subunit of the high affinity immunoglobulin E receptor. *Br. J. Dermatol.* **138**: 182–187.

Cox, N., Beaty, T., Rich, S., *et al.* (1998) Genome screen for asthma susceptibility loci in US populations. *Am. J. Respir. Crit. Care Med.* **157**(3): A855.

CSGA (1997) A genome-wide search for asthma susceptibility loci in ethnically diverse populations. *Nature Genet.* **15**: 389–392.

Cullinan, P. and Newman-Taylor, A.J. (1994) Asthma in children: environmental factors. *Br. Med. J.* **308**: 1585–1586.

Daniels, S., Bhattacharrya, S., James, A., *et al.* (1996) A genome-wide search for quantitative trait loci underlying asthma. *Nature* **383**: 247–250.

Davies, J., Kawaguchi, Y., Bennett, S. *et al.* (1994) A genome-wide search for human type I diabetes susceptibility loci. *Nature* **371**: 130–136.

Dawkins, H., Windon, R. and Eagleson, G. (1989) Eosinophil responses in sheep selected for high and low responsiveness to *Trichostrongylus colubriformis*. *Int. J. Parasitol.* **19**: 199–205.

De Sanctis, G.T., Merchant, M., Beier, D.R. *et al.* (1995) Quantitative locus analysis of airway hyperresponsiveness in A/J and C57BL/6J mice. *Nature* Genet **11**(2): 150–154.

Dizier, M., Hill, M., James, A., *et al.* (1995) Detection of a recessive major gene for high IgE levels acting independently of specific response to allergens. *Genet. Epidemiol.* **12**: 93–105.

Dizier, M.H., Sandford, A., Walley, A.J. *et al.* (1999) Genome screen for asthma and related phenotypes in the French EGEA study. *Am. J. Respir. Crit. Care Med.* **159**(3): A649.

Dold, S., Wjst, M., von Mutius, E., Reitmeir, P. and Stiepel, E. (1992) Genetic risk for asthma, allergic rhinitis, and atopic dermatitis. *Arch. Dis. Child.* **67**: 1018–1022.

Drazen, J.M., Yandava, C.N., Dube, L., *et al.* (1999) Pharmacogenetic association between ALOX5 promoter genotype and the response to anti-asthma treatment. *Nat. Genet.* **22**(2): 168–170.

Duffy, D.L., Martin, N.G., Battistutta, D., Hopper, J.L. and Mathews, J.D. (1990) Genetics of asthma and hay fever in Australian twins. *Am. Rev. Respir. Dis.* **142**: 1351–1358.

Edfors-Lubs, M. (1971) Allergy in 7000 twin pairs. *Acta Allergol.* **26**: 249–285.

Elston, R. (1995a) The genetic dissection of multifactorial traits. *Clin. Exp. Allergy* **2**(Suppl. 2): 103–106.

Elston, R.C. (1995b) 'Twixt cup and lip: how intractable is the ascertainment problem? [editorial; comment]. *Am. J. Hum. Genet.* **56**(1): 15–17.

Ewens, W. and Spielman, R. (1995) The transmission/disequilibrium test: history, subdivision, and admixture. *Am. J. Hum. Genet.* **57**: 455–464.

Ferguson, A. and Wong, F. (1989) Bronchial hyperresponsiveness in asthmatic children: correlation with macrophages and eosinophils in broncholavage fluid. *Chest* **96**: 988–991.

Folkerts, G. and Nijkamp, F.P. (1995) Virus-induced airway hyperresponsiveness. Role of inflammatory cells and mediators. *Am. J. Respir. Crit. Care Med.* **151**(5): 1666–1673; discussion 1673–1674.

Freidhoff, L., Ehrlich-Kautzky, E., Meyers, D., Ansari, A., Bias, W. and Marsh, D. (1988) Association of HLA-DR3 with human immune response to Lol p I and Lol p II allergens in allergic subjects. *Tissue Antigens* **31**: 211–219.

Fujii, H., Kondo, N., Agata, H., *et al.* (1995) Genetic analysis of IgE and the IGHE, IGHEP1 and IGHEP2 genes in atopic families. *Int. Arch. Allergy Immunol.* **106**(1): 62–68.

George, V. and Elston, R. (1987) Testing the association between polymorphic markers and quantitative traits in pedigrees. *Genet. Epidemiol.* **4**: 193–201.

Gerrard, J., Rao, D. and Morton, N. (1978) A genetic study of Immunoglobulin E. *Am. J. Hum. Genet.* **30**: 46–58.

Godfrey, S. (1993) Airway inflammation, bronchial reactivity and asthma. *Agents and Actions – Supplements* **40**: 109–143.

Green, S., Turki, J., Innis, M. and Liggett, S. (1994) Amino-terminal polymorphisms of the human b2-adrenergic receptor impart distinct agonist-promoted regulatory properties. *Biochemistry* **33**: 9414–9419.

Grundbacher, F.J. (1975) Causes of variation in serum IgE levels in normal populations. *J. Allergy Clin. Immunol.* **56**(2): 104–111.

Hall, I.P. (1996) Beta 2 adrenoceptor polymorphisms: are they clinically important? [editorial]. *Thorax* **51**(4): 351–353.

Halonen, M., Barbee, R., Lebowitz, M. and Burrows, B. (1982) An epidemiologic study of the interrelationships of total serum immunoglobin E, allergy skin test reactivity and eosinophilia. *J. Allergy Clin. Immunol.* **69**: 221–228.

Hanson, B., McGue, M., Roitman-Johnson, B., Segal, N.L., Bouchard, T.J. and Blumenthal, M.N. (1991) Atopic disease and immunoglobulin E in twins reared apart and together. *Am. J. Hum. Genet.* **48**: 873–879.

Haseman, J. and Elston, R. (1972) The investigation of linkage between a quantitative trait and a marker locus. *Behav. Genet.* **2**: 3–19.

Hasstedt, S., Meyers, D. and Marsh, D. (1983) Inheritance of IgE. *Am. J. Med. Genet.* **14**: 61–66.

Hay, J.G., Danel, C., Chu, C.S. and Crystal, R.G. (1995) Human CC10 gene expression in airway epithelium and subchromosomal locus suggest linkage to airway disease. *Am. J. Physiol.* **268**(4 Pt 1): L565–575.

Hegele, R.G., Hayashi, S., Hogg, J.C. and Pare, P.D. (1995) Mechanisms of airway narrowing and hyperresponsiveness in viral respiratory tract infections. *Am. J. Respir. Crit. Care Med.* **151**(5): 1659–1664; discussion 1664–1665.

Hershey, G.K.K., Friedrich, M.F., Esswein, L.A., Thomas, M.L. and Chatila, T.A. (1997) The association of atopy with a gain-of-function mutation in the alpha subunit of the interleukin-4 receptor. *N. Engl. J. Med.* **337**(24): 1720–1725.

Hill, M. and Cookson W., (1996) A new variant of the beta subunit of the high-affinity receptor for immunoglobin E (FC-epsilon-RI-beta E237G) – associations with measures of atopy and bronchial hyper-responsiveness. *Hum. Mol. Genet.* **5**(7): 959–962.

Hill, M., James, A., Faux, J., *et al.* (1995) FcεRI-β polymorphism and risk of atopy in a general population sample. *Br. Med. J.* **311**: 776–779.

Hirshman, C., Downes, H. and Veith, L. (1984) Airway responses in offspring of dogs with and without airway hyperreactivity. *J. Appl. Physiol.* **56**: 1272–1277.

Hizawa, N., Collins, G., Rafnar, T., *et al.* (1998) Linkage analysis of *dermatophagoides pteronyssinus*-specific Ige responsiveness with polymorphic markers on chromosome 6p21 (Hla-D Region) in caucasian families by the transmission/disequilibrium test. *J. Allergy Clin. Immunol.* **102**(3): 443–448.

Hizawa, N., Yamaguchi, E., Furuya, K., *et al.* (1995) Association between high serum total IgE levels and D11S97 on chromosome 11q13 in Japanese subjects. *J. Med. Genet.* **32**: 363–369.

Hizawa, N., Yamaguchi, E., Ohe, M., *et al.* (1992) Lack of linkage between atopy and locus 11q13. *Clin. Exp. Allergy* **22**: 1065–1069.

Holberg, C., Elston, R., Halonen, M., *et al.* (1996) Segregation analysis of physician-diagnosed asthma in hispanic and non-hispanic white families. *Am. J. Respir. Crit. Care Med.* **154**: 144–150.

Holgate, S. (1996) The inflammatory basis of asthma and its implications for drug treatment. *Clin. Exp. Allergy* **26**(Supplement 4): 1–4.

Holt, P. (1996) Infections and the development of allergy. *Toxicol. Lett.* **86**(2–3): 205–210.

Hopkin, J. (1995) Genetics of atopy. *Pediatr. Allergy Immunol.* **6**: 139–144.

Hopp, R., Bewtra, A., Watt, G., Nair, N. and Townley, R. (1984) Genetic analysis of allergic disease in twins. *J. Allergy Clin. Immunol.* **73**: 265–270.

Hopp, R., Bewtra, A., Biven, R., Nair, N. and Townley, R. (1988) Bronchial reactivity pattern in nonasthmatic parents of asthmatics. *Ann. Allergy* **61**: 184–186.

Hopp, R., Townley, R., Biven, R., Bewtra, R. and Nair, N. (1990) The presence of airway reactivity before the development of asthma. *Am. Rev. Respir. Dis.* **141**: 2–8.

Hopper, J. (1983) The utility of a multivariate normal model for studying familial patterns in medical and psychiatric data. *Aust. NZ J. Psychiat.* **17**: 342–348.

Hopper, J., Jenkins, M., Carlin, J. and Giles, G. (1995) Increase in the self-reported prevalence of asthma and hay fever in adults over the last generation: a matched parent–offspring study. *Aust. J. Public Health* **19**: 120–124.

Howarth, P.H. (1995) The airway inflammatory response in allergic asthma and its relationship to clinical disease. *Allergy* **50**(Suppl. 22): 13–21.

Howell, W. and Holgate, S. (1995) HLA genetics and allergic disease. *Thorax* **50**: 815–818.

Hyde, J., Werner, P., Kumar, C. and Moore, B. (1979) Protease inhibitor variants in children and young adults with chronic asthma. *Ann. Allergy* **43**: 8–13.

Hyer, R., Julier, C., Buckley, J. *et al.* (1991) High-resolution linkage mapping for susceptibility genes in human polygenic disease: insulin-dependent diabetes mellitus and chromosome 11q. *Am. J. Hum. Genet.* **48**: 243–257.

Jenkins, M., Hopper, J.L., Bowes, G., Carlin, J.B., Flander, L.B. and Giles, G.G. (1994) Factors in childhood as predictors of asthma in adult life. *BMJ* **309**(6947): 90–93.

Jenkins, M., Hopper, J. and Giles, G. (1997) Regressive logistic modelling of familial aggregation for asthma in 7,394 population-based nuclear families. *Genet. Epidemiol.* **14**: 317–332.

Kartasamita, C.B., Rosmayudi, O., Demedts, M. and T.R.D.W. Group (1994) Total serum IgE and eosinophil count in children with and without a history of asthma, wheezing, or atopy in an urban community in Indonesia. *J. Allergy Clin. Immunol.* **94**(6 Pt 1): 981–988.

Katz, R., Lieberman, J. and Siegel, S. (1976) Alpha-1-antitrypsin levels and the prevalence of Pi variant phenotypes in asthmatic children. *J. Allergy Clin. Immunol.* **57**: 41–45.

Kauffmann, F., Frette, C., Pham, Q., Nafissi, S., Bertrand, J. and Oriol, R. (1996) Associations of blood group-related antigens to fev(1), wheezing, and asthma. *Am. J. Respir. Crit. Care Med.* **153**(1): 76–82.

Khoury, M., Beaty, T. and Cohen, B. (1993) *Fundamentals of Genetic Epidemiology*. Oxford University Press, Oxford.

Kjellman, N.-I.M. and Johansson, S.G.O. (1976) IgE and atopic allergy in newborns and infants with a family history of atopic disease. *Acta Paediatr. Scand.* **65**: 601–607.

Koumantaki, Y., Giziaki, E., Linos, A., *et al.* (1997) Family history as a risk factor for rheumatoid arthritis: a case–control study. *J. Rheumatol.* **24**(8): 1522–1526.

Kuehr, J., Karmaus, W., Forster, J., Frischer, T., *et al.* (1993) Sensitization to four common inhalant allergens within 302 nuclear families. *Clin. Exp. Allergy* **23**(7): 600–605.

Lammas, D., Mitchell, L. and Wakelin, D. (1988) Genetic control of eosinophilia in parasitic infections: responses of mouse strains to treatment with cyclophosphamide and parastite antigen. *Int. J. Parasit.* **18**(8): 1077–1085.

Lander, E. and Schork, N. (1994) Genetic dissection of complex traits. *Science* **265**: 2037–2048.

Lange, K. (1986) Cohabitation, convergence, and environmental covariances. *Am. J. Med. Genet.* **24**: 483–491.

Laprise, C. and Boulet, L. (1996) Airway responsiveness and atopy in families of patients with asthma. *Clin. Invest. Med.* **19**(6): 461–469.

Lawrence, S., Beasley, R., Doull, I., *et al.* (1994) Genetic analysis of atopy and asthma as quantitative traits and ordered polychotomies. *Ann. Hum. Genet.* **58**(Pt 4): 359–368.

Leung, R. and Jenkins, M. (1994) Asthma, allergy and atopy in southern Chinese school students. *Clin. Exp. Allergy* **24**(4): 353–358.

Levitt, R. and Mitzner, W. (1988) Expression of airway hyperreactivity to acetylcholine as a simple autosomal recessive trait in mice. *FASEB J.* **2**: 2605–2608.

Liebermann, J. and Colp, C. (1990) A role for intermediate, heterozygous alpha1-antitrypsin deficiency in obstructive lung disease. *Chest* **98**: 522–523.

Longo, G., Strinati, R., Poli, F. and Fumi, F. (1987) Genetic factors in non-specific bronchial hyperreactivity: an epidemiologic study. *Am. J. Dis. Child.* **141**: 331–334.

Lympany, P., Welsh, K.I., Cochrane, G.M., Kemeny, D.M. and Lee, T.H. (1992) Genetic analysis of the linkage between chromosome 11q and atopy. *Clin. Exp. Allergy* **22**: 1085–1092.

Lynch, N., Lopez, R., Isturiz, G. and Tenias-Salazar, E. (1983) Allergic reactivity and helminthic infection in Amerindians of the Amazon basin. *Int. Arch. Allergy. Appl. Immunol.* **72**: 369–372.

Marsh, D. and Huang, S. (1991) Molecular genetics of human immune responsiveness to pollen allergens. *Clin. Exp. Allergy* **21**(suppl 1): 168–172.

Marsh, D.G. and Meyers, D.A. (1992) A major gene for allergy – fact or fancy? *Nature Genet.* **2**: 252–254.

Marsh, D., Bias, W. and Ishizaka, K. (1974) Genetic control of basal serum immunoglobulin E level and its effect on specific reaginic sensitivity. *Proc. Natl Acad. Sci. USA* **71**: 3588–3592.

Marsh, D., Neely, J., Breazeale, D., *et al.* (1994) Linkage analysis of IL4 and other chromosome 5q31.1 markers and total serum immunoglobin E concentrations. *Science* **264**: 1152–1156.

Martinez, F.D. (1995) Viral infections and the development of asthma. *Am. J. Respir. Crit. Care Med.* **151**(5): 1644–1647; discussion 1647–1648.

Martinez, F., Holberg, C., Halonen, M., Morgan, W., Wright, A. and Taussig, L. (1994) Evidence for Mendelian inheritance of serum IgE levels in Hispanic and non-Hispanic white families. *Am. J. Hum. Genet.* **55**: 555–565.

Massicot, J. and Cohen, S. (1986) Epidemiologic and socioeconomic aspects of allergic diseases. *J. Allergy Clin. Immunol.* **78**(5(2)): 954–958.

McNally, N.J., Phillips, D.R. and Williams, H.C. (1998) The problem of atopic eczema: aetiological clues from the environment and lifestyles. *Soc. Sci. Med.* **46**(6): 729–741.

McNichol, K., Williams, H., Allan, J. and McAndrew, I. (1973) Spectrum of asthma in children – III, psychological and social components. *BMJ.* **4**: 16–20.

Mennie, M., Gilfillan, A., Brock, D. and Liston, W. (1995) Heterozygotes for the ΔF508 cystic fibrosis allele are not protective against bronchial asthma. *Nature Med.* **1**: 978–979.

Mensinga, T., Schouten, J., Weiss, S. and vanderLende, R. (1992) Relationship of skin test reactivity and eosinophilia to level of pulmonary function in a community-based population study. *Am. Rev. Respir. Dis.* **146**: 638–643.

Meyers, D., Bias, W. and Marsh, D. (1982) A genetic study of total IgE levels in the Amish. *Hum. Hered.* **32**: 15–23.

Meyers, D., Beaty, T., Freidhoff, L. and Marsh, D. (1987) Inheritance of serum IgE (basal levels) in man. *Am. J. Hum. Genet.* **41**: 51–62.

Meyers, D., Postma, D., Panhuysen, C., *et al.* (1994) Evidence for a locus regulating total serum IgE levels mapping to chromosome 5. *Genomics* **23**: 464–470.

Moffatt, M. and Cookson, W. (1997) Tumour necrosis factor haplotypes and asthma. *Hum. Mol. Genet.* **6**(4): 551–554.

Moffatt, M., Hill, M., Cornelius, F., *et al.* (1994) Genetic linkage of T-cell receptor α/δ complex to specific IgE responses. *Lancet* **343**: 1597–1600.

Moffatt, M.F. and Cookson, W.O. (1998) The genetics of asthma. Maternal effects in atopic disease. *Clin. Exp. Allergy* **28** Suppl 1: 56–61; discussion 65–66.

Moffatt, M.F., Sharp, P.A., Faux, J.A., Young, R.P., Cookson, W.O.C.M. and Hopkin, J.M. (1992) Factors confounding genetic linkage between atopy and chromosome 11q. *Clin. Exp. Allergy* **22**: 1046–1051.

Morgan, W.J. and Martinez, F.D. (1992) Risk factors for developing wheezing and asthma in childhood. *Pediatr. Clin. North Am.* **39**(6): 1185–1203.

Moro-Furlani, A. and Krieger, H. (1992) Familial analysis of eosinophilia caused by helminthic parasites. *Genet. Epidemiol.* **9**: 185–190.

Morton, N.E. (1992) Major loci for atopy? *Clin. Exp. Allergy* **22**: 1041–1043.

Nieminen, M., Kaprio, J. and Koskenvuo, M. (1991) A population-based study of bronchial asthma in twin pairs. *Chest* **100**: 71–75.

NIH (1995) Global Initiative for Asthma. Global Strategy for Asthma Management and Prevention. NHLBI/WHO Workshop Report. National Institutes of Health, National Heart, Lung and Blood Institute, Washington.

Ober, C., Cox, N.J., Abney, M., *et al.* (1998) Genome-wide search for asthma susceptibility loci in a founder population. The Collaborative Study on the Genetics of Asthma. *Hum. Mol. Genet.* **7**(9): 1393–1398.

O'Connor, G. and Weiss, S. (1994) Clinical and symptom measures. *Am. J. Respir. Crit. Care Med.* **149**(2 Pt 2): S21–28; discussion S29–30.

O'Connor, G., Sparrow, D., Taylor, D., Segal, M. and Weiss, S. (1987) Analysis of dose-response curves to methacholine: an approach suitable to population studies. *Am. Rev. Respir. Dis.* **136**: 1412–1417.

Ownby, D.R. (1990) Environmental factors versus genetic determinants of childhood inhalant allergies. *J. Allergy Clin. Immunol.* **86**(3(1)): 279–287.

Oxelius, V.-A. (1990) Correlation between atopy and Gm allotypes. *Int. Arch. Allergy Appl. Immunol.* **91**: 54–57.

Palmer, L.J. (1998) The genetic epidemiology of asthma: An investigation of a complex human disease. PhD Thesis. Department of Paediatrics, University of Western Australia, Perth.

Palmer, L., Pare, P., Faux, J., *et al.* (1997) Fc(Epsilon)R1-beta polymorphism and total serum Ige levels in endemically parasitized Australian aborigines. *Am. J. Hum. Genet.* **61**(1): 182–188.

Palmer, L., Daniels, S., Rye, P., *et al.* (1998a) Linkage of chromosome 5q and 11q gene markers to asthma-associated quantitative traits in Australian children. *Am. J. Respir. Crit. Care Med.* **158**: 1825–1830.

Palmer, L.J., Burton, P.R. and Cookson, W.O.C.M. (1998b) Power to detect linkage to binary and quantitative asthma-associated traits using sibpair methods. *Am. J. Respir. Crit. Care Med.* **157**(3): A771.

Panhuysen, C., Meyers, D., Postma, D. and Bleecker, E. (1995) The genetics of asthma and atopy. *Allergy* **50**: 863–869.

Paoletti, F., Carrozzi, L., Viegi, G., *et al.* (1995) Distribution of bronchial responsiveness in a general population: effect of sex, age, smoking, and level of pulmonary function. *Am. J. Respir. Crit. Care Med.* **151**: 1770–1777.

Pattemore, P., Asher, M., Harrison, A., Mitchell, E., Rea, H. and Stewart, A. (1990) The interrelationship among bronchial responsiveness, the diagnosis of asthma and asthma symptoms. *Am. Rev. Respir. Dis.* **142**: 549–554.

Pauwels, R., VanDerStraeten, M., Weyne, J. and Bazin, H. (1985) Genetic factors in non-specific bronchial reactivity in rats. *Eur. Respir. J.* **66**: 98–104.

Peat, J., Britton, W., Salome, C. and Woolcock, A. (1987) Bronchial hyperresponsiveness in two populations of Australian school children. II: relative importance of associated factors. *Clin. Allergy* **17**: 283–290.

Platts-Mills, T. (1995) The role of indoor allergens in asthma. Allergy **50**(suppl 22): 5–12.

Postma, D.S., Bleecker, E.R., Amelung, P.J., *et al.* (1995) Genetic susceptibility to asthma – bronchial hyperresponsiveness coinherited with a major gene for atopy. *N. Engl. J. Med.* **333**(14): 894–900.

Pritchard, D.I., Quinnell, R.J., and Walsh, E.A. (1995) Immunity in humans to *Necator americanus*: IgE, parasite weight and fecundity. *Parasite Immunol.* **17**(2): 71–75.

Rich, S.S., Roitman-Johnson, B., Greenberg, B., Roberts, S. and Blumenthal, M.N. (1992) Genetic analysis of atopy in three large kindreds: no evidence of linkage to D11S97. *Clin. Exp. Allergy* **22**: 1070–1076.

Rijcken, B., Shouten, J., Weiss, S., Speizer, F. and Lende, R.V.d. (1988) The relationship between airway responsiveness to histamine and pulmonary function in a random population sample. *Am. Rev. Respir. Dis.* **137**: 826–832.

Risch, N. (1990) Linkage strategies for genetically complex traits. I. Multilocus models. *Am. J. Hum. Genet.* **46**: 222–228.

Risch, N. and Zhang, H. (1995) Extreme discordant sib pairs for mapping quantitative trait loci in humans. *Science* **268**: 1584–1589.

Rodrigues, V., Abel, L., Piper, K. and Dessein, A. (1996) Segregation analysis indicates a major gene in the control of interleukine-5 production in humans infected with *Schistosoma mansoni*. *Am. J. Hum. Genet.* **59**(2): 453–461.

Roorda, R.J., Gerritsen, J., van Aalderen, W.M., *et al.* (1994) Follow-up of asthma from childhood to adulthood: influence of potential childhood risk factors on the outcome of pulmonary function and bronchial responsiveness in adulthood. *J. Allergy Clin. Immunol.* **93**(3): 575–584.

Rosenwasser, L.J. (1997) Genetics of atopy and asthma: promoter-based candidate gene studies for IL-4. *Int. Arch. Allergy Immunol.* **113**(1–3): 61–64.

Rosenwasser, L., Klemm, D., Dresback, J., *et al.* (1995) Promoter polymorphisms in the chromosome 5 gene cluster in asthma and atopy. *Clin. Exp. Allergy* **25**(Suppl 2): 74–78.

Rothwell, T., Abeydeera, L. and Geczy, A. (1988) Relationship between basophils and eosinophils in cutaneous basophil hypersensitivity reactions in guinea pigs and susceptibility to *Trichostrongylus colubriformis* infection. *Int. J. Parasit.* **18**(3): 347–351.

Sandford, A.J., Shirakawa, T., Moffatt, M.F., *et al.* (1993) Localisation of atopy and β subunit of high-affinity IgE receptor (FCεRI) on chromosome 11q. *The Lancet* **341**: 332–334.

Sandford, A., Weir, T. and Pare, P. (1996) The genetics of asthma. *Am. J. Respir. Crit. Care Med.* **153**: 1749–1765.

Sapin, C., Hirsch, F., Delaporte, J., Bazin, H. and Druet, P. (1984) Polyclonal IgE increase after $HgCl_2$ injections in BN and LEW rats: a genetic analysis. Immunogenetics **20**: 227–236.

Sarafino, E.P. and Goldfedder, J. (1995) Genetic factors in the presence, severity, and triggers of asthma. *Arch. Dis. Child.* **73**(2): 112–116.

Sasazuki, T., Nichimura, Y., Muto, M. and Ohta, N. (1983) HLA-linked genes controlling the immune response and disease susceptibility. *Immunol. Rev.* **70**: 51–75.

Schroeder, S., Gaughan, D. and Swift, M. (1995) Protection against bronchial asthma by CFTR ΔF508 mutation: a heterozygote advantage in cystic fibrosis. *Nature* Med **1**: 703–705.

Sears, M.R. (1991) Epidemiological trends in bronchial asthma. In: *Asthma: Its Pathology and Treatment.* (eds M.A. Kalines, P.J. Barnes and C.G.A. Persson) Marcel Dekker, New York, pp. 1–49.

Sears, M.R., Burrows, B., Herbison, G.P., Holdaways, M.D. and Flannery, E.M. (1993a) Atopy in childhood. II. Relationship to airway responsiveness, hay fever and asthma. *Clin. Exp. Allergy* **23**: 949–956.

Sears, M.R., Burrows, B., Herbison, G.P., Flannery, E.M. and Holdaways, M.D. (1993b) Atopy in childhood. III. Relationship with pulmonary function and airway responsiveness. *Clin. Exp. Allergy* **23**: 957–963.

Seaton, A., Godden, D.J. and Brown, K. (1994) Increase in asthma: a more toxic environment or a more susceptible population? *Thorax* **49**(2): 171–174.

Shibasaki, M., Tajima, K., Morikawa, A., Mitsuhashi, M., Sumazaki, R. and Tokuyama, K. (1988) Relation between frequency of asthma and IgE antibody levels against *Dermatophagoides farinae* and total serum IgE levels in schoolchildren. *J. Allergy Clin. Immunol.* **82**(1): 86–94.

Shirakawa, T. and Morimoto, K. (1991) Lifestyle effect on total IgE. Lifestyles have a cumulative impact on controlling total IgE levels. *Allergy* **46**(8): 561–569.

Shirakawa, T., Hashimoto, T., Furuyama, J. and Morimoto, K. (1994a) Linkage between severe atopy and chromosome 11q13 in Japanese families. *Clin. Genet.* **46**(3): 228–232.

Shirakawa, T., Li, A., Dubowitz, M., *et al.* (1994b) Association between atopy and variants of the β subunit of the high-affinity immunoglobin E receptor. *Nature Genet.* **7**: 125–130.

Shirakawa, T., Mao, X.-Q., Sasaki, S., *et al.* (1995) Association between atopy and FceRIB variants in the Japanese population. In: *Recent Advances in Prediction and Prevention of Childhood Allergy.* (eds S. Sasaki, T. Miyamoto and J. Hopkin) Churchill Livingstone, London, pp. 29–31.

Shirakawa, T., Hayakawa, K., Shimizu, T. and Morimoto, K. (1996a) Association of life style with high risk of hyperimmunity and of immunosuppression mediated by IgE. *J. Clin. Epidemiol.* **49**(9): 1059–1065.

Shirakawa, T., Mao, X., Sasaki, S., *et al.* (1996b) Association between atopic asthma and a coding variant of FCeRIβ in a Japanese population. *Hum. Mol. Genet.* 5(8): 1129–1130.

Shirakawa, T., Mao, X.-Q., Sasaki, S., Kawai, M., Morimoto, K. and Hopkin, J. (1996c) Association between FceRIβ and atopic disorder in a Japanese population. *Lancet* **347**: 394–395.

Sibbald, B., Horn, M., Brain, E. and Gregg, I. (1980) Genetic factors in childhood asthma. *Thorax* **35**: 671–674.

Sly, R.M. (1999) Changing prevalence of allergic rhinitis and asthma. *Ann. Allergy Asthma Immunol.* **82**(3): 233–248; quiz 248–252.

Speight, A. (1978) Is childhood asthma being underdiagnosed and undertreated? *BMJ* **2**: 331–332.

Spielman, R., McGinnis, R. and Ewens, W. (1994) The transmission/disequilibrium test detects cosegregation and linkage [letter]. *Am. J. Hum. Genet.* **54**: 559–560.

Sporik, R., Chapman, M.D. and Platts-Mills, T.A. (1992) House dust mite exposure as a cause of asthma. *Clin. Exp. Allergy* **22**(10): 897–906.

Spry, C. (1988) *Eosinophils: A Comprehensive Review and Guide to the Scientific and Medical Literature.* Oxford University Press, Oxford.

Suarez, B., VanEerewegh, P. and Hampe, C. (1991) *Detecting Loci for Oligogenic Traits by Linkage Analysis.* Proceedings of the International Congress of Human Genetics.

Tariq, S.M., Matthews, S.M., Hakim, E.A., Stevens, M, Arshad, S.H. and Hide, D.W. (1998) The prevalence of and risk factors for atopy in early childhood: a whole population birth cohort study. *J. Allergy Clin. Immunol.* **101**(5): 587–593.

Taylor, K. and Luksza, A. (1987) Peripheral blood eosinophil counts and bronchial responsiveness. *Thorax* **42**: 452–456.

The International Study of Asthma and Allergies in Childhood (ISAAC) Steering Committee (1998) Worldwide variation in prevalence of symptoms of asthma, allergic rhinoconjunctivitis, and atopic eczema: ISAAC. *Lancet* **351**: 1225–1232.

Toelle, B., Peat, J., Mellis, C. and Woolcock, A. (1995) The cost of childhood asthma to Australian families. *Pediatr. Pulmonol.* **19**(6): 330–335.

Townley, R., Bewtra, A., Wilson, A., *et al.* (1986) Segregation analysis of bronchial response to methacholine inhalation challenge in families with and without asthma. *J. Allergy Clin. Immunol.* **77**: 101–107.

Tregouet, D., Ducimeteire, P., and Tiret, L. (1997) Testing association between candidate-gene markers and phenotype in related individuals, by use of estimating equations. *Am. J. Hum. Genet.* **61**: 189–199.

Turner, K.J., Rosman, D.L. and O'Mahony, J. (1974) Prevalence and familial association of atopic disease and its relationship to serum IgE levels in 1,061 school children and their families. *Int. Arch. Allergy* **47**: 650–664.

vanHerwerden, L., Harrap, S., Wong, Z., *et al.* (1995) Linkage of high-affinity IgE receptor gene with bronchial hyperreactivity, even in the absence of atopy. *Lancet* **346**: 1262–1265.

vonMutius, E. (1996) Progression of allergy and asthma through childhood to adolescence. *Thorax* **51** (Suppl 1): S3–S6.

Vorehovsky, I., Webster, A., Plebani, A. and Hammarstrom, L. (1999) Genetic linkage of IgA deficiency to the major histocompatibility. *Am. J. Hum. Genet.* **64**: 1096–1109.

Wakelin, D. (1985) Genetic control of immunity to helminth infections. *Parasit. Today* **1**: 17–23.

Wardlaw, A., Dunnette, S., Gleich, G., Collins, J. and Kay, A. (1988) Eosinophils and mast cells in bronchoalveolar lavage and mild asthma; relationship to bronchial responsiveness. *Am. Rev. Respir. Dis.* **137**: 62–69.

Warner, J., Gotz, M., Landau, L., *et al.* (1989) Management of asthma: A consensus statement. *Arch. Dis. Child.* **64**: 1065–1069.

Warram, J.H., Krolewski, A.S., Gottlieb, M.S. and Kahn, C.R. (1984) Differences in risk of insulin-dependent diabetes in offspring of diabetic mothers and diabetic fathers. *N. Engl. J. Med.* **311**(3): 149–152.

Watson, M., Lawrence, S., Collins, A. *et al.* (1995) Exclusion from proximal 11q of a common gene with megaphenic effects on atopy. *Ann. Hum. Genet.* **59**: 403–411.

Weeks, D. and Lange, K. (1988) The affected-pedigree-member method of linkage analysis. *Am. J. Hum. Genet.* **42**: 315–326.

Weeks, D. and Lathrop, G. (1995) Polygenic disease: methods for mapping complex disease traits. *TIG* **11**(12): 513–519.

Weiss, K.B., Gergen, P.J. and Hodgson, T.A. (1992) An economic evaluation of asthma in the United States. *N. Engl. J. Med.* **326**: 862–866.

Weiss, S.T. (1994) The origins of childhood asthma. *Monaldi Arch. Chest Dis.* **49**(2): 154–158.

Wilkinson, J., Thomas, S., Lio, P., Holgate, S. and Morton, N. (1996) Evidence for linkage for atopy and asthma to markers on chromosome 12q. *Eur. Respir. J.* **9**(S23): 435s.

Wittig, H., McLaughlin, E., Leifer, K. and Belloit, J. (1978) Risk factors for the development of allergic disease: analysis of 2190 patient records. *Ann. Allergy* **41**: 84–88.

Wjst, M., Fischer, G., Immervoll, T. *et al.* (1999) A genome-wide search for linkage to asthma. *Genomics* **58**(1): 1–8.

Woolcock, A.J. (1991) Worldwide trends in asthma morbidity and mortality. Explanation of trends. *Bulletin of the International Union Against Tuberculosis and Lung Disease* **66**(2–3): 85–89.

Xu, J., Levitt, R.C., Panhuysen, C.I., *et al.* (1995) Evidence for two unlinked loci regulating total serum IgE levels. *Am. J. Hum. Genet.* **57**(2): 425–430.

Young, R., Sharp, P., Lynch, J. *et al.* (1992) Confirmation of genetic linkage between atopic IgE responses and chromosome 11q13. *J. Med. Genet.* **29**: 236–238.

Young, R., Dekker, J., Wordsworth, B., *et al.* (1994) HLA-DR and HLD-DP genotypes and immunoglobin E responses to common major allergens. *Clin. Exp. Allergy* **24**: 431–439.

Young, S., Le Souëf, P.N., Geelhoed, G.C., Stick, S.M., Turner, K.J. and Landau, L.I. (1991) The influence of a family history of asthma and parental smoking on airway responsiveness in early infancy. *N. Engl. J. Med.* **324**: 1168–1173.

Zamel, N., McClean, P., Sandell, P., *et al.* (1996) Asthma on Tristan da Cunha: Looking for the genetic link. *Am. J. Respir. Crit. Care Med.* **153**(6): 1902–1906.

12

Rheumatoid arthritis

Jennifer H. Barrett and Sally John

1. Introduction

1.1 Diagnosis

Rheumatoid arthritis (RA) is one of the most common autoimmune diseases. It is characterized by chronic inflammation of the synovial lining of multiple joints, often leading to progressive joint destruction and disability. A diagnosis of RA covers a wide clinical spectrum, varying from very mild to severely disabling disease. The course of the disease is highly variable. Hand and wrist joints are usually the first to become affected. Presentation and diagnosis are often succeeded by a period of improvement, perhaps induced by treatment, followed by gradual deterioration. Some patients develop extra-articular features such as rheumatoid nodules. There is no cure for RA, and once joint damage occurs it is irreversible. Treatment with (potentially toxic) drugs such as methotrexate and sulphasalazine may slow down or halt the progress of the disease, especially if given in the early stages. Steroids and nonsteroidal anti-inflammatory drugs are also used to relieve pain and reduce inflammation.

Pathology is characterized by infiltration of T lymphocytes, macrophages and active B lymphocytes into the joint. Cartilage and bone destruction appears to be mediated by cytokine induction of destructive enzymes such as matrix metalloproteinases. The synovial membrane thickens and develops blood vessels, a structure known as the pannus, which is a major source of degradative enzymes.

There is no gold standard for the diagnosis of RA, and many of the clinical manifestations are common to other rheumatic diseases. The most commonly used diagnostic criteria are those adopted by the American College of Rheumatology in 1987 (Arnett *et al.*, 1988). A patient is classified as having RA if at least four of the following seven criteria are satisfied:

(i) morning stiffness lasting at least 1 hour;
(ii) arthritis of three or more joint areas;
(iii) arthritis of hand joints;
(iv) symmetric arthritis;
(v) rheumatoid nodules;
(vi) presence of serum rheumatoid factor (RF);
(vii) radiographic changes in hand or wrist joints.

Analysis of Multifactorial Disease, edited by T. Bishop and P. Sham.

An alternative application of the criteria was also suggested, which uses them in a classification tree. For example a patient satisfying criteria (ii) and (vii) (arthritis of three or more joint areas and radiographic changes) is classified as having RA, whereas one satisfying neither (ii) nor (vi) (arthritis in less than three joint areas and absence of RF) does not have RA. For patients in neither of these two categories, others of the above criteria are examined, and classification is made accordingly. Further definitions of the seven criteria, and details of the classification tree, are given elsewhere (Arnett *et al.*, 1988). Different methods of RA classification were compared by MacGregor *et al.* (1994), who concluded that the above criteria work well in epidemiological studies, provided they incorporate past as well as current evidence of disease activity.

1.2 Epidemiology

The prevalence of RA in most adult populations is estimated to be between 0.5% and 1%, and there is some evidence that incidence rates are declining (Hochberg and Spector, 1990). The disease is almost three times more common in women than in men. In a recent population-based study in Norfolk, England, the annual incidence was estimated as 36/100 000 for women and 14/100 000 for men (Symmons *et al.*, 1994). In women the incidence rate was highest (approximately 60/100 000) in those aged 45–75 years, but was also fairly high (36/100 000) in the 35–44 year age group. In men, onset before the age of 45 years was very rare, and thereafter the incidence rate increased with age.

Little is known about the etiology of RA. It is thought that the disease may develop in a genetically susceptible host in response to an environmental trigger. Several potential environmental risk factors have been suggested, although none has been established. These include exposure to infection (Harris, 1990), immunization (Symmons and Chakravarty, 1993), diet (Shapiro *et al.*, 1996), smoking, blood transfusion and obesity (Symmons *et al.*, 1997).

Several observations, including the higher prevalence of RA in women, have led to interest in the role of hormones (Brennan and Silman, 1995). Some studies have found oral contraceptive use to be protective against RA, but there is no consensus about this or the role of estrogen replacement therapy in postmenopausal women. Nulliparity, probably associated with infertility rather than choice, has been found to increase the risk of RA. Studies have consistently found that men with RA have lower than average levels of testosterone, a hormone with immunosuppressive properties.

The most striking evidence for hormonal involvement is found in the effect of pregnancy on RA. It has long been observed that during pregnancy women with RA commonly experience remission followed by postpartum relapse (Barrett *et al.*, 1999; Hench, 1935). Furthermore the risk of RA onset is increased in the postpartum period, especially in those who breast-feed (Brennan and Silman, 1994), and recent evidence suggests that postpartum relapse in women with RA may also be related to breast-feeding (Barrett *et al.*, 2000). A possible role for prolactin has been suggested to explain the association with lactation.

2. Familial aggregation

Various studies have assessed familial aggregation of RA, yielding varying estimates of the ratio of the risk to first degree relatives of cases, to the population risk. A recent UK study based on 134 RA cases sampled from the population found a risk ratio of only 2.2 (Jones *et al.*, 1996), compared with risk ratios typically of 3–5 in studies based on hospital RA patients (Koumantaki *et al.*, 1997, Thomas *et al.*, 1983). Most studies are consistent in showing an increased risk of a modest size. Although higher values of 8–10 are sometimes quoted for RA, these have not often been substantiated. Variation in the estimates can at least partly be accounted for by differences in case ascertainment and are consistent with the hypothesis that the risk is greater to relatives of cases with more severe disease (Deighton *et al.*, 1992).

It has also been shown that RA is associated with other autoimmune diseases within families. Thomas *et al.* (1983) demonstrated an increased prevalence of both insulin-dependent diabetes mellitus (IDDM) and thyroid disease in first and second degree relatives of RA patients.

There have been several studies of RA twin pairs which have demonstrated a higher concordance rate in monozygotic (MZ) than dizygotic (DZ) twins. A study of 91 MZ and 112 DZ twin pairs in the UK, each with at least one RA-affected member, showed 15% disease concordance for the MZ and 4% concordance for the DZ pairs (Silman *et al.*, 1993).

These epidemiological studies demonstrate that there is a genetic component to RA, but the 15% concordance rate in MZ twins sets a limit to the genetic contribution to disease susceptibility.

No Mendelian subforms of RA are known, although some rare related conditions follow a clear Mendelian form of inheritance. These include familial Hibernian fever (autosomal dominant; McDermott *et al.*, 1998) and familial Mediterranean fever (autosomal recessive; French FMF Consortium, 1997), both of which are characterized by periodic fever accompanied by musculoskeletal manifestations such as joint inflammation and stiffness. These are however clinically clearly distinct diseases from RA, having more in common with the systemic form of juvenile idiopathic arthritis.

3. Major histocompatibility complex

To date the only indisputable genetic association with RA is with human leukocyte antigen (HLA) DRB1 in the major histocompatibility complex (MHC) on chromosome 6. The association with HLA was first described over 20 years ago (Stastny, 1978) and has since been established by case-control studies in many different populations (Reveille, 1998).

Although the HLA association with RA represents one of the longest established genetic risk factors for a complex disease and has been the subject of a great deal of study, much is still unknown about the role of HLA in the development of RA. In most populations HLA DRB1 *04 alleles have been found to be more frequent in RA patients than in controls, although in some studies certain DRB1

*01 alleles were associated with RA. With advances in DNA technology it was noted that some HLA DRB1 alleles (*0101, *0102, *0401, *0404, *0405, *0408, *0409, *0410, *1001 and *1402), including those which had previously shown an association with RA, share a common coding sequence. It has been proposed that a direct role for this shared sequence may explain the observed association between certain HLA DRB1 alleles and RA; this is the shared epitope (SE) hypothesis (Gregersen *et al.*, 1987). However, it should be noted that at least one quarter of RA patients do not carry the SE, and conversely many subjects with the SE do not develop RA.

Most of the available data from epidemiological studies are broadly compatible with the SE hypothesis (Ollier and Thomson, 1992), although the role of the HLA DRB1 gene in RA is clearly complex. There is some evidence that particular SE alleles and combinations of them (especially the *0401/*0404 genotype) are associated with greater risk than others (Wordsworth *et al.*, 1992). Furthermore different SE alleles show the strongest association in different populations, and this may be only partly explained by the considerable differences between populations in background frequencies of HLA polymorphisms. Differential effects between males and females may complicate matters further.

Linkage to the HLA region has also been confirmed in many family studies, commonly based on the analysis of HLA haplotype-sharing in affected sibling pairs (ASPs). More recently the transmission disequilibrium test (TDT) has also been applied (Marlow *et al.*, 1997). This examines which marker alleles are transmitted and not transmitted from a heterozygous parent to an affected child. A distortion from the expected transmission rates will only be observed in the presence of both association and linkage between the marker and a disease gene.

A variety of different analytic methods have been used to investigate the mode of inheritance of the HLA RA susceptibility gene. One approach is the antigen genotype frequency among patients (AGFAP) method (Thomson, 1983), which compares observed genotype frequencies in cases with expected frequencies under different inheritance models. For example in 180 community-based female RA patients the numbers with zero, one and two SE alleles were analyzed using AGFAP, and a recessive mode of inheritance was supported (Mu *et al.*, 1998). Another approach which has been used (Genin *et al.*, 1998) is the marker association segregation chi-square method (Clerget-Darpoux *et al.*, 1988), which allows simultaneous analysis of the distribution of genotypes in unrelated affected individuals and haplotype sharing between two affected sibs. Again observed genotype frequencies are compared to those expected under different genetic models. Other researchers have investigated the possibility that the association between RA and HLA DRB1 is due to a gene in linkage disequilibrium with HLA DRB1. For example Hasstedt *et al.* (1994) used likelihood methods to compare different genetic models in a combined segregation and linkage analysis of 28 pedigrees with multiple affected members.

Although most studies of the RA-HLA relationship support a recessive mode of inheritance for DRB1 or another HLA-linked locus, there are many conflicting results. This could partly be due to the complexity of the relationship between RA and HLA DRB1. A further possible factor accounting for the lack of agreement between studies is that the HLA gene may be associated with severity of RA rather than susceptibility. Studies examining outcome in RA patients have found that disease severity increases with the number of SE alleles (Reveille, 1998). Most reports of an association between the SE and RA have used hospital-based cases, who are likely to have established and relatively severe disease. A study of newly diagnosed RA cases from the community (Thomson *et al.*, 1998) found only a modest association with the presence of an SE allele (odds ratio 1.8, 95% confidence interval (1.4, 2.4)). This raises the possibility that the SE may not be related to disease onset, but rather a predictor of disease persistence and severity.

Associations between RA and other MHC class II genes have been described. However, these may not be independent of the HLA DRB1 association, since strong linkage disequilibrium exists across this region, especially between HLA DRB1, DQA1 and DQB1.

MHC class III genes include those at the tumor necrosis factor (TNF) locus. The evidence for involvement of TNFα in RA is compelling. TNF is a potent pro-inflammatory cytokine and is up-regulated in synovial tissue from RA patients. In animal models of induced arthritis TNF is produced early in the inflammatory process, and depleting TNF by using anti-TNF antibodies leads to amelioration of the disease. Anti-TNF antibodies have also been used to treat human RA with some success. However, interpretation of results at the TNF locus is again difficult, as any effect may be due to linkage disequilibrium with HLA DRB1.

Two microsatellite markers have been mapped to the TNFα locus. These are in linkage disequilibrium with two single nucleotide polymorphisms in the TNFα promoter region at positions –238 and –308 from the transcription start site. Both linkage and association studies have been carried out at this locus, and results have been conflicting. Hajeer *et al.* (1996) found that associated TNF microsatellite alleles associated with RA were not independent of class II HLA associations. Another recent study by Field *et al.* (1997) examined markers in 98 cases and 91 controls. They also found that, although there was an association with TNFα alleles, they were carried on extended MHC haplotypes known to be associated with RA. These authors also concluded that TNFα alleles were unlikely to make an independent contribution to RA. A larger study by Brinkman *et al.* (1997) of 283 cases and 116 controls found that the –238GA genotype was associated with reduced radiographic progression independently of the presence of HLA DRB1 *04. This was supported by another recent study that found the –238GA genotype to be associated with a lower rate of joint damage independent of HLA DRB1 *04 (Kaijzel *et al.*, 1998). Different TNF associations have also been shown in male compared to female patients (Hajeer *et al.*, 1997).

Despite considerable evidence for involvement of the protein in RA pathology, it is still unclear whether different TNF alleles influence disease susceptibility or progression. It is noteworthy however that the most promising results have come from studying clinical features of disease, such as erosions and the number of joints involved. Disease heterogeneity may be a considerable obstacle to identifying genetic factors in RA, and it may be important to identify more homogeneous subgroups in order to detect loci involved in disease susceptibility or severity.

The relative risk of sibs attributable to the HLA locus has been estimated to be about 1.6 (Rigby *et al.*, 1993). Hence, the HLA region accounts for less than half of the genetic or familial contribution to RA. Attention has therefore also been focused on the search for genes outside this region.

4. The search for non-MHC genes using linkage and association methods

The vast majority of published studies investigating genetic factors in RA have been association studies using RA cases and controls. Both the genetic epidemiology of RA and its clinical features have influenced this choice of study design. RA is a common condition, and clinically well characterized cases are relatively easy and cheap to recruit. In contrast the modest genetic component of RA, the relatively late age at onset, the absence of any obvious inheritance pattern and the relapsing-remitting nature of disease make it difficult to identify families with multiple-affected members. In addition, the known pathology and epidemiology of RA implicate a number of potential candidate genes. Most investigators have therefore chosen to test whether alleles of a single polymorphic candidate gene are associated with RA. Despite the difficulties of collecting RA families, in recent years more linkage studies have been published.

4.1 Candidate genes in RA

The evidence suggesting a particular gene as a candidate for investigation varies enormously from one gene to another. Whilst on the one hand it is clearly of interest to investigate any gene which may be implicated in the pathology of the disease, doing so will inevitably lead to a high false-positive rate unless very stringent significance levels are applied.

Both the epidemiology and pathology of RA suggest a number of potential candidate genes. Cytokines are a key group of inflammatory molecules, which are upregulated in RA synovial tissue. Once an inflammatory reaction has started, many molecules will be produced as part of a cascade reaction. It is therefore often difficult to determine whether altered levels of cytokine production in disease are 'cause or effect'. If an association between RA and polymorphisms of a candidate gene could be demonstrated, this would provide evidence for a role in etiology. The main problem with this type of approach in RA is that there are a great many potential candidate genes which have been shown to be differentially expressed in patients. It is also clear that the immune system does not act in isolation and interacts with other systems such as the neuroendrocine system. A number of observations have led researchers in RA to investigate hormones as candidate genes in RA (Masi *et al.*, 1996). The role of sex hormones in particular is supported by epidemiological studies (Section 1.2).

Immunological genes such as the T-cell receptor and immunoglobulin genes have been investigated in a number of studies with conflicting results. Interpretation of results from studies of these genes is particularly difficult, as research is often done on genomic DNA and these genes undergo extensive rearrangement to produce the protein products.

Polymorphisms have been published within many of the cytokine and cytokine receptor genes such as interleukin 1α (IL1α), IL2 and IL10, and many of these proteins have been shown to present in abnormal levels in synovial tissue. A number of studies have shown positive association or linkage to either RA as a single definition or to certain aspects of RA. Several examples exist in RA of studies where no overall difference in allele frequencies has been shown between cases and controls, but where a significant difference is seen in a subgroup.

More recently some attention has focused on candidate genes in hormone pathways. The female excess in RA cases, and evidence that estrogens affect the immune system directly, have led investigators to study genes involved in estrogen production and signaling. Polymorphisms within the estrogen receptor gene are currently being investigated in RA. A recent linkage study has also shown some preliminary evidence of linkage to a marker in the estrogen synthase locus (the gene encoding the enzyme that converts androgens to estrogens) in RA (John *et al.*, 1999). There is increasing evidence that patients with RA have a defective cortisol response to stress (Harbuz and Jessop, 1999). Linkage and association have also been shown to the corticotropin-releasing hormone (CRH) locus (unpublished data). CRH is produced by the hypothalamus in response to inflammatory cytokines such as TNFα and IL6 and leads to an increase in cortisol concentration, the body's main natural immunosuppressant. It should be noted that these studies are preliminary, and that linkage results could be due to other genes in the region.

Animal models can provide useful insights into the genetic epidemiology of a disease, as breeding experiments can be carried out between genetically resistant and susceptible strains. A form of arthritis with similar features to RA can be induced in genetically susceptible rats by injection with collagen, oil or adjuvant. A number of whole genome screens using these autoimmune experimental models have recently been published. Although no gene has yet been identified using this approach, a number of regions have been detected in more than one experimental model, some of which contain plausible candidate genes. Loci in humans syntenic to the linked animal loci are therefore obvious candidates for investigation in RA. An additional benefit of animal models is that the different stages of disease can be more easily studied. Different loci were found to be associated with the genetic control of onset, severity and chronicity of pristane-induced arthritis in rats (Vingsbo-Lundberg *et al.*, 1998). Studying the genetics of disease severity in human RA is problematic, as most measures of disease activity are affected by treatment, making interpretation of such results difficult.

This is by no means an extensive list of candidate genes which have been investigated in RA. To date however, no gene other than DRB1 has been confirmed to have a role in RA. It is likely that RA susceptibility genes will have quite small genetic relative risks and be of very low penetrance (associated HLA genotypes are only approximately 5% penetrant). This means that most published studies are very under-powered to detect genes.

4.2 Whole genome screens

One of the standard methods of genetic analysis of complex disease entails genotyping several hundred ASPs for a panel of 300–400 informative microsatellite markers spaced at approximately 10 cM intervals throughout the genome.

To date, two whole genome screens (WGS) of RA ASPs have been published. The European Consortium on Rheumatoid Arthritis Families (ECRAF) performed a WGS on a collection of 97 RA families (114 ASPs) (Cornelis *et al.*, 1998).

The HLA region showed the strongest evidence for linkage, and 19 other markers in 14 regions reached nominal significance. It was of interest that four of these regions had previously been identified in IDDM. The authors only followed up two regions in the original publication, providing further support for a region on chromosome 3 ($P = 0.002$) in a further 164 families (194 ASPs), although overall the evidence did not reach genome-wide significance. Clearly this number of ASPs lacks power to detect linkage at genome-wide significance levels, particularly in view of the fact that RA has a modest genetic component and individual genes may have very small effects. A WGS on 41 multiplex Japanese RA families has also been recently published (Shiozawa *et al.*, 1998). In contrast to the European study, HLA was not the most significant result; markers on chromosomes 1 and 8 had lod scores greater than 3. The small sample size suggests that these results should be treated with caution and require further confirmation. ECRAF have recently published a confirmation of a locus on chromosome 1, close to one of the loci identified in the Japanese study.

Further studies will be required before any loci can be identified with certainty. Other large family collections do exist, and WGSs are currently underway. The Arthritis Research Council's National Repository for DNA (Worthington *et al.*, 1994) now holds 400 RA ASP families, and the North American Rheumatoid Arthritis Consortium (NARAC) are committed to collecting 1000 RA ASP families. This will lead to a collection of nearly 2000 families. Combining data from these families may give sufficient power to detect loci with small relative risks. However, pooling of data may be hampered by the use of different markers by different groups. It is also important to consider disease heterogeneity, genetic heterogeneity and the possibility that different criteria may have been used to diagnose disease.

5. Discussion

The relatively slow progress there has been in dissecting the genetic epidemiology of RA, and the lack of consistency in the results that have been published, can be attributed to various factors. Some of these factors are common to most other multifactorial diseases, and others may be more specific to RA.

The first problem lies in the complexity of the disease etiology. It seems likely that several genes are involved in the pathogenesis of RA, each with a relatively small effect, possibly interacting with each other and with unknown environmental factors. As can be seen from the example of HLA DRB1, even when a gene has been identified, its effect on disease susceptibility may be complex and difficult to characterize. The complex pathology of RA suggests many genes as possible candidates for involvement, but if their effects are small and inter-dependent, current methods may not be adequate to their study.

Second, RA is a highly heterogeneous disease, and it is possible that subforms exist with distinct genetic etiologies. Related to this, diagnosis is difficult, and there are undoubtedly inconsistencies across studies of both diagnosis and case ascertainment. This is a problem shared by many complex diseases, notably psychiatric disorders, but is less of a problem for diseases whose diagnosis can be confirmed histologically. Because of the heterogeneous nature of the disease course, it is likely that some genes associated with RA actually influence severity rather than susceptibility.

A further problem arises through genetic heterogeneity in different populations. This has contributed to the difficulty in studying the effect of HLA on RA, and is likely to also hamper the study of other genes.

Difficulties in interpreting the available evidence are compounded by the fact that many studies conducted so far have been small, lacking adequate power to detect small effects. The large number of studies has clearly also produced a number of false-positive results.

The most common form of study design in the search for genes involved in RA has been the case-control approach. In addition to the practical advantages mentioned earlier (Section 4), such studies are also relatively powerful to detect modest effects. There has been some success in their use, notably in identifying and beginning to unravel the role of HLA in RA. However, unless controls are chosen very carefully, association studies can lead to spurious results. Linkage studies using ASPs avoid this problem, but they are relatively under-powered. The WGSs recently reported and currently in progress have the advantage of potentially identifying genes that have not been studied as potential candidates for involvement in RA.

Further progress is likely to be made in RA by conducting larger studies, and investigating more complex disease models, for example allowing for the interaction between promising candidate genes and HLA DRB1. Because of disease heterogeneity and late age at onset, it may not be sufficient to simply categorize subjects as affected or unaffected, as in most studies to date. It may be more fruitful to assemble collections of well-characterized cases and controls, and to incorporate factors such as age, gender and disease course into the analysis. To avoid the problem of population stratification, there is an increasing move to use other family members as controls, and flexible methods can be used to analyze such data (Schaid and Rowland, 1998).

Although few studies are able to provide convincing evidence for or against the involvement of a particular gene in RA, each study contributes to the body of evidence. The final proof of genetic involvement will come when associated polymorphisms can be shown to make a functional difference to disease pathology.

References

Arnett, F.C., Edworthy, S.M., Bloch, D.A., *et al.* (1988) The American Rheumatism Association 1987 revised criteria for the classification of rheumatoid arthritis. *Arthritis Rheum.* **31:** 315–324.

Barrett, J.H., Brennan, P., Fiddler, M. and Silman, A.J. (1999) Does rheumatoid arthritis remit during pregnancy and relapse postpartum? *Arthritis Rheum.* **42**: 1219–1227.

Barrett, J.H., Brennan, P., Fiddler, M. and Silman, A.J. (2000) Breast-feeding and postpartum relapse in women with rheumatoid and inflammatory arthritis. *Arthritis Rheum.* **43**: 1010–1015.

Brennan, P. and Silman, A. (1994) Breast-feeding and the onset of rheumatoid arthritis. *Arthritis Rheum.* **37:** 808–813.

Brennan, P. and Silman, A. (1995) Why the gender difference in susceptibility to rheumatoid arthritis? *Ann. Rheum. Dis.* **54:** 694–695.

Brinkman, B.M., Huizinga, T.W., Kurban, S.S., *et al.* (1997) Tumour necrosis factor alpha gene polymorphism in rheumatoid arthritis: association with, susceptibility to, or severity of, disease? *Br. J. Rheumatol.* **36:** 516–521.

Clerget-Darpoux, F., Babron, M.C., Prum, B., Lathrop, G.M., Deschamps, I. and Hors, J. (1988) A new method to test genetic models in HLA associated disease: The Masc method. *Ann. Hum. Genet.* **52:** 247–258.

Cornelis, F., Faure, S., Martinez, M., *et al.* (1998) New susceptibility locus for rheumatoid arthritis suggested by a genome wide linkage study. *Proc. Natl Acad. Sci. USA* **95:** 10746–10750.

Deighton, C.M., Roberts, D.F. and Walker, D.J. (1992) Effect of disease severity on RA concordance in same sexed siblings. *Ann. Rheum. Dis.* **51:** 943–945.

Field, M., Gallagher, G., Eskdale, J., *et al.* (1997) Tumor necrosis factor locus polymorphisms in rheumatoid arthritis. *Tissue Antigens* **50:** 303–307.

French FMF Consortium, The (1997) A candidate gene for familial Mediterranean fever. *Nature Genet.* **17:** 25–31.

Genin, E., Babron, M.C., McDermott, M.F., *et al.* (1998) Modelling the major histocompatibility complex susceptibility to RA using the MASC method. *Genet. Epidemiol.* **15:** 419–430.

Gregersen, P.K., Silver, J. and Winchester, R.J. (1987) The shared epitope hypothesis: an approach to understanding the molecular genetics of susceptibility to rheumatoid arthritis. *Arthritis Rheum.* **30:** 1205–1213.

Hajeer, A.H., Worthington, J., Silman, A.J. and Ollier, W.E.R., *et al.* (1996) Association of tumor necrosis factor microsatellite polymorphisms with HLA-DRB1*04-bearing haplotypes in rheumatoid arthritis patients. *Arthritis Rheum.* **3:** 1109–1114.

Hajeer, A.H., John, S., Ollier, W.E.R., *et al.* (1997) Tumor necrosis factor microsatellite haplotypes are different in male and female patients with RA. *J. Rheumatol.* **24:** 217–219.

Harbuz, M.S. and Jessop, D.S. (1999) Is there a defect in cortisol production in rheumatoid arthritis? *Rheumatol.* **38:** 298 –302.

Harris, E.D. (1990) Mechanisms of disease: rheumatoid arthritis – pathophysiology and implications for therapy. *N. Engl. J. Med.* **322:** 1277–1289.

Hasstedt, S.J., Clegg, D.O., Ingles, L., Ward, R.H. (1994) HLA-linked rheumatoid arthritis. *Am. J. Hum. Genet.* **55:** 738–746.

Hench, P.S. (1935) The ameliorating effect of pregnancy on chronic atrophic (infectious rheumatoid) arthritis, fibrositis and intermittent hydrarthritis. *Mayo Clin. Proc.* **13:** 161–167.

Hochberg, M.C. and Spector, T.D. (1990) Epidemiology of rheumatoid arthritis: update. *Epidemiol. Rev.* **12:** 247–252.

John, S., Myerscough, A., Eyre, S., *et al.* (1999) Linkage of a marker in intron D of the estrogen synthase locus to rheumatoid arthritis. *Arthritis Rheum.* **42:** 1617–1620.

Jones, M.A., Silman, A.J., Whiting, S., Barrett, E.M. and Symmons, D.P.M. (1996) Occurrence of rheumatoid arthritis is not increased in the first degree relatives of a population based inception cohort of inflammatory polyarthritis. *Ann. Rheum. Dis.* **55:** 89–93.

Kaijzel, E.L., van-Krugten, M.V., Brinkman, B.M., *et al.* (1998) Functional analysis of a human tumor necrosis factor alpha (TNF-alpha) promoter polymorphism related to joint damage in rheumatoid arthritis. *Molecular Medicine.* **4:** 724–733.

Koumantaki, Y., Giziaki, E., Linos, A., *et al.* (1997) Family history as a risk factor for rheumatoid arthritis: a case-control study. *J. Rheumatol.* **24:** 1522–1526.

MacGregor, A.J., Bamber, S. and Silman, A.J. (1994) A comparison of the performance of different methods of disease classification for rheumatoid arthritis. Results of an analysis from a nationwide twin study. *J. Rheumatol.* **21:** 1420–1426.

Marlow, A., John, S., Hajeer, A., Ollier, W.E.R., Silman, A.J. and Worthington, J. (1997) The sensitivity of different analytical methods to detect disease susceptibility genes in rheumatoid arthritis sibling pair families. *J. Rheumatol.* **24:** 208–211.

Masi, A.T., Da Silva, J.A.P. and Cutulo, M. (1996) Perturbations of hypothalamic-pituitary-gonadal (HPG) axis and adrenal androgen (AA) functions in rheumatoid arthritis. *Baillières Clin. Rheumatol.* **10:** 295 –332.

McDermott, M.F., Ogunkolade, B.W., McDermott, E.M., *et al.* (1998) Linkage of familial Hibernian fever to chromosome 12p13. *Am. J. Hum. Genet.* **62:** 1446–1451.

Mu, H., King, M.-C. and Criswell, L.A. (1998) Relative predispositional effects and mode of inheritance of HLA-DRB1 alleles among community-based Caucasian females with rheumatoid arthritis. *Genet. Epidemiol.* **15:** 123–134.

Ollier, W. and Thomson, W. (1992) Population genetics of rheumatoid arthritis. In: *Rheumatic Diseases Clinics of North America* (ed. B. Dudlick). WB Saunders, Philadelphia, pp. 741–759.

Reveille J.D. (1998) The genetic contribution to the pathogenesis of rheumatoid arthritis. *Curr. Opin. Rheumatol.* **10:** 187–200.

Rigby, A.S., Voelm, L. and Silman, A.J. (1993) Epistatic modelling in rheumatoid arthritis: an application of the Risch theory. *Genet. Epidemiol.* **10:** 311–320.

Schaid, D.J. and Rowland, C. (1998) Use of parents, sibs, and unrelated controls for detection of associations between genetic markers and disease. *Am. J. Hum. Genet.* **63:** 1492–1506.

Shapiro, J.A., Koepsell, T.D., Voigt, L.F., Dugowson, C.E., Kestin, M. and Nelson, J.L. (1996) Diet and rheumatoid arthritis in women: a possible protective effect of fish consumption. *Epidemiology* **7:** 256–263.

Shiozawa, S., Hayashi, S., Tsukamoto, Y., *et al.* (1998) Identification of the gene loci that predispose to rheumatoid arthritis. *Int. Immunol.* **10:** 1891–1895.

Silman, A.J., MacGregor, A.J., Thomson, W., *et al.* (1993) Twin concordance rates for rheumatoid arthritis: results from a nationwide study. *Br. J. Rheumatol.* **32:** 903–907.

Stastny, P. (1978) Association of the B-cell alloantigen DRw4 with rheumatoid arthritis. *N. Engl. J. Med.* **298:** 869–871.

Symmons, D.P.M. and Chakravarty, K. (1993) Can immunisation trigger RA? *Ann. Rheum. Dis.* **52:** 843–844.

Symmons, D.P.M., Barrett, E.M., Bankhead, C.R., Scott, D.G.I. and Silman, A.J. (1994) The incidence of rheumatoid arthritis in the United Kingdom: results from the Norfolk Arthritis Register. *Br. J. Rheumatol.* **33:** 735–739.

Symmons, D.P.M., Bankhead, C.R., Harrison, B.J., *et al.* (1997) Blood transfusion, smoking, and obesity as risk factors for the development of rheumatoid arthritis. *Arthritis Rheum.* **40:** 1955–1961.

Thomas, D.J.B., Young, A., Gorsuch, A.N., Bottazzo, G.F. and Cudworth, A.G. (1983) Evidence for an association between rheumatoid arthritis and autoimmune endocrine disease. *Ann. Rheum. Dis.* **42:** 297–300.

Thomson, G. (1983) Investigation of the mode of inheritance of the HLA associated diseases by the method of antigen genotype frequencies among diseased individuals. *Tissue Antigens* **21:** 81–108.

Thomson, W., Harrison, B., Ollier, B., *et al.* (1998) Quantifying the exact role of HLA-DRB1 alleles in susceptibility to inflammatory polyarthritis. *Arthritis Rheum.* **42:** 757–762.

Vingsbo-Lundberg, C., Nordquist, N., Olofsson, P., *et al.* (1998) Genetic control of arthritis onset, severity and chronicity in a model for rheumatoid arthritis in rats. *Nature Genetics.* **20:** 401–404.

Wordsworth, P., Pile, K.D., Buckely, J.D., *et al.* (1992) HLA heterozygosity contributes to susceptibility to rheumatoid arthritis. *Am. J. Hum. Genet.* **51:** 585–591.

Worthington, J., Ollier, W.E.R., Leach, M.K., *et al.* (1994) The Arthritis and Rheumatism Council's National Repository of Family Material: pedigrees from the first 100 rheumatoid arthritis families containing affected sibling pairs. *Br. J. Rheumatol.* **33:** 970–976.

13

Genes and environment in eating disorders and obesity

David A. Collier and Janet Treasure

1. Introduction

'They are as sick that surfeit with too much, as they that starve with nothing.'

William Shakespeare, The Merchant of Venice

The genetics of body composition encompass diseases ranging from extreme underweight (anorexia nervosa) to extreme overweight (morbid obesity), as well as normal variation in body composition. Obesity is a chronic disease characterized by an excess of adipose tissue, whereas other body composition disorders affect multiple tissues and organs, for example, in anorexia nervosa, adipose tissue, skeletal and heart muscle, bone and brain mass are all affected. Body composition phenotypes may also be specific; mineral bone density, osteoporosis and muscle mass all show significant and specific heritability. Of body composition disorders, obesity is the most prevalent in industrialized societies and has the largest overall impact on health, society and the economy. Anorexia nervosa (AN) has the highest mortality, reaching 16%.

1.1 Epidemiology

The current epidemic of obesity in industrialized countries originates from a culture that promotes excessive eating and discourages physical activity (Hill and Peters, 1998). Almost one third of people in the EC-15 countries are overweight and one in ten is obese, ranging from 8% in France to 19% in Germany. This is a total of almost 50 million people, most of whom are still gaining weight. In the USA, the figures are even more striking, with 63% of men and 55% of women having a body mass index (BMI) over 25 (Must *et al.*, 1999). As in Europe, the prevalence and severity of obesity continues to increase. Binge eating disorder (BED)

Analysis of Multifactorial Disease, edited by T. Bishop and P. Sham.

has emerged as a clinical subtype of obesity; about 30% of obese subjects attending eating disorder clinics have BED.

Little is known about the causes of anorexia nervosa (AN) and bulimia nervosa (BN). Eating disorders are uncommon in men, with women accounting for 90–98% of cases. Between 0.1% and 0.8% of teenage women have AN, and between 1.2% and 2.8% have BN (Bushnell *et al.*, 1990; Gotestam and Agras, 1995; Hoek *et al.*, 1995; Kendler *et al.*, 1991; Lucas *et al.*, 1991; Rastam *et al.*, 1989; Rooney *et al.*, 1995; Soundy *et al.*, 1995; Wade *et al.*, 1996; Walters and Kendler, 1995). In addition, about 3.8% of women have partial syndromes and almost 11% sub-clinical eating disorders (Patton *et al.*, 1999; Cotrufo *et al.*, 1998; Morande *et al.*, 1999). The age-adjusted incidence per annum for AN is between 15 and 26 cases per 100 000 population per year overall (prevalence per 100 000, about 270 for women and 22.5 for males), rising to 51 per 100 000 in teenage women (Lucas *et al.*, 1991; Pawluck and Gorey, 1998; Turnbull *et al.*, 1996).

Views on whether the prevalence of eating disorders is changing are conflicting. Reviews of population studies indicate that rates of AN (Hoek *et al.*, 1995; Pawluck and Gorey, 1998) and BN (Fombonne, 1996) are not increasing significantly overall. However, several studies argue for an increase in overall rates (Eagles *et al.*, 1999; Munk-Jorgensen *et al.*, 1996) or for an increase restricted to women in their 20s (Lucas *et al.*, 1991; Pawluck and Gorey, 1998). Observed increases may reflect greater disease awareness and the referral of less severe cases. Contrary to popular belief, the prevalence of AN also appears to be stable across cultures, with comparable rates observed in Curacao (Hoek *et al.*, 1998), Japan (Kuboki *et al.*, 1996), Egypt (Nasser, 1994), Singapore (Kok and Tian, 1994; Ung *et al.*, 1997) and Hong Kong (Lee and Lee, 1996). Perceived differences in prevalence across cultures may simply be because of a paucity of epidemiological analyses in non-industrialized countries, and lower awareness of eating disorders.

1.2 Health risks

Body weight has a major impact on health. At a body mass index (BMI) below 20 kg/m^2 and above 25 kg/m^2, there is an increase in mortality from all causes in a classical curvilinear relationship (Bray, 1997a). The overweight individual has increased risk of diabetes mellitus, cancer, heart disease, high blood pressure, pulmonary abnormalities, gall bladder and joint disease. For example, the risk of developing type II diabetes more than doubles for every 20% of excess body weight, whereas for a BMI of 20 kg/m^2, there is essentially no risk. For the severely obese (BMI > 40) the prevalence ratio (using normal weight individuals as a baseline) was 18.1 (male) and 12.6 (female) for type II diabetes, and 21.1 and 5.6 for gall bladder disease (Must *et al.*, 1999).

The underweight individual also has increased risk of all-cause mortality. There are specific and severe risks related to anorexia nervosa, which include osteoporosis and associated fractures, heart failure from loss of muscle mass, brain shrinkage and suicide. Depression and anxiety are very common (Fairburn *et al.*, 1999a). The 20-year mortality from anorexia nervosa is 16% (Crisp *et al.*, 1992; Moller-Madsen *et al.*, 1996; Nielsen *et al.*, 1998; Norring and Sohlberg, 1993; Theander, 1985), the highest for any psychiatric disorder.

While weight is often unremarkable in BN and mortality much lower than AN

(Keel *et al.*, 1999), the health risks are still high. Almost 50% of bulimics report suicide attempts or suicidal ideation (Favaro and Santonastaso, 1997). Severe fluid and electrolyte abnormalities are common, including hypovolemia (low plasma volume) depletion of total body potassium and cardiac arrhythmias (Comerci, 1990). BN is also increased in women with diabetes (Verrotti *et al.*, 1999), although no causative interrelationship has been established. However, the combined effects of BN and diabetes result in greatly increased morbidity and mortality (Nielsen and Molbak, 1988).

2. Environment

2.1 Environment in obesity

Obesity is strongly influenced by environmental factors which affect energy intake and usage. In the general population, it is clearly associated with increased food intake, and obesity is most common in societies where food availability is high and portion control is poor (Hill and Peters, 1998). A high level of dietary fat is also a significant contributory factor. In animals, the proportion of body fat increases in line with the fat content of the diet (Salmon and Flatt, 1985), although palatable calorie-dense foods such as sucrose have a similar effect (Rothwell and Stock, 1984). The 'American Paradox' is that obesity prevalence is rising in line with the increased use of reduced-fat foods (Heini and Weinsier, 1997). This may result from increased consumption of calorie-dense refined carbohydrate, but alternative explanations are underreporting of fat consumption and clustering of obesity risk factors in subgroups of the population (Astrup, 1998).

Low levels of physical activity are also associated with obesity, since volitional exercise mitigates the effects of a high-fat diet on obesity, and there is good evidence of energy balance being determined by interaction between diet and exercise. An example of the role of environment in obesity can be observed in Pima Indians, who split into two geographic groups several centuries ago. The Arizona Pimas consume a high-fat, industrialized diet and have the highest incidence of obesity in the world. They are on average 26 kg heavier than their Mexican Pima kindred, who, unlike their Americans counterparts, engage in extensive physical work and live on a diet of grains and vegetables (Gibbs, 1996). The 'thrifty' genotype of the Pima Indians is well suited to the traditional environment, but not to a high-fat, high-calorie diet. The penetrance of susceptibility genes for obesity will increase as exposure to environmental risk rises. Thus, despite the clear role for cultural environment in obesity, physical correlates such as body mass index and percentage fat are highly heritable, especially in the young.

2.2 NEAT: The cause of differential weight gain?

The genetic component to differential weight gain has been clearly demonstrated by overfeeding studies in identical twins, where there is about three times more variability between unrelated twins than within twin pairs (Bouchard *et al.*, 1990; Tremblay *et al.*, 1997). This concurs with the commonly expressed view that some people can eat what they want and not gain weight, whereas others on the same diet will become obese. This variation can be put down to differences in resting metabolic rate (RMR), voluntary exercise, elevated thermic effects of food, and increased fidgeting, or spontaneous physical activity (SPA) (Jequier, 1999). This last factor is also known as NEAT, or nonexercise activity thermogenesis (Levine *et al.*, 1999). High 24-hour respiratory quotient (RQ) and high insulin sensitivity (IS) have also been shown to predict weight gain (Ravussin and Swinburn, 1993).

RMR, although variable between humans, only accounts for differences of about 100 kcal/day in energy expenditure, and hence can only explain some differences in weight gain (Ravussin *et al.*, 1988). Likewise, although voluntary exercise has a major effect on body composition, most people in Western societies are sedentary, and exercise accounts for less than 25% of energy expenditure (Jequier and Schutz, 1988). Defects in the thermogenic effects of food have been reported, but this effect is self-limiting as weight gain is accompanied by an increase in RMR and the energy cost of physical exercise, and can only account for differences of a few kilograms of fat mass. The answer may lie with SPA or NEAT, as proposed by Ravussin *et al.* in 1986. SPA includes activities such as strolling and limb movements and can account for interindividual differences in energy usage of up to 800 kcal/day. Levine *et al.* (1999) made a complete assessment of energy balance in individuals fed 1000 kcal/day over their normal intake; the ten-fold difference in fat storage they found was mainly accounted for by differences in NEAT ranging from –98 to +692 kcal/day. The heritability of NEAT is unknown. In the eating disorders, voluntary exercise may have a much greater effect on body composition. Women with anorexia and bulimia use intense physical exercise as a method of controlling weight, and in bulimia, as compensatory behavior for binge eating.

2.3 Risk factors for eating disorders

Unlike obesity, eating disorders are regarded as having strong psychosocial risk factors. These risk factors are well described elsewhere. Many of these factors are nonspecific in that they are shared with other forms of psychopathology; for example, preterm birth and birth trauma are associated with anorexia (Cnattingius *et al.*, 1999), and premorbid levels of physical and sexual abuse are similar to those seen in affective disorders (Fairburn *et al.*, 1998, 1999a; Wonderlich *et al.* 1997). History of sexual abuse is strongest in women with comorbid bulimia and substance dependence (Deep *et al.*, 1999). Comorbidity for clinical depression, generalized anxiety, alcohol abuse, panic disorder as well as low self-esteem, are seen in both AN and BN (Fairburn *et al.*, 1999; Walters and Kendler, 1995).

Factors linked to dieting, such as premorbid obesity, family history of obesity and teasing or bullying about weight or shape are specifically linked to BED and BN, but not AN. Contrary to popular belief, dieting is not a major risk factor for AN once other risk factors are taken into account (Fairburn *et al.*, 1999b). BN is also distinguished from AN by early menarche and parental psychiatric disorder. These findings were echoed by Karwautz *et al.*, (2000) who also found parental pressure to succeed was a significant risk factor. Overall, eating disorders can be thought of as disorders of low self-esteem. They may lie on a continuum with obesity, with AN associated with propensity to leanness and BN associated with propensity to obesity.

3. Heritability

3.1 The obesity phenotype

The aim of phenotypic measurement in the present context is to select the most robust and heritable traits for use in genetic epidemiology and molecular genetic studies. Although obesity is relatively straightforward to measure, there are a number of related phenotypes that arise from the different distribution of fat and muscle between people, and the underlying metabolic and endocrinological traits that lead to, or are associated with obesity. The most commonly used measure is BMI, formally defined as weight in kilograms divided by body area, although in practice usually estimated by dividing weight by height squared. This method is reproducible, reliable and inexpensive, although it does not account for the proportion of fat to lean tissue in the body; people with a high proportion of muscle may have the same BMI as someone much fatter (Michels *et al.*, 1998). Percentage fat mass (%FAT) and lean body mass (LBM), both of which can be measured by bioimpedance, the sum of skinfold thickness at up to eight body sites (SF8) measured using calipers, and visceral adipose tissue might be more precise measures of obesity. There is also increasing attention to body shape because of the differential health risks associated with variation in fat distribution (Wickelgren, 1998), such as waist-to-hip ratio (WHR).

There are also a number of intermediary physiological indices of obesity-related traits that are more complex to measure but which may have a more direct relationship with underlying genetic variation, such as the rate of energy expenditure. Total energy expenditure (TEE) in humans consists of RMR, the thermic effect of food and the energy cost of voluntary and spontaneous physical activity. Sixty to seventy percent of TEE is derived from RMR, which is almost entirely a result of the energy needed to sustain life, with an additional 10% contributed by neural activity during the waking state. RMR is correlated with LBM and total body fat (TBF) and is higher in men than women (Rice *et al.*, 1996). Respiratory quotient (RQ) reflects the ratio of energy derived from carbohydrate as opposed to fat metabolism. High RQ is indicative of a low ratio of fat to carbohydrate metabolism (Zurlo *et al.*, 1990; Ravussin and Swinburn, 1993).

Both insulin sensitivity, fasting insulin and leptin levels are also commonly used as a phenotype for obesity studies, since they are highly correlated with fat mass and share common features as adiposity signals. Their circulating concentrations are proportional to adiposity and acute changes in energy balance; both enter the CNS and bind to receptors in the hypothalamus, both reduce food intake in a dose-dependent manner and both induce responses leading to reduction in body fat stores when centrally administered.

3.2 Twin, family and adoption studies in obesity

Twin, family and adoption studies are all consistent with a major role of heritability in susceptibility to obesity. In adults from industrialized countries, twin studies have yielded heritability estimates ranging from 50% to 80% for BMI (Allison *et al.*, 1994, 1996; Borecki et al. 1998; Pietilainen *et al.*, 1999; Price and Gottesman, 1991; Stunkard *et al.*, 1986). However, estimates of heritability are lower for adoption studies (30–40%) and family studies give intermediate estimates (reviewed in Comuzzie and Allison, 1998). Part of this variation appears to come from age-specific variation in the effects of genes on BMI as well as gender differences (Korkeila *et al.*, 1991). Other measures of obesity also have a genetic influence; SF8 is about 34% heritable (Rice *et al.*, 1997) and adjusted subscapular skinfold thickness is up to 60% heritable (Brook *et al.*, 1975; Selby *et al.*, 1990; Slattery *et al.*, 1988). There is evidence from analysis of familial clustering that total body fat and fat distribution have different genetic and environmental factors (Rice *et al.*, 1999; Selby *et al.*, 1990). For example, WHR shows lower heritability at 31% and is influenced by cigarette smoking (Selby *et al.*, 1990), especially in men (Visser *et al.*, 1999). Segregation analysis indicates that at least two major loci are involved, one for extreme overweight and one for variations within the normal range (Borecki *et al.*, 1998).

Twin studies in children and adolescents show the highest heritability, of about 80% for BMI (Pietilainen *et al.*, 1999). Longitudinal studies support the idea that the effects of genes on BMI vary with age, with genes having a greater effect when subjects are young. Fat has different characteristics at each age, and gender differences are also evident from birth to adulthood (Korkeila *et al.*, 1991), with men tending to have more abdominal fat ('apple-shape') whereas women tend to have more fat on their hips and buttocks ('pear-shape'). Although any gender differences are likely to be sex-limited (e.g., hormonal), there is also evidence from twin studies that some genetic factors contributing to variation in BMI are separable between men and women.

Physiological measures are also heritable. Circulating leptin levels and insulin sensitivity are correlated with fat mass and have similar levels of heritability to BMI. The heritability of RMR, estimated from both twin and family studies, is at least 40% (Bouchard *et al.*, 1989; Rice *et al.*, 1996), and the effect appears to be polygenic. Respiratory quotient has a familial component (Toubro *et al.*, 1998) but there is little information on its heritability. The heritability of NEAT has not been measured.

3.3 The eating disorder phenotype

The eating disorders encompass AN and the more recently described BN and BED. Eating disorders are more difficult to define than obesity since they cannot be defined on the basis of body composition alone and require the presence of certain behaviors (e.g., episodic overconsumption, compensatory behavior such as self-induced vomiting, excessive exercise), psychiatric manifestations (e.g., fear of weight gain) and physiological abnormalities (e.g., amenorrhea). Diagnosis is also made more difficult by the need for longitudinal information, since BN often manifests itself initially as an anorexic illness and up to five years may be required for bulimic symptoms to emerge.

Anorexia nervosa is not a new disorder. Early Christian ascetic women practised restricted eating, in part because the suspension of menstruation was regarded as a sign of purity (Garnsey, 1999). It has been reported as an illness for over 300 years, and by 1874 was clearly defined (Vandereycken and Van Deth, 1994; Walsh and Devlin, 1998). The key diagnostic features of anorexia are maintenance of body weight at below normal level, intense fear of gaining weight or becoming fat, and amenorrhea in females. Anorexia can be divided into the restricting subtype (RAN) where there is no compensatory behavior, and binge–purging anorexia (AN/BP). The physiological consequences of starvation are common in anorexia.

Bulimia nervosa was defined as an illness much more recently (Russell, 1979), and is characterized by uncontrolled eating binges, compensatory behavior (use of laxatives, self-induced vomiting, excessive exercise) and low self-esteem, which is linked to body image. Most of those with BN are of unremarkable weight. BN can be divided into purging and nonpurging types; the purging type may be a more severe phenotype (Hay and Fairburn, 1998). Nonpurging BN is similar to BED, but BED appears less severe and has a different outcome. Binge eating disorder was first formally defined in 1992 (Spitzer *et al.*, 1993), but was first described in Japan ('kibarashi-gui') in 1977 (Nogami and Yabana, 1977). It is characterized by binge eating without compensatory behavior, but with marked guilt and distress. It is associated with obesity (Fairburn *et al.*, 1998), and about 30% of attendees at weight loss clinics have BED. Heritability estimates for BED are high at 82% (Bulik *et al.*, 1998). Eating disorders which do not fit into the above diagnostic categories are termed EDNOS (eating disorder, not otherwise specified). EDNOS includes subjects with a milder phenotype which is unstable over time (Hay and Fairburn, 1998).

The phenotypic characterization of eating disorders can be extended beyond diagnostic classification by examining comorbidity, personality and behavior (Lilenfeld *et al.*, 1998, 1999). By examining 237 affected relative pairs with anorexia, Lilenfeld *et al.* (1999) found that, consistent with earlier analyses, women with anorexia display behavioral rigidity, emotional restraint, perfectionism and obsessionality. The most influential combination derived from factor analysis was one of trait anxiety, harm avoidance, perfectionism, obsessionality and diminished self-directedness. The aim of this analysis is to define continuous traits which are characteristic of eating disorders and can thus be used to search for quantitative trait loci and to define genetically homogeneous sub-groups among the population.

3.4 Heritability of eating disorders

Although the contribution of genetic factors to the etiology of eating disorders has not been as thoroughly investigated as in obesity, and the results of twin studies are inconsistent, a 'best-guess' is moderate heritability of about 50%. Family studies have confirmed the familial aggregation of eating disorders, and provide some insight into the relationship between anorexia and bulimia. Strober *et al.* (1999) performed a controlled family study of eating disorders and obesity in which they also reviewed previous family studies. In nine out of ten family studies of anorexia, cases were found among relatives of AN probands but not of controls. In five out of six studies which included BN, the frequency of BN was elevated in the relatives of AN probands, indicating that some etiological factors may be shared. Likewise, BN was elevated in the relatives of BN probands in three out of four family studies. Strober *et al.* (2000) examined the rates of eating disorders in 1831 relatives of 504 probands, and found relative risks of 11.1 in female relatives of AN and 11.9 in female relatives of BN probands. Stein *et al.* (1999) assessed lifetime eating disorders in relatives of probands with BN and found very strong familial aggregation, with 43% of sisters and 26% of mothers receiving a diagnosis, most often EDNOS.

Fairburn *et al.* (1999a) has reviewed twin studies in eating disorders in detail and concludes that heritability is uncertain. For example, the Virginia Twin Registry study (Walters and Kendler, 1995) found only 11 subjects meeting DSMIV criteria for anorexia nervosa and 24 with definite or probable AN based on clinical criteria, out of over 1000 female twin pairs, and only six concordant pairs. There was no evidence for an excess of MZ over DZ concordance; however, the power of the sample was insufficient to reliably measure heritability. Targeted twin studies, although potentially biased, may be more appropriate for the study of eating disorders because of the low prevalence of cases in population studies. Using twin registers, advertisement and clinic populations, Treasure and Holland (1994) found 67% concordance for AN in 31 MZ twins, and 0% in 28 DZ twins, giving an estimated heritability of 70%. A reanalysis of clinically ascertained twin pairs (Bulik *et al.*, 2000) found heritability of 88% (95%CI 33–97) resulting from additive genetic effects.

Several subsequent studies have examined the heritability of BN in the Virginia Twin sample, the first of which (Kendler *et al.*, 1991) found an excess of MZ (23%) over DZ (9%) concordance, consistent with a heritability of 50% or more. Once again, however, the power of this study was limited because of the small sample size. Two subsequent studies examined bulimia and its relationship with other psychiatric disorders, the first finding similar heritability at about 50% (Walters *et al.*, 1992) and the second 31% (Kendler *et al.*, 1995). A re-evaluation of the data over five years (Bulik *et al.*, 1998) found more substantial evidence for heritability (83%). As Fairburn *et al.* (1999a) and Bulik *et al.* (2000) conclude, the extent and nature of the genetic and environmental contribution to eating disorders remains unclear. New twin studies in progress will clarify the role of genes in these disorders.

4. Human genetic studies of obesity

4.1 Mendelian syndromal forms of obesity

There are currently five single gene forms of obesity (Chagnon *et al.*, 2000a), involving mutations in prohormone convertase 1 gene (Jackson *et al.*, 1997), the leptin (Montague *et al.*, 1997; Strobel *et al.*, 1998) or leptin receptor genes (Clement *et al.*, 1998), proopiomelanocortin (POMC; Krude *et al.*, 1998), and the melanocortin-4 receptor (MC4R) (Hinney *et al.*, 1999; Vaisse *et al.*, 1998; Yeo *et al.*, 1998). There are also currently over 26 single gene syndromes with obesity as a component (Chagnon *et al.*, 2000a). The majority of these genes appear to have an effect restricted to rare single gene forms of obesity, although the loci for some have also been tentatively implicated as causes of general population obesity. A comprehensive review of genetic findings in obesity, including human single gene obesity and related phenotypes, linkage and association in the general population, mouse single genes and mouse quantitative trait loci (QTLs) can be found in Chagnon *et al.* (2000a) (www.obesity.chair.ulaval.ca/genes.html).

4.2 Linkage genome scans of obesity

From the public health perspective, common forms of obesity, i.e. those which have a complex origin from both genes and environment, are of considerable importance. Obesity is likely to have multiple contributions from genes that affect appetite, the physiological parameters of food usage and food-related behavior. As for all complex diseases, two convergent approaches have been used: family and sib-pair linkage and candidate gene analysis.

While the analysis of candidate genes selected for their role in metabolic pathways related to weight control has been very successful, the causes of obesity will also involve unknown or unexpected genes. The most robust approach for the localization of these is genetic linkage analysis, which has the additional advantage in obesity that quantitative trait loci (QTL) methods can easily be used, increasing the power to detect linkage several fold. Linkage scans also have the advantage of placing any positive findings in a genome-wide context, allowing the rough estimation of the impact of a particular locus in the population under test. Several genome-wide scans have now been completed (*Table 1*). The most significant of these loci are discussed below.

Chromosome 2p21: a leptin QTL? Comuzzie *et al.* (1997) performed a genome-wide linkage scan in Mexican-Americans using leptin levels as a quantitative trait, and found a maximum lod score of 4.95 at chromosome 2p21. Further support for this locus comes from the genome scan of Hager *et al.* (1998a) in a French population, who also found suggestive evidence for linkage with leptin levels at the same locus on chromosome 2p. This suggests that 2p21 may be an obesity-related locus which acts independently of ethnic and cultural factors. The proopiomelanocortin gene (POMC) is the strongest candidate gene within this region.

Chromosome 10: a major locus for obesity? Hager *et al.* (1998a) analysed 158 French nuclear families for linkage to obesity using BMI and leptin levels as their main phenotypic measures. Their strongest linkage finding was with chromosome 10 and BMI, with a maximum model-free multipoint lod score of 4.85. The region of linkage contains no obvious candidate genes for obesity. This locus may represent a major gene for obesity since estimates for λ_S, the sibling recurrent risk, are about 1.66 for the marker D10S197 compared to a total λ_S of 4.15 for the French population. Calculation of the log ratios of these measures indicate that the locus accounts for as much as 36% of liability to obesity under a non additive and 21% under an additive genetic model.

Chromosome 8p and the β-3 adrenergic receptor. Mitchell *et al.* (1999) performed a QTL analysis in Mexican-American families, ascertained as part of the San Antonio Family Heart Study. The study population was not ascertained for

Table 1. Genome wide scans for linkage in human populations.

Population	Phenotype	Locus	lod	*P*-value	Candidate genes?	Reference
Mexican Americans	Serum leptin	2p21	4.95	9×10^{-7}	*GCKR, POMC*	Comuzzie *et al.*, 1997
Pima Indians	% body fat	11q21–q22	2.1	0.001	*DRD2*	Norman *et al.*, 1998
Pima Indians	% body fat	18q21	2.3	0.0006	*MC4R, MC5R*	Norman *et al.*, 1998
Pima Indians	24 h TEE	11q23–q24	2.0	0.001	*APOC3, PC8*	Norman *et al.*, 1998
Pima Indians	24 h RQ	1p31–p21	2.8	0.0003	*LEPR*	Norman *et al.*, 1998
Pima Indians	24 h RQ	20q11.2	3.0	0.0006	*ASIP*	Norman *et al.*, 1998
French	BMI	10p	4.85	–	–	Hager *et al.*, 1998a
French	Serum leptin	2p21	2.68	–	*GCKR, POMC*	Hager *et al.*, 1998
French	Serum leptin	5p	2.93	–	–	Hager *et al.*, 1998
Pima Indians	BMI	11q	3.6	–	*DRD2, APOC3, PC8*	Hanson *et al.*, 1998
Mixed USA	BMI $\geqslant$ 30[a]	20q13	3.17	–	*MC3R, ASIP, ADA, C/EBPβ*	Lee *et al.*, 1999b
Mexican-American	BMI	8p12–q23.3	3.21	0.0002	*ADRB3*	Mitchell *et al.*, 1999
Canadian	Fat free mass	7p15.3	2.72	–	–	Chagnon *et al.*, 2000b
Canadian	Fat free mass	15q25–q26	3.56	–	*IGFR1*	Chagnon *et al.*, 2000b
Canadian	% body fat	18q12–q21	2.3	–	*MC4R, MC5R*	Chagnon *et al.*, 2000b
Canadian	Fat free mass	18q12–q21	3.58	–	*MC4R, MC5R*	Chagnon *et al.*, 2000b

[a]A variety of other measures provided evidence for linkage.

obesity, but included the relatives of probands aged 40–60 years with large families from a low income area. A genome scan at 10 cM average intervals identified a QTL for BMI on chromosome 8, with a multipoint lod score of 3.21 close to the β-3 adrenergic receptor locus. This linkage is particularly interesting since the same group have previously observed association with the gene in the same San Antonio population.

Chromosome 20q11–q13. Two genome scans have implicated chromosome 20q in obesity, using %FAT, WHR and 24 h RQ (Norman *et al.*, 1998), or BMI and %FAT (Lee *et al.*, 1999b). This locus is supported by other linkage findings, including a mouse QTL for obesity on a region syntenic to chromosome 20q (Lembertas *et al.*, 1997; Mehrabian *et al.*, 1998; Pomp, 1997; Taylor and Philips, 1997; York *et al.*, 1996), and targeted linkage of the region using %FAT, BMI and fasting insulin measurements (Lembertas *et al.*, 1997). Maturity onset diabetes of the young, caused by mutations in the *4HNF4* (hepatic nuclear factor) gene, has also been localized nearby at 20q12 region (Yamagata *et al.*, 1996). There is also linkage to noninsulin-dependent diabetes mellitus (NIDDM) (Bowden *et al.*, 1997; Ji *et al.*, 1997) at 20q13 and a second NIDDM locus may lie near to the phosphoenolpyruvate locus (PCK1), which is more centromeric (Zouali *et al.*, 1997). The most likely linkage interval in humans is about 10 cM wide, although linkage extends for about 16 cM. The region contains a number of candidate genes, including the adenosine deaminase, the melanocortin 3 receptor (MC3R), the agouti signaling protein (ASIP) and the CAAT/enhancer binding protein beta (C/EBPβ) (Lee *et al.*, 1999; Lembertas *et al.*, 1997).

4.3 Candidate genes for obesity

The physiology of energy balance and adiposity is relatively well understood, and a variety of candidate genes identified from human or animal physiology and single-gene disorders have been examined for association with obesity (Eichwald, 1999). Many of the genes identified from human and animal studies as being involved in single-gene forms of obesity do not appear relevant to obesity in the general human population. However in some cases, evidence from linkage studies points to a role for some ‘Mendelian’ obesity genes, such as POMC and MC4R, in common human obesity. A full description of candidate gene studies in obesity can be found in Pérusse *et al.* (1999) (www.obesity.chair.ulaval.ca/genes.html). This section concentrates on the ‘lipostat’ system in the hypothalamus, which is the central mediator of the regulation of body composition. Most single-gene defects and transgenic mouse models of obesity map to this system, and there is also evidence that the adipostat may determine common forms of human obesity.

The agouti pathway and the ‘lipostat’. The agouti pathway (*Figure 1*) was first uncovered through the yellow obese or agouti mouse (A^y) which has ectopic expression of the endogenous agouti protein ligand, normally an antagonist of melanocortin 1 receptors (MC1R) (Bultman *et al.*, 1992; Cone *et al.*, 1996; Huszar *et al.*, 1997). Production of agouti in the brain of A^y mice results in the aberrant antagonism of the melanocortin 4 receptor (MC4R) and consequent obesity,

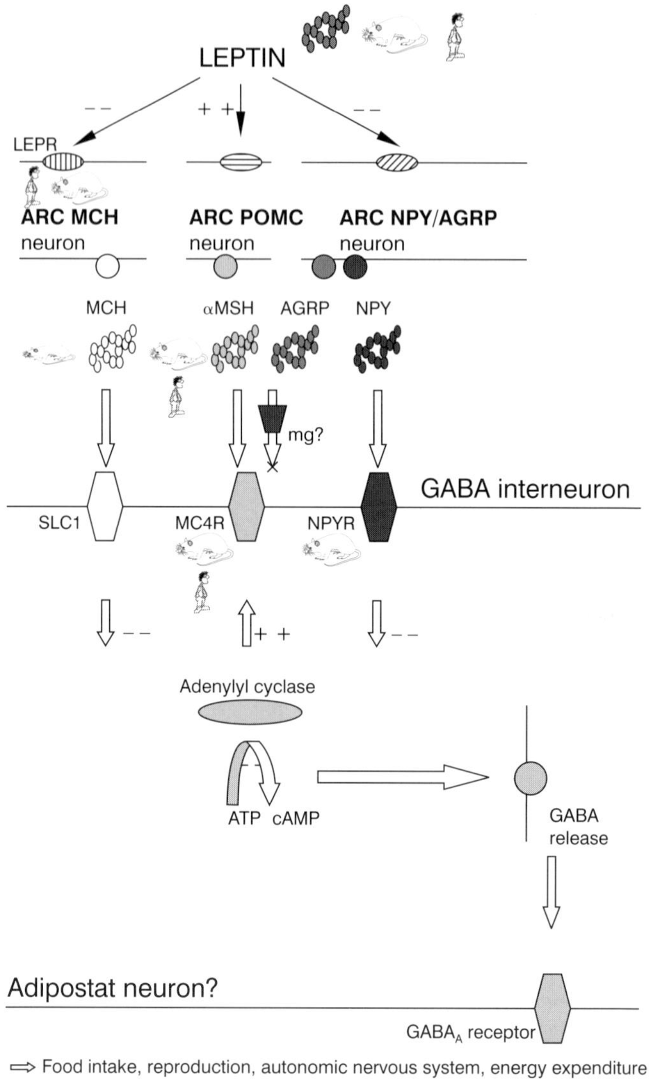

Figure 1. Schematic representation of components of the putative 'lipostat' feeding control pathway in the hypothalamus. Circulating leptin is transported into the CNS where it inhibits MCH and NPY/AGRP expression, and stimulates POMC (α-MSH) expression. MCH and α-MSH are functional antagonists of each other but act at separate receptors. Signaling at GABA interneurons occurs principally by modulation of adenylyl cyclase and subsequent alteration in cAMP levels, which directs inhibitory GABA release. Other direct and indirect signaling occurs, including via peptides such as galanin, which are not shown. Human single forms of obesity are shown by the human figure, transgenic mouse obesity or leanness by the mouse. LEPR, leptin receptors; MCH, melanin concentrating hormone; POMC, pre-proopiomelanocortin; AGRP, agouti-related peptide; NPY, neuropeptide Y; mg, mahogany; SLC1, MCH receptor; MC4R, melanocortin 4 receptor; NPYR, NPY receptors Y1, Y2 and Y5. A more detailed model of the adipostat can be found in Cowley *et al.* (1999).

whereas antagonism of the MC1R in the skin results in a yellow coat. Agouti has a brain homolog, agouti-related protein (AGRP) which antagonizes both the MC3R and MC4R (Graham *et al.*, 1997; Ollmann *et al.*, 1997). Overexpression of AGRP in transgenic mice also results in obesity (Huszar *et al.*, 1997).

The endogenous agonists for the MCRs are the melanocyte stimulating hormones, α-, β- and γ-MSH. γ-MSH is involved in blood pressure and heart function, whereas α-MSH is important in the control of body weight. All are encoded by the *POMC* gene. Recent evidence indicates that the agouti pathway, together with neuropeptide Y, participate in an integrated system in the hypothalamic paraventricular nucleus (Cowley *et al.*, 1999), which detects the amount of energy stored in the body and attempts to hold adiposity constant. This concept was historically formulated as the adipostat or lipostat (Kennedy, 1953).

The proopiomelanocortin gene: a common QTL for leptin levels and a single-gene cause of obesity? The *POMC* gene encodes a number of peptides involved in diverse functions, including pigmentation, adrenocortical function, the regulation of energy stores, the immune system and central and peripheral nervous systems. *POMC* encodes ligands for the melanocortin receptors, adrenocorticotrophin (ACTH), α-, β- and γ-MSH as well as the endogenous opioid β-endorphin. A role for α-MSH in weight control was first identified in mice (Lu et al., 1994), and a mouse model of POMC deficiency develops obesity which responds to peripheral melanocortin (Yaswen et al., 1999). α-MSH is a high affinity agonist for MC3 and MC4 receptors, where its binding inhibits feeding (Fan *et al.*, 1997), the opposite effect to the MCR antagonists AGRP and ASIP, both of which stimulate feeding and lead to obesity.

The first indications of a role for POMC in human obesity came from linkage analysis, which implicated the POMC locus at chromosome 2p21 as a leptin QTL (Comuzzie *et al.*, 1997), and mutational analysis in early-onset obesity with adrenal insufficiency and red hair, which identified two cases with truncated or abolished POMC expression (Krude *et al.*, 1998). Thus, POMC may lead to both common and single-gene forms of human obesity. A number of additional studies have evaluated the *POMC* gene as a candidate for obesity and eating disorders. Hinney *et al.* (1998a) found *POMC* variants in obese, underweight and anorexic subjects, none of which could be correlated with a phenotype. Echwald *et al.* (1999) analyzed the gene and its putative promoter in normal and obese subjects; they likewise found no association with obesity, %FAT or leptin levels. However, Hixson *et al.* (1999) found association ($P = 0.001$) between a haplotype of POMC polymorphisms and leptin levels in Mexican-Americans. The most likely explanation for this finding is that an as yet unknown *POMC* polymorphism (for example in distal promoter elements) is the determinant of leptin levels at this locus, or there is another gene close by which is responsible.

Melanocortin receptors. The melanocortin receptors (MCRs) are a family of seven transmembrane G protein-coupled receptors involved in pigmentation and the regulation of body weight in humans. There are five known MCRs (MC1R–MC5R) which have distinct functions; MC1R alleles are associated with skin pigmentation and vulnerability to UV-induced damage whereas MC3R and

MC4R are involved in body weight regulation and MC5R in exocrine gland function. Melanocortin receptors have been implicated as genetic risk factors for obesity. Transgenic mice with a targeted disruption of MC4R develop maturity onset obesity, hyperinsulinemia and hyperglycemia (Huszar *et al.*, 1997), analogous to maturity-onset human obesity. Transgenic mice lacking the MCR ligand α-MSH also become obese, and MC4R null mice do not respond to MSH-like antagonists (Marsh *et al.*, 1999), indicating that α-MSH inhibits feeding mainly by activating MC4R signal transduction.

MC4R mutations have been identified in dominant forms of human obesity. Two frameshift variants of MC4R, either a deletion (ΔCTCT311) (Yeo *et al.*, 1998) or insertion (InsGATT732) (Vaisse *et al.*, 1998), which disrupt the gene, were found to segregate with obesity in a single family with a proband from a panel of obese subjects. Hinney *et al.* (1999) screened a large cohort of obese children and adolescents ($N = 306$) and in addition to ΔCTCT311, found a nonsense mutation (Tyr35Stop) and a variety of missense mutations exclusively in obese probands. Gu *et al.* (1999) also found three additional variants from 190 obese probands, one of which (Ile137Thr) shows severely impaired ligand binding. Thus, mutations in MC4R, although rare in the general population, may account for up to 3% of morbid obesity in humans.

Evidence that MCR receptors may have a more general effect on human obesity comes from linkage analysis in the Quebec family study (Chagnon *et al.*, 1997; 2000a), where the MC5R (human chromosome 18p11.2) and MC4R (18q21.3) loci showed linkage and association to a variety of obesity phenotypes, with the strongest effect seen for MC5R on BMI in females. Linkage with the MC4R locus was also observed in a Finnish sib-pair linkage study of obesity (Ohman *et al.*, 1999), but no mutations were found within the gene in obese subjects from the linkage panel. A role for MC4R is also supported by studies in the pig, where a missense mutation in a highly conserved region of MC4R has been associated with fatness, growth, and feed intake traits in several different pig lines (Kim *et al.*, 2000).

The MC3R locus has also been linked to obesity or related phenotypes in both mice (*Pfat2, Mob5*; chromosome 2; Lembertas *et al.*, 1997; Mehrabian *et al.*, 1998; Taylor and Phillips, 1997) and the syntenic region in humans (chromosome 20q13; Lee *et al.*, 1999; Lembertas *et al.*, 1997). There are a number of other candidate genes for obesity in this region, including the MC3R/MC4R antagonist human agouti signaling protein (ASIP), the adipocyte differentiation regulator CAAT/enhancer-binding protein beta, and the mahogany gene, *mg*.

Mahogany. Mahogany (*mg*; Lane and Green, 1960) is also a component of the agouti pathway, since mutations can suppress the phenotype (including obesity) of the A^y mouse (Miller *et al.*, 1997; Dinulescu *et al.*, 1998). However, *mg* does not suppress obesity resulting from disruption of the MC4R or other mouse obesity alleles. Its most likely mode of action is by presenting antagonist ligands such as AGRP to the MC4R receptor (Nagle *et al.*, 1999; Gunn *et al.*, 1999), although it could be a functional receptor in its own right. Thus, in A^y mice, *mg* acts in a dose-dependent manner to compensate for agouti over-expression, most likely through failure to present agonist. The mahogany gene also acts as a suppresser of diet-induced obesity, since transgenic mice lacking *mg* do not gain weight on a high-fat diet in comparison to controls (Nagle *et al.*, 1999). The mahogany gene is thus involved in the control of metabolic rate and feeding behavior independently of its interaction with agouti. No human mutations in mahogany have as yet been identified.

Melanin concentrating hormone and SLC-1. Melanin concentrating hormone (MCH) is a cyclic peptide which shows increased expression on fasting, is up-regulated in genetically obese mice, and stimulates feeding if administered centrally. Mice lacking MCH have reduced body weight, leanness due to hypophagia, and increased metabolic rate despite having reduced amounts of leptin and POMC in the actuate nucleus (Shimada *et al.*, 1998). MCH is thus an orexigenic peptide which is a critical regulator of feeding and energy balance, and which acts downstream of the leptin system. It has functional and mutually antagonistic action with αMSH, mediated by a separate receptor. The MCH receptor (SLC-1) is a 353-amino acid G protein-coupled receptor on chromosome 22q, which acts by increasing protein kinase A signaling. This is in contrast to NPY and AGRP, which act through their receptors to reduce protein kinase A signaling.

Neuropeptide Y. Neuropeptide Y is a 36-amino acid peptide which modulates a variety of physiological processes, including feeding, blood pressure, circadian rhythm and anxiety. It is widely expressed in the central and peripheral nervous systems, including the hypothalamus, where it regulates energy balance by stimulating food intake. It binds to a series of at least six receptors, NPY-Y1 to NPY-Y6 (Blomqvist and Herzog, 1997). Its role in stimulating food intake has catalyzed interest in the gene and its receptors as candidates for genetic causes of obesity. The effect of NPY on feeding is thought to lie with the Y1, Y2 and Y5 receptors; transgenic mice lacking these genes develop obesity and hyperinsulinemia (Y1; Kushi *et al.*, 1998; Pedrazzini *et al.*, 1998) mild late-onset obesity (Y5; Marsh *et al.*, 1999) and obesity, increased food intake and fat deposition (Y2; Naveilhan *et al.*, 1999). Transgenic mice lacking NPY itself, however, have a normal pattern of feeding behavior, although they are more sensitive to the action of leptin (Erickson *et al.*, 1996). This may be because of compensation for loss of NPY with AGRP, since they are coexpressed in the same neurons (Cowley *et al.*, 1999).

NPY is also involved in reproduction. Obesity and eating disorders produce a variety of alterations in the reproductive system, particularly in women (Bray, 1997b). For example, in obese women, menarche is earlier, and in anorexic women menarche may either be delayed or amenorrhea occur. NPY appears to play an important role in regulating the control of positive feedback exerted by ovarian steroids (Barreca *et al.*, 1998). Estradiol stimulates leptin production (Demerath *et al.*, 1999) which in turn suppresses central NPY expression (Bray, 1997b).

The NPY peptide gene has been screened by Karvonen *et al.* (1998) who found a Leu7Pro amino acid change in the signal peptide region of the NPY-precursor. This was not associated with obesity, but with elevated cholesterol, particularly in obese subjects. The Y1 (Roche *et al.*, 1997) and Y5 (Rosenkranz *et al.*, 1998) receptors have also been screened in obesity and eating disorders, but no associations have been found.

Leptin. Leptin is a peripheral signal of obesity which is produced by adipocytes in proportion to adiposity (Loftus, 1999). It was originally discovered through the obese (ob/ob) and diabetic (db/db) mice, which are massively obese, hyperphagic and have insulin insensitivity. The ob/ob mice lack leptin (Zhang *et al.*, 1994) whereas the db/db and other obese mice lack the leptin receptor (Chua *et al.*, 1996; Lee and Lee, 1996). Leptin has the effect of reducing food intake and specifically reducing adipose tissue mass, it prevents the reduction in energy expenditure associated with fasting, and also acts as a signal for reproduction. Leptin actions are also linked to insulin, since both molecules are secreted in proportion to adiposity, they influence the production of each other, and insulin signaling in the brain may be dependent on leptin action (Loftus, 1999).

The marked reduction in leptin seen in conditions of extreme leanness is a cause of amenorrhea (Laughlin and Yen, 1997). Leptin concentrations differ between adolescent men and women, with the latter having higher leptin levels before, during and after puberty, independent of the level of adiposity (Demerath *et al.*, 1999). This is because estrogen has a stimulatory effect on leptin, whereas testosterone has a negative influence.

Leptin and leptin receptor mutations are rare single-gene causes of obesity in humans (Clement *et al.*, 1998; Elmquist *et al.*, 1998; Montague *et al.*, 1997; Strobel *et al.*, 1998). In genome-wide linkage scans of obesity, one study found a suggestive lod score of 2.8 at the leptin receptor locus for 24-h RQ (Norman *et al.*, 1998), but other studies are negative. Single-point linkage analysis of the LEPR gene found significant linkage between a Glu223Arg polymorphism and fat free mass (FFM) in the Quebec family study (Chagnon *et al.*, 1999), although there were no significant differences between lower and higher BMI subjects. Candidate gene association analysis of the leptin gene or its receptor have not provided evidence for population effects of pathogenic genetic variants in these genes (Gotoda *et al.*, 1997; Hinney *et al.* 1998b; Lonnqvist *et al.*, 1999; Oksanen *et al.*, 1997). However, some studies have observed haplotypic association with the leptin receptor which may indicate the presence of as yet uncharacterized causative genetic variants (Reed *et al.*, 1996; Roth *et al.*, 1998; Thompson *et al.*, 1997), and one study found association between a polymorphism in the 5′ untranslated region of the leptin gene and circulating leptin levels (Hager *et al.*, 1998b).

Leptin may also play an indirect role in common human obesity, through physiologically low levels of leptin in some obese individuals (Ravussin *et al.*, 1997) and the development of leptin resistance (Caro *et al.*, 1996), analogous to insulin insufficiency or resistance in diabetes mellitus. Evidence for an indirect genetic basis for this comes from both animal and human studies. A cross between the KK type 2 diabetes-prone mouse and C57BL/6J revealed two QTLs for leptin levels on chromosome 9 and chromosome 7 (Taylor *et al.*, 1999). A human QTL for leptin levels maps to chromosome 2p21 (Comuzzie *et al.*, 1997; Hager *et al.*, 1998b) and analysis of the *POMC* gene, which maps within the 95% confidence interval for linkage, indicates that its polymorphisms may be associated with variation in leptin levels in normal humans (Hixson *et al.*, 1999).

The 'adipostat': integration of orexigenic and anorexigenic pathways in the PVH? The lipostat or adipostat is a classical model for the regulation of food intake which involves the detection of the amount of stored energy in adipose tissue by the CNS, and the use of this information to regulate fat stores at a constant level (Cowley *et al.*, 1999; Kennedy, 1953). There is a body of literature to support the notion of a lipostat. For example, rats held on a restricted diet so that they lose 10–15% of body weight regain this amount when fed *ad libitum* (Mitchel and Keesey, 1977), and the converse applies to rats after rapid weight gain (Steffens, 1975). Two hypotheses have been put forward: the 'set-point' hypothesis and the 'settling point' hypothesis. In the former, alterations in body fat stores will elicit compensatory behaviors and a return to the pre-set level (Kennedy, 1953), whereas in the latter the set-point can be reset by environmental factors, such as increased availability of high-fat food (Bennett, 1995). The etiology of obesity and eating disorders may involve dysfunction of the adiposity set point.

The circuitry for maintaining energy homeostasis should sense the amount of energy stored in adipose tissue, sense and integrate the multiple opposing signals regarding nutritional state and provide output regulating energy intake and expenditure to maintain energy homeostasis. The hypothalamus is the most likely site for the adipostat, since it is the main location for the CNS action of leptin, and its systems for energy homeostasis are highly integrated and elicit robust compensatory responses in feeding behavior and energy expenditure. Anatomical and pharmacological data suggest that hypothalamic integration of NPY, AGRP and POMC signaling may form the cellular basis of the adiposat. Neuropeptide Y and POMC mRNA colocalize within the arctuate nucleus of the hypothalamus (Cheung *et al.*, 1997). The hypothalamus also contains a unique NPY cell, NPY/AGRP neurons, which can activate feeding through postsynaptic NPY receptors and inhibit melanocortin anorexigenic systems (Cowley *et al.*, 1999; Hahn *et al.*, 1998). In the paraventricular nucleus of the hypothalamus (PVH), NPY/AGRP and melanocortin/POMC projections are largely convergent, and neurons exist which receive both α-MSH and NPY projections and integrate the orexigenic and anorexigenic information they provide. Microinjection of the potent MC4R/MC3R agonist MTII into the PVH increases resting oxygen consumption and suppresses the effect of NPY. Furthermore, MTII and NPY have opposing actions on GABA-evoked currents within individual neurons. In the adipostat model, leptin regulates *POMC, AGRP* and *NPY* gene expression in the PVH; their neurons project to GABA interneurons in the PVH, which then regulate GABA release and provide inhibitory input to the hypothetical 'adipostat' neuron. This demonstrates the possibility of a neuroanatomical substrate for cellular integration of opposing physiological inputs to energy homeostasis, which fulfils the three criteria for the adipostat: sensing adipose mass, integrating opposing nutritional signals, and providing output regulating energy intake and expenditure.

Adipose tissue and lipid metabolism. Adipose tissue is designed to act as an energy store for use during fasting. This important mechanism has evolved to allow humans to survive the intermittent food supplies and frequent prolonged famine during evolution; humans can live for months without food. The regulation of adipose cell differentiation and fatty acid metabolism is a key feature of this process, governing energy metabolism and fatty acid signaling (Bernlohr *et al.*, 1999).

Fatty acids are used in adipose tissue to regulate transcriptional control, membrane structure, cellular metabolism and energy storage. The trafficking of fatty acids is under the control of proteins involved in transport, intracellular lipid binding and control at the level of transcription. Fatty acids are transported to adipose cells as lipoprotein-bound triglycerides circulating in the plasma, from where they cross the adipocyte plasma membrane after binding to serum albumin, probably through an active transporter such as fatty acid translocase (FAT) or transport protein (FATP) (*Figure 2*). Inside the cell, fatty acids are bound by abundant intracellular lipid binding proteins, which make up 1–3% of soluble adipose tissue protein.

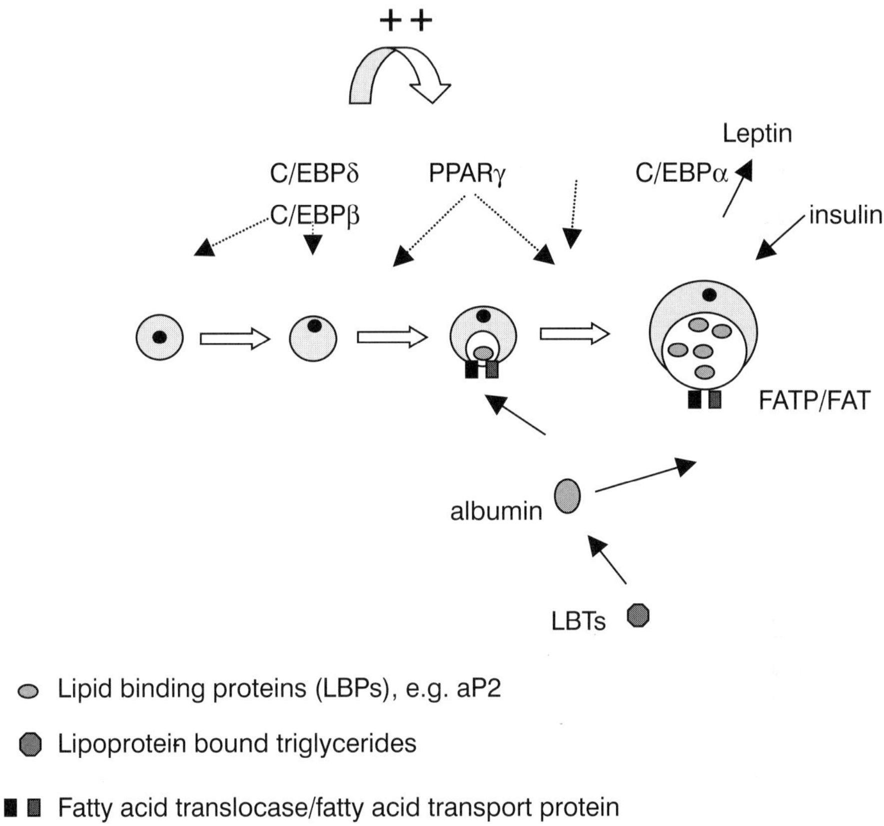

Figure 2. Adipocyte differentiation and fatty acid uptake. The process of adipocyte differentiation begins with the induction of C/EPBβ and δ expression by unknown signals, which may include glutocorticoids and growth factors. Subesequent strong induction of PPARγ expression by C/EPBβ and δ, as well as PPARγ ligands, leads to positive feedback of PPARγ and increase in expression of C/EPBα, maintaining both at high levels and leading to differentiation of mature adipocytes. As differentiation occurs, expression of fatty acid binding protein genes such as aP2 increases, and fatty acids are brought into the cell by fatty acid translocase (FAT) and fatty acid transport protein (FATP). These mature adipocytes are insulin sensitive and produce leptin. PPARγ mutations and polymorphisms in humans are associated with reduced leptin levels for a given BMI, obesity and lower BMI and improved insulin sensitivity. The role of PPARγ in the process is described in Lowell (1999).

A central system in the regulation of fatty acids and adipose tissue are the peroxisome proliferator-activated receptors (PPARs), which, together with the retinoid X receptor, activate many genes by binding to peroxisome proliferator responsive elements (PPREs). There are three isoforms, α, β and γ, with α and γ predominantly found in adipose tissue. Regulated genes involved in fatty acid metabolism include the transport proteins FAT, FATP and the adipocyte lipid binding protein aP2 (Bernlohr *et al.*, 1999).

Adipocyte differentiation, C/EBPα and PPARγ. An understanding of the mechanisms which lead to the differentiation of pre-adipocytes and stem cells to adipocytes is essential to the understanding of the mechanisms of obesity, since adipocytes are not passive storage vesicles for fat but rather play an active role in complex signaling pathways. The process of unraveling adipocyte differentiation began in the 1970s when it was found that differentiated adipocytes possess many mRNA transcripts, such as the fatty acid-binding protein aP2, CCAAT/enhancer binding protein C/EBPα and the peroxisome proliferator-activated receptor γ (PPARγ), which are not seen in preadipocytes (Djian *et al.*, 1985). During the early stages of the differentiation process, C/EBPβ and C/EBPδ expression induces PPARγ and C/EBPα, which induces adipocytes to terminal differentiation (*Figure 2*). Adipose tissue mass in transgenic mice lacking both C/EBPβ and C/EBPδ is normal but reduced by 70% (Tanaka *et al.*, 1997), but aP2, PPARγ and C/EBPα expression in the remaining tissue is normal, indicating that other factors, such as the sterol transcription factor ADD-1/SREBP-1 (Kim *et al.*, 1998; Tontonoz *et al.*, 1993; Wang *et al.*, 1995) must also induce their expression. It is cleaved upon activation (for example, by low cholesterol levels) and probably acts by promoting synthesis of endogenous ligands for PPARγ, with which it has a synergistic effect on adipocyte differentiation.

Transgenic mice lacking C/EBPα fail to develop normal adipose tissue, but instead have adipocytes which lack triglyceride and fail to become insulin sensitive (Wang *et al.*, 1995). Failure of PPARγ induction in these mice indicates that while C/EBPα is not required for adipogenesis, it is required for the fully functional differentiated adipocyte. Genetic studies of PPARγ in mice indicate that it has a more profound function in adipogenesis, however. Because of placental lethality of PPARγ null mice, both placental rescue (Barak *et al.*, 1999) and chimeric null mice (Rosen *et al.*, 1999) have been generated. Both of these studies, combined with analyses of embryos and ES cells (Kubota *et al.*, 1999; Rosen *et al.*, 1999) confirm that while lack of PPARγ does not affect expression of C/EBPβ and C/EBPδ, levels of C/EBPα are low, both aP2 and leptin expression are absent, and there is no lipid accumulation. Thus, PPARγ is required for adipogenesis *in vivo* (Lowell, 1999).

PPARγ is thus a key protein in mature adipocytes, but its exact function is not fully known. It forms a heterodimeric protein complex with the retinoid X receptor (RXR), forming the AFR6 protein complex which interacts with peroxisome proliferator responsive elements (PPREs or adipocyte response elements) in the promoter of a variety of genes (Cowherd *et al.*, 1999). The protein has two alternatively sliced isoforms, γ1 and γ2 (Fajas *et al.*, 1999), the latter of which has an additional 28 amino acids which may enhance ligand binding by fatty acids and eicosanids such as prostaglandin J2. Transcriptional activity of PPARγ is induced by natural ligands, including prostaglandin J2.

A major role for PPARγ, along with C/EBPα, is in leptin gene expression. While C/EBPα is a potent transactivator of leptin expression (He, 1995), PPARγ inhibits leptin expression without binding to the promoter (Hollenberg, 1997), possibly by competing for shared limiting cofactors. Transgenic mice heterozygous for PPARγ are partially resistant to diet-induced obesity and are strongly protected against insulin resistance (Kubota *et al.*, 1999). The mechanism for this appears to be release of leptin gene expression from inhibition by PPARγ, thus raising leptin levels, reducing appetite, increasing energy expenditure and increasing insulin sensitivity. A twist to this tale is provided by drugs such as thiazolidinediones, which improve insulin resistance by *simulating* PPARγ. This apparent paradox may be explained by the ability of PPARγ–TZD interaction to induce adipogenesis, and hypertrophy of mature adipocytes resulting from reduced expression of PPARγ. Both of these have the effect of reducing the number of large adipocytes and consequently ameliorating insulin resistance (Lowell, 1999).

The central role of PPARγ and other PPARs in adipocyte function has resulted in several efforts to associate genetic variation in these genes with disease (Kersten *et al.*, 2000), and has led to the identification of alleles which may have a significant impact on BMI and insulin sensitivity. Meirhaeghe *et al.* (1998) analyzed a silent C→T polymorphism in the PPARγ gene in a population from Northern France and found that the leptin level for a given BMI was lower in obese subjects with at least one T allele, indicating that other (functional) alleles in *PPARγ* may modify the relationship between leptin and BMI. Ristow *et al.* (1998) found such an allele (Pro115Glu), which leads to defective phosphorylation of a serine at position 114, in about 5% of obese German subjects, but in none of the controls. Deeb *et al.* (1998) found a relatively common (~12%) Pro12Ala substitution in the gene which was associated with lower BMI and improved insulin sensitivity in middle-aged and elderly Finns. This variant was also functional, since the Ala12 allele showed decreased binding activity to cognate promoter elements and reduced ability to transactivate responsive promoters. A role for PPARγ is supported by Cole *et al.* (2000) who found associaton of this polymorphism with measures of fatness in Mexican-Americans. However, other studies do not find evidence for a role of PPARγ in morbid obesity (Evans *et al.*, 2000), but instead it appears important in the control of insulin sensitivity, glucose homeostasis and blood pressure, since dominant negative mutations in humans cause type II diabetes and hypertension (Barroso *et al.*, 1999).

Adrenoreceptors. Substantial effort has focused on the genetics of adrenoreceptors in obesity because of their role in adipocyte differentiation, energy mobilization and utilization (Arner and Hoffstedt, 1999; Lafontan *et al.*, 1998). The lipolytic response of subcutaneous adipocytes to catecholamines is modulated by a functional balance between inhibitory α2- and stimulatory β-adrenoreceptors (Arner and Hoffstedt, 1999; Lafontan and Berlan, 1995), which act via cAMP production to modulate hormone-sensitive lipase (Holm *et al.*, 1997). All three known β-adrenoreceptors can modulate lipolysis in human adipose tissue (Barbe *et al.*, 1996; Enocksson *et al.*, 1995). In addition, there is evidence that β-adrenoreceptors can increase metabolic rate (Connacher *et al.*, 1992; Lamont *et al.*, 1997; Yamashita *et al.*, 1994), at least in part through inceasing the expression of the uncoupling proteins (UCPs) (Riquier *et al.*, 1986).

A functional polymorphism (Trp64Arg) in the β3-adrenoreceptor (ADRB3) gene has been associated with both obesity and insulin phenotypes (Walston *et al.*, 1995; Widen *et al.*, 1995) in several studies. However, the number of negative findings in other studies have led to skepticism of its role in obesity (Mauriege and Bouchard, 1996). Furthermore, two recent meta-analyses provided conflicting results, with one finding association with BMI across diverse populations (Fujisawa *et al.*, 1998) and the other not (Allison *et al.*, 1998). Epigenetic effects are possible: Fogelholm *et al.* (1998) found an interaction between polymorphisms in the brown adipose tissue-expressed *UCP-1* gene and the *ADRB3* Trp64Arg polymorphism and the rate of weight gain after a very low calorie diet. Support for a role of the *ADRB3* gene in obesity comes from two other areas, functional genomics and linkage analysis. Studies of the effect of the Arg64 allele on lipolysis indicate that it is associated with lower activity and decreased agonist sensitivity (Hoffstedt *et al.*, 1999; Umekawa *et al.*, 1999), and a third study reported an effect on ligand-mediated cAMP accumulation (Pietri-Rouxel *et al.*, 1999). As discussed above, a family study reported strong linkage between BMI and the ADRB3 locus on chromosome 8p (Mitchell *et al.*, 1999). Overall, there is convincing but not compelling evidence that the β3AR locus is involved in obesity. However a direct role for the Trp64Arg polymorphism is less certain. Association with this polymorphism could represent the effect of other alleles in the gene or promoter, since additional ADRB3 variants in linkage disequilibrium with Trp64Arg have been identified (Hoffstedt *et al.*, 1999), or a gene nearby which also plays a role in weight homeostasis (Mitchell *et al.*, 1999).

In the β2 adrenoreceptor (*ADRB2*) gene there are two common variants, Arg16Gly and Gln27Glu which alter receptor function in recombinant cell lines, with the Gly16 allele in particular showing a five-fold-higher agonist sensitivity (Green *et al.*, 1995). The Gln27Glu polymorphism has associated with a relative risk for obesity of about 7 in a Swedish population (Large *et al.*, 1997) and has been replicated in Japan (Ishiyama-Shigemoto *et al.*, 1998). However, some studies have failed to find association (Echwald *et al*, 1998) or linkage (Takami *et al.*, 1999). Meirhaeghe *et al.* (1999) examined this polymorphism in a Northern French population and found that the Gln27Glu polymorphism had no effect in men who took regular physical exercise, but in men homozygous for the Gln27 variant who did not take exercise there was an association with body weight, BMI and WHR. The Arg16Gly polymorphism was initially not associated with obesity (Large *et al.*, 1997), but subsequent analysis in Japan (Sakane *et al.*, 1997) found that reduction in body weight was greater in women with the Gly16 allele after a combined calorie reduction and exercise regime. Two ADR2B 5′-leader cistron polymorphisms have also been identified, one of which alters cysteine to arginine in the leader peptide (Yamada *et al.*, 1999). These were associated with greater BMI and serum triglycerides in Japanese subjects. More recently, a three-amino-acid deletion from a run of 12 glutamic acid repeats in the alpha-2B gene (ADRB2) was identified (Heinonen *et al.*, 1999). This repeat array appears to be important in agonist dependent receptor desensitization and was associated with a 5.6% lower resting metabolic rate. ADR1B polymorphisms have also been identified (Maqbool *et al.*, 1999), but their role in obesity is unknown.

The precise role of β-adrenergic receptors in obesity remains controversial, but Mauriege *et al.* (1999) provide some interesting insights into the regional and gender-induced differences in diet-induced weight loss. They found that moderate weight loss leads to a higher catecholamine-induced lipolytic efficiency, which was associated with increased β2- and decreased α2-receptor sensitivity. These changes were regional, with greater subcutaneous abdominal compared to femoral sensitivity, presumably because α2- and β2-adrenergic receptors are more highly expressed in subcutaneous abdominal fat compared to femoral fat. This reinforces the notion that 'female' femoral fat is more resistant to dieting than 'male' abdominal fat, and may account for some of the male–female differences in associations seen between obesity phenotypes and ADR receptor genes.

Uncoupling proteins, RQ and metabolic rate. Metabolic measures such as RQ and RMR are amenable to genome-wide analysis such as quantitative trait linkage. However, recent advances in the physiology of energy expenditure have led to the identification of good candidate genes for regulation of metabolic rate, such as the mitochondrial uncoupling proteins (UCPs) (Warden, 1999). UCPs were first discovered in the brown fat tissue of hibernating animals. They produce heat by dissociating the reaction between energy production from food and the synthesis of ATP by creating a proton leak in the mitochondrial membrane; this raises metabolic rate and generates heat (Gura, 1998). However, the exact physiological role of the UCPs is not fully known, since, paradoxically, muscle UCP expression increases during starvation along with a shift to the utilization of lipids as fuel (Samec *et al.*, 1998). This may mean that UCPs are involved in lipid fuel handling rather than just thermogenesis, since refeeding on a high-fat diet leads to greater UCP expression in muscle than low-fat refeeding (Samec *et al.*, 1999).

However, only newborn humans have significant brown fat, to which expression of *UCP1* gene is restricted. The discovery of two novel *UCP* genes in humans, *UCP2* (Fleury *et al.*, 1997) and *UCP3* (Boss *et al.*, 1997; Gong *et al.*, 1997; Vidal-Puig *et al.*, 1997), which are widely expressed in muscle, fat, immune tissue and CNS neurons, including the hypothalamus and SCN, led to investigations of the role of these proteins in metabolic rate and obesity. *UCP2* and *UCP3* are only a few kb apart on chromosome 11q13 and most likely arose by gene duplication. Using RMR as a quantitative trait, Bouchard *et al.* (1997) analyzed the UCP2/3 locus by linkage analysis in the Quebec Family Study, and found significant evidence for a role of this locus in RMR, BMI and %FAT. However, linkage in obese type II diabetes patients was negative (Elbein *et al.*, 1997). Detailed analysis of the genes themselves showed association between a UCP2 exon 8 polymorphic variant and RMR, and with BMI when older subjects (>age 45) were analyzed (Walder *et al.*, 1998). One study has replicated this finding twice in South Indian populations (Cassell *et al.*, 1999), but not in British subjects, although in obese women the exon 8 variant was associated with fasting leptin levels. In obese NIDDM patients, however, no association with UCP2 has been found (Otabe *et al.*, 1998; Urhammer *et al.*, 1997), and overall association studies with the Ala55Val polymorphism in UCP2 have been negative (Warden, 1999). Reduction in UCP2 muscle expression of 28%

in obese subjects has been found, suggesting that impaired expression of the protein is a risk factor for obesity (Nordfors *et al.*, 1998). UCP-1 has a polymorphism in the distal promoter (Esterbauer *et al.*, 1998) which may account for up to 20% of interindividual variation in UCP1 expression. This region binds an unknown adipocyte-derived transcription factor. However no effect of this variant on transcription factor binding was seen, nor was any association with obesity found (Schaffler *et al.*, 1999). Recently, transgenic mice overexpressing UCP3 were found to be hyperphagic and lean, indicating that muscle UCP3 can influence metabolic rate and glucose homeostasis (Chapman *et al.*, 2000).

Other systems; galanin and orexins. Many other genes encoding peptides, receptors and transport systems are involved in regulating feeding, such as galanin, the orexins (hypocretins) and apolipoproteins, and are not considered in detail within this review. Some of these proteins, although shown to regulate feeding, have different primary functions. For example, the orexins are involved in narcolepsy and regulating the sleep-wake cycle (Hagan *et al.*, 1999; Siegel 1999), and their role in circadian and short-term aspects of feeding may be secondary to this. Galanin appears to be important in relation to fat intake; although serum levels of galanin do not vary between obese, normal and lean individuals (Invitti *et al.*, 1995), galanin expression is specifically induced by a high-fat diet (Leibowitz *et al.*, 1998) and may be involved in the development of obesity on a high fat diet. Battleboro rats, which overexpress galanin 1 receptors in the hypothalamus, have a dietary preference for fat at 45% of daily consumption (Odorizzi *et al.*, 1999). Expression of galanin receptors in the paraventricular nucleus of the hypothalamus indicates that they may also participate in the adipostat (Gundlach and Burazin, 1998).

5. Genetic findings in eating disorders

The molecular genetics of eating disorders (anorexia nervosa, bulimia nervosa and binge eating disorder) is in its infancy. Unlike obesity, the eating disorders have been regarded as psychiatric disorders and until recently, significant heritability was dismissed. However, there is increasingly substantial evidence in favor of heritability and for measurable biological traits in the eating disorders which might lead to the identification of genetic susceptibility factors.

5.1 Trait or state? Biological markers for eating disorders

Serotonin (5-HT) has been implicated in the regulation of eating behaviors, meal size and body weight, and in particular in regulating carbohydrate preference. Catecholamine neurons are under dietary control (Wurtman *et al.*, 1980); carbohydrate ingestion stimulates the release of hypothalamic 5-HT, which serves to control the size of carbohydrate-rich meals (Leibowitz and Alexander, 1998), and postsynaptic 5-HT receptor agonists have the same effect (Blundell *et al.*, 1995; Simansky, 1996). Tryptophan depletion decreases brain serotonin (Sahakian *et al.*, 1979) and moderate dieting triggers 5-HT2c receptor supersensitivity (Cowen *et al.*, 1996), which may be a response to falling tryptophan and brain 5-HT levels.

Mice lacking the 5-HT2c receptor develop chronic hyperphagia and late-onset obesity (Nonogaki *et al.*, 1998) by a mechanism which is independent of leptin.

Studies in animals and humans suggest a role for altered monoamine transmission in eating disorders (Morley and Blundell, 1988). People with BN have neuroendocrine abnormalities (Gwirtsman *et al.*, 1983), and blunted prolactin response after the administration of either 5-HT (Goldbloom *et al.*, 1996) or the 5-HT receptor agonist m-CPP (Brewerton *et al.*, 1992; Levitan *et al.*, 1997) and d-fenfluramine (Jimerson *et al.*, 1997). A disorganized behavioral response to mCPP has also been reported (Kaye *et al.*, 1998).

Overall, the evidence suggests that lowered brain 5-HT neurotransmission contributes to the pathology of BN. This reduction appears to be a trait rather than a state-associated index of disordered eating, since symptomatic relapse, in the form of lowering of mood, loss of dietary restraint and increased concern for body image, is seen after acute tryptophan depletion in BN (Weltzin *et al.*, 1995; Smith *et al.*, 1999).

In women with AN, neuroendocrine abnormalities are less clear cut, but may also involve serotonin and dopamine. Women with AN, particularly those with the restricting subtype, show stereotyped and hyperactive motor behavior (Beumont *et al.*, 1994; Kron *et al.*, 1978), reduced motivation and reward, and reduced novelty seeking (Sohlberg and Strober, 1994) consistent with the neuropharmacology of dopamine. Mice lacking tyrosine hydroxylase in dopaminergic neurons are born normally, but do not eat enough food to survive (Szczypka *et al.*, 1999) and dopamine is required for hyperphagia in ob/ob mice (Sczcypka *et al.*, 2000). In AN, homovanillic acid, a major metabolite of dopamine, is decreased in the CSF of underweight subjects (Jimerson *et al.*, 1992) and in recovered restricting AN subjects (Kaye *et al.*, 1999). Abnormalities of noradrenergic activity have also been observed (Prike, 1996).

5.2 Animal models of eating disorders

Unlike obesity, there are no ideal animal models of eating disorders. While there are genetic models of hypophagia and leanness, such as transgenic mice lacking melanin concentrating hormone, these show no gender bias, as seen for eating disorders. However, mouse and animal breeding experiments have provided two useful models which might provide insight into the etiology of eating disorders, the anorexia (*anx*) mouse and thin sow syndrome.

The *anx* mouse is an autosomal recessive lethal mutation, characterized by an emaciated appearance, abnormal behavior including head weaving and body tremors, uncoordinated gait and hyperactivity (Maltais *et al.*, 1984). Mice die because they fail to eat enough to survive. While the *anx* mouse has features of a neurological disorder, there are some interesting effects on feeding, especially involving serotonin, which is thought to be involved in suckling response and the emerging control of food intake (Spear and Ristine, 1982). The mouse anx locus maps to chromosome 2, in a region syntenic with human chromosome 15q15. Neurobiological analysis of these mice reveals drastic and selective serotonergic hyperinnervation of the hypothalamus (Son *et al.*, 1994), which may account for the feeding phenotype of these mice. In particular, these mice appear to have altered processing of neuropeptide Y, with increased expression in arcuate cell bodies, but decreased expression in the arcuate nucleus (Broberger *et al.*, 1997). Jahng *et al.* (1998) also observed a failure to upregulate NPY in these mice in comparison with wild-type littermates and heterozygotes, and found that this effect was paralleled by serotonergic hyperinnervation of the arcuate nucleus. These mice also have other neurochemical abnormalities, including fewer POMC neurons and decreased NPY Y1, Y2 and Y5 receptor expression, indicating that degeneration of the actuate POMC system is a feature of the *anx* mouse (Broberger *et al.*, 1999). Jahng *et al.* (1998b) found that the *anx* mutation and 24-h food deprivation of wild type mice resulted in differential effects on the expression of *TH, NET*, and *5-HTT* genes. Decreased *5-HTT* expression in the *anx* RN is consistent with upregulation of serotonergic neurotransmission that may accompany *5-HT* hyperinnervation. The *anx* mutation has not yet been cloned.

A second animal model of anorexia is thin sow syndrome. Pigs bred for leanness can develop irreversible self-starvation and emaciation (Maclean, 1968; Martinsson *et al.*, 1978) and a similar syndrome can be seen in sheep and goats. The prevalence is from 6–30% (reviewed in Owen and Treasure, 1997). Like human anorexia nervosa, thin sow syndrome appears to have a psychosocial aspect, with early maternal separation and mixing with unfamiliar animals acting as a trigger. The subsequent bullying as a new social order is established results in some animals losing their appetite. There is also a genetic component, since some strains are more vulnerable (Maclean, 1968). The illness may involve the serotonin system, since 5-HT_2 antagonist drugs such as amperozide are reported to prevent and treat the symptoms (Kyriakis *et al.*, 1990).

5.3 Candidate gene studies in eating disorders

The molecular genetics of eating disorders is an emerging field. At the time of writing, genome-wide scans for linkage are underway, but have not been completed (Kaye *et al.*, 2000); however, a number of candidate gene studies have been performed. These have focused on two areas – the examination of genes implicated in obesity, on the basis that many women with eating disorders (especially BN and BED) have a positive family history, and genes involved in serotoninergic neurotransmission, because of their involvement in dietary restraint and some of the behaviors associated with eating disorders (such as obsessive compulsive symptoms and perfectionism). No associations have been reported with genes primarily involved in controlling feeding behavior.

Serotoninergic candidate genes in eating disorders. A variety of serotonin neurotransmitters have been examined in eating disorders, particularly 5-HT_{2C} and 5-HT_{2A}. Studies in both BN and AN found no evidence for involvement of polymorphisms in the *5-HT_{2C}* gene (Lentes *et al.*, 1997; Burnet *et al.*, 1999), including a Cys23Ser variant (Lappalainen *et al.*, 1995) which may have functional consequences for the receptor (Lappalainen *et al.*, 1999; Quested *et al.*, 1999). Analysis of the *5-HT_{2A}* gene is more interesting; Collier *et al.* (1997) reported association between a silent polymorphism in the promoter of the *5-HT_{2A}* gene and AN (-1348G/A) in which the A allele appeared to confer increased risk under a recessive model. This finding was subsequently replicated in AN and control populations from the USA and Italy (Enoch *et al.*, 1998; Sorbi *et al.*, 1998), but not in UK and German subjects (Campbell *et al.*, 1998; Hinney *et al.*, 1998). A meta-analysis of all published data (Ziegler *et al.*, 1999) showed no overall statistically significant association with AN and the $5\text{-}HT_{2A}$ gene, although a positive odds ratio of 1.26 (1.00–1.90) was observed. Differences between samples may be accounted for by diagnosis; strongest association was seen with the restricting subtype of AN in one study; AN usually begins as a restricting illness and then a proportion of cases switch to the binge/purging subtype. A three- to five-year period may be needed to be sure of correctly defining anorexia into its two subtypes, and most studies have not followed subjects for this length of time.

6. Conclusions

The genetics of obesity has moved at an astonishing pace over the last five years. This has been driven by the use of molecular and cell biology to clone and characterize the genes involved in regulation of body weight. Studies in transgenic mice have provided insight into the consequences of the loss of many components of the feeding system, with the majority of mouse knockouts, such as leptin, the leptin receptor and melanocortin receptors, showing obesity. Detailed analysis of the expression and pharmacology of these genes has shown how multiple input signals on the state of nutrition can be integrated to provide a coherent output, through the hypothetical 'adipostat' in the hypothalamus (see *Figure 1*). However, the current model of the adipostat is only partial, since many components and additional regulatory mechanisms, such as corticotrophin releasing hormone and galanin, need to be incorporated.

Many of the polymorphisms and mutations associated with human obesity either lie within this emerging CNS adipostat pathway, are proteins regulated within adipose tissue, such as PPARγ and adrenoreceptors, or are directly involved in energy expenditure, such as the UCPs. All of these systems are linked through a complex interdependent regulatory network governed by the CNS. The prospect of an almost complete understanding of the physiology of feeding and adiposity is a tempting prospect. The real challenge will be to use this information to treat obesity. While it is likely that highly effective anti-obesity drugs will be developed based on genetics and physiology, the long-term answer must lie in re-educating people in their eating habits in populations who have become accustomed to freely available high-fat food.

Eating disorders are a more complex problem. AN and BN have been regarded as disorders with a psychosocial origin and have largely been treated with psychotherapy. However, this has, at best, only been partially successful. The recognition that eating disorders have a heritable component and are biological in origin has catalyzed a new interest in molecular genetic analysis; however, this work is in its infancy. Eating disorders are psychiatric illnesses which rely in part on the patients' descriptions of their own behavior and, consequently, they cannot be defined simply. As seen with schizophrenia, unraveling the genetics of eating disorders might be a long and difficult process. However, the field does promise the advent of novel pharmacological therapies, which are urgently needed to reduce mortality.

References

Allison, D.B., Heshka, S., Neale, M.C., Lykken, D.T. and Heymsfield, S.B. (1994) A genetic analysis of relative weight among 4020 twin pairs, with an emphasis on sex effects. *Health Psychol.* **13**: 362–365.

Allison, D.B., Kaprio, J., Korkeila, M., Koskenvuo, M., Neale, M.C. and Hayakawa, K. (1996) The heritability of body mass index among an international sample of monozygotic twins reared apart. *Int. J. Obes. Relat. Metab. Disord.* **20**: 501–506.

Allison, D.B., Heo, M., Faith, M.S. and Pietrobelli, A. (1998) Meta-analysis of the association of the Trp64 Arg polymorphism in the beta3 adrenergic receptor with body mass index. *Int. J. Obes. Relat. Metab. Disord.* **22**: 559–566.

Arner, P. and Hoffstedt, J. (1999) Adrenoceptor genes in human obesity. *J. Intern. Med.* **245**: 667–672.

Astrup, A. (1998) The American paradox: the role of energy-dense fat-reduced food in the increasing prevalence of obesity. *Curr. Opin. Clin. Nutr. Metab. Care* **1**: 573–577.

Barak, Y., Nelson, M.C., Ong, E.S., Jones, Y.Z., Ruiz-Lozano, P., Chien, K.R., *et al.* (1999) PPAR gamma is required for placental, cardiac, and adipose tissue development. *Mol. Cell.* **4**: 585–595.

Barbe, P., Millet, L., Galitzky, J., Lafontan, M. and Berlan, M. (1996) In situ assessment of the role of the β1-, β2- and β3-adrenoceptors in the control of lipolysis and nutritive blood flow in human subcutaneous adipose tissue. *Br. J. Pharmacol.* **117**: 907–913.

Barreca, A., Valli, B., Cesarone, A., Arvigo, M., Balasini, M., Battista La Sala, G., *et al.* (1998) Effects of the neuropeptide Y on estradiol and progesterone secretion by human granulosa cells in culture. *Fertil. Steril.* **70**: 320–325.

Barroso, I., Gurnell, M., Crowley, V.E.F., Agostini, M., Schwabe, J.W., Soos, M.A., *et al.* (1999) Dominant negative mutations in human PPARγ associated with severe insulin resistance, diabetes mellitus and hypertension. *Nature* **402**: 880–883.

Bennett, W.I., (1995) Beyond overeating, *N.Engl. J. Med.* **332**: 673 674.

Bernlohr, D.A., Coe, N.R. and LiCata, V.J. (1999) Fatty acid trafficking in the adipocyte. *Semin. Cell Dev. Biol.* **10**: 43–49.

Beumont, P.J., Arthur, B., Russell, J.D. and Touyz, S.W. (1994) Excessive physical activity in dieting disorder patients: proposals for a supervised exercise program. *Int. J. Eat. Disord.* **15**: 21–36.

Blomqvist, A.G. and Herzog, H. (1997) Y-receptor subtypes – how many more? *Trends Neurosci.* **20**: 294–298.

Blundell, J.E., Lawton, C.L. and Halford, J.C. (1995) Serotonin, eating behavior, and fat intake. *Obes. Res.* **3**: (Suppl. 4) 471S–476S.

Borecki, I.B., Blangero, J., Rice, T., Perusse, L., Bouchard, C. and Rao, D.C. (1998) Evidence for at least two major loci influencing human fatness. *Am. J. Hum. Genet.* **63**: 831–838.

Boss, O., Samec, S., Paoloni-Giacobino, A., Rossier, C., Dulloo, A., Seydoux, J. *et al.* (1997) Uncoupling protein-3: a new member of the mitochondrial carrier family with tissue-specific expression. *FEBS Lett.* **408**: 39–42.

Bouchard, C. and Tremblay, A. (1997) Genetic influences on the response of body fat and fat distribution to positive and negative energy balances in human identical twins. *J. Nutr.* **127**(5 Suppl.): 943S–947S.

Bouchard, C., Tremblay, A., Nadeau, A., Despres, J.P., Theriault, G., Boulay, M.R., *et al.* (1989) Genetic effect in resting and exercise metabolic rates. *Metabolism* **38**: 364–370.

Bouchard, C., Tremblay, A., Despres, J.P. *et al.* (1990) The response to long-term overfeeding in identical twins. *N. Engl. J. Med.* **322**: 1477–1482.

Bouchard, C., Perusse, L., Chagnon, Y.C., Warden, C. and Ricquier, D. (1997) Linkage between markers in the vicinity of the uncoupling protein 2 gene and resting metabolic rate in humans. *Hum. Mol. Genet.* **6**: 1887–1889.

Bowden, D.W., Sale, M., Howard, T.D., Qadri, A., Spray, B.J., Rothschild, C.B. *et al.* (1997)

Linkage of genetic markers on human chromosomes 20 and 12 to NIDDM in Caucasian sib pairs with a history of diabetic nephropathy. *Diabetes* **46**: 882–886.

Bray, G.A. (1997a) Overweight is risking fate. Definition, classification, prevalence and risks. *Ann. NY. Acad. Sci.*, **499**: 14–28.

Bray, G.A. (1997b) Obesity and reproduction. *Hum. Reprod.* **12**(Suppl. 1): 26–32.

Brewerton, T.D., Mueller, E.A., Lesem, M.D., Brandt, H.A., Quearry, B., George, D.T. *et al.* (1992) Neuroendocrine responses to m-chlorophenylpiperazine and L-tryptophan in bulimia. *Arch. Gen. Psychiatry* **49**: 852–861.

Broberger, C., Johansen, J., Schalling, M. and Hokfelt, T. (1997) Hypothalamic neurohistochemistry of the murine anorexia (*anx*/*anx*) mutation: altered processing of neuropeptide Y in the arcuate nucleus. *J. Comp. Neurol.* **387**: 124–135.

Broberger, C., Johansen, J., Brismar, H., Johansson, C., Schalling, M. and Hokfelt, T. (1999) Changes in neuropeptide Y receptors and pro-opiomelanocortin in the anorexia (*anx*/*anx*) mouse hypothalamus. *J. Neurosci.* **19**: 7130–7139.

Brook, C.G., Huntley, R.M. and Slack, J. (1975) Influence of heredity and environment in determination of skinfold thickness in children. *Br. Med. J.* **2**: 719–721.

Bulik, C.M., *et al.* (1998) Heritability of binge-eating and broadly defined bulimia nervosa. *Biol. Psychiatry* **44**: 1210–1218.

Bulik, C.M., Sullivan, P.F., Wade, T.D. and Kendler, K.S. (2000) Twin studies of eating disorders: a review. *Int. J. Eat. Disord.* **27**: 1–20.

Bultman, S.J., Michaud, E.J. and Woychik, R.P. (1992) Molecular characterization of the mouse agouti locus. *Cell* **71**: 1195–1204.

Burnet, P.W., Smith, K.A., Cowen, P.J., Fairburn, C.G. and Harrison, P.J. (1999) Allelic variation of the 5-HT2C receptor (HTR2C) in bulimia nervosa and binge eating disorder. *Psychiatr. Genet.* **9**: 101–104.

Bushnell, J.A., Wells, J.E., Hornblow, A.R., Oakley-Browne, M.A. and Joyce, P. (1990) Prevalence of three bulimia syndromes in the general population. *Psychol. Med.* **20**: 671–680.

Campbell, D.A., Sundaramurthy, D., Markham, A.F. and Pieri, L.F. (1998) Lack of association between 5-HT_{2a} gene promoter polymorphism and susceptibility to anorexia nervosa. *Lancet* **351**: 499.

Caro, J.F., Kolaczynski, J.W., Nyce, M.R., Ohannesian, J.P., Opentanova, I., Goldman, W.H., *et al.* (1996) Decreased cerebrospinal fluid/serum leptin ratio in obesity: a possible mechanism for leptin resistance. *Lancet* **348**: 159–161.

Cassell, P.G., Neverova, M., Janmohamed, S., Uwakwe, N., Qureshi, A., McCarthy, M.I. *et al.* (1999) An uncoupling protein 2 gene variant is associated with a raised body mass index but not type II diabetes. *Diabetologia* **42**: 688–692.

Chagnon, Y.C., Chen, W.J., Perusse, L., Chagnon, M., Nadeau, A., Wilkison, W.O. and Bouchard, C. (1997) Linkage and association studies between the melanocortin receptors 4 and 5 genes and obesity-related phenotypes in the Quebec Family Study. *Mol. Med.* **3**: 663–674.

Chagnon, Y.C., Chung, W.K., Perusse, L., Chagnon, M., Leibel, R.L. and Bouchard, C. (1999) Linkages and associations between the leptin receptor (*LEPR*) gene and human body composition in the Quebec Family Study. *Int. J. Obes. Relat. Metab. Disord.* **23**: 278–286.

Chagnon, Y.C., Pérusse, L., Weisnagel, S.J., Rankinen, T. and Bouchard, C. (2000a) The Human Obesity Gene Map: the 1999 update. *Obesity Res.* **8**: 89–117.

Chagnon, Y.C., Borecki, I.B., Pérusse, L., Roy, S., Lacaille, M. and Chagnon, M. *et al.* (2000b) Genes related to the fat-free body mass results from a genome-wide search in the Québec Family Study. *Metabolism* **49**: 491–496.

Cheung, C.C., Clifton, D.K. and Steiner, R.A. (1997) Proopiomelanocortin neurons are direct targets for leptin in the hypothalamus. *Endocrinology* **138**: 4489–4492.

Chua, S.C. Jr, Chung, W.K., Wu-Peng, X.S., Zhang, Y., Liu, S.M., Tartaglia, L. and Leibel, R.L. (1996) Phenotypes of mouse diabetes and rat fatty due to mutations in the OB (leptin) receptor. *Science*. **271**: 994–996.

Clapham, J.C., Arch, J.R.S., Chapman, H. *et al.* (2000) Mice over-expressing human uncoupling protein 3 are hyperphagic and lean. *Nature* **406**: 415–418.

Clement, K., Vaisse, C., Lahlou, N., Cabrol, S., Pelloux, V. and Cassuto, D. *et al.* (1998) A mutation in the human leptin receptor gene causes obesity and pituitary dysfunction. *Nature* **392**: 398–401.

Cnattingius, S., Hultman, C.M., Dahl, M. and Sparen, P. (1999) Very preterm birth, birth trauma, and the risk of anorexia nervosa among girls. *Arch. Gen. Psychiatry* **56**: 634–638.

Cole, S.A., Mitchell, B.D., Hsueh, W.C., Pineda, P., Beamer, B.A. and Shuldiner, A.R. *et al.* (2000) The Pro12Ala variant of peroxisome proliferator-activated receptor-γ2 (PPAR-γ2) is associated with measures of obesity in Mexican Americans. Int. *J. Obes. Relat. Metab. Disord.* **24**: 522–524.

Collier, D.A., Arranz, M.J., Li, T., Mupita, D., Brown, N. and Treasure, J. (1997) Association between 5-HT_{2a} gene promoter polymorphism and anorexia nervosa. *Lancet* **350**: 412.

Comerci, G.D. (1990) Medical complications of anorexia nervosa and bulimia nervosa. *Med. Clin. North. Am.* **74**: 1293–1310.

Comuzzie, A.G. and Allison, D.B. (1998) The search for human obesity genes. *Science* **280**: 1374–1377.

Comuzzie, A.G., Hixson, J.E., Almasy, L., Mitchell, B.D., Mahaney, M.C. and Dyer, T.D. *et al.* (1997) A major quantitative trait locus determining serum leptin levels and fat mass is located on human chromosome 2. *Nat. Genet.* **15**: 273–276.

Cone, R.D., Lu, D., Koppula, S., Vage, D.I., Klungland, H. and Boston, B., *et al.* (1996) The melanocortin receptors: agonists, antagonists, and the hormonal control of pigmentation. *Recent Prog. Horm. Res.* **51**: 287–317.

Connacher, A.A., Bennet, W.M., Jung, R.T. and Rennie, M.J. (1992) Metabolic effects of three weeks administration of the β-adrenoceptor agonist BRL 26830A. *Int. J. Obes. Relat. Metab. Disord.* **16**: 685–694.

Cotrufo, P., Barretta, V., Monteleone, P. and Maj, M. (1998) Full-syndrome, partial-syndrome and subclinical eating disorders: an epidemiological study of female students in Southern Italy. *Acta. Psychiatr. Scand.* **98**: 112–115

Cowen, P.J., Clifford, E.M., Walsh, A.E., Williams, C. and Fairburn, C.G. (1996) Moderate dieting causes 5-HT_{2C} receptor supersensitivity. *Psychol. Med.* **26**: 1155–1159.

Cowherd, R.M., Lyle, R.E. and McGehee, R.E. Jr (1999) Molecular regulation of adipocyte differentiation. *Semin. Cell. Dev. Biol.* **10**: 3–10.

Cowley, M.A., Pronchuk, N., Fan, W., Dinulescu, D.M., Colmers, W. and Cone, R.D. (1999) Integration of NPY, AGRP, and melanocortin signals in the hypothalamic paraventricular nucleus: evidence of a cellular basis for the adiposat. *Neuron* **24**: 155–163.

Crisp, A.H., Callender, J.S., Halek, C. and Hsu, L.K. (1992) Long-term mortality in anorexia nervosa. A 20-year follow-up of the St George's and Aberdeen cohorts. *Br. J. Psychiatry*. **161**: 104–107.

Deeb, S.M., Fajas, L., Nemoto, M., Pihlajamaki, J., Mykkanen, L. and Kuusisto, J. *et al.* (1998) A Pro12Ala substitution in PPARγ2 associated with decreased receptor activity, lower body mass index and improved insulin sensitivity. *Nature Genetics* **20**: 284–287.

Deep, A.L., Lilenfeld, L.R., Plotnicov, K.H., Pollice, C. and Kaye, W.H. (1999) Sexual abuse in eating disorder subtypes and control women: the role of comorbid substance dependence in bulimia nervosa. *Int. J. Eat. Disord.* **25**: 1–10.

Demerath, E.W., Towne, B., Wisemandle, W., Blangero, J., Chumlea,W.C. and Siervogel, E.M. (1999) Serum leptin concentration, body composition, and gonadal hormones during puberty. *Int. J. Obes. Relat. Metab. Disord.* **23**: 678–685.

Dinulescu, D.M., Fan, W., Boston, B.A., McCall, K., Lamoreux, M.L., Moore, K.J., Montagno, J. and Cone, R.D. (1998) Mahogany (mg) stimulates feeding and increases basal metabolic rate independent of its suppression of agouti. *Proc. Natl. Acad. Sci. USA* **95**: 12707–12712.

Djian, P., Phillips, M. and Green, H. (1985) The activation of specific gene transcription in the adipose conversion of 3T3 cells. *J. Cell. Physiol.* **124**: 554–556.

Eagles, J.M., Easton, E.A., Nicoll, K.S., Johnston, M.I. and Millar, H.R. (1999) Changes in the presenting features of females with anorexia nervosa in Northeast Scotland, 1965–1991. *Int. J. Eat. Disord.* **26**: 289–294.

Echwald, S.M. (1999) Genetics of human obesity: lessons from mouse models and candidate genes. *J. Intern. Med.* **245**: 653–666.

Echwald, S.M., Sørensen, T.I., Tybjaerg-Hansen, A., Andersen, T. and Pedersen, O. (1998) Gln27Glu variant of the human β2-adrenoreceptor gene is not associated with early-onset obesity in Danish men. *Diabetes* **47**: 1657–1658.

Echwald, S.M., Sørensen, T.I., Andersen, T., Tybjaerg-Hansen, A., Clausen, J.O. and Pedersen, O.

(1999) Mutational analysis of the proopiomelanocortin gene in Caucasians with early onset obesity. *Int. J. Obes. Relat. Metab. Disord.* **23**: 293–298.

Elbein, S.C., Leppert, M. and Hasstedt, S. (1997) Uncoupling protein 2 region on chromosome 11q13 is not linked to markers of obesity in familial type 2 diabetes. *Diabetes* **46**: 2105–2107.

Elmquist, J.K., Maratos-Flier, E., Saper, C.B. and Flier, J.S. (1998) Unraveling the central nervous system pathways underlying responses to leptin. *Nat. Neurosci.* **1**: 445–450.

Enoch, M., Kaye, W.H., Rotondo, A., Greenberg, B.D., Murphy, D.L. and Goldman, D. (1998) 5-HT_{2a} promoter polymorphism – 1438G/A, anorexia nervosa, and obsessive-compulsive disorder. *Lancet* **351**: 1785–1786.

Enocksson, S., Shimizu, M., Lonnqvist, F., Nordenstrom, J. and Arner, P. (1995) Demonstration of an in vivo functional β3-adrenoceptor in man. *J. Clin. Invest.* **95**: 2239–2245.

Erickson, J.C., Clegg, K.E. and Palmiter, R.D. (1996) Sensitivity to leptin and susceptibility to seizures in NPY deficient mice. *Nature* **381**: 415–421.

Esterbauer, H., Oberkofler, H., Liu, Y.M., Breban, D., Hell, E., Krempler, F. and Patsch, W.J. (1998) Uncoupling protein-1 mRNA expression in obese human subjects: the role of sequence variations at the uncoupling protein-1 gene locus. *Lipid Res.* **39**: 834–844.

Evans, D., Mann, W.A., Heer, J., Michel, U., Wendt, D., Kortner, B., Wolf, A. and Beisiegel, U. (2000) Variation in the gene for human peroxisome proliferator activated receptor γ (PPARγ) does not play a major role in the development of morbid obesity. *Int. J. Obes. Relat. Metab. Disord.* **24**: 647–651.

Fairburn, C.G., Doll, H.A., Welch, S.L., Hay, P.J., Davies, B.A. and O'Connor, M.E. (1998) Risk factors for binge eating disorder: a community-based, case-control study. *Arch. Gen. Psychiatry* **55**: 425–432.

Fairburn, C.G., Cowen, P.J. and Harrison, P.J. (1999a) Twin studies and the etiology of eating disorders. *Int. J. Eat. Disord.* **26**: 349–358.

Fairburn, C.G., Cooper, Z., Doll, H.A. and Welch, S.L. (1999b) Risk factors for anorexia nervosa: three integrated case-control comparisons. *Arch. Gen. Psychiatry* **56**: 468–476.

Fajas, L., Schoonjans, K., Gelman, L., Kim, J.B., Najib, J. and Martin, G. *et al.* (1999) Regulation of peroxisome proliferator-activated receptor γ expression by adipocyte differentiation and determination factor 1/sterol regulatory element binding protein 1: implications for adipocyte differentiation and metabolism. *Mol. Cell. Biol.* **19**: 5495–5503.

Fan, W., Boston, B.A., Kesterson, R.A., Hruby, V.J. and Cone, R.D. (1997) Role of melanocortinergic neurons in feeding and the agouti obesity syndrome. *Nature* **385**: 165–168.

Favaro, A. and Santonastaso, P. (1997) Suicidality in eating disorders: clinical and psychological correlates. *Acta Psychiatr. Scand.* **95**: 508–514.

Fleury, C., Neverova, M., Collins, S., Raimbault, S., Champigny, O. and Levi-Meyrueis, C. *et al.* (1997) Uncoupling protein-2: a novel gene linked to obesity and hyperinsulinemia. *Nature Genet.* **15**: 223–224.

Fogelholm, M., Valve, R., Kukkonen-Harjula, K., Nenonen, A., Hakkarainen, V., Laakso, M. and Uusitupa, M.J. (1998) Additive effects of the mutations in the β3-adrenergic receptor and uncoupling protein-1 genes on weight loss and weight maintenance in Finnish women. *Clin. Endocrinol. Metab.* **83**: 4246–4250.

Fombonne, E. (1996) Is bulimia nervosa increasing in frequency? *Int. J. Eat. Disord.* **19**: 287–296.

Fujisawa, T., Ikegami, H., Kawaguchi, Y. and Ogihara, T.J. (1998) Meta-analysis of the association of Trp64Arg polymorphism of β3-adrenergic receptor gene with body mass index. *Clin. Endocrinol. Metab.* **83**: 2441–2444.

Garnsey, P. (1999) *Food and Society in Classical Antiquity*. Cambridge University Press, Cambridge, UK.

Gibbs, W.W. (1996) Gaining on fat. *Scientific American* **275**: 88–94.

Goldbloom, D.S., Garfinkel, P.E., Katz, R. and Brown, G.M.J. (1996) The hormonal response to intravenous 5-hydroxytryptophan in bulimia nervosa. *Psychosom. Res.* **40**: 289–297.

Gong, D.W., He, Y., Karas, M. and Reitman, M. (1997) Uncoupling protein-3 is a mediator of thermogenesis regulated by thyroid hormone, β3-adrenergic agonists, and leptin. *J. Biol. Chem.* **272**: 24129–24132.

Gotestam, K.G. and Agras, W.S. (1995) General population-based epidemiological study of eating disorders in Norway. *Int. J. Eat. Disord.* **18**: 119–126.

Gotoda, T., Manning, B.S., Goldstone, A.P., Imrie, H., Evans, A.L., Strosberg, A.D., McKeigue,

P.M., Scott, J. and Aitman, T.J. (1997) Leptin receptor gene variation and obesity: lack of association in a white British male population. *Hum. Mol. Genet.* **6**: 869–876.

Graham, M., Shutter, J.R., Sarmiento, U., Sarosi, I. and Stark, K.L. (1997) Overexpression of Agrt leads to obesity in transgenic mice. *Nat. Genet.* **17**: 273–274.

Green, S.A., Turki, J., Bejarano, P., Hall, I.P. and Liggett, S.B. (1995) Influence of β2-adrenergic receptor genotypes on signal transduction in human airway smooth muscle cells. *Am. J. Respir. Cell. Mol. Biol.* **13**: 25–33.

Gu, W., Tu, Z., Kleyn, P.W., Kissebah, A., Duprat, L., Lee, J. *et al.* (1999) Identification and functional analysis of novel human melanocortin-4 receptor variants. *Diabetes* **48**: 635–639.

Gundlach, A.L. and Burazin, T.C. (1998) Galanin-galanin receptor systems in the hypothalamic paraventricular and supraoptic nuclei. Some recent findings and future challenges. *Ann. N.Y Acad. Sci.* **863**: 241–251.

Gunn, T.M., Miller, K.A., He, L., Hyman, R.W., Davis, R.W., Azarani, A., Schlossman, S.F., Duke-Cohan, J.S. and Barsh, G.S. (1999) The mouse mahogany locus encodes a transmembrane form of human attractin. *Nature* **398**: 152–156.

Gura, T. (1998) Uncoupling proteins provide new clue to obesity's causes. *Science* **280**: 1369–1370.

Gwirtsman, H.E., Roy-Byrne, P., Yager, J. and Gerner, R.H. (1983) Neuroendocrine abnormalities in bulimia. *Am. J. Psychiatry* **140**: 559–563.

Hagan, J.J., Leslie, R.A., Patel, S., Evans, M.L., Wattam, T.A., Holmes, S. *et al.* (1999) Orexin A activates locus coeruleus cell firing and increases arousal in the rat. *Proc. Natl Acad. Sci. USA* **96**: 10911–10916.

Hager, J., Dina, C., Francke, S., Dubois, S., Houari, M., Vatin, V. *et al.* (1998a) A genome-wide scan for human obesity genes reveals a major susceptibility locus on chromosome 10. *Nature Genet.* **20**: 304–308.

Hager, J., Clement, K., Francke, S., Dina, C., Raison, J., Lahlou, N. *et al.* (1998b) A polymorphism in the 5′ untranslated region of the human *ob* gene is associated with low leptin levels. *Int. J. Obes. Relat. Metab. Disord.* **22**: 200–205.

Hahn, T.M., Breininger, J.F., Baskin, D.G. and Schwartz, M.W. (1998) Coexpression of Agrp and NPY in fasting-activated hypothalamic neurons. *Nat. Neurosci.* **1**: 271–272.

Hanson, R.L., Ehm, M.G., Pettitt, D.J. *et al.* (1998) An autosomal genomic scan for loci linked to type II diabetics mellitus and body-mass index in Pima Indians. *Am. J. Hum. Genet.* **63**: 1130–1138

Hay, P. and Fairburn, C. (1998) The validity of the DSM-IV scheme for classifying bulimic eating disorders. *Int. J. Eat. Disord.* **23**: 7–15.

He, Y., Chen, H., Quon, M.J. and Reitman, M. (1995) The mouse obese gene. Genomic organization, promoter activity, and activation by CCAAT/enhancer-binding protein alpha. *J. Biol. Chem.* **270**: 28887–28891.

Heinonen, P., Koulu, M., Pesonen, U., Karvonen, M.K., Rissanen, A., Laakso, M., Valve, R., Uusitupa, M. and Scheinin, M. (1999) Identification of a three-amino acid deletion in the α2B-adrenergic receptor that is associated with reduced basal metabolic rate in obese subjects. *J. Clin. Endocrinol. Metab.* **84**: 2429–2433.

Heini, A.F. and Weinsier, R.L. (1997) Divergent trends in obesity and fat intake patterns: the American paradox. *Am. J. Med.* **102**: 259–264.

Hill, J.O. and Peters, J.C. (1998) Environmental contributions to the obesity epidemic. *Science* **280**(5368): 1371–1374.

Hinney, A., Ziegler, A., Nothen, M.M., Remschmidt, H. and Hebebrand, J. (1997) 5-HT_{2A} receptor gene polymorphisms, anorexia nervosa, and obesity. *Lancet* **350**: 1324–1325.

Hinney, A., Becker, I., Heibult, O., Nottebom, K., Schmidt, A., Ziegler, A. (1998a) Systematic mutation screening of the pro-opiomelanocortin gene: identification of several genetic variants including three different insertions, one nonsense and two missense point mutations in probands of different weight extremes. *J. Clin. Endocrinol. Metab.* **83**: 3737–3741.

Hinney, A., Bornscheuer, A., Depenbusch, M., Mierke, B., Tolle, A., Middeke, K. *et al.* (1998b) No evidence for involvement of the leptin gene in anorexia nervosa, bulimia nervosa, underweight or early onset extreme obesity: identification of two novel mutations in the coding sequence and a novel polymorphism in the leptin gene linked upstream region. *Mol. Psychiatry* **3**: 539–543.

Hinney, A., Schmidt, A., Nottebom, K., Heibult, O., Becker, I., Ziegler, A. *et al.* (1999) Several mutations in the melanocortin-4 receptor gene including a nonsense and a frameshift mutation associated with dominantly inherited obesity in humans. *J. Clin. Endocrinol. Metab.* **84**: 1483–1486.

Hixson, J.E., Almasy, L., Cole, S., Birnbaum, S., Mitchell, B.D., Mahaney, M.C. (1999) Normal variation in leptin levels in associated with polymorphisms in the proopiomelanocortin gene, *POMC. J. Clin. Endocrinol. Metab.* **84**: 187–191.

Hoek, H.W., Bartelds, A.I., Bosveld, J.J., van der Graaf, Y., Limpens, V.E., Maiwald, M. and Spaaij, C.J. (1995) Impact of urbanization on detection rates of eating disorders. *Am. J. Psychiatry* **152**: 1272–1278.

Hoek, H.W., van Harten, P.N., van Hoeken, D. and Susser, E. (1998) Lack of relation between culture and anorexia nervosa – results of an incidence study on Curacao. *N. Engl. J. Med.* **338**: 1231–1232.

Hoffstedt, J., Poirier, O., Thorne, A., Lonnqvist, F., Herrmann, S.M., Cambien, F. and Arner, P. (1999) Polymorphism of the human beta3-adrenoceptor gene forms a well-conserved haplotype that is associated with moderate obesity and altered receptor function. *Diabetes* **48**: 203–205.

Hollenberg, A.N., Susulic, V.S., Madura, J.P., Zhang, B., Moller, D.E., Tontonoz, P., Sarraf, P., Spiegelman, B.M. and Lowell, B.B. (1997) Functional antagonism between CCAAT/Enhancer binding protein-α and peroxisome proliferator-activated receptor-γ on the leptin promoter. *J. Biol. Chem.* **272**: 5283–5290.

Holm, C., Langin, D., Manganiello, V., Belfrage, P. and Degerman, E. (1997) Regulation of hormone-sensitive lipase activity in adipose tissue. *Methods Enzymol.* **286**: 45–67.

Huszar, D., Lynch, C.A., Fairchild-Huntress, V., Dunmore, J.H., Fang, Q., Berkemeier, L.R., *et al.* (1997) Targeted disruption of the melanocortin-4 receptor results in obesity in mice. *Cell* **88**: 131–141.

Invitti, C., Brunani, A., Pasqualinotto, L., Dubini, A., Bendinelli, P., Maroni, P. and Cavagnini, F. (1995) Plasma galanin concentrations in obese, normal weight and anorectic women. *Int. J. Obes. Relat. Metab. Disord.* **19**: 347–349.

Ishiyama-Shigemoto, S., Yamada, K., Yuan, X., Ichikawa, F. and Nonaka, K. (1999) Association of polymorphisms in the beta 2-adrenergic receptor gene with obesity, hypertriglyceridaemia, and diabetes mellitus. *Diabetologia* **42**: 98–101.

Jackson, R.S., Creemers, J.W., Ohagi, S., Raffin-Sanson, M.L., Sanders, L., Montague, C.T., Hutton, J.C. and O'Rahilly, S. (1997) Obesity and impaired prohormone processing associated with mutations in the human prohormone convertase 1 gene. *Nat. Genet.* **16**: 303–306.

Jahng, J.W., Houpt, T.A., Kim, S.J., Joh, T.H. and Son, J.H. (1998a) Neuropeptide Y mRNA and serotonin innervation in the arcuate nucleus of anorexia mutant mice. *Brain Res* **790**: 67–73.

Jahng, J.W., Houpt, T.A., Joh, T.H. and Son, J.H. (1998b) Differential expression of monoamine oxidase A, serotonin transporter, tyrosine hydroxylase and norepinephrine transporter mRNA by anorexia mutation and food deprivation. *Brain. Res. Dev. Brain Res.* **107**: 241–246.

Jequier, E. and Schutz, Y. (1988) Energy expenditure in obesity and diabetes. *Diabetes Metab. Rev.* **4**: 583–593.

Jequier, E. (1999) Response to and range of acceptable fat intake in adults. *Eur. J. Clin. Nutr.* **53** (Suppl 1): S84–88, discussion S88–93.

Ji, L., Malecki, M., Warram, J.H., Yang, Y., Rich, S.S. and Krolewski, A.S. (1997) New susceptibility locus for NIDDM is localized to human chromosome 20q. *Diabetes* **46**: 876–881.

Jimerson, D.C., Lesem, M.D., Kaye, W.H. and Brewerton, T.D. (1992) Low serotonin and dopamine metabolite concentrations in cerebrospinal fluid from bulimic patients with frequent binge episodes. *Arch. Gen. Psychiatry* **49**: 132–138.

Jimerson, D.C., Wolfe, B.E., Metzger, E.D., Finkelstein, D.M., Cooper, T.B. and Levine, J.M. (1997) Decreased serotonin function in bulimia nervosa. *Arch. Gen. Psychiatry* **54**: 529–534.

Karwautz, A., Rabe-Hesketh, S., Hu, X., Zhao, J.H., Sham, P., Collier, D.A. and Treasure, J.L. (2000) Individual-specific risk factors for anorexia nervosa: a pilot study using a discordant sister-pair design. *Psych. Med.* (in press).

Karvonen, M.K., Pesonen, U., Koulu, M., Niskanen, L., Laakso, M., Rissanen, A. *et al.* (1998) *Nat. Med.* **4**: 1434–1437.

Kaye, W., Gendall, K. and Strober, M. (1998) Serotonin neuronal function and selective serotonin reuptake inhibitor treatment in anorexia and bulimia nervosa. *Biol. Psychiatry* **44**: 825–838.

Kaye, W.H., Frank, G.K. and McConaha, C. (1999) Altered dopamine activity after recovery from restricting-type anorexia nervosa. *Neuropsychopharmacology* **21**: 503–506.

Kaye, W.H., Lilenfeld, L.R., Berrettini, W.H. *et al.* (2000) **47**: 794–803.

Keel, P.K., Mitchell, J.E., Miller, K.B., Davis, T.L. and Crow, S.J. (1999) Long-term outcome of bulimia nervosa. *Arch. Gen. Psychiatry* **56**: 63–69.

Kendler, K.S., MacLean, C., Neale, M., Kessler, R., Heath, A. and Eaves, L. (1991) The genetic epidemiology of bulimia nervosa. *Am. J. Psychiatry* **148**: 1627–1637.

Kendler, K.S., Walters, E.E., Neale, M.C., Kessler, R.C., Heath, A.C. and Eaves, L.J. (1995) The structure of the genetic and environmental risk factors for six major psychiatric disorders in women. Phobia, generalized anxiety disorder, panic disorder, bulimia, major depression, and alcoholism. *Arch. Gen. Psychiatry* **52**: 374–383.

Kennedy, G.C. (1953) The role of depot fat in the hypothalamic control of food intake in the rat. *Proc. Roy. Soc. Lond. B. Biol. Sci.* **140**: 579–592.

Kersten, S., Desvergne, B. and Wahli, W. (2000) Roles of PPARs in health and disease. *Nature* **405**: 421–424.

Kim, J.B., Sarraf, P., Wright, M., Yao, K.M., Mueller, E., Solanes, G., Lowell, B.B. and Spiegelman, B.M. (1998) Nutritional and insulin regulation of fatty acid synthetase and leptin gene expression through ADD1/SREBP1. *J. Clin. Invest.* **101**: 1–9.

Kim, K.S., Larsen, N., Short, T., Plastow, G. and Rothschild, M.F. (2000) A missense variant of the porcine melanocortin-4 receptor (MC4R) gene is associated with fatness, growth, and feed intake traits. *Mamm. Genome* **11**: 131–135.

Kok, L.P. and Tian, C.S. (1994) Susceptibility of Singapore Chinese schoolgirls to anorexia nervosa – Part I (Psychological factors). *Singapore Med. J.* **35**: 481–485.

Korkeila, M., Kaprio, J., Rissanen, A. and Koskenvuo, M. (1991) Effects of gender and age on the heritability of body mass index. *Int. J. Obes.* **15**: 647–654.

Kron, L., Katz, J.L., Gorzynski, G. and Weiner, H. (1978) Hyperactivity in anorexia nervosa: a fundamental clinical feature. *Compr. Psychiatry* **19**: 433–440.

Krude, H., Biebermann, H., Luck, W., Horn, R., Brabant, G. and Gruters, A. (1998) Severe early-onset obesity, adrenal insufficiency and red hair pigmentation caused by POMC mutations in humans. *Nature Genet.* **19**: 155–157.

Kuboki, T., Nomura, S., Ide, M., Suematsu, H. and Araki, S. (1996) Epidemiological data on anorexia nervosa in Japan. *Psychiatry Res.* **62**: 11–16. (Published erratum appears in *Psychiatry Res.* 1996 **62**(3): 285–286.

Kubota, N., Terauchi, Y., Miki, H., Tamemoto, H., Yamauchi, T., Komeda, K. *et al.* (1999) PPAR gamma mediates high-fat diet-induced adipocyte hypertrophy and insulin resistance. *Mol. Cell* **4**: 597–609.

Kushi, A., Sasai, H., Kuizumi, H., Takeda, N., Yokoyama, M. and Nakamura, M. (1998) Obesity and mild hyperinsulinemia found in neuropeptide Y-Y1 receptor-deficient mice. *Proc. Natl. Acad. Sci.* **95**: 15659–15664.

Kyriakis, S.C., Martinsson, K., Olsson, N.G. and Bjork, A. (1990) Thin sow syndrome (TSS): the effect of amperozide. *Br. Vet. J.* **146**: 463–467.

Lafontan, M. and Berlan, M. (1995) Fat cell α2-adrenoceptors: the regulation of fat cell function and lipolysis. *Endocr. Rev.* **16**: 716–738.

Lafontan, M., Betuing, S., Saulnier-Blache, J.S., Valet, P., Bouloumie, A., Carpene, C., Galitzky, J. and Berlan, M. (1998) Regulation of fat-cell function by α2-adrenergic receptors. *Adv. Pharmacol.* **42**: 496–498.

Lamont, L.S., Romito, R.A., Finkelhor, R.S. and Kalhan, S.C. (1997) Beta 1-adrenoreceptors regulate resting metabolic rate. *Med. Sci. Sports Exerc.* **29**: 769–774.

Lane, P.W. and Green, M.C. (1960) Mahogany, a recessive color mutation in linkage group V of the mouse., *J. Hered.* **51**: 228–230.

Lappalainen, J., Long, J.C., Virkkunen, M., Ozaki, N., Goldman, D. and Linnoila, M. (1999) HTR2C Cys23Ser polymorphism in relation to CSF monoamine metabolite concentrations and DSM-III-R psychiatric diagnoses. *Biol. Psychiatry* **46**: 821–826.

Large, V., Hellstrom, L., Reynisdottir, S., Lonnqvist, F., Eriksson, P., Lannfelt, L. and Arner, P. (1997) Human β-2 adrenoceptor gene polymorphisms are highly frequent in obesity and associate with altered adipocyte β-2 adrenoceptor function. *J. Clin. Invest.* **100**: 3005–3013.

Laughlin, G.A. and Yen, S.S. (1997) Hypoleptinemia in women athletes: absence of a diurnal rhythm with amenorrhea. *J. Clin. Endocrinol. Metab.* **82**: 318–321.

Lee, A.M. and Lee, S. (1996) Disordered eating and its psychosocial correlates among Chinese adolescent females in Hong Kong. *Int. J. Eat. Disord.* **20**: 177–183.

Lee, F.Y., Li, Y., Yang, E.K., Yang, S.Q., Lin, H.Z., Trush, M.A., Dannenberg, A.J. and Diehl, A.M. (1999a) Phenotypic abnormalities in macrophages from leptin-deficient, obese mice. *Am. J. Physiol.* **276**(2 Pt 1): C386–394.

Lee, J.H., Reed, D.R., Li, W.D., Xu, W., Joo, E.J., Kilker, R.L. *et al.* (1999b) Genome scan for human obesity and linkage to markers in 20q13. *Am. J. Hum. Genet.* **64**: 196–209.

Leibowitz, S.F. and Alexander, J.T. (1998) Hypothalamic serotonin in control of eating behavior, meal size and body weight. *Biol. Psychiatry* **44**: 851–864.

Leibowitz, S.F., Akabayashi, A. and Wang, J. (1998) Obesity on a high-fat diet: role of hypothalamic galanin in neurons of the anterior paraventricular nucleus projecting to the median eminence. *J. Neurosci.* **18**: 2709–2719.

Lembertas, A.V., Perusse, L., Chagnon, Y.C., Fisler, J.S., Warden, C.H., Purcell-Huynh, D.A. *et al.* (1997) Identification of an obesity quantitative trait locus on mouse chromosome 2 and evidence of linkage to body fat and insulin on the human homologous region 20q. *J. Clin. Invest.* **100**: 1240–1247.

Lentes, K.U., Hinney, A., Ziegler, A., Rosenkranz, K., Wurmser, H., Barth, N. *et al.* (1997) Evaluation of a Cys23Ser mutation within the human 5-HT_{2C} receptor gene: no evidence for an association of the mutant allele with obesity or underweight in children, adolescents and young adults. *Life Sci.* **61**: PL9–16.

Levine, J.A., Eberhardt, N.L. and Jensen, M.D. (1999) Role of nonexercise activity thermogenesis in resistance to fat gain in humans. *Science* **283**: 212–214.

Levitan, R.D., Kaplan, A.S., Joffe, R.T., Levitt, A.J. and Brown, G.M. (1997) Hormonal and subjective responses to intravenous meta-chlorophenylpiperazine in bulimia nervosa. *Arch. Gen. Psychiatry* **54**: 521–527.

Lilenfeld, L.R., Kaye, W.H., Greeno, C.G., Merikangas, K.R., Plotnicov, K., Pollice, C., *et al.* (1998) A controlled family study of anorexia nervosa and bulimia nervosa: psychiatric disorders in first-degree relatives and effects of proband comorbidity. *Arch. Gen. Psychiatry* **55**: 603–610.

Loftus, T.M. (1999) An adipocyte–central nervous system regulatory loop in the control of adipose homeostasis. *Semin. Cell Dev. Biol.* **10**: 11–18.

Lonnqvist, F., Nordfors, L. and Schalling, M. (1999) Leptin and its potential role in human obesity. *J. Intern. Med.* **245**: 643–652.

Lowell, B.B. (1999) PPARg: An essential regulator of adipogenesis and modulator of fat function. *Cell* **99**: 239–242.

Lu, D., Willard, D., Patel, I.R., Kadwell, S., Overton, L., Kost, T. *et al.* (1994) Agouti protein is an antagonist of the melanocyte-stimulating-hormone receptor. *Nature* **371**: 799–802.

Lucas, A.R., Beard, C.M., O'Fallon, W.M. and Kurland, L.T. (1991) 50-year trends in the incidence of anorexia nervosa in Rochester, Minn.: a population-based study. *Am. J. Psychiatry* **148**: 917–922.

Maclean, C.W. (1968) The thin sow problem. *Veterinary Record* **83**: 308–316.

Maltais, L.J., Lane, P.W. and Beamer, W.G. (1984) Anorexia, a recessive mutation causing starvation in preweanling mice. *J. Hered.* **75**: 468–472.

Maqbool, A., Hall, A.S., Ball, S.G. and Balmforth, A.J. (1999) Common polymorphisms of β1-adrenoceptor: identification and rapid screening assay. *Lancet* **353**: 897.

Marsh, D.J., Hollopeter, G., Kafer, K.E. and Palmiter, R.D. (1998) Role of the Y5 neuropeptide Y receptor in feeding and obesity. *Nat. Med.* **4**: 718–721.

Marsh, D.L., Hollopeter, G., Huszar, D., Laufer, R., Yagaloff, K.A., Fisher, S.L., Burn, P. and Palmiter, R.D. (1999) Response of melanocortin-4 receptor-deficient mice to anorectic and orexigenic peptides. *Nature Genet.* **21**: 119–122.

Martinsson, K., Ekman, L., Lofstedt, M., Figueiras, H. and Jonsson, L. (1978) Organ weights and concentrations of zinc in different tissues of wasting pigs and pigs with regional ileitis. *Zentralbl. Veterinarmed.* [A] **25**: 570–578.

Mauriege, P. and Bouchard, C. (1996) Trp64Arg mutation in beta 3-adrenoceptor gene of doubtful significance for obesity and insulin resistance. *Lancet* **348**: 698–699.

Mauriege, P., Imbeault, P., Langin, D., Lacaille, M., Almeras, N., Tremblay, A. and Despres, J.P. (1999) Regional and gender variations in adipose tissue lipolysis in response to weight loss. *J. Lipid Res.* **40**: 1559–1571.

Mehrabian, M., Wen, P.Z., Fisler, J., Davis, R.C. and Lusis, A.J. (1998) Genetic loci controlling body fat, lipoprotein metabolism, and insulin levels in a multifactorial mouse model. *J. Clin. Invest.* **101**: 2485–2496.

Meirhaeghe, A., Fajas, L., Helbecque, N., Cottel, D., Lebel, P., Dallongeville, J., Deeb, S., Auwers, J. and Amouyel, P. (1998) A genetic polymorphism of the peroxisome proliferator-activated receptor gamma gene influences plasma leptin levels in obese humans. *Hum. Mol. Genet.* 7: 435–440.

Meirhaeghe, A., Helbecque, N., Cottel, D. and Amouyel, P. (1999) Beta2-adrenoceptor gene polymorphism, body weight, and physical activity. *Lancet* **353**: 896.

Michels, K.B., Greenland, S. and Rosner, B.A. (1998) Does body mass index adequately capture the relation of body composition and body size to health outcomes? *Am. J. Epidemiol.* **147**: 167–172.

Miller, K.A., Gunn, T.M., Carrasquillo, M.M., Lamoreux, M.L., Galbraith, D.B. and Barsh, G.S. (1997) Genetic studies of the mouse mutations mahogany and mahoganoid. *Genetics* **146**: 1407–1415.

Mitchel, J.S. and Keesey, R.E. (1977) Defense of a lowered weight maintenance level by lateral hypothamically lesioned rats: evidence from a restriction-refeeding regimen. *Physiol. Behav.* **18**: 1121–1125.

Mitchell, B.D., Cole, S.A., Comuzzie, A.G., Almasy, L., Blangero, J., MacCluer, J.W. and Hixson, J.E. (1999) A quantitative trait locus influencing BMI maps to the region of the beta-3 adrenergic receptor. *Diabetes* **48**: 1863–1867.

Moller-Madsen, S., Nystrup, J. and Nielsen, S. (1996) Mortality of anorexia nervosa in Denmark 1970–1987. *Acta. Psychiatr. Scand.* **94**: 454–459.

Montague, C.T., Farooqi, I.S., Whitehead, J.P., Soos, M.A., Rau, H., Wareham, N.J. ***et al.*** (1997) Congenital leptin deficiency is associated with severe early-onset obesity in humans. *Nature* **387**: 903–908.

Morande, G., Celada, J. and Casas, J.J. (1999) Prevalence of eating disorders in a Spanish school-age population. *J. Adolesc. Health* **24**: 212–219.

Morley, J.E. and Blundell, J.E. (1988) The neurobiological basis of eating disorders: some formulations. *Biol. Psychiatry* **23**: 53–78.

Munk-Jorgensen, P., Moller-Madsen, S.M., Nielsen, S. and Nystrup, J. (1996) Incidence of eating disorders in psychiatric health services in Denmark 1970–1993. *Ugeskr. Laeger* **158**: 5785–5789.

Must, A., Spadano, J., Coakley, E.H., Field, A., Colditz, G. and Dietz, W.H. (1999) The disease burden associated with overweight and obesity. *JAMA* **282**: 1523–1529.

Nagle, D.L., McGrail, S.H., Vitale, J., Woolf, E.A., Dussault, B.J. Jr, DiRocco, L. ***et al.*** (1999) The mahogany protein is a receptor involved in suppression of obesity. *Nature* **398**: 148–152.

Nasser, M. (1994) Screening for abnormal eating attitudes in a population of Egyptian secondary school girls. *Soc. Psychiatry Psychiatr. Epidemiol.* **29**: 25–30.

Naveilhan, P., Hassani, H., Canals, J.M., Ekstrand, A.J., Larefalk, A., Chhajlani, V. ***et al.*** (1999) Normal feeding behavior, body weight and leptin response require the neuropeptide Y Y2 receptor. *Nat. Med.* 5: 1188–1193.

Nielsen, S. and Molbak, A.G. (1998) Eating disorders and type 1 diabetes: Overview and summing up. *Eur. Eat. Dis. Rev.* **6**: 4–27.

Nielsen, S., Moller-Madsen, S., Isager, T., Jorgensen, J., Pagsberg, K. and Theander, S. (1998) Standardized mortality in eating disorders – a quantitative summary of previously published and new evidence. *J. Psychosom. Res.* **44**: 413–434.

Nogami, Y. and Yabana, F. (1977) On Kibarashi-gui (binge eating). *Folia Psychiatr. Neurol. Jpn.* **31**: 294–295.

Nonogaki, K., Strack, A.M., Dallman, M.F. and Tecott, L.H. (1998) Leptin-independent hyperphagia and type 2 diabetes in mice with a mutated serotonin 5-HT_{2C} receptor gene. *Nat. Med.* **4**: 1152–1156.

Nordfors, L., Hoffstedt, J., Nyberg, B., Thörne, A., Arner, P., Schalling, M. and Lönnqvist, F. (1998) Reduced gene expression of UCP2 but not UCP3 in skeletal muscle of human obese subjects. *Diabetologia* **41**: 935–939.

Norman, R.A., Tataranni, P.A., Pratley, R., Thompson, D.B., Hanson, R.L., Prochazka, M. ***et al.*** (1998) Autosomal genomic scan for loci linked to obesity and energy metabolism in Pima Indians. *Am. J. Hum. Genet.* **62**: 659–668.

Norring, C.E. and Sohlberg, S.S. (1993) Outcome, recovery, relapse and mortality across six years in patients with clinical eating disorders. *Acta. Psychiatr. Scand.* **87**: 437–444.

Odorizzi, M., Max, J.P., Tankosic, P., Burlet, C. and Burlet, A. (1999) Dietary preferences of Brattleboro rats correlated with an overexpression of galanin in the hypothalamus. *Eur. J. Neurosci.* **11**: 3005–3014.

Ohman, M., Oksanen, L., Kainulainen, K., Jöanne, O.A., Kaprio, J., Koskenvuo, M., Mustajoki, P., Kontula, K. and Peltonen, L., (1999) Testing of human homologues of murine obesity genes as candidate regions in Finnish obese sib pairs. *Eur. J. Hum. Genet.* 7: 17–24.

Oksanen, L., Kainulainen, K., Heiman, M., Mustajoki, P., Kauppinen-Makelin, R. and Kontula, K. (1997) Novel polymorphism of the human ob gene promoter in lean and morbidly obese subjects. *Int. J. Obes. Relat. Metab. Disord.* **21**: 489–494.

Ollmann, M.M., Wilson, B.D., Yang, Y.K., Kerns, J.A., Chen, Y., Gantz, I. and Barsh, G.S. (1997) Antagonism of central melanocortin receptors in vitro and in vivo by agouti-related protein. *Science* **278**: 135–138.

Otabe, S., Clement, K., Rich, N., Warden, C., Pecqueur, C., Neverova, M. ***et al.*** (1998) Mutation screening of the human UCP 2 gene in normoglycemic and NIDDM morbidly obese patients: lack of association between new UCP 2 polymorphisms and obesity in French Caucasians. *Diabetes* **47**: 840–842.

Patton, G.C., Selzer, R., Coffey, C., Carlin, J.B. and Wolfe, R. (1999) Onset of adolescent eating disorders: population based cohort study over 3 years. *BMJ* **318**: 765–768.

Pawluck, D.E. and Gorey, K.M. (1998) Secular trends in the incidence of anorexia nervosa: integrative review of population-based studies. *Int. J. Eat. Disord.* **23**: 347–352.

Pedrazzini, T., Seydoux, J., Kunstner, P., Aubert, J.F., Grouzmann, E., Beermann, F. and Brunner, H.R. (1998) Cardiovascular response, feeding behavior and locomotor activity in mice lacking the NPY Y1 receptor. *Nat. Med.* **4**: 722–726.

Perusse, L., Chagnon, Y.C., Weisnagel, J. and Bouchard, C. (1999) The Human Obesity Gene Map: the 1998 update. *Obesity Res.* **7**: 111–129.

Pietilainen, K.H., Kaprio, J., Rissanen, A., Winter, T., Rimpela, A., Viken, R.J. and Rose, R.J. (1999) Distribution and heritability of BMI in Finnish adolescents aged 16y and 17y: a study of 4884 twins and 2509 singletons. *Int. J. Obes. Relat. Metab. Disord.* **23**: 107–115.

Pietri-Rouxel, F., St John Manning, B., Gros, J. and Strosberg, A.D. (1997) The biochemical effect of the naturally occurring Trp64 → Arg mutation on human beta3-adrenoceptor activity. *Eur. J. Biochem.* **247**: 1174–1179.

Pomp, D. (1997) Genetic dissection of obesity in polygenic animal models. *Behav. Genet.* **27**: 285–306.

Price, R.A. and Gottesman, I.I. (1991) Body fat in identical twins reared apart: roles for genes and environment. *Behav. Genet.* **21**: 1–7.

Prike, K.M. (1996) Central and peripheral noradrenalin regulation in eating disorders. *Psychiatry Res.* **62**: 43–49.

Quested, D.J., Whale, R., Sharpley, A.L., McGavin, C.L., Crossland, N., Harrison, P.J. and Cowen, P.J. (1999) Allelic variation in the 5-HT2C receptor (HTR2C) and functional responses to the 5-HT2C receptor agonist, m-chlorophenylpiperazine. *Psychopharmacology (Berl.)* **144**: 306–307.

Rastam, M., Gillberg, C. and Garton, M. (1989) Anorexia nervosa in a Swedish urban region. A population-based study. *Br. J. Psychiatry* **155**: 642–646.

Ravussin, E. and Swinburn, B.A. (1993) Metabolic predictors of obesity: cross-sectional versus longitudinal data. *Int. J. Obes. Relat. Metab. Disord.* **17**: S28–31: S42–2.

Ravussin, E., Lillioja, S., Anderson, T.E., Christin, L. and Bogardus, C. (1986) Determinants of 24-hour energy expenditure in man. Methods and results using a respiratory chamber. *J. Clin. Invest.* **78**: 1568–1578.

Ravussin, E., Lillioja, S., Knowler, W.C., Christin, L., Freymond, D., Abbott, W.G., Boyce, V., Howard, B.V. and Bogardus, C. (1988) Reduced rate of energy expenditure as a risk factor for body-weight gain. *N. Engl. J. Med.* **318**: 467–472.

Ravussin, E., Pratley, R.E., Maffei, M., Wang, H., Friedman, J.M., Bennett, P.H. and Bogardus, C. (1997) Relatively low plasma leptin concentrations precede weight gain in Pima Indians. *Nat. Med.* **3**: 238–240.

Reed, D.R., Ding, Y., Xu, W., Cather, C., Green, E.D. and Price, R.A. (1996) Extreme obesity may be linked to markers flanking the human OB gene. *Diabetes* **45**: 691–694.

Rice, T., Tremblay, A., Deriaz, O., Perusse, L., Rao, D.C. and Bouchard, C. (1996) Genetic pleiotropy for resting metabolic rate with fat-free mass and fat mass: the Quebec Family Study. *Obes. Res.* **4**: 125–131. (Published erratum appears in *Obes. Res.* **5**: 385.)

Rice, T., Perusse, L., Bouchard, C. and Rao, D.C. (1999) Familial aggregation of body mass index and subcutaneous fat measures in the longitudinal Quebec family study. *Genet. Epidemiol.* **16**: 316–334.

Rice, T., Daw, E.W., Gagnon, J., Bouchard, C., Leon, A.S., Skinner, J.S., Wilmore, J.H. and Rao, D.C. (1997) Familial resemblance for body composition measures: the HERITAGE. *Family Study Obes. Res.* 5: 557–562.

Ricquier, D., Bouillaud, F., Toumelin, P., Mory, G., Bazin, R., Arch, J. and Penicaud, L. (1986) Expression of uncoupling protein mRNA in thermogenic or weakly thermogenic brown adipose tissue. Evidence for a rapid beta-adrenoreceptor-mediated and transcriptionally regulated step during activation of thermogenesis. *J. Biol. Chem.* **261**: 13905–13910.

Ristow, M., Muller-Wieland, D., Pfeiffer, A., Krone, W. and Kahn, C.R. (1998) Obesity associated with a mutation in a genetic regulator of adipocyte differentiation. *N. Engl. J. Med.* **339**: 953–959.

Roche, C., Boutin, P., Dina, C., Gyapay, G., Basdevant, A., Hager, J., Guy-Grand, B., Clement, K. and Froguel, P. (1997) Genetic studies of neuropeptide Y and neuropeptide Y receptors Y1 and Y5 regions in morbid obesity. *Diabetologia* **40**: 671–675.

Rothwell, N.J. and Stock, M.J. (1984) The development of obesity in animals: the role of dietary factors. *Clin. Endocrinol. Metab.* **13**: 437–449.

Rooney, B., McClelland, L., Crisp, A.H. and Sedgwick, P.M. (1995) The incidence and prevalence of anorexia nervosa in three suburban health districts in south west London, UK. *Int. J. Eat. Disord.* **18**: 299–307.

Rosen, E.D., Sarraf, P., Troy, A.E., Bradwin, G., Moore, K., Milstone, D.S., Spiegelman, B.M. and Mortensen, R.M. (1999) PPAR gamma is required for the differentiation of adipose tissue in vivo and in vitro. *Mol. Cell.* **4**: 611–617.

Rosenkranz, K., Hinney, A., Ziegler, A., von Prittwitz, S., Barth, N., Roth, H., *et al.* (1998) Screening for mutations in the neuropeptide Y Y5 receptor gene in cohorts belonging to different weight extremes. *Int. J. Obes. Relat. Metab. Disord.* **22**: 57–63.

Roth, H., Korn, T., Rosenkranz, K., Hinney, A., Ziegler, A., Kunz, J. *et al.* (1998) Transmission disequilibrium and sequence variants at the leptin receptor in extremely obese German children and adolescents. *Hum. Genet.* **103**: 540–546.

Russell, G. (1979) Bulimia nervosa: an ominous variant of anorexia nervosa. *Psychol. Med.* **9**: 429–448.

Sahakian, B.J., Wurtman, R.J., Barr, J.K., Millington, W.R. and Chiel, H.J. (1979) Low tryptophan diet decreases brain serotonin and alters response to apomorphine. *Nature* **279** (5715): 731–732.

Sakane, N., Yoshida, T., Umekawa, T., Kogure, A., Takakura, Y. and Kondo, M. (1997) Effects of Trp64Arg mutation in the beta 3-adrenergic receptor gene on weight loss, body fat distribution, glycemic control, and insulin resistance in obese type 2 diabetic patients. *Diabetes Care* **20**: 1887–1890.

Salmon, D.M. and Flatt, J.P. (1985) Effect of dietary fat content on the incidence of obesity among ad libitum fed mice. *Int. J. Obesity* **9**: 443.

Samec, S., Seydoux, J. and Dulloo, A.G. (1998) Role of UCP homologues in skeletal muscles and brown adipose tissue: mediators of thermogenesis or regulators of lipids as fuel substrate? *FASEB J.* **12**: 715–724.

Samec, S., Seydoux, J. and Dulloo, A.G. (1999) Post-starvation gene expression of skeletal muscle uncoupling protein 2 and uncoupling protein 3 in response to dietary fat levels and fatty acid composition: a link with insulin resistance. *Diabetes* **48**: 436–441.

Schaffler, A., Palitzsch, K.D., Watzlawek, E., Drobnik, W., Schwer, H., Scholmerich, J. and Schmitz, G. Frequency and significance of the A→G (–3826) polymorphism in the promoter of the gene for uncoupling protein-1 with regard to metabolic parameters and adipocyte transcription factor binding in a large population-based Caucasian cohort. *Eur. J. Clin. Invest.* **29**: 770–709.

Siegel, J.M. (1999) Narcolepsy: a key role for hypocretins. *Cell* **98**: 409–412.

Selby, J.V., Newman, B., Quesenberry, C.P. Jr, Fabsitz, R.R., Carmelli, D., Meaney, F.J. and Slemenda, C. (1990) Genetic and behavioral influences on body fat distribution. *Int. J. Obes.* **14**: 593–602.

Shimada, M., Tritos, N.A., Lowell, B.B., Flier, J.S. and Maratos-Flier, E. (1998) Mice lacking melanin-concentrating hormone are hypophagic and lean. *Nature* **396**: 670–674.

Simansky, K.J. (1996) Serotonergic control of the organization of feeding and satiety. *Behav. Brain Res.* **73**: 37–42.

Slattery, M.L., Bishop, D.T., French, T.K., Hunt, S.C., Meikle, A.W. and Williams, R.R. (1988) Lifestyle and blood pressure levels in male twins in Utah. *Genet. Epidemiol.* 5: 277–287.

Smith, K.A., Fairburn, C.G. and Cowen, P.J. (1999) Symptomatic relapse in bulimia nervosa following acute tryptophan depletion. *Arch. Gen. Psychiatry* **56**: 171–176.

Sohlberg, S. and Strober, M. (1994) Personality in Anorexia nervosa: an update and a theoretical integration. *Acta. Psychiatr. Scand. Suppl.* **378**: 1–15.

Son, J.H., Baker, H., Park, D.H. and Joh, T.H. (1994) Drastic and selective hyperinnervation of central serotonergic neurons in a lethal neurodevelopmental mouse mutant, Anorexia (*anx*). *Brain Res. Mol. Brain Res.* 25: 129–134.

Sorbi, S., Nacmias, B., Tedde, A., Ricca, V., Mezzani, B. and Rotella, C.M. (1998) 5-HT_{2a} promoter polymorphism in anorexia nervosa. *Lancet* **351**: 1785.

Soundy, T.J., Lucas, A.R., Suman, V.J. and Melton, L.J. 3rd (1995) Bulimia nervosa in Rochester, Minnesota from 1980 to 1990. *Psychol. Med.* 25: 1065–1071.

Spear, L.P. and Ristine, L.A. (1982) Suckling behavior in neonatal rats: psychopharmacological investigations. *J. Comp. Physiol. Psychol.* **96**: 244–255.

Spitzer, R.L., Yanovski, S., Wadden, T., Wing, R., Marcus, M.D., Stunkard, A. *et al.* (1993) Binge eating disorder: its further validation in a multisite study. *Int. J. Eat. Disord.* **13**: 137–153.

Steffens, A.B. (1975) Influence of reversible obesity on eating behavior, blood glucose, and insulin in the rat. *Am. J. Physiol.* **228**: 1738–1744.

Stein, D., Lilenfeld, L.R., Plotnicov, K., Pollice, C., Rao, R., Strober, M. and Kaye, W.H. (1999) Familial aggregation of eating disorders: results from a controlled family study of bulimia nervosa. *Int. J. Eat. Disord.* **26**: 211–215.

Strobel, A., Issad, T., Camoin, L., Ozata, M. and Strosberg, A.D. (1998) A leptin missense mutation associated with hypogonadism and morbid obesity. *Nat. Genet.* **18**: 213–215.

Strober, M., Freeman, R., Lampert, C., Diamond, M.S.W. and Kaye, W. (2000) Controlled family study of anorexia nervosa and bulimia nervosa: evidence of shared liability and transmission of partial syndromes. *Am. J. Psychiatry* **157**: 393–401.

Stunkard, A.J., Foch, T.T. and Hrubec, Z. (1986) A twin study of human obesity. *JAMA*, **256**: 51–54.

Szczypka, M.S., Rainey, M.A., Kim, D.S., Alaynick, W.A., Marck, B.T., Matsumoto, A.M. and Palmiter, R.D. (1999) Feeding behavior in dopamine-deficient mice. *Proc. Natl Acad. Sci. USA* **96**: 12138–12143.

Szczypka, M.S., Rainey, M.A. and Palmiter, R.D. (2000) Dopamine is required for hyperphagia in Lep(ob/ob) mice. *Nature Genet.* 25: 102–104.

Takami, S., Wong, Z.Y., Stebbing, M. and Harrap, S.B. (1999) Linkage analysis of glucocorticoid and beta2-adrenergic receptor genes with blood pressure and body mass index. *Am. J. Physiol.* **276**: H1379–1384.

Tanaka, T., Yoshida, N., Kishimoto, T. and Akira, S. (1997) Defective adipocyte differentiation in mice lacking the C/EBPβ and/or C/EBPδ gene. *EM30 J.* **16**: 7432–7443.

Taylor, B.A. and Phillips, S.J. (1997) Obesity QTLs on mouse chromosomes 2 and 17. *Genomics* **43**: 249–257.

Taylor, B.A., Tarantino, L.M. and Phillips, S.J. (1999) Gender-influenced obesity QTLs identified in a cross involving the KK type II diabetes-prone mouse strain. *Mamm. Genome* **10**: 963–968.

Theander, S. (1985) Outcome and prognosis in anorexia nervosa and bulimia nervosa: some results of previous investigations compared with those of a Swedish long term study. *J. Psychiatric Res.* **9**: 493–508.

Thompson, D.B., Ravussin, E., Bennett, P.H. and Bogardus, C. (1997) Structure and sequence variation at the human leptin receptor gene in lean and obese Pima Indians. *Hum. Mol. Genet.* **6**: 675–679.

Tontonoz, P., Kim, J.B., Graves, R.A. and Spiegelman, B.M. (1993) ADD1: a novel helix-loop-helix transcription factor associated with adipocyte determination and differentiation. *Mol. Cell. Biol.* **13**: 4753–4759.

Toubro, S., Sorensen, T.I., Hindsberger, C., Christensen, N.J. and Astrup, A. (1998) Twenty-four-hour respiratory quotient: the role of diet and familial resemblance. *J. Clin. Endocrinol. Metab.* **83**: 2758–2764.

Treasure, J.L. and Holland, A.J. (1994) Genetic factors in eating disorders. In: *Handbook of Eating Disorders* (eds Szmukler, G. *et al.*). John Wiley & Sons Ltd, Chichester.

Treasure, J. and Owen, J.B. (1997) Intriguing links between animal behavior and anorexia nervosa. *Int. J. Eat. Disord.* **21**: 307–311.

Tremblay, A., Poehlman, E.T., Despres, J.P., Theriault, G., Danforth, E. and Bouchard, C. (1997) Endurance training with constant energy intake in identical twins: changes over time in energy expenditure and related hormones. *Metabolism* **46**: 499–503.

Turnbull, S., Ward, A., Treasure, J., Jick, H. and Derby, L. (1996) The demand for eating disorder care. An epidemiological study using the general practice research database. *Br. J. Psychiatry* **169**: 705–712.

Umekawa, T., Yoshida, T., Sakane, N., Kogure, A., Kondo, M. and Honjyo, H. (1999) Trp64Arg mutation of beta3-adrenoceptor gene deteriorates lipolysis induced by beta3-adrenoceptor agonist in human omental adipocytes. *Diabetes* **48**: 17–20.

Ung, E.K., Lee, S. and Kua, E.H. (1997) Anorexia nervosa and bulimia – a Singapore perspective. *Singapore Med. J.* **38**: 332–335.

Urhammer, S.A., Dalgaard, L.T., Sorensen, T.I., Moller, A.M., Andersen, T., Tybjaerg-Hansen, A., *et al.* (1997) Mutational analysis of the coding region of the uncoupling protein 2 gene in obese NIDDM patients: impact of a common amino acid polymorphism on juvenile and maturity onset forms of obesity and insulin resistance. *Diabetologia* **40**: 1227–1230.

Vaisse, C., Clement, K., Guy-Grand, B. and Froguel, P. (1998) A frameshift mutation in human MC4R is associated with a dominant form of obesity. *Nat. Genet.* **20**: 113–114.

Vandereycken, W. and Van Deth, R. (1994) From Fasting Saints to Anorexic Girls. New York University Press, New York.

Verrotti, A., Catino, M., De Luca, F.A., Morgese, G. and Chiarelli, F. (1999) Eating disorders in adolescents with type 1 diabetes mellitus. *Acta Diabetol.* **36**: 21–25.

Vidal-Puig, A., Solanes, G., Grujic, D., Flier, J.S. and Lowell, B.B. (1997) UCP3: an uncoupling protein homologue expressed preferentially and abundantly in skeletal muscle and brown adipose tissue. *Biochem. Biophys. Res. Commun.* **235**: 79–82.

Visser, M., Launer, L.J., Deurenberg, P., Deeg, D.J.J and Gerontol, A. (1999) Past and current smoking in relation to body fat distribution in older men and women. *Biol. Sci. Med. Sci.* **54**: M293–298.

Wade, T., Heath, A.C., Abraham, S., Treloar, S.A., Martin, N.G. and Tiggemann, M. (1996) Assessing the prevalence of eating disorders in an Australian twin population. *Aust. NZ J. Psychiatry* **30**: 845–851.

Walder, K., Norman, R.A., Hanson, R.L., Schrauwen, P., Neverova, M., Jenkinson, C.P., *et al.* (1998) Association between uncoupling protein polymorphisms (UCP2–UCP3) and energy metabolism/obesity in Pima Indians. *Hum. Mol. Genet.* **7**: 1431–1435.

Walsh, B.T. and Devlin, M.J. (1998) Eating disorders: progress and problems. *Science* **280**: 1387–1390.

Walston, J., Lowe, A., Silver, K., Yang, Y., Bodkin, N.L., Hansen, B.C. and Shuldiner, A.R. (1995) The beta3-adrenergic receptor in the obesity and diabetes prone rhesus monkey is very similar to human and contains arginine at codon 64. *Gene* **188**: 207–213.

Walters, E.E. and Kendler, K.S. (1995) Anorexia nervosa and anorexic-like syndromes in a population-based female twin sample. *Am. J. Psychiatry* **152**: 64–71.

Walters, E.E., Neale, M.C., Eaves, L.J., Heath, A.C., Kessler, R.C. and Kendler, K.S. (1992) Bulimia nervosa and major depression: a study of common genetic and environmental factors. *Psychol. Med.* **22**: 617–622.

Wang, N.D., Finegold, M.J., Bradley, A., Ou, C.N., Abdelsayed, S.V., Wilde, M.D., Taylor, L.R., Wilson, D.R. and Darlington, G.J. (1995) Impaired energy homeostasis in C/EBP alpha knockout mice. *Science* **269**: 1108–1112.

Warden, C. (1999) Genetics of uncoupling proteins in humans 23. *Int. J. Obes. Relat. Metab. Disord.* **6** (Suppl.): S46–48.

Weltzin, T.E., Fernstrom, M.H., Fernstrom, J.D., Neuberger, S.K. and Kaye, W.H. (1995) Acute tryptophan depletion and increased food intake and irritability in bulimia nervosa. *Am. J. Psychiatry* **152**: 1668–1671.

Wickelgren, I. (1998) Obesity: how big a problem? *Science* **280**: 1364–1367.

Widen, E., Lehto, M., Kanninen, T., Walston, J., Shuldiner, A.R. and Groop, L.C. (1995) Association of a polymorphism in the beta 3-adrenergic-receptor gene with features of the insulin resistance syndrome in Finns. *N. Engl. J. Med.* **10**: 348–351.

Wonderlich, S.A., Brewerton, T.D., Jocic, Z., Dansky, B.S. and Abbott, D.W. (1997) Relationship of childhood sexual abuse and eating disorders. *J. Am. Acad. Child Adolesc. Psychiatry* **36**: 1107–1115.

Wurtman, R.J., Hefti, F. and Melamed, E. (1980) Precursor control of neurotransmitter synthesis. *Pharmacol. Rev.* **32**: 315–335.

Yamada, K., Ishiyama-Shigemoto, S., Ichikawa, F., Yuan, X., Koyanagi, A., Koyama, W. and Nonaka, K. (1999) Polymorphism in the 5′-leader cistron of the beta2-adrenergic receptor gene associated with obesity and type 2 diabetes. *J. Clin. Endocrinol. Metab.* **84**: 1754–1757.

Yamagata, K., Furuta, H., Oda, N., Kaisaki, P.J., Menzel, S., Cox, N.J., *et al.* (1996) Mutations in the hepatocyte nuclear factor-4alpha gene in maturity-onset diabetes of the young. *Nature* **384**: 458–460.

Yamashita, J., Onai, T., York, D.A. and Bray, G.A. (1994) Relationship between food intake and metabolic rate in rats treated with beta-adrenoceptor agonists. *Int. J. Obes. Relat. Metab. Disord.* **18**: 429–433.

Yaswen, L., Diehl, N., Brennan, M.B. and Hochgeschwender, U. (1999) Obesity in the mouse model of pro-opiomelanocortin deficiency responds to peripheral melanocortin. *Nat. Med.* **5**: 1066–1070.

Yeo, G.S., Farooqi, I.S., Aminian, S., Halsall, D.J., Stanhope, R.G. and O'Rahilly, S. (1998) A frameshift mutation in MC4R associated with dominantly inherited human obesity [letter]. *Nat. Genet.* **20**: 111–112.

York, B., Lei, K. and West, D.B. (1996) Sensitivity to dietary obesity linked to a locus on chromosome 15 in a CAST/Ei × C57BL/6J F2 intercross. *Mamm. Genome* **7**: 677–681.

Ziegler, A., Hebebrand, J., Gorg, T., Rosenkranz, K., Fichter, M., Herpertz-Dahlmann, B., Remschmidt, H. and Hinney, A. (1999) Further lack of association between the 5-HT2A gene promoter polymorphism and susceptibility to eating disorders and a meta-analysis pertaining to anorexia nervosa. *Mol. Psychiatry* **4**: 410–412.

Zhang, Y., Proenca, R., Maffei, M., Barone, M., Leopold, L. and Friedman, J.M. (1994) Positional cloning of the mouse obese gene and its human homologue. *Nature* **372**: 425–432. (Published erratum appears in *Nature* 1995, **374**: 479.)

Zouali, H., Hani, E.H., Philippi, A., Vionnet, N., Beckmann, J.S., Demenais, F. and Froguel, P. (1997) A susceptibility locus for early-onset non-insulin dependent (type 2) diabetes mellitus maps to chromosome 20q, proximal to the phosphoenolpyruvate carboxykinase gene. *Hum. Mol. Genet.* **6**: 1401–1408.

Zurlo, F., Lillioja, S., Esposita-Del Puente, A., Nyomba, B.L., Raz, I., Sand, M.F. *et al.* (1990) Low ratio of fat to carbohydrate oxidation as predictor of weight gain: study of 24-h RQ. *Am. J. Physiol.* **259**: E650–657.

14

Depression and anxiety

Janette Moore, Gin S. Malhi and Peter McGuffin

1. Introduction

A wide range of disorders is characterized by pathological anxiety and mood. Contemporary classification systems such as the Diagnostic and Statistical Manual-IV (DSM-IV; American Psychiatric Association, 1994) and the International Classification of Diseases-10 (ICD-10; World Health Organization, 1992) adopt a phenomenological approach, which attempts to be atheoretical with regard to etiology (Kahn, 1995). This reflects the lack of definitive tests in the diagnosis of these disorders as well as our current limited state of knowledge about the factors that are of importance in causation.

The classification of mood disorders has been a controversial issue for most of this century. In current classification systems a distinction is usually drawn between *unipolar* (e.g. major depression, MD) and bipolar disorders, on the basis of whether a manic episode has ever occurred. Characteristic features of a unipolar depressive disorder in addition to depressed mood include disturbance of sleep and appetite, loss of energy and interest in usual activities and often thoughts of suicide. The lifetime risk for major depressive disorder has been estimated at 17% (Blazer *et al.*, 1994) and there is some evidence that it may be becoming more common (Klerman and Weissman, 1989).

Pathological anxiety is characterized by a feeling of apprehension, uncertainty and fear, which is out of proportion to any apparent cause and usually accompanied by physiological changes such as tachycardia, sweating or tremor. Several overlapping subcategories of anxiety disorder are recognized. In generalized anxiety disorder (GAD) the individual suffers mainly from 'free-floating' or generalized pathological anxiety or worry, to the extent that it disrupts everyday life. Panic disorder (PD) is characterized by episodic, recurrent, and unpredictable anxiety, with physiological symptoms typically predominating. In contrast, phobic disorders are characterized by anxiety evoked by specific situations, the degree of anxiety being excessive given the objective contextual risks. For instance, social phobia describes a marked and persistent fear of interactive situations in which there is exposure to the scrutiny of others (Bell, 1995), whilst agoraphobia refers to an overlapping cluster of phobias including fears of leaving home, entering public or crowded places and traveling alone.

The overall lifetime prevalence of GAD is about 5%, PD 3–4% and phobias

Analysis of Multifactorial Disease, edited by T. Bishop and P. Sham.

including social phobia approximately 25% (Kessler *et al.*, 1994). Both depressive disorders and anxiety disorders have been shown to occur more frequently in women. Furthermore anxiety disorders are the most common form of mental illness, and although not generally as debilitating as major depression, are associated with considerable disability and dysfunction.

Until the introduction of operationalized diagnostic classification, the relationship between symptoms and diagnostic categories in psychiatry was ill defined and led to substantial differences in diagnostic practice between countries (Cooper *et al.*, 1972). The introduction and widespread adoption of explicit diagnostic systems and the use of standardized structured or semi-structured interviews such as the Present State Examination (Wing *et al.*, 1974) or the Schedule for Affective Disorders and Schizophrenia (Endicott and Spitzer, 1978), has greatly improved the reliability of diagnosis and enhanced comparability of research studies (Tsuang and Faraone, 1990). However, difficulties remain in that these diagnostic concepts are still somewhat 'elastic' and open to varying interpretations (McGuffin *et al.*, 1994a). Epidemiological studies are complicated further in that they commonly assess whether or not a subject has had a psychiatric disorder on the basis of a single interview. The reliability of diagnostic conclusions drawn depend upon an individual's ability to recall past episodes of psychiatric symptoms. An associated difficulty arises in the investigation of families. Indirect family history methods substantially underestimate the amount of psychiatric disorder in comparison with the use of personal interviews for all family members (the family study method) (Tsuang and Faraone, 1990).

Although standardized classification systems increase reliability there is, however, no guarantee that the diagnostic concepts they contain are valid. In particular, current classifications of depression and anxiety phenotypes may not result in categories that are most informative for genetic studies. It has been argued that the currently accepted nosology does not capture the natural tendency of these psychiatric disorders to co-occur (Sullivan and Kendler, 1998). In the National Comorbidity Study (NCS) eight out of 10 individuals with a lifetime diagnosis of GAD also had a comorbid mood disorder at some time in their life (Judd *et al.*, 1998).

There is much debate about whether a categorical approach to phenotypic definition is most appropriate. We can all relate to transient feelings of low mood and anxiety and these are considered normal responses to various life events. In those people diagnosed with depression and anxiety disorders, individuals also vary in the severity and frequency of their symptoms. A continuum from normal to abnormal therefore seems possible, with symptoms increasing in a continuous fashion and individuals with a diagnosis representing the extreme end of the dimension. Some authors have suggested that categorical classification systems such as DSM-IV are merely diagnostic conventions imposed on such a continuum of symptoms (Kendler and Gardner, 1998a). Whether the dimensional or categorical approach to classification is correct is not yet clear and awaits clarification of the degree to which common etiological factors explain both normal experiences and abnormal psychopathology. It is clear, however, that the fundamental problem of phenotypic definition has crucial implications both for the interpretation and the validity of genetic research in psychiatry.

2. Family studies

2.1 Depression

It has long been recognized that mood disorder tends to run in families (Slater and Cowie, 1971). Angst (1966) found that among the relatives of probands with unipolar illness there was an increased risk of unipolar disorder, however, among the relatives of bipolar probands there was an excess of both unipolar and bipolar disorders. A review of 12 subsequent family studies found that most research has replicated this pattern (McGuffin and Katz, 1986). The average risk of bipolar and unipolar illness in the first degree relatives of bipolar probands was estimated to be 7.8% and 11.4% respectively; whereas the familial risk in relatives of unipolar probands was 0.6% for bipolar and 9.1% for unipolar disorder. As these studies focused on severe affective disorder that often required hospitalization, these figures should be compared to population risks of approximately 3% for unipolar disorder (Sturt *et al.*, 1984) and less than 1% for bipolar disorder. These findings have led to the suggestion that bipolar affective disorder is a more severe disorder than unipolar depression (McGuffin and Katz, 1986) and that this explains why the 'familial risk is greater for bipolar depression, why bipolar probands have an excess of unipolar relatives and why unipolar probands do not have many relatives with bipolar disorder' (Plomin *et al.*, 1997a). However, an alternative explanation may be that the majority of unipolar depression is etiologically distinct from bipolar disorder (Sham, 1996).

2.2 Anxiety

Anxiety disorders in general also show evidence of increased familial risk (Marks, 1986). Family studies suggest that panic disorder and generalized anxiety disorder are not related, although both appear to be familial (Noyes *et al.*, 1987). First degree relatives of GAD probands show an approximate 20% risk for GAD, about five times the risk in control families, whereas family studies of severe PD have estimated the risk of PD to first degree relatives as 17.3% compared with 1.8% for controls (Crowe *et al.*, 1983). The relationship between agoraphobia and panic disorder has also been explored by Noyes *et al.* (1986). Agoraphobia, like PD, also appears not to be associated with an increased risk of GAD in first degree relatives. A high risk of PD (7%) is, however, evident in the families of agoraphobic probands. Interestingly there is no increased risk of agoraphobia seen in the families of PD probands. Similar to unipolar and bipolar depression, therefore, it has been suggested that agoraphobia may represent a more severe variant of PD (Noyes *et al.*, 1986). Taken together these studies suggest that GAD and PD are both familial but have separate etiologies, whereas agoraphobia and PD share common familial and possibly genetic etiological factors (McGuffin *et al.*, 1994b). Similar significant familial effects are shown for specific phobias, with a risk to first degree relatives of 31% compared with 11% for controls (Fryer *et al.*, 1990), and social phobia where relative risk for generalized social phobia among the first degree relatives of probands is markedly higher than expected (Reich and Yates, 1988).

3. Twin studies

Familial aggregation does not imply that a disorder is genetically influenced. Twin and adoption studies attempt to untangle the relative contribution of genes from other familial factors. However, the validity of the twin method depends on the equal environments assumption: that monozygotic (MZ) and dizygotic (DZ) twins are equally similar in their exposure to shared environmental factors relevant to the illness of study. For MD and anxiety disorders there is evidence that this assumption is unlikely to introduce substantial bias (Kendell and Gardner, 1998b; Lytton, 1977; McGuffin *et al.*, 1996; Scarr, 1968). The use of the twin design in the investigation of depression and anxiety disorders has not only focused on estimating heritabilities, but has also used multivariate analyses to explore the covariance between these traits. This has begun to clarify the role of genetic factors in comorbidity and the opportunity for 'identification of the most heritable constellations (comorbidity) and components (homogeneity)' (Plomin *et al.*, 1997b) of these phenotypes.

Multivariate analyses of family data are also being used to explore a variety of other questions. Is there continuity in the influence of genetic and environmental factors at different stages of the disorder? Are the genetic and environmental factors that affect depressive and anxiety disorders the same as factors that affect variability in these experiences within the normal range? What is the relationship between genetic and environmental factors such as life events? And what are the genetic correlations between behavior and biological processes?

Recent studies using twin data have used structural equation modeling such as path analysis to facilitate comparisons between competing causal biometrical models. Each causal model considers a number of latent genetic or environmental variables, for example additive genetic, common environmental or nonshared environmental influences. Although these latent variables cannot be measured directly, their shared effect can be predicted to induce a particular pattern of correlation, between twins, in the variables observed. By comparing predicted and observed correlations it is possible to validate a causal model and to estimate the relative contributions of the latent variables to the variability of the observed phenotypes (Sham, 1996). In the case of categorically defined illness, observed twin concordances and the population prevalence of the disorder can be used to

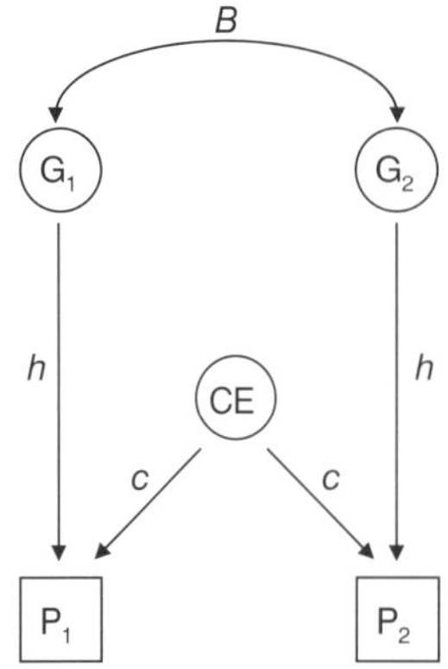

Figure 1. A simple path model of the sources of resemblance between twins or pairs of siblings. G_1 and G_2 are genotypes with correlation B, CE is common environment, P_1 and P_2 are phenotypes and h and c are path coefficients.

estimate a correlation in liability to the disorder assuming a liability threshold model (Falconer, 1965; Sham, 1996; Smith, 1974).

3.1 Depression

Data from a compilation of older twin studies reported by Gershon *et al.* (1976) strongly suggests a genetic influence in the etiology of affective disorders. A review of older studies (Zerbin–Rudin, 1979), in which the unipolar/bipolar distinction was considered, found a strong tendency for twin pairs concordant for affective disorder to be also concordant for subtype. This was confirmed by Bertelsen *et al.* (1977); however, it is of note that even among MZ twins homotypia was not complete. Thus in nine out of 45 pairs concordant for affective illness, the proband showed one subtype of disorder while the co-twin showed the other. The estimated MZ/DZ concordance ratio for any affective disorder in the co-twin of a proband with unipolar disorder was between 2 and 3; less than that for a bipolar disorder proband which was estimated at 4. Further evidence for the role of genetic factors in unipolar depression was provided by Torgersen (1986) and McGuffin *et al.* (1991).

In the latter study, based on a sample from the Maudsley Hospital twin register, formal biometrical model fitting suggested the importance of both additive genetic and common environmental factors in the transmission of unipolar disorder. However, subsequent and more complete analyses of these data applying the diagnostic criteria in a more restrictive fashion, corresponding to a population lifetime risk of major depression of around 4–8%, failed to find evidence for a common environmental factor but estimated heritability to be between 48% and 75% (McGuffin *et al.*, 1996). A large population based sample of female twins (Kendler *et al.*, 1992a) similarly failed to find evidence for a common environmental contribution. However in this study, estimates of heritability were initially much lower, being between 24% and 45% depending on the definition of major depression used. This is possibly related to a much more inclusive interpretation of the DSM-IIIR diagnostic criteria, which resulted in lifetime prevalence estimates for MD ranging from 12% to more than 30%. Such findings are consistent with the hypothesis that more severe forms of depression show greater genetic influence. However measurement errors are included in standard biometrical models as nonshared or 'residual' environment. Hence if the error rate is high heritability may be falsely deflated. In a later analysis based on the results of a second wave of interviews, Kendler *et al.* (1993a) showed that reassessment improved detection of lifetime depression and resulted in heritability estimates close to those from the hospital based study (McGuffin *et al.*, 1996).

Several recent large twin registry studies have also explored the influence of genetic factors on MD in male twins (Kendler and Prescott, 1999; Lyons *et al.*, 1998) suggesting similar estimates of heritability (36–39%) as the population based findings for females. Kendler and Prescott (1999) explored further the genetic correlation in the liability to MD in the two sexes and found this to be +0.57. They concluded that although MD is equally heritable in men and women and most genetic risk factors influence liability to MD similarly in both sexes, genes may exist that act differently on the risk for the disorder between the two sexes.

3.2 Anxiety

Twin studies of anxiety disorders have been less consistent in their evidence for a genetic influence. Older studies investigating broad concepts of anxiety neurosis indicated a marked difference in the concordance rates between MZ and DZ twins, implying a genetic effect (Slater and Shields, 1969; Torgersen, 1983). A more recent population based study using an Australian twin registry failed to find significant differences between MZ and DZ twins for GAD, PD or agoraphobia with panic (Andrews *et al.*, 1990). This conflict in findings may be related to the severity of illness under investigation: population based studies perhaps being more likely to come in contact with milder forms of disorder, which possibly show less genetic influence. There is, however, evidence from other population based studies that even broadly defined anxiety disorders may be heritable (Kendler *et al.*, 1992b). Twin studies focusing specifically on panic disorder and minor phobias are also indicative of a genetic etiology for these disorders (Perna *et al.*, 1997; Rose and Ditto, 1983; Torgersen, 1979; Weissman, 1993). Taken together the evidence indicates a genetic contribution to anxiety disorders, although this is more consistent for PD.

4. Adoption studies

There have been few adoption studies exploring depression and anxiety disorders. It has been suggested that adoptees and their families are generally not representative of the population as a whole, both in the prevalence and nature of the psychopathology experienced and in their social environment (Moore and Fombonne, 1999). Further complications arise with the possibility of selective placement, with children being matched with adoptive parents on the basis of background, and in relation to age at adoption, which affects the degree to which genetic and environmental influences can be disentangled. Despite these problems several studies focusing on affective disorder have supported the evidence from twin studies for a genetic effect. In an adoptees' family design, Wender *et al.* (1986) reviewed hospital records of the biological relatives of adoptees with affective illness and compared these both with adoptive relatives and the biological families of matched controls. Their findings indicated an eight-fold increase in unipolar depression and a 15-fold increase in suicide among the biological relatives of affected adoptees. Similar indications for a genetic influence were found by Cadoret (1978) who, using an adoptees' study method, showed that the adopted away offspring of patients with affective illness had higher rates of affective disorder than those adoptees whose natural parents were psychiatrically well. The findings, however, have not all been consistent. Von Knorring *et al.* (1983), in a study which reviewed health insurance records, found little support for specific genetic influences in affective disorder.

5. Heterogeneity and comorbidity

Several investigators have explored the possibility of etiological heterogeneity (the existence of a mixture of illness forms in which different causative mechanisms are operating). It has been suggested that there is little evidence for a genetic/nongenetic distinction for MD (Andrew *et al.*, 1998). However, the possibility that certain clinical features such as number of depressive episodes, duration of episode and age of onset as indicators of more heritable subtypes has been raised (Kendler *et al.*, 1999a; McGuffin *et al.*, 1996). In studies of panic disorder it has been suggested that hypersensitivity to CO_2 (Perna *et al.*, 1996) and early onset of disorder (Goldstein *et al.*, 1997) might define subtypes of panic disorder which are associated with greater familial loading.

Merikangas *et al.* (1994), in a family study design, explored the relationship between depression, anxiety and alcoholism and demonstrated evidence for the cotransmission of depression and anxiety within families. Weaker associations between depression and alcoholism and between anxiety and alcoholism were also seen. This work builds on older studies such as Weissman *et al.* (1984) and Angst *et al.* (1990) which have consistently shown the co-occurrence of anxiety and depression within families to be common. The genetic relationship between these disorders has been explored further in several twin studies. Population studies based both on self-report measures (Kendler *et al.*, 1987) and on personal interview (Kendler *et al.*, 1992b) suggest a substantial degree of genetic overlap between anxiety disorder and MD. As noted previously for depression, shared environment appeared to play no role in the etiology of either condition but genetic factors were important for both. It is interesting to note that the genetic factors contributing to both disorders appeared to be perfectly correlated. This therefore suggests that nonoverlapping, nonshared environmental factors may account for the existence of two apparently different syndromes. This multivariate approach to the investigation of comorbidity in psychiatric disorders has been extended to include PD, phobia, bulimia and alcoholism (Kendler *et al.*, 1995a). Two sets of shared genetic factors have been suggested by this analysis. The first contributes to MD and GAD and the second to phobia, PD and bulimia: genetic effects for alcoholism appear to be specific in nature. It is not clear, however, whether these findings can be extrapolated to clinical samples given the somewhat more liberal interpretation of the diagnostic criteria that suggest that the lifetime prevalence of MD and GAD is 31% and 24% respectively. There is also evidence that the covariation of anxiety and depressive symptoms in childhood (maternally rated) may also be explained by a common set of genes (Thapar and McGuffin, 1997).

6. The role of environment

It is clear from the findings presented above that twin studies provide strong evidence for the role of environmental factors in the liability to MD. It is well established that certain forms of adversity, such as life events of severe or moderately severe threat, have a strong association with the onset of such illness (Bebbington, 1985; Brown and Harris, 1978). In the past it has been common practice, particularly in Europe, to think of depression as either 'reactive' to adversity or endogenous, arising 'out of the blue' without an obvious precipitant (McGuffin and Katz, 1993). It was felt that this distinction, which could be made on the basis of the symptomatology experienced by the patient, was likely to be indicative of different etiological mechanisms: perhaps the endogenous type represents a more innate (possibly genetic) predisposition whereas the reactive form may be precipitated by stress. If this is valid then it would predict a strong relationship between exposure to life events/adversity and the type of depressive subtype manifest. In the Camberwell Collaborative Depression Study (Bebbington *et al.*, 1888) this hypothesis was tested using a sample of 130 probands who had attended the Maudsley Hospital and received a diagnosis of unipolar depression. The hypothesis that the endogenous group would be relatively independent of prior life events was, however, not confirmed. Furthermore, when the study was extended to include the first degree relatives of unipolar probands, it was found that depression was no more common among the relatives of those probands whose illness was not related to prior adversity than among the relatives of probands whose depression clearly was (McGuffin *et al.*, 1988a). This indicated a lack of the predicted inverse relationship between the presence of familial loading and reactivity to stress.

It was particularly interesting that not only was there an increased rate of current depression among relatives of all depressive probands (McGuffin *et al.*, 1988b) but the frequency of reported life events in these families was also significantly higher than in a sample from the general population (McGuffin *et al.*, 1988a). This finding was still evident when life events related to the proband's illness were excluded from the analysis and suggested that adverse life experiences also aggregate within these families. The authors further suggested that part of the association between life events and depression may be due to this familial aggregation.

These findings suggest the possibility that familial factors may be influencing the liability to MD indirectly, by predisposing an individual to a more adversive environment (Kendler, 1998). There is further evidence that some individuals are more prone to report or to experience adverse life events (Andrews, 1981) and several studies have indicated that this may be due to a genetic influence (Kendler *et al.*, 1993b; Lyons *et al.*, 1993; Plomin *et al.*, 1990). Kendler *et al.* (1993b) categorized life events, according to their likely range of impact, into 'personal' events which focused primarily on the proband and 'network' events where the impact extended to other individuals in the proband's social circle. Model fitting using twin data suggested three separate classes of life event: network events in which twin resemblance appeared to result purely from common environmental effects, personal events (being robbed /assaulted, experiencing financial or marital problems, personal injury or illness) in which twin resemblance appeared to be due to genetic influence only, and interpersonal difficulties where both genetic and common environmental factors appeared to operate. The distinct familial origins of different categories of events is underlined by a recent sib-pair study (Farmer *et al.*, 2000). This showed positive correlations between siblings for events classified as 'independent' (that is those events not likely to result from the subject's deliberate actions) but that this was almost entirely explained by 'shared' events such as death or illness of a parent. Nevertheless it appears that some life events (or how they are reported) are influenced by genetic factors. Further to this there is evidence that the genetic factors operating in depression and in the predisposition to life events are positively correlated, thereby implying an

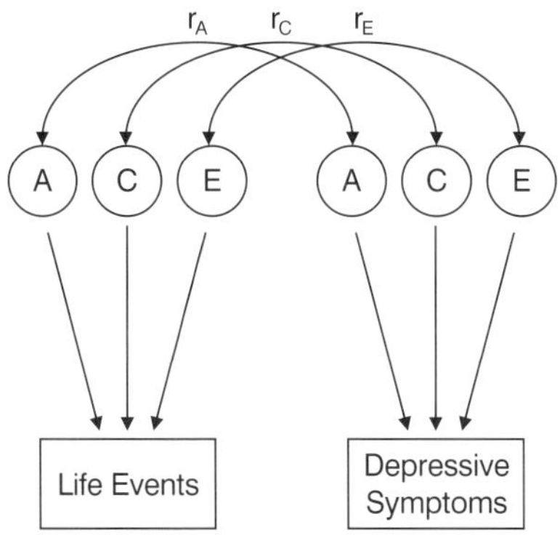

Figure 2 A bivariate model for the covariation of life events and depressive symptoms. A, additive genes; C, shared environment; E, nonshared environment; r_A, genetic correlation; r_C, shared environment correlation; r_E, nonshared environment correlation (Thapar *et al.*, 1998).

association via a common set of genes (Kendler and Karowski-Shuman, 1997; Silberg *et al.*, 1999; Thapar *et al.*, 1998) (*Figure 2*). The degree to which the relationship between stressful life events and depression is due to a causal relationship, with life events contributing directly to the onset of MD, was explored by Kendler *et al.* (1999b) in their population based sample of female twins. They found strong evidence that the observed association was, at least in part, causal.

There is evidence that a genetic liability to depression may also increase 'an individual's sensitivity to the depressogenic effects' of life events (Kendler, 1998). Evidence of such a gene–environment interaction has come from studies such as that of Kendler *et al.* (1995b), in which onset of MD was shown to be predicted not only by separate main effects for genetic factors and severe life events, but also by an interaction between the two. The absolute increase in risk for MD given a severe life event was found to be twice as high in those at high genetic risk compared with those at low risk. Similar gene–environment interactions have been demonstrated. For example, it has been shown that a marital relationship can act in reducing the impact of inherited liability to depressive symptoms (Heath *et al.*, 1998).

7. Molecular genetic studies

Having demonstrated a genetic contribution to anxiety and depressive disorders, the next stage of the investigation draws on molecular genetic methods in an attempt to locate and identify the genes involved. Depression and anxiety do not follow Mendelian segregation patterns, thereby indicating a complex mode of inheritance. It is likely that the genetic influences on these disorders are multigenic in nature, acting in concert with environmental factors. Segregation analyses have been unable to confirm specific modes of inheritance.

7.1 Linkage studies

Linkage studies of affective disorder have focused primarily on the bipolar phenotype (see Chapter 10). There have been few recent linkage studies of MD. Older studies, predating the availability of highly polymorphic DNA markers, made use of classical genetic markers such as blood group antigens and protein polymorphisms, yielding inconclusive results (Cox *et al.*, 1989; Goldin *et al.*, 1983; Tanna *et al.*, 1976, 1977; Wilson *et al.*, 1991). More recently, analyses using DNA polymorphisms have concentrated on excluding linkage to chromosomal regions known to contain possible candidate genes, such as those involved in neuroendocrine or neurotransmitter systems (Neiswanger *et al.*, 1998; Wang *et al.*, 1992), or those implicated in bipolar disorder (Balciunine *et al.*, 1998). Within the anxiety disorders, linkage studies have focused predominantly on PD. It has been suggested that family segregation patterns for PD are consistent with a single autosomal locus (Crowe, 1990). Linkage studies using classical markers (Crowe *et al.*, 1987) and candidate gene polymorphisms (Crowe *et al.*, 1997; Mutchler *et al.*, 1990) have, however, not resulted in definitive findings.

Many of the above linkage studies of depression (Balciunine *et al.*, 1998; Cox *et al.*, 1989; Goldin *et al.*, 1983; Tanna *et al.*, 1977; Wang *et al.*, 1992; Wilson *et al.*, 1991) and PD (Crowe *et al.*, 1987, 1997; Mutchler *et al.*, 1990) used parametric linkage methods. Such conventional LOD score linkage analysis requires precise knowledge of the disease mode of transmission and assumes that a gene of major effect is operating (McGuffin and Owen, 1996). There is little evidence for rare single gene families for MD and, as stated previously, the mode of inheritance is unknown. This strategy, which usually requires the assumption of a monogenic model of inheritance, is therefore open to the possibility of error. An alternative linkage strategy requiring no information about the mode of inheritance, and useful in oliogenic or polygenic disease, is the affected sib-pair method. Although such an approach has been used in several of the above linkage studies of depression (Wilson *et al.*, 1991) and panic disorder (Crowe *et al.*, 1987; Knowles *et al.*, 1998), these studies have generally involved only a small number of pedigrees. When combined with a genomewide screen the sib-pair method offers the possibility of a systematic search for linkage in these disorders and a recent genome scan for PD has been reported (Knowles *et al.*, 1998).

7.2 Association studies

Given that the individual gene effects in complex disorders may be modest and therefore not detectable by feasible linkage studies unless sample sizes are large (Risch and Merikangas, 1996), some investigators have explored the use of association studies to locate and identify susceptibility genes in depression and anxiety phenotypes. However, as linkage disequilibrium occurs over very short distances this strategy has typically been restricted to the testing of candidate genes.

Candidate genes for MD and anxiety have been suggested mainly by the molecular mechanisms implicated in the action of therapeutic drugs and the putative pathophysiology for these disorders. Of particular interest are genes associated with neurotransmitter systems. Many of the association studies associated with these phenotypes have focused on the serotonergic system and in particular the serotonin transporter gene located on chromosome 17 (17q11.1–17q12). Many antidepressant and anxiolytic drugs, such as tricyclic antidepressants and the newer serotonin selective reuptake inhibitors, influence the transporter facilitated uptake of serotonin. Several polymorphisms of the serotonin transporter (SERT) gene have been described.

A 44 bp insertion/deletion polymorphism in the transcriptional control region of the gene has been described by Heils *et al.* (1996). The short version of this promoter polymorphism has been shown to be associated with decreased transcriptional efficiency resulting in reduced serotonin transporter expression. Since an original report of association between this genetic variant and anxiety related personality traits (Lesch *et al.*, 1996) various studies have explored this further. Despite Katsuragi *et al.* (1999) replicating this finding in a Japanese sample, several studies have failed to do so (Ball *et al.*, 1997; Ebstein *et al.*, 1997; Gelernter *et al.*, 1998). Ohara *et al.* (1998a) described a tendency for the shorter variant of this polymorphism to be more frequent in patients diagnosed with an anxiety disorder using DSM-IV criteria. Associations of this polymorphism with affective disorders have also been described (Collier *et al.*, 1996a). Similar associations have been shown for both unipolar and bipolar subtypes separately (Furlong *et al.*, 1998), although this too has not been replicated (Ohara *et al.*, 1998b).

A further variable number tandem repeat polymorphism, in the second intron of the serotonin transporter gene, has also been described (Lesch *et al.*, 1994). Several variants of this polymorphism have been found to be associated with affective disorder (Collier *et al.*, 1996b; Rees *et al.*, 1997). In particular an association of the nine copy variant with unipolar depression has been reported (Battersby *et al.*, 1996; Ogilvie *et al.*, 1996). However, this has not been replicated in independent studies (Kunugi *et al.*, 1997a). Gelerneter *et al.* (1999) investigated the range of allele frequency variation for both the promoter and second intron polymorphisms of the serotonin transporter gene in seven different populations and found significant global variation. Linkage disequilibrium varied among the populations, suggesting a large potential for population stratification in association studies using population controls – as the majority of studies exploring depressive and anxiety related phenotypes have done. Other positive association findings for this gene, using family controls, have been described for autism and obsessive compulsive disorder (Cook *et al.*, 1997; McDougle *et al.*, 1998). Taken as a whole, therefore, the various association findings suggest that polymorphic variation of this gene results in some phenotypic effect but it is unclear at present what phenotypic construct best captures this (Gelerneter *et al.*, 1998).

Other candidate gene association studies in depression and anxiety have investigated the dopamine receptor genes (Serretti *et al.*, 1998a), the tryosine hydroxylase gene (Serretti *et al.*, 1998b) genes related to the γ-aminobutryic acid system (Oruc *et al.*, 1997) and the catechol-O-methyltransferase gene (Kunugi *et al.*, 1997b; Ohara *et al.*, 1998c) among others. Candidate gene approaches are limited in that they are only as good as the current understanding of the disease pathophysiology (Craddock and Owen, 1996). However, the possibility of a genomewide search for association, similar to that for linkage studies, is increasingly feasible with the development of high throughput genotyping techniques such as DNA pooling (Daniels *et al.*, 1998).

7.3 Future and prospects for functional genomics

As there is something uniquely human about psychological disturbance it is difficult to find plausible animal models for depression and anxiety. Despite the difficulties inherent in this there are several models that are worthy of further investigation. It has long been suggested, for example, that the activity of mice exposed to a brightly lit, open field environment can be used as an empirical measure of 'emotionality' or fearfulness (Hall *et al.*, 1938). Similarly particular mouse strains, which vary in their susceptibility to develop 'learned helplessness' (Henn and Edwards, 1994), may provide a model for depression. Such mouse models are potentially useful as it has been suggested that mice are likely to have a similar genetic architecture to humans for complex behaviors due to extensive synteny homology; the similar ordering of genetic loci in chromosomal regions. Mapping of relevant mouse genes can therefore augment gene finding approaches in humans by the identification of candidate regions for further investigation. For example, mouse strains developed for both high and low levels of activity in an open field environment have helped to demonstrate linkage for this behavior to several chromosomal regions (Flint *et al.*, 1995; Talbot *et al.*, 1999), by use of quantitative trait locus analyses.

8. Conclusion

Quantitative genetic methods have been instrumental in disentangling the relative contributions of genetic and environmental influences involved in depression and anxiety. It seems likely that such strategies will continue to play an important role in the elucidation of such etiological factors. In particular the identification of the more heritable components and combinations of these phenotypes and the clarification of gene and environment interaction will be increasingly informative for molecular genetic studies. Systematic gene finding approaches have not yet been applied in the study of MD and most anxiety disorders. However, such methods are being used increasingly and genes for these disorders are likely to be located in the foreseeable future. With the convergence of quantitative and molecular genetic methods, the exploration of etiological processes, which result in depression and anxiety disorders, is intensifying. Increasingly this partnership is also incorporating the social and environmental perspective leading to a more complex and intriguing understanding of how risk factors for these disorders interrelate.

References

American Psychiatric Association (1994) *Diagnostic and Statistical Manual of Mental Disorders*, 4th Edn. American Psychiatric Association, Washington, DC.

Andrew, M., McGuffin, P. and Katz, R. (1998) Genetic and non-genetic subtypes of major depressive disorder. *Br. J. Psychiatry* **173**: 523–526.

Andrews, G. (1981) A prospective study of life events and psychological symptoms. *Psycholog. Med.* **11**: 795–801.

Andrews, G., Stewart, G., Allen, R. and Henderson, A.S. (1990) The genetics of six neurotic disorders: a twin study. *J. Affective Disord.* **19**: 23–29.

Angst, J. (1966) Zur Atiologie und Nosologie endogener depressiver Psychosen. *Monographen ans der Neurologie und Psychiatrie,* No. 112. Springer, Berlin.

Angst, J., Vollrath, M., Merikangas, K., Ernst, C. (1990) Comorbidity of anxiety and depression in the Zurich Cohort study of young adults. In: *Comorbidity of Mood and Anxiety Disorders* (eds J.D. Maser and C.R. Cloninger). American Psychiatric Press, Washington DC, pp. 123–137.

Balciuniene, J., Yuan, Q.P., Engstrom, C., Lindblad, K., Nylander, P.O., Sundvall, M., Schalling, M., Pettersson, U., Adolfsson, R. and Jazin, E.E. (1998) Linkage analysis of candidate loci in families with recurrent major depression. *Mol. Psychiatry* **3**: 162–168.

Ball, D., Hill, L., Freeman, B., Eley, T.C., Strelau, J., Riemann, R., Spinath, F.M., Angleitner, A. and Plomin, R. (1997) The serotonin transporter gene and peer-rated neuroticism. *Neuroreport* **8**: 1301–1304.

Battersby, S., Ogilvie, A.D., Smith, C.A.D., Blackwood, D.H.R., Muir, W.J., Quinn, J.P., Fink, G., Goodwin, G.M. and Harmar, A.J. (1996) Structure of a variable number tandem repeat of the serotonin transporter gene and association with affective disorder. *Psychiatric Genet.* **6**: 177–181.

Bebbington, P. (1985) Psychosocial etiology of schizophrenia and affective disorders. In: *Psychiatry* (ed. R. Michels). Lippincott, Philadelphia.

Bebbington, P., Brugha, T., McCarthy, B., Potter, J., Sturt, E., Wykes, T., Katz, R. and McGuffin, P. (1988) The Camberwell Collaborative Depression Study I. Depressed probands: adversity and the form of depression. *Br. J. Psychiatry* **152**: 754–765.

Bell, J. (1995) Social phobias and specific phobias. In: *Psychiatric Secrets* (ed. J.L. Jacobson and A.M. Jacobson). Hanley and Belfus, Philadelphia, pp. 84–87.

Bertelsen, A., Harvald, B. and Hauge, M. (1977) A Danish twin study of manic–depressive disorders. *Br. J. Psychiatry* **130**: 330–351.

Blazer, D.G., Kessler, R.C., McGonagle, K.A. and Swartz, M.S. (1994) The prevalence and distribution of major depression in a national community sample: The national comorbidity survey. *Am. J. Psychiatry* **151**: 979–986.

Brown, G. W. and Harris, T. (1978) The social origins of depression: A study of psychiatric disorder in women. Tavistock, London.

Cadoret, R.J. (1978) Evidence of genetic inheritance of primary affective disorder in adoptees. *Am. J. Psychiatry* **135**: 463–466.

Collier, D.A., Stober, G., Li, T., *et al.* (1996a) A novel functional polymorphism within the promoter of the serotonin transporter gene: possible role in the susceptibility to affective disorder. *Mol. Psychiatry* **1**: 453–460.

Collier, D.A., Arranz, M.J., Sham, P,. *et al.* (1996b) The serotonin transporter is a potential susceptibility factor for bipolar affective disorder. *Neuroreport* **7**: 1675–1679.

Cook, E.H., Courchesne, R., Lord, C., Cox, N.J., Yan, S., Lincoln, A., Haas, R., Courchesne, E. and Leventhal, B.L. (1997) Evidence of linkage between the serotonin transporter and autistic disorder. *Mol. Psychiatry* **2**: 247–250.

Cooper, J.E., Kendell, R.E., Gurland, B.J., Sharpe, L., Copeland, J.R.M. and Simon R. (1972) *Psychiatric Diagnosis in New York and London*, Maudsley Monograph 20. Oxford University Press, London.

Cox, N., Reich, T., Rice, J., Elston, R., Schober, J. and Keats, B. (1989) Segregation and linkage analyses of bipolar and major depressive illnesses in multigenerational pedigrees. *J. Psychiatr. Res.* **23**: 109–123.

Craddock, N. and Owen, M.J. (1996) Modern molecular genetic approaches to psychiatric disease. *Br. Med. Bull.* **52**: 434–452.

Crowe, R.R. (1990) Panic disorder: genetic considerations. *J. Psychiatr. Res.* **24(suppl. 2)**: 129–134.

Crowe, R.R., Noyes, R., Pauls, D.L. and Slyman, D. (1983) A family study of panic disorder. *Arch. Gen. Psychiatry* **40**: 1065–1069.

Crowe, R.R., Noyes, R., Wilson, A.F., Elston, R.C. and Ward, L.J. (1987) A linkage study of panic disorder. *Arch. Gen. Psychiatry* **44**: 933–937.

Crowe, R.R., Wang, Z., Noyes, R., Albrecht, B.E, Darlison, M.G., Bailey. M.E.S., Johnson, K.J. and Zoega, T. (1997) Candidate gene study of eight GABA-sub(A) receptor subunits in panic disorder. *Am. J. Psychiatry* **154**: 1096–1100.

Daniels, J., Holmans, P., Williams, N., Turic, D., McGuffin, P., Plomin, R., and Owen, M.J. (1998)

A simple method for analyzing microsatellite allele image patterns generated from DNA pools and its application to allelic association studies. *Am. J. Hum. Genet.* **62**: 1189–1197.

Ebstein, R.P., Gritsenko, I., Nemanov, L., Frisch, A., Osher, Y. and Belmaker, R.H. (1997) No association between the serotonin transporter gene regulatory region polymorphism and the Tridimensional Personality Questionnaire (TPQ) temperament of harm avoidance. *Mol. Psychiatry* **2**: 224–226.

Endicott, J. and Spitzer, R.L. (1978) A diagnostic interview. The schedule for affective disorders and schizophrenia. *Arch. Gen. Psychiatry* **35**: 837–844.

Falconer, D.S. (1965) The inheritance of liability to certain diseases, estimated from the incidence among relatives. *Ann. Hum. Genet.* **29**: 51–76.

Farmer, A., Harris, T., Redman, K., Sadler, S., Mahmood, A. and McGuffin, P. (2000) The Cardiff Depression Study – a sib pair study of life events and familiality in major depression. *Br. J. Psychiatry* **176**: 150–155.

Flint, J., Corley, R., DeFries, J.C., Fulker, D.W., Gray, J.A., Miller, S. and Collins, A.C. (1995) A simple genetic basis for a complex psychological trait in laboratory mice. *Science* **269**: 1432–1435.

Fryer, A.J., Mannuzza, S., Gallops, M.S., Martin, L.Y., Aaronson, C., Gorman, J.M., Liebowitz, M.R. and Klein, D.F. (1990) Familial transmission of simple phobias and fears. A preliminary report. *Arch. Gen. Psychiatry* **47**: 252–256.

Furlong, R.A., Ho, L., Walsh, C., Rubinsztein, J.S., Jain, S., Paykel, E.S., Easton, D.F. and Rubinsztein, D.C. (1998) Analysis and meta-analysis of two serotonin transporter gene polymorphisms in bipolar and unipolar affective disorders. *Am. J. Med. Genet.* **81**: 58–63.

Gelernter, J., Kranzler, H., Coccaro, E.F., Siever, L.J. and New, A.S. (1998) Serotonin transporter protein gene polymorphism and personality measures in African American and European American subjects. *Am. J. Psychiatry* **155**: 1332–1338.

Gelernter, J., Cubells, J.F., Kidd, J.R., Pakstis, A.J. and Kidd, K.K. (1999) Population studies of polymorphisms of the serotonin transporter protein gene. *Am. J. Med. Genet.* **88**: 61–66.

Gershon, E.S., Bunney, W.E., Leckman, J.F., Eerdewegh, M. and DeBauche, B.A. (1976) The inheritance of affective disorders: a review of data and hypotheses. *Behavior Genet.* **6**: 227–261.

Goldin, L.R., Gershon, E.S., Targum, S.D., Sparkes, R.S. and McGinniss, M. (1983) Segregation and linkage analyses in families of patents with bipolar, unipolar, and schizoaffective mood disorders. *Am. J. Hum. Genet.* **35**: 274–287.

Goldstein, R.B., Wickramaratne, P.J., Horwath, E. and Weissman, M.M. (1997) Familial aggregation and phenomenology of 'early'-onset (at or before age 20 years) panic disorder. *Arch. Gen. Psychiatry* **54**: 271–278.

Hall, C.S. (1938) The inheritance of emotionality. *Sigma Xi Quart.* **26**: 17–27.

Heath, A.C., Eaves, L.J. and Martin, N.G. (1998) Interaction of marital status and genetic risk for symptoms of depression. *Twin Res.* **1**: 119–122.

Heils, A., Teufel, A., Petri, S., Stober, G., Riederer, P., Bengel, D. and Lesch, K.P. (1996) Allelic variation of human serotonin transporter gene expression. *J. Neurochem.* **66**: 2621–2624.

Henn, F.A. and Edwards, E. (1994) Animal models in the study of genetic factors in human psychopathology. In: *Genetic Studies in Affective Disorders. Overview of Basic Methods, Current Directions, and Critical Research Issues* (eds D.F. Papolos and H.M. Lachman). John Wiley, New York.

Judd, L.L., Kessler, R.C., Paulus, M.P., Zeller, P.V., Wittchen, H.U. and Kunovac, J.L. (1998) Comorbidity as a fundamental feature of generalized anxiety disorders: results from the National, Comorbidity Study (NCS). *Acta Psychiatr. Scand. Suppl.* **393**: 6–11.

Kahn, M. (1995) Introduction to DSM IV. In: *Psychiatric Secrets* (ed. J.L. Jacobson and A.M. Jacobson). Hanley and Belfus, Philadelphia, pp. 19–20.

Katsuragi, S., Kunugi, H., Sano, A., Tsutsumi, T., Isowaga, K., Nanko, S. and Akiyoshi, J. (1999) Association between serotonin transporter gene polymorphism and anxiety-related traits. *Biol. Psychiatry* **45**: 368–370.

Kendler, K. (1998) Major depression and the environment: a psychiatric genetic perspective. *Pharmacopsychiatry* **31**: 5–9.

Kendler, K.S. and Gardner, C.O. (1998a) Boundaries of major depression: an evaluation of DSM-IV criteria. *Am. J. Psychiatry* **155**: 172–177.

Kendler, K.S. and Gardner, C.O. (1998b) Twin studies of adult psychiatric and substance dependence disorders: are they biased by differences in the environmental experiences of monozygotic and dizygotic twins in childhood and adolescence? *Psycholog. Med.* **28**: 625–633.

Kendler, K.S. and Karkowski-Shuman, L. (1997) Stressful life events and genetic liability to major depression: genetic control of exposure to the environment? *Psycholog. Med.* **27**: 539–547.

Kendler, K.S. and Prescott, C.A. (1999) A population-based twin study of lifetime major depression in men and women. *Arch. Gen. Psychiatry* **56**: 39–44.

Kendler, K.S., Heath, A.C., Martin, N.G. and Eaves, L.J. (1987) Symptoms of anxiety and symptoms of depression. Same genes, different environments? *Arch. Gen. Psychiatry* **44**: 451–457.

Kendler, K.S., Neale, M.C., Kessler, R.C., Heath, A.C. and Eaves, L.J. (1992a) A population based twin study of major depression in women. The impact of varying definitions of illness. *Arch. Gen. Psychiatry* **49**: 257–266.

Kendler, K.S., Neale, M.C., Kessler, R.C., Heath, A.C. and Eaves, L.J. (1992b) Major depression and generalised anxiety disorder. Same genes, (partly) different environments? *Arch. Gen. Psychiatry* **49**: 716–722.

Kendler, K.S., Neale, M.C., Kessler, R.C., Heath, A.C. and Eaves, L.J. (1993a) The lifetime history of major depression in women. Reliability of diagnosis and heritability. *Arch. Gen. Psychiatry* **50**: 863–870.

Kendler, K.S., Neale, M., Kessler, R., Heath, A. and Eaves, L. (1993b) A twin study of recent life events and difficulties. *Arch. Gen. Psychiatry* **50**: 789–796.

Kendler, K.S., Walters, E.E., Neale, M.C., Kessler, R.C., Heath, A.C. and Eaves, L.J. (1995a) The structure of the genetic and environmental risk factors for six major psychiatric disorders in women. Phobia, generalised anxiety disorder, panic disorder, bulimia, major depression and alcoholism. *Arch. Gen. Psychiatry* **52**: 374–383.

Kendler, K.S., Kessler, R.C., Walters, E.E., MacLean, C., Neale, M.C., Heath, A.C. and Eaves, L.J. (1995b) Stressful life events, genetic liability and onset of an episode of major depression in women. *Am. J. Psychiatry* **152**: 833–842.

Kendler, K.S., Gardner, C.O. and Prescott, C.A. (1999a) Clinical characteristics of major depression that predict risk of depression in relatives. *Arch. Gen. Psychiatry* **56**: 322–327.

Kendler, K.S., Karkowski, L.M. and Prescott, C.A. (1999b) Causal relationship between stressful life events and the onset of major depression. *Am. J. Psychiatry* **156**: 837–841.

Kessler, R.C., McGonagle, K.A., Zhao, S., Nelson, C.B., Hughes, M., Eshleman, S., Wittchen, H.U. and Kendler, K.S. (1994) Lifetime and 12-month prevalence of DSM-III-R psychiatric disorders in the United States: Results from the National Comorbidity Study. *Arch. Gen. Psychiatry* **51**: 8–19.

Klerman, G.L. and Weissman, M.M. (1989) Increasing rates of depression. *J. Am. Med. Assoc.* **261**: 2229–2235.

Knowles, J.A., Fyer, A.J., Vieland, V.J., *et al.* (1998) Results of a genome-wide genetic screen for panic disorder. *Am. J. Med. Genet.* **81**: 139–147.

Kunugi, H., Hattori, M., Kato, T., Tatsumi, M., Sakai, T., Sasaki, T., Hirose, T. and Nanko, S. (1997a) Serotonin transporter gene polymorphisms: ethnic difference and possible association with bipolar affective disorder. *Mol. Psychiatry* **2**: 457–462.

Kunugi, H., Vallada, H.P., Hoda, F., Kirov, G., Gill, M., Aitchison, K.J., Ball, D., Arranz, M.J., Murray, R.M. and Collier, D.A. (1997b) No evidence for an association of affective disorders with high or low activity allele of catechol-o-methyltransferase gene. *Biol. Psychiatry* **42**: 282–285.

Lesch, K.P., Balling, U., Gross, J., Strauss, K., Wolozin, B.L., Murphy, D.L. and Riederer, P. (1994) Organisation of the human serotonin transporter gene. *J. Neural Trans.* **95**: 157–162.

Lesch, K.P., Bengel, D., Heils, A., Sabol, S.Z., Greenberg, B.D., Petri, S., Benjamin, J., Muller, C.R., Hamer, D.H. and Murphy, D.L. (1996) Association of anxiety-related traits with a polymorphism in the serotonin transporter gene regulatory region. *Science* **274**: 1527–1531.

Lyons, M.J., Goldberg, J., Eisen, S.A., True.W., Tsuang, M.T., Meyer, J.M. and Henderson, W.G. (1993) Do genes influence exposure to trauma? A twin study of combat. *Am. J. Med. Genet.* **48**: 22–27.

Lyons, M.J., Eisen, S.A., Goldberg, J., True, W., Lin, N., Meyer, J.M., Toomey, R., Faraone, S.V., Merla-Ramos, M. and Tsuang, M.T. (1998) A registry-based twin study of depression in men. *Arch. Gen. Psychiatry* 55: 468–472.

Lytton, H. (1977) Do parents create, or respond to, differences in twins? *Dev. Psychol.* **13**: 456–459.

Marks, I.M. (1986) Genetics of fear and anxiety disorders. *Br. J. Psychiatry* **149**: 406–418.

McDougle, C.J., Epperson, C.N., Price, L.H. and Gelernter, J. (1998) Evidence for linkage disequilibrium between serotonin transporter protein gene (SLC6A4) and obsessive compulsive disorder. *Mol. Psychiatry* **3**: 270–273.

McGuffin, P. and Katz, R. (1986) Nature, nurture and affective disorder. In: *The Biology of Depression* (ed. J.W.F. Deakin). Gaskell, London, pp. 26–51.

McGuffin, P. and Katz, R. (1993) Genes, adversity and depression. In: *Nature, Nurture and Psychology* (eds R. Plomin & G.E. McClearn). American Psychological Press, Washington DC, pp. 217–230.

McGuffin, P. and Martin, N. (1999) Behaviour and Genes. *Br. Med. J.* **319**: 37–40.

McGuffin, P. and Owen, M.J. (1996) *Molecular Genetic Studies of Schizophrenia*. Cold Harbor Symposia on Quantitative Biology Vol. **61**. Cold Spring Harbor Laboratory Press, New York, pp. 815–822.

McGuffin, P., Katz, R. and Bebbington, P. (1988a) The Camberwell Collaborative Depression Study. III. Depression and adversity in the relatives of depressed probands. *Br. J. Psychiatry* **152**: 775–782.

McGuffin, P., Katz, R., Aldrich, J. and Bebbington, P. (1988b) The Camberwell Collaborative Depression Study. II. Investigation of family members. *Br. J. Psychiatry* **152**: 766–774.

McGuffin, P., Katz, R. and Rutherford, J. (1991) Nature, nurture and depression: a twin study. *Psychol. Med.* **21**: 329–335.

McGuffin, P., Owen, M., O'Donovan, M.C., Thapar, A. and Gottesman, I.I. (1994a) Affective disorders. In: *Seminars in Psychiatric Genetics* (eds P. McGuffin, M. Owen, M.C. O'Donovan, A. Thapar and I.I. Gottesman). Gaskell, The Royal College of Psychiatrists, London, pp 110–127.

McGuffin, P., Owen, M., O'Donovan, M.C., Thapar, A. and Gottesman, I.I. (1994b) Neurotic disorders. In: *Seminars in Psychiatric Genetics* (eds P. McGuffin, M. Owen, M.C. O'Donovan, A. Thapar and I.I. Gottesman). Gaskell, The Royal College of Psychiatrists, London, pp. 128–145.

McGuffin, P., Katz, R., Watkins, S. and Rutherford, J. (1996) A hospital-based twin register of the heritability of DSM-IV unipolar depression. *Arch. Gen. Psychiatry* **53**: 129–136.

Merikangas, K.R., Risch, N.J. and Weissman, M.M. (1994) Comorbidity and co-transmission of alcoholism, anxiety and depression. *Psychol. Med.* **24**: 69–80.

Moore, J. and Fombonne, E. (1999) Psychopathology in adopted and nonadopted children: A clinical sample. *Am. J. Orthopsychiatry* **69**: 403–409.

Mutchler, K., Crowe, R.R., Noyes, R. and Wesner, R.W. (1990) Exclusion of the tryosine hydroxylase gene in 14 panic disorder pedigrees. *Am. J. Psychiatry* **147**: 1367–1369.

Neiswanger, K., Zubenko, G.S., Giles, D.E., Frank, E., Kupfer, D.J. and Kaplan, B.B. (1998) Linkage and association analysis of chromosomal regions containing genes related to neuroendocrine or serotonin function in families with early-onset, recurrent major depression. *Am. J. Med. Genet.* **81**: 443–439.

Noyes, R., Crowe, R.R., Harris, E.L., Hamra, B.J., McChesney, C.M. and Chaudhry, D.R. (1986) Relationship between panic disorder and agoraphobia. A family study. *Arch. Gen. Psychiatry* **43**: 227–232.

Noyes, R., Clarkson, C., Crowe, R.R., Yates, W.R. and McChesney, C.M. (1987) A family study of generalised anxiety disorder. *Am. J. Psychiatry* **144**: 1019–1024.

Ogilvie, A.D., Battersby, S., Bubb, V.J., Fink,G., Harmar, A.J., Goodwin, G.M. and Smith, C.A.D. (1996) Polymorphism in serotonin transporter gene associated with susceptibility to major depression. *Lancet* **347**: 731–733.

Ohara, K., Nagai, M., Suzuki, Y., Ochiai, M. and Ohara, K. (1998a) Association between anxiety disorders and a functional polymorphism in the serotonin transporter gene. *Psychiatry Res.* **81**: 277–279.

Ohara, K., Nagai, M., Tsukamoto, T., Tani, K., Suzuki, Y. and Ohara, K. (1998b) Functional polymorphism in the serotonin transporter promoter at the SLC6A4 locus and mood disorders. *Biol. Psychiatry* **44**: 550–554.

Ohara, K., Nagai, M., Suzuki, Y., Ochiai, M. and Ohara, K. (1998c) No association between anxiety disorders and catechol-o-methyltransferase polymorphism. *Psychiatry Res.* **80**: 145–148.

Oruc, L., Verheyen, G.R., Furac, I., Ivezic, S., Jakovljevic, M., Raeymaekers, P. and Van Broeckhoven, C. (1997) Positive association between the GABRA5 gene and unipolar recurrent major depression. *Neuropsychobiology* **36**: 62–64.

Perna, G., Bertani, A., Caldirola, D. and Bellodi, L. (1996) Family history of panic disorder and hypersensitivity to CO_2 in patients with panic disorder. *Am. J. Psychiatry* **153**: 1060–1064.

Perna, G., Cladirola, D., Arancio, C. and Bellodi, L. (1997) Panic attacks: a twin study. *Psych. Res.* **66**: 69–71.

Plomin, R., Lichtensein, P., Pedersen, N.L., McClearn, G.E. and Nesselroade, J.R. (1990) Genetic influence on life events during the last half of the life span. *Psycholog. Ageing* 5: 25–30.

Plomin, R., De Fries, J.C., McClearn, G. and Rutter, M. (1997a) Psychopathology. In: *Behavioral Genetics* (eds R. Plomin, J.C. DeFries, G. McCleam and M. Rutter). W.H. Freeman and Company, New York, pp. 169–193.

Plomin, R., De Fries, J.C., McClearn, G. and Rutter, M. (1997b) Behavioral Genetics is the Twenty-first century. In: *Behavioral Genetics* (eds R. Plomin, J.C. DeFries, G. McCleam and M. Rutter). W.H. Freeman and Company, New York, pp. 274–279.

Rees, M., Norton, N., Jones, I., *et al.* (1997) Association studies of bipolar disorder at the human serotonin transporter gene (hSERT; 5HTT). *Mol. Psychiatry* **2**: 398–402.

Reich, J. and Yates, W. (1988) Family history of psychiatric disorders in social phobia. *Comp. Psychiatry* **29**: 72–75.

Risch, N. and Merikangas, K. (1996) The future of genetic studies of complex human diseases. *Science* **273**: 1516–1517.

Rose, R.J. and Ditto, W.B. (1983) A developmental-genetic analysis of common fears from early adolescence to early adulthood. *Child Dev.* **54**: 361–368.

Scarr, S. (1968) Environmental bias in twin studies. *Eugenics Q.* **15**: 34–40.

Serretti, A., Macciardi, F., Cusin, C., Lattuada, E., Lilli, R. and Smeraldi, E. (1998a) Dopamine receptor D4 gene is associated with delusional symptomatology in mood disorders. *Psychiatry Res.* **80**: 129–136.

Serretti, A., Macciardi, F., Verga, M., Cusin, C., Pedrini, S. and Smeraldi, E. (1998b) Tyrosine hydroxylase gene associated with depressive symptomatology in mood disorder. *Am. J. Med. Genet.* **81**: 127–130.

Sham, P. (1996) Genetic epidemiology. *Br. Med. Bull.* 52: 408–433.

Silberg, J., Pickles, A., Rutter, M., Hewitt, J., Simonoff, E., Maes, H., Carbonneau, R., Murrelle, L., Foley, D. and Eaves, L. (1999) The influence of genetic factors and life stress on depression among adolescent girls. *Arch. Gen. Psychiatry* **56**: 225–232.

Slater, E. and Cowie, V. (1971) *The Genetics of Mental Disorder*. Oxford University Press, London.

Slater, E. and Shields, J. (1969) Genetic aspects of anxiety. In: *Studies of Anxiety* (ed. M.H. Lader). Royal Medico-Psychological Association, London, pp. 62–71.

Smith, C. (1974) Concordance in twins: methods and interpretation. *Am. J. Hum. Genet.* **26**: 454–466.

Stein, M.B., Chartier, M.J., Hazen, A.L., Kozak, M.V., Tancer, M.E., Lander, S., Furer, P., Chubaty, D. and Walker, J.R. (1998) A direct-interview family study of generalized social phobia. *Am. J. Psychiatry* **155**: 90–97.

Sturt, E., Kumakura, N. and Der, G. (1984) How depressing life is – lifelong morbidity risk for depressive disorder in the general population. *J. Affect. Disord.* **7**: 109–122.

Sullivan, P.F. and Kendler, K.S. (1998) Typology of common psychiatric syndromes. An empirical study. *Br. J. Psychiatry* **173**: 312–319.

Talbot, C.J., Nicod, A., Cherny, S.S., Fulker, D.W., Collins, A.C. and Flint, J. (1999) High-resolution mapping of quantitative trait loci in outbred mice. *Nature Genet.* **21**: 305–308.

Tanna, V.L., Winokur, G., Elston, R.C. and Go, R.C. (1976) A linkage study of pure depressive disease: The use of the sib-pair method. *Biol. Psychiatry* **11**: 767–771.

Tanna, V.L., Go, R.C., Winokur, G. and Elston, R.C. (1977) Possible linkage between group-specific component (Gc protein) and pure depressive disease. *Acta Psychiatr. Scand.* **55**: 111–115.

Thapar, A. and McGuffin, P. (1997) Anxiety and depressive symptoms in childhood – a genetic study of comorbidity. *J. Child Psychol. Psychiatry Allied Discipl.* **38**: 651–656.

Thapar, A., Harold, G. and McGuffin, P. (1998) Life events and depressive symptoms in childhood – shared genes or shared adversity? A research note. *J. Child Psychol. Psychiatry* **39**: 1153–1158.

Torgersen, S. (1979) The nature and origin of common phobic fears. *Br. J. Psychiatry* **134**: 343–351.

Torgersen, S. (1983) Genetic factors in anxiety disorders. *Arch. Gen. Psychiatry* **40**: 1085–1089.

Torgersen, S. (1986) Genetic factors in moderately severe and mild affective disorders. *Arch. Genet. Psychiatry* **43**: 222–226.

Tsuang, M. and Faraone, S.D. (1990) Diagnostic and methodological issues. In: *The Genetics of Mood Disorders* (eds M. Tsuang and S.D. Faraone). Johns Hopkins University Press, Baltimore, pp. 1–30.

Von Knorring, A.L., Cloninger, C.R., Bohman, M. and Sigvardsson, S. (1983) An adoption study of depressive disorders and substance abuse. *Arch. Gen. Psychiatry* **40**: 943–950.
Wang, Z., Crowe, R.R., Tanna, V.L. and Winokur, G. (1992) Alpha 2 adrenergic receptor subtypes in depression: a candidate gene study. *J. Affect. Disord.* **25**: 191–196.
Weissman, M.M. (1993) Family genetic studies of panic disorder. *J. Psychiatr. Res.* **27(suppl. 1)**: 69–78.
Weissman, M.M., Leckman, J.F., Merikangas, K.R., Gammon, G.D. and Prusoff, B.A. (1984) Depression and anxiety disorders in parents and children. Results from the Yale Family Study. *Arch. Gen. Psychiatry* **41**: 845–852.
Wender, P.H., Kety, S.S., Rosenthal, D., Schulsinger, F., Ortmann, J. and Lunde, I. (1986) Psychiatric disorders in the biological and adoptive families of adopted individuals with affective disorders. *Arch. Gen. Psychiatry* **43**: 923–929.
Wilson, A.F., Elston, R.C., Mallott, D.B., Tran, L.D. and Winokur, G. (1991) The current status of genetic linkage studies of alcoholism and unipolar depression. *Psychiatric Genet.* **2**: 107–124.
Wing, J.K., Cooper, J.E. and Sartorius, N. (1974) *Measurement and Classification of Psychiatric Symptoms; an Instruction Mmanual for the PSE and Catego Program*. Cambridge University Press, Cambridge.
World Health Organization (1992) *The ICD-10 Classification of Mental and Behavioural Disorders: Clinical Descriptions and Diagnostic Guidelines*. World Health Organization, Geneva.
Zerbin-Rudin, E. (1979) Genetics of affective disorders. In: *Origin, Prevention and Treatment of Affective Disorders* (eds M. Schou & E. Stromgren). Academic Press, London, pp. 185–197.

15

Genetic strategies for common complex disorders: ischemic heart disease as an example

Steve E. Humphries

1. Introduction

Since the identification of the first polymorphisms in the gene for apolipoprotein (apo) CIII more than 15 years ago (Rees *et al.*, 1985), there has been a vast amount of research carried out with the aim of identifying candidate genes (and specific functional mutations in these genes) that are involved in determining risk of ischemic heart disease (IHD). As with all molecular genetic analysis of human disorders, there have been two major goals of this research. The first of these was the expectation that the identification of disease-causing mutations would allow the different etiologies and pathophysiologies of IHD to be identified. The detection of association between a mutation in a particular enzyme and plasma levels of an end product or IHD risk identifies the enzyme (and the pathway that the enzyme is involved in) as a pharmocological target. Few would argue that this aspect of the use of molecular genetics has made significant contributions to understanding the pathological processes that are involved in determining IHD. By contrast there has been essentially no progress in achieving the second major goal of molecular genetic research, the development of DNA-based tests to identify individuals who are at increased risk of IHD because of their genetic make-up. Although this has led to disillusionment, a fully informed understanding of the nature of a multifactorial disease such as IHD, and a detailed understanding of the way common genetic variations of modest effect interact with environmental factors, suggested that this goal was always going to be difficult and technically challenging to achieve, and further discussion of this area would be useful

Analysis of Multifactorial Disease, edited by T. Bishop and P. Sham.

(Khoury and Wagener, 1995). This chapter focuses on genetic approaches that have proved useful in recent years to identify candidate genes for IHD and for IHD risk factor 'intermediate phenotypes', and to highlight our current understanding of the mechanisms by which these mutations interact with environmental factors to predispose an individual to IHD.

2. Risk factors for IHD

There are many personal characteristics that show strong population associations with the risk of developing IHD, including increasing age, male sex, use of cigarettes, presence of hypertension, diabetes and obesity. As shown in *Figure 1*, three major systems are involved in cardiovascular disease: the enzymes, apoproteins and receptors that are important in maintaining homeostasis of lipids as the chief energy source for cells; the proteins of the coagulation system which is important in maintaining blood flow; and proteins expressed by cells in the vessel wall which acts not simply as an inert barrier and conduit, but has a key role in maintaining adequate vessel tone and thus blood pressure. With this understanding, a number of biochemical IHD 'intermediate phenotypes' have been identified, with greater risk being associated with elevated plasma levels of cholesterol, triglycerides (Hokanson and Austin, 1996), lipoprotein(a) [Lp(a)] and fibrinogen (Meade *et al.*, 1986), and low levels of high density lipoprotein (HDL)-cholesterol with elevated levels of homocysteine (Frosst *et al.*, 1995) and serum positivity for *Chlamydia pneumonii* (Danesh *et al.*, 1997) attracting recent interest. It is now becoming clear that the inflammatory system occupies a central position in the

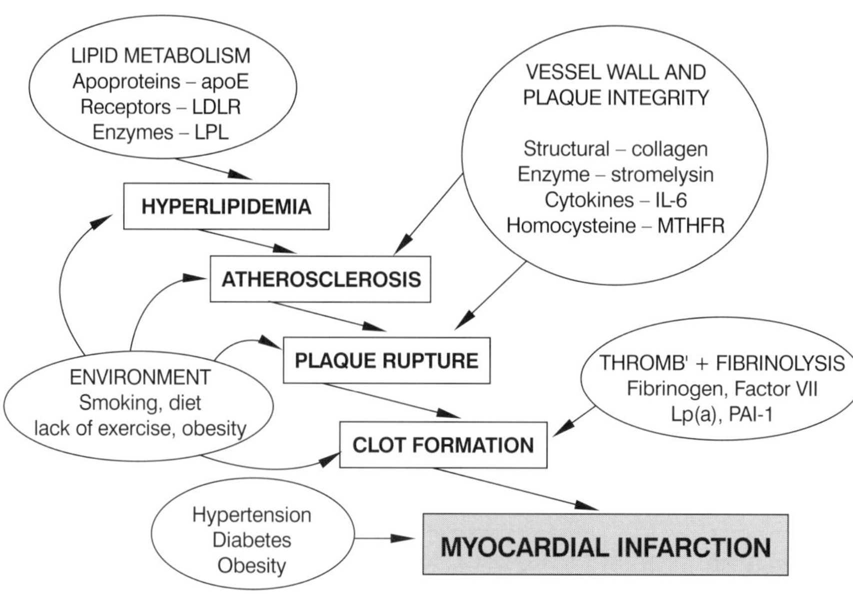

Figure 1. Model showing effect of genetic variations and environmental variations on determining levels of 'intermediate' phenotypes and risk of IHD.

atherosclerotic process (Ross, 1999), which may explain the role of *Chlamydia* as a risk factor and also why elevated levels of markers of the acute phase response system such as C-reactive protein (CRP) (Ridker *et al.*, 1997), as well as cytokines such as interleukin (IL)-6, are also elevated in IHD patients.

For any of these biochemical traits the overall distribution observed in the general population is a result of the combined and interactive effects of inherited factors, environmental factors and the impact of chance events. For each trait the relative contribution of genes and environment to the population variance will be different, and it will also be different in populations of different ethnic background and environments (e.g. depending on dietary homogeneity). The major goal of research is first to identify and then to quantify these variables. The hypothesis is that a more complete identification and description of these variables will result in a better ability to predict these individuals who, because of their position in the trait distribution, are at particular risk of developing disease. The framework for this is the determination for all key genes, the frequency of all functional mutations in the population under study, the magnitude of the effect of the mutation on the IHD intermediate phenotype in the individuals who carry it, and the range of expression of the mutations as they are influenced by other genetic and environmental factors. As discussed below, this gene–environment interaction appears, for IHD, to be the major area for further research. Firstly we need better analytical tools with which to recognize its presence with a high degree of statistical certainty, and secondly we need to be able to tease apart the molecular mechanisms which explain it. This will give us the understanding necessary to be able to give genotype specific life-style advice, or to tailor clinical decisions.

3. Heritability of IHD risk traits

The classical approach to estimating the relative contribution of genetic and environmental factors for a particular trait has been to use biometrical methods to compare individuals who share genes and environment to different extents, such as identical and nonidentical twins, sibling pairs and parents and children. These methods have demonstrated a significant impact of genetic variation (*Table 1*) in determining the plasma levels of cholesterol and lipoproteins such as low density lipoprotein (LDL) and HDL, with reported heritabilities in the range of 0.5–0.6, and a similar high heritability for plasma levels of apolipoproteins (apo) B and apoAI (Hamsten *et al.*, 1986). Family studies and complex segregation analyses have found evidence for a major gene determining levels of apoB and apo AI, with both environmental factors and genes of small or intermediate effect making a contribution (Hasstedt *et al.*, 1984; Moll *et al.*, 1989; Pairitz *et al.*, 1988). To date, the major genes having these effects have not been identified. A risk factor that has received much attention in recent years is elevated levels of small dense LDL particles (Austin *et al.*, 1993), for which strong evidence for the influence of major genes (Austin *et al.*, 1988) and good evidence for a multifactorial or at least multigenic involvement (Austin *et al.*, 1998; Rotter *et al.*, 1996) have been obtained. For other traits, such as fasting plasma levels of triglycerides, there is evidence for a

Table 1. Reported estimates of genetic heritability (h^2) for plasma lipid, lipoprotein and apolipoprotein traits from selected studies (for further references see Humphries *et al.*, 1999)

Trait	Heritability	Reference
Total cholesterol	~ 0.50	Hamsten *et al.*, 1986
Triglyceride	~ 0.30	Friedlander and Kark, 1987
LDL/ApoB	~ 0.60	Pairitz *et al.*, 1988
HDL/ApoAI	~ 0.60	Hasstedt *et al.*, 1984
Small dense LDL	~ 0.60	Austin *et al.*, 1988
Lp(a)	> 0.90	Lamon-Fava, 1991
Fibrinogen	0.3–0.5	Berg and Kierulf, 1989; Hamsten *et al.*, 1987; Reed *et al.*, 1994
Plasminogen activator inhibitor I	0.42	Hong *et al.*, 1997
Platelet number	> 0.50	Whitfield and Martin, 1985
Homocysteine	~ 0.55	Berg *et al.*, 1992

much greater effect of environmental factors, resulting in a low heritability of 0.3–0.4. However, the trait where genetic variation has the largest contribution is undoubtedly in determining levels of the lipoprotein Lp(a). Genetic variation determines in excess of 90% of the extraordinary degree of variability seen in the levels of this lipoprotein, mostly due to variation at the (*a*) gene locus itself. This subject has been reviewed extensively (Utermann *et al.*, 1995) and will not be covered here.

There is considerably less published data regarding the heritability of clotting factors. The extent to which genetic factors may determine plasma fibrinogen level is unclear, and published estimates range between 0.30 and 0.50 (Berg and Kierulf, 1989; Hamsten *et al.*, 1987; Reed *et al.*, 1994). Recently, complex segregation analysis has suggested that there is a major gene determining fibrinogen levels but with strong evidence for environmental factors such as age and sex interacting with the major gene effect in determining levels (Friedlander *et al.*, 1995). The obvious candidate for this major gene is the fibrinogen gene cluster itself, but no studies have yet addressed this possibility and other genes are also likely to be important. There is also good evidence for a high heritability for platelet number and related traits (Whitfield and Martin, 1985), as well as for more complex traits in which environmental factors are expected to play a major role; for example, obesity, in which heritability has been estimated to be in the range 0.3–0.5 (Bouchard, 1997). Finally, twin studies also suggest a high heritability for age at death from IHD (Marenberg *et al.*, 1994), confirming the importance of a family history of early heart disease as a major predictive risk factor.

As for many of the IHD intermediate phenotypes, this heritability is of the order of 0.5–0.6, the task of the genetic epidemiologist is to partition the 50–60% into the amounts explained by variation at key genetic loci, whereas the task of the classical epidemiologist is to partition the 'environmental' component into different causes (e.g. diet, lack of exercise, obesity, etc.). The simplest hypothesis for this genetic and environmental dissection would be that a complete description of the measurable environmental factors plus a knowledge of all mutations present in the population in the genes for the known enzymes, receptors and apoproteins already described in lipid synthesis, metabolism and catabolism, would partition the genetic component of heritability completely, although in the final analysis, as discussed above, interaction terms between genetic and environmental factors will also need to be included. The failure to explain all of the heritability would suggest the presence of genes encoding unidentified key proteins in the process, which could then be sought using additional candidate genes or analytical strategies.

4. Rare mutations of major effect on IHD

Severe hyperlipidemias are often familial, and confer a high risk of early coronary disease: the most common is familial combined hyperlipidemia with a prevalence of perhaps 1 in 100, the most severe is familial hypercholesterolemia (FH), with an estimated prevalence of 1 in 500 (Goldstein and Brown, 1989). It is characterized by hypercholesterolemia due to elevated plasma LDL levels, xanthomas, premature IHD and a family history of one or more of these. Angina, heart attacks or death typically occur in men aged between 30 and 50 years, and in women aged between 50 and 70 years (Goldstein and Brown, 1989), and those who smoke, are hypertensive or have other risk factors are at particularly high risk. FH is present in 5–10% of individuals who develop IHD under the age of 55 years (Patterson and Slack, 1972). FH is an autosomal dominant inherited disorder and is usually caused by a mutation in the LDL receptor gene (*LDLR*). More than 500 mutations in *LDLR* have been reported (Hobbs *et al.*, 1992; Day *et al.*, 1997, http://www.ucl.ac.uk/fh) and different types of mutation result in different residual functional LDL-receptor activity, varying from < 2% to 55% (Hobbs *et al.*, 1992). Patients with FH show considerable benefit in reduction of IHD risk by lipid-lowering therapy, such as the statin group of drugs (Neil *et al.*, 1999) which have been demonstrated to reduce mortality and morbidity in subjects with high plasma cholesterol (Shepherd *et al.*, 1995) and to induce regression of coronary artery stenosis in FH patients (Thompson *et al.*, 1995).

An *LDLR* mutation is not found in all patients with a clinical diagnosis of FH, however, and in approximately 3% of FH patients in the UK, northern Europe and the USA a defect has been detected in the apoB-100 gene (*apoB*), the ligand for LDL receptor (Tybjaerg Hansen and Humphries, 1992). This disorder has been designated familial defective apoB-100 (FDB) (Myant, 1993). Recently two reports of a third distinct genetic cause of monogenic FH have also been published (Haddad *et al.*, 1999; Varret *et al.*, 1999) but no specific gene has yet been identified.

For familial combined hyperlipidemia (FCH) the genetic inheritance pattern is not so clear-cut, and a major gene determining this phenotype has so far been identified only by linkage in a few families from Finland (Pajukantam *et al.*, 1998). FCH is likely to be more 'multifactorial' than FH but the identification of the gene(s) involved will be of interest whether the disorder is caused by a 'major gene' or the interplay of several. To date, apart from some extremely rare familial disorders, no other monogenic IHD families have been reported. On the coagulation side, although familial monogenic disorders have been identified they are mostly extremely rare. Mutations in the gene for clotting Factor V (*R506Q* designated Factor V Lieden) have been identified (Price and Ridker, 1997) with a carrier frequency of 2–3%, but this mutation increases risk mainly of venous thrombosis and has little effect on arterial thrombosis and risk of IHD.

5. Common mutations of small effect on IHD

By definition, for a multifactorial disease any single gene (and in particular any single mutation in that gene) can at best be having only a modest effect on causing the disease. The theoretical basis for this is that almost all IHD intermediate phenotypes show a normal distribution of levels in the general population. If a mutation has a substantial effect on the phenotype, for example by doubling levels, and this mutation is common (e.g. present in 5–10% of the population), the resulting distribution would be bimodal. For plasma cholesterol levels, mutations in the LDL receptor gene that cause FH have this magnitude of effect, but the frequency of FH is 1 in 500, and the population cholesterol curve shows no evidence for bimodality. By contrast, if a mutation has a modest effect on cholesterol levels (say ± 10%), carriers may be relatively common and still not result in bimodality. It is now generally accepted that the causes of the normal distribution of IHD risk factor traits/intermediate phenotypes are the result of the interplay of environment, chance factors and the combined contribution of common mutations of modest effect, rather than to rare mutations of large effect, like FH.

A good example of this is the gene encoding apoE. ApoE is synthesized in both the liver and the intestine and is found in the plasma associated with the triglyceride rich lipoproteins. It is the ligand for removal of these particles and thus occupies a central role in determining the metabolic fate of these lipoproteins. There are three common isoforms of the protein each differing from the other by one charged amino acid (reviewed in Davignon *et al.*, 1988). The apoprotein designated apoE3 is the most common allele in populations (frequency about 0.77) with apo E4 being present at a frequency of 0.15 and apoE2 at a frequency of 0.08 in western Caucasian populations (Davignon *et al.*, 1988). The frequency of this polymorphism differs within and between the different racial groups which reflects both population drift and the possible effect of selection. It is of interest that carriers of the apoE4 isoform have increased risk and apoE2 carriers reduced risk, of developing Alzheimer's disease (Mayeaux, 1998), although this will not be associated with 'fitness' in a Darwinian sense. Carriers of the apoE2 isoform, who represent approximately 12% of the population, have cholesterol levels that are, on average, 10% lower than those of E3E3 individuals, whereas apoE4 carriers represent approximately 20% of the population and have average cholesterol levels 5% higher than those of E3E3 individuals (Davignon *et al.*, 1988). However, because of the multifactorial nature of IHD, it is predicted that mutations of modest impact on IHD intermediate phenotypes such as apoE will have an even smaller impact on IHD risk itself. The largest effect of the apo E protein polymorphism is on the plasma levels of apo E itself, and around 20% of the variance in apoE levels in the population are determined solely by this protein polymorphism (Davignon *et al.*, 1988). The polymorphism has a smaller effect on levels of plasma apoB, with its effect on plasma cholesterol being smaller still (4–5%). Using the estimate of the effect on cholesterol levels associated with the polymorphism, it is expected that variation at the E locus explains approximately 2–3% of the risk in myocardial infarction and this prediction is supported by several recent studies (Hixson *et al.*, 1991).

A second good example of this is the G/A mutation at position –455 in the promoter of the beta fibrinogen gene (Thomas *et al.*, 1991). The three fibrinogen genes (α-, β- and γ-) are in a cluster of less than 50 kb on the long arm of chromosome 4, and each chain is synthesized as a separate mRNA, with the levels of all three mRNAs being coordinately controlled. The rate limiting step in the production of the mature fibrinogen molecule in the human hepatoma cell line HepG2 is the synthesis of the Bβ-polypeptide chain (Roy *et al.*, 1990), which in turn is influenced by the amount of its mRNA available. It is therefore likely that an alteration in the level of synthesis of the Bβ-chain may have an effect on the amount of mature fibrinogen secreted by the liver. The β-gene promoter has been well studied, and the region from –150 base pairs to the start of transcription appears to contain all the information required to act as a promoter in HepG2 cells and has been shown to bind proteins from a HepG2 cell nuclear extract. The sequence from –89 to –76 contains a conserved liver specific transcription element which binds hepatic nuclear factor 1, and deletion mapping shows that just upstream lies an IL-6 responsive element (Dalmon *et al.*, 1993), which has been identified in other genes as the motif CTGGGA. It is therefore likely that sequence changes in this region of the gene would have a direct effect on the rate of transcription and thus on plasma fibrinogen levels.

For the G/A sequence variation at position –455, the A is common, with an allele frequency of approximately 0.20 (95% confidence interval 0.18–0.22) in European populations (reviewed in Humphries *et al.*, 1997). In more than eight independent studies from 10 different laboratories and with samples of more than 20 000 healthy individuals the A-455 allele has been consistently associated with higher fibrinogen levels (e.g. Tybjaerg Hansen *et al.*, 1997), with those with one or more copies of the A-455 allele having on average 0.28 g l^{-1} higher fibrinogen levels than those with the genotype G/G (weighted average in healthy men, see Humphries *et al.*, 1997). The molecular mechanism of the effect has not been fully elucidated, but is likely to be due to a greater rate of transcription of the beta fibrinogen gene A allele compared with the G allele, leading to higher levels of secretion of the mature protein from the liver (Lane *et al.*, 1993). As elevated fibrinogen levels are an independent risk factor for IHD (Meade *et al.*, 1986), A allele carriers would be expected to have a higher risk of disease, and the size of this risk increase can be estimated from the prospective Northwick Park Heart Study (NPHS) in which an 84% higher risk of IHD was associated with 0.64 g l^{-1} increase in fibrinogen levels (Meade *et al.*, 1986). This suggests that, on average, men carrying the A allele would have a risk of IHD that would be 20–40% greater than men with the genotype GG. To put this genotype associated risk effect in context, in a large population based study of over 9000 men and women from Copenhagen (Tybjaerg Hansen *et al.*, 1997), factors such as age, sex, smoking and body mass index (BMI) explained 12 times as much of the sample variance in plasma fibrinogen levels as the G/A genotype (24% versus 2%). To put this risk impact another way, men who smoke had on average 20% higher fibrinogen levels than nonsmokers, whereas men with the genotype AA had levels that were only 9% higher than GG men, implying that smoking is associated with a much greater fibrinogen related risk of IHD than the G/A genotype. As will be discussed below in more detail, to detect such a modest increase in relative risk (RR) with statistical certainty requires a very large study and it is therefore not surprising that, despite the consistent effect on fibrinogen levels, most studies have not found a significant effect on IHD risk associated with this allele (e.g., Tybjaerg Hansen *et al.*, 1997).

6. Study design for intermediate traits and IHD risk

6.1 Linkage

Many of the recent advances in the genetic dissection of human disorders have relied on the collection of multi-generation families – often large pedigrees with many affected individuals. A linkage approach has then been applied, using highly polymorphic markers on every chromosome to carry out a 'genomewide scan' for the locus or loci determining the 'affected' status. Classically, this linkage strategy has been most powerful for single gene disorders of high penetrance, and there is little doubt that the locus for the LDL-receptor on chromosome 19 would have been detected by this approach in FH families. However, one of the problems with this approach is that of genetic heterogeneity, that is, where the phenotype is caused by mutations in two or more different unlinked genes. If a linkage study were to be carried out in a group of clinically defined FH patients, while in the majority the hypercholesterolemia would segregate with markers on chromosome 19, a proportion of families would have FDB (Myant, 1993) and would thus show linkage to chromosome 2 (*apo B* is on 2p 23,24), while others would show linkage to chromosome 1 (Varret *et al.*, 1999) leading to potential confounding of the linkage approach. There are statistical strategies with which to explore for potential genetic heterogeneity in linkage analysis, but it can be seen that as the degree of heterogeneity increases (i.e., three or more genes) the size of the sample needed to detect linkage with statistical confidence will increase; that is, if the disorder is caused by three different genes, a sample of at least three or more times the size will be needed.

The candidate gene approach turns linkage on its head and asks if there is evidence for or against variation at the gene locus determining the disorder or trait. Thus a recombination event (an affected relative without the haplotype or an unaffected relative with the haplotype) means the hypothesis of linkage is rejected, and the phenotype is not caused by a mutation in the candidate gene. Success with this approach has been reported recently by using a wide screen of candidate genes to analyze the disorder polycystic ovary disease (Urbanek *et al.*, 1999). However, for a common multifactorial disorder there are several complicating problems (reviewed in Greenberg, 1993; Moll, 1993). Firstly, the phenotype may not develop until middle age (e.g. clinically detectable atherosclerosis, angina or heart attack). Secondly, where the phenotype is affected by environmental factors, such as diet or obesity, an unaffected individual whose exposure to these environmental factors is reduced may not develop the phenotype at all. With a common disease such as hyperlipidemia, the probability of a spouse bringing in a defective allele causing the phenotype is high, and could result in affected offspring who do not carry the particular haplotype. Conversely, spouses may pass on genes that protect a particular individual, who has inherited the defective gene, from developing the phenotype.

An example of these problems can be seen in the report of a family with the disorder hypobetalipoproteinemia, where individuals have levels of apoB below the fifth centile. This disorder is inherited in a codominant fashion, and parents of severely affected individuals have intermediate or usually low levels of apoB. An obvious candidate gene for this disorder is the *apoB* gene itself, and study of a large kindred found evidence for cosegregation between a particular haplotype of the *apoB* gene and the phenotype of levels of apoB below the fifth centile (Leppert *et al.*, 1988). However, within this kindred there are some individuals with low levels of *apoB* but who had not inherited the haplotype, and some individuals who had inherited the haplotype but who had normal apoB levels. Further analysis shows that the first group of individuals had the apoE phenotype E2E2, which is associated with low plasma cholesterol and apoB levels, whereas the second group had inherited a defect in the LDL receptor, and therefore compensate for low plasma *apoB* levels by slow removal of plasma LDL. Thus, even the study of a rare mutation with a major effect on the levels of plasma apoB was complicated by variations in other genes that protect from, or predispose to, the development of the trait.

The use of the linkage approach has another problem where intermediate traits such as levels of cholesterol or fibrinogen are to be analyzed, rather than 'affected' or 'nonaffected' status. Linkage is a statistically powerful approach when the gene being sought is having a relatively large effect on the trait, for example if associated with higher (or lower) levels of the trait in carriers by two standard deviations from that of noncarriers. For effects that are smaller than this the sample sizes required become unfeasibly large. Thus, as an example, although the common apoE isoforms (E2, E3 and E4) are associated with consistent differences in plasma levels of total cholesterol, LDL cholesterol and apoE, it is likely that a linkage approach would require > 1000 families to have a reasonable power to detect an effect of $\pm$ 10% on the trait level.

6.2 Affected sibs

One method which overcomes the problems of the linkage and the population association approach is the affected sib-pair method (Penrose, 1953; Wilson *et al.*, 1991). This is based on the relative frequencies of pairs of sibs to be alike or unalike for two traits the linkage of which is to be investigated. For example, if sibs with IHD or with hyperlipidemia share a common candidate gene haplotype more often than expected by chance alone, this would implicate defects in the candidate gene in the development of the disorder. The method has the advantage that it can detect linkage in the presence of genetic heterogeneity. This approach has been very useful in analysing multifactorial traits such as diabetes, using

HLA markers. Furthermore, a quantitative sib-pair analysis can be used to analyze quantitative traits such as plasma cholesterol or triglycerides (Kruglyak and Lander, 1995; Xu and Gessler, 1998) and has been successful for some lipid traits. The trait of levels of small dense LDL was originally analyzed by linkage in families using a dichotomous variable (i.e. levels above or below a cut-off point) and the lod score method requiring specification of the model of inheritance; this rejected linkage with the *apoB* gene (Austin *et al.*, 1991). This analysis may also have been confounded by genetic heterogeneity and the influence of other modifying genes (or environmental factors) as discussed above. However, using the quantitative trait of LDL particle size and a nonparametric sib-pair approach, strong evidence of linkage of the *apoB* gene with the trait has been reported using dizigotic twins (essentially sibs) (Austin *et al.*, 1988). Modifications and enhancements of these approaches have been suggested using both discordant and concordant sibs (Gu *et al.*, 1996); however, so far no reports on the identification of novel genes involved in IHD risk have been published using the sib-pair approach.

6.3 Association in unrelated subjects

The approach that has proved most useful over the last 15 years to identify genes involved in risk of IHD is to search for population associations between a (possibly 'neutral') genetic variation detected as a DNA polymorphism (historically a restriction fragment length polymorphism) and a functionally significant genetic variation in the same or a nearby gene. The simplest design using unrelated subjects is a case–control comparison of genotype or allele frequency. Cases are selected, defined as subjects with a myocardial infarction (MI) or other cardiovascular disease, or with levels of an intermediate phenotype trait above or below a chosen threshold (i.e. triglycerides > 2.0 mmol l^{-1} or fibrinogen > 3.0 g dl^{-1}). They are 'matched' with a group of healthy controls, with matching criteria designed to avoid potential confounders such as sex or age. The frequency of carriers of a mutation in a candidate gene is determined, and if it differs in the two groups this is taken as evidence for the involvement of the mutation in determining the disease or trait. Formally, the mutation itself may not be causal in this, but could be acting as a genetic marker for another mutation elsewhere at the gene locus. Therefore, using two to three genetic variants within and flanking the candidate gene, this approach can be effectively used for a rapid screen, to see if there is any evidence for involvement of a gene locus in the trait. Once a case–control difference in frequency is observed, detailed sequencing to identify functional changes can be carried out, followed by *in vitro* expression of any mutations identified to confirm this effect.

The power of association studies for complex multifactorial disorders compared to classical linkage appoaches has been more appreciated recently (Risch and Merikangas, 1996). There are, however, a number of potential problems with this approach which renders the interpretation of results difficult. For an MI case comparison, only survivors are available, and these subjects may be genetically and biochemically different from the whole MI group. To give an unbiased estimate of allele frequency in healthy subjects the sample must be selected with strict criteria to ensure that the sample is indeed representative of the general population from which the 'cases' are drawn. Thus, for example, subjects recruited as controls through having had coronary angiography showing minimal atherosclerosis cannot be regarded as being a healthy population based sample.

IHD risk factors tend to cluster, so that cases are usually more obese and show a higher proportion of smokers, hypertension and diabetes than controls. Biochemical risk factors also cluster, so that such comparisons may overestimate the relative importance of some factors and underestimate others. However, most importantly, there may be undetected genetic stratification, so that the cases may, by chance alone, contain a different proportion of individuals with a different genetic background than the controls. If the frequency of the studied mutation is higher in subjects from this different genetic background, there will be an observed difference in frequency between cases and controls that will not be causal. To some extent this problem of misinterpretation of results because of genetic stratification can be overcome by replicating the study in different countries or centers, and the failure to confirm some exciting initial frequency studies may be explained by this problem (e.g. Weiss *et al.*, 1996).

Many of these problems are overcome by prospective studies of IHD risk, and studies where the impact of mutations on intermediate traits is examined in a random sample of the healthy population. Such epidemiologically based studies need to be large, and for IHD risk to be estimated subjects must be followed for several years; and both of these have cost implications. However, such studies are less susceptible to confounding by genetic stratification, and are also able to measure environmental factors in detail and thus to look for gene–environment interaction. Because of their size they are also well suited for detection of the modest effects on traits that is predicted from the multifactorial model. DNA samples from large prospective cohorts are now available and being analyzed (such as the US Physicians Health Study (n = 20 000; Ridker *et al.*, 1997), the Scottish study WOSCOPs (n = 6000; Shepherd *et al.*, 1995) and the UK healthy men in NPHS-II (n = 3000; Talmud *et al.*, 1999). The availability of several such studies allows rapid replication of any novel observations.

7. Power calculations

For both case–control and prospective studies the most important consideration in study design is the power of the study to detect a statistically significant effect. Many published studies that report that they 'fail' to detect IHD risk, or an effect on an intermediate phenotype associated with a candidate gene mutation, simply did not have the power to detect a biologically plausible effect. Epidemiologists are very aware of this problem, but most molecular geneticists are not, and if journals had a policy of insisting on the inclusion of power calculation in the methods or discussion sections this would clarify the field considerably. The power of a study to detect an effect on risk is dependent on the size of the sample and the frequency of the mutation being studied. *Figure 2* shows the relative risk (RR) detectable with a study of one case and three controls at conventional levels of statistical probability for mutation of different carrier frequency.

It can be seen that for a mutation of low carrier frequency of 5%, a sample of 200 cases would only have the power to detect a RR of > 2.5-fold, which is implausibly high for a single mutation in a single gene in a multifactorial disorder. Even for a common variant such as the insertion/deletion polymorphism in the gene for angiotensin converting enzyme (ACE), the sample only has the power to detect an RR of 1.7. For ACE, a recent meta analysis (Samani *et al.*, 1996) has suggested an RR of 1.3 for the *DD* genotype (24% of the general population), and to obtain a statistically significant difference would thus require a sample of at least 800 cases and 2400 controls. Similar calculations are shown in *Figure 2b* for the power to detect protective effects. Formally, such genetic protection should be as common

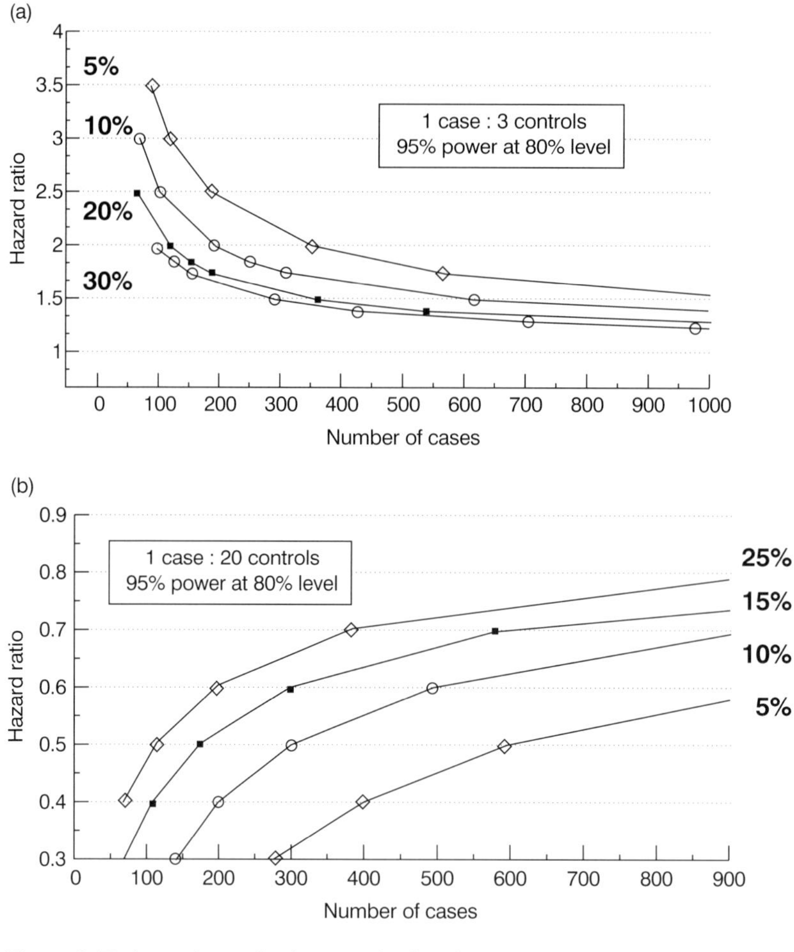

Figure 2. Estimated sample sizes required to detect a RR of IHD with a particular statistical power for mutations of different rare allele frequency. (a) The rare allele is associated with increased risk of IHD. (b) The rare allele is associated with reduced risk of IHD. Data were calculated using the unmatched case–control package in the Epi program [Epi Info (version 6.02) Centers for Disease Control and Prevention, USA] with $P = 0.05$ and 80% power, and assuming three controls to each case.

as genetic predisposition, and certainly, from the known effects on lipid traits or clotting factors, the apoE2 mutation which lowers cholesterol levels (Davignon *et al.*, 1988), the LPL447X mutation which lowers triglyceride and raises HDL levels (see Fisher *et al.*, 1997) and the FactorVII Q353 mutation which lowers levels of clotting factor VII (Green *et al.*, 1991) should all be protective, and some studies supporting this have been reported (e.g. Luc *et al.*, 1994, Humphries *et al.*, 1998, and Iacoviello *et al.*, 1998 respectively). However, the size of the sample needed to detect a 30–40% protective effect, which clinically would be extremely beneficial, is of the order of 400–500 cases for each of these rare alleles.

For association studies with intermediate phenotypes the power is also easily estimated, based on the known mean and standard deviation of the trait in the sample being studied and the frequency of the mutation. As a general rule a sample of 17 common homozygotes and 17 rare homozygotes has an 80% power at the $P = 0.05$ level to detect a difference in a trait of standard deviation (e.g. 0.64 g l^{-1} for fibrinogen, or 0.6 mmol l^{-1} total cholesterol). For this type of study subjects can be selected matched for age, sex, BMI or other relevant characteristics, thus reducing the problem of confounding. For a larger sample, drawn to be representative of the healthy general population, appropriate statistical adjustment (e.g. by regression) can be used to take the effect of these differing characteristics into account. Even for relatively rare polymorphisms a sample of 200 subjects can detect a 10% difference in trait levels, though with greater power for a trait such as HDL cholesterol, where for metabolic reasons the standard deviation is smaller than for triglycerides where variation is large.

For both the case–control and association with trait approaches it is easy to factor into the calculation the potential problem of gene–environment interaction, providing the prevalence of the environmental factor is known (e.g. sex, smoking, noninsulin dependent diabetes mellitus (NIDDM), obesity) and the sample size required to detect interactions can be increased proportionately.

8. Gene–environment interaction

A key factor in the identification and study of genes that determine IHD risk is that an individual carrying such mutations will be at risk only if and when they enter the high risk environment. Thus the mutation will cause high plasma cholesterol or high fibrinogen only in the presence of this environmental challenge. This classical 'lack of penetrance' of a mutation will cause analytical problems and misphenotyping which will be particularly problematic with some sampling analytical designs. The problem that this 'context dependency' of a mutation creates from an analytical point of view (i.e. gene × environment effect), has largely been overlooked in the field until recently. The problem occurs when a second study fails to confirm a reported association between a candidate gene polymorphism and levels of intermediate traits or with risk of IHD. Although the failure of a second study to reproduce an association found may cast doubt on the validity of the first report, it may also reflect the presence of a potentially interesting gene–environment interaction that requires further exploration. If, for example, smoking is the key environmental stress that amplifies or diminishes the genotype effect, then the different proportion of smokers in the samples being studied (and possibly the number of cigarettes being smoked) will have a major effect on the power of the sample to detect the association.

As we predict that for common mutations, their single impact on IHD will be very modest (increasing RR by 20–40% at most) the important question from a clinical point of view is whether there are subgroups of the population where the risk associated with a mutation is considerably greater than average, that is, to the point where the RR is of clinical relevance. To be clinically useful in any risk algorithm, we might require any factor to have RR $\geqslant 2$; which is, for example, the RR estimated to be associated with smoking habit in middle-aged men (Meade *et al.*, 1987). There are two obvious ways to search for such subgroups, either by genetic stratification, by comparing the impact on a trait of a mutation in one gene in carriers and noncarriers of a mutation at a second gene (gene–gene interaction), or by stratification by the presence of an environmental risk factor such as smoking, diet, obesity, NIDDM, IHD (gene–environment interaction). To date there are few good examples of gene–gene interactions, although an example of this has been reported between the ACE gene insertion/deletion variant and the angiotensinogen II type I receptor gene (*A1166C*) polymorphism on risk of IHD (Tiret *et al.*, 1994), although this has not been confirmed or taken forward in terms of molecular mechanisms. However, many gene–environment interactions have already been identified and some of these are shown in *Table 2*. Interacting environmental factors identified to date include obesity, sex, age, presence or absence of IHD (possibly a marker of inflammation), exercise, dietary fat, and alcohol consumption and smoking. Until proven otherwise for a particular mutation it would be safer to use an analytical strategy that is based on the expectation of such potential interaction. From a practical point of view, this means that large samples are required for interactions to be statistically significant and that the characteristics of the sample being studied must be carefully recorded and compared between samples.

8.1 Gene–environment interaction and the lipoprotein lipase gene

As an example of gene–environment interaction common mutations in the LPL gene are particularly informative (see review by Fisher *et al.*, 1997). The plasma level of any IHD intermediate phenotypes, such as plasma triglyceride, will be due to the balance between the rate of production [in this case very LDL (VLDL) secretion from the liver] and rate of removal. The key step in VLDL removal is its metabolism in adipose and muscle tissue by LPL. The resulting LDL remnants are cleared via apoE acting as a ligand by receptor mediated processes in the liver, while some remnants are further metabolized to LDL. Any mutation that results in a partial deficiency of LPL would thus be predicted to result in a modest increase in plasma triglyceride levels, with the increase being proportional to the degree of deficiency caused by the mutation. This prediction is confirmed in population based samples of healthy subjects (Fisher *et al.*, 1995; Gerdes *et al.*, 1997; Mailly *et al.*, 1995), where carriers of the more severe LPL mutation Asn291Ser (which has approximately 50% of wild type LPL activity; Zhang *et al.*, 1996) have higher triglyceride and lower HDL than carriers of the milder LPL Asp9Asn

Table 2. Reported examples of gene–environment interaction with candidate genes and IHD risk traits and IHD risk

Gene	Environment	Trait	Reference
ApoAI	Sex/smoking	AI/HDL	Sigurdsson *et al.*, 1992
CETP	Alcohol/statin/ sex	AI/IHD	Fumeron *et al.*, 1995; Gudnason *et al.*, 1997;
Fibrinogen	Gender/IHD	Fibrinogen	Humphries *et al.*, 1995 Tybjaerg Hansen *et al.*, 1997
LPL-D9N	Obesity	Triglyceride	Fisher *et al.*, 1997
ACE	Exercise	LVH	Montgomery *et al.*, 1997
FVII	Triglyceride	FVII	Humphries *et al.*, 1994; Mennen *et al.*, 1997
PAI-I	Triglyceride	PAI-1	Mansfield *et al.*, 1995; Panahloo *et al.*, 1995

LVH = Left ventrical hypertrophy; FVII = clotting factor VII; PAI-1 = plasminogen activator inhibitor-1

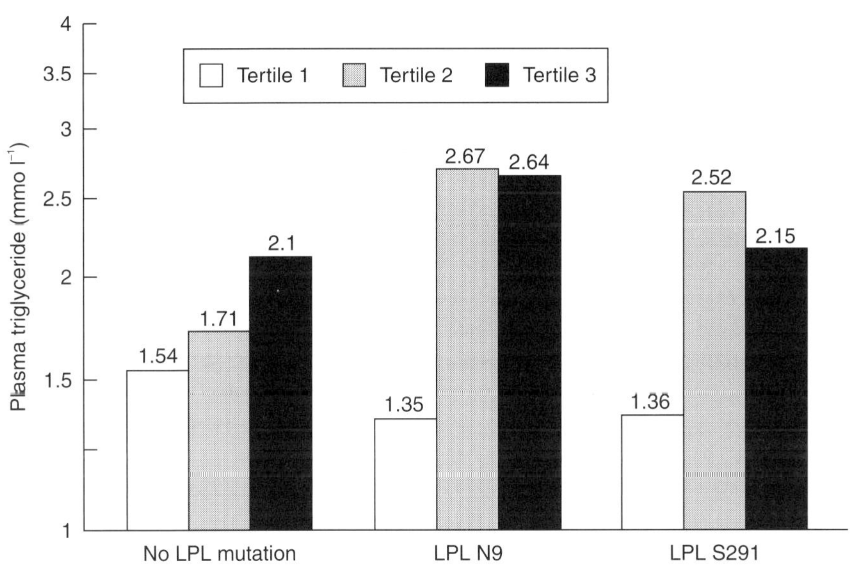

Figure 3. Mean plasma triglyceride concentration according to tertiles of BMI in healthy noncarriers and carriers of the Asn9 (n = 26) or Ser291 (n = 24) LPL mutations. For carriers of either mutation the test for interaction between carriers/noncarriers and tertiles of BMI was $P = 0.006$. To test differences in triglyceride, values were log-transformed prior to statistical analysis (data from Fisher *et al.*, 1997).

mutation which have around 70% of wild type activity (Mailly *et al.*, 1995, Zhang *et al.*, 1996). However, this partial deficiency of lipolytic activity will not be rate-limiting if VLDL production is low, and may become important only in situations where VLDL production is high, leading to the prediction that in environmental situations of VLDL overproduction, such as increasing obesity, or in genetic situations that predispose to increased VLDL secretion (e.g. variants in the genes involved in hepatic triglyceride lipoprotein assembly such as microsomal triglyceride transfer protein (Karpe *et al.*, 1998), carriers of these LPL mutations could have considerably higher triglyceride levels. This prediction of obesity × LPL interaction has been borne out in several studies (reviewed in Fisher *et al.*, 1997) and is shown in *Figure 3*. The prediction would thus be that the impact on IHD risk associated with such an LPL mutation (acting through effects on triglycerides and HDL), would be negligible in thin individuals, but might be considerable in obese individuals when the rate limiting nature of the defect is apparent. To date no study has reported such an obesity–LPL interaction on IHD risk, although most studies have not had the power to investigate this in detail.

8.2 Gene–environment interaction and the fibrinogen gene

Fibrinogen is an acute phase protein, and its plasma level is raised following infection or injury and, because of its sensitivity to environmental factors, the within individual variation of fibrinogen levels is high. It is possible that in the general population some individuals may have a 'plastic' genotype that responds easily to environmental factors, and that such individuals would experience a greater than average increase in plasma fibrinogen levels in response to a moderate environmental stimulus. These individuals would then be at greater than average risk of a thrombotic event at the peak of fibrinogen levels, and thus have a greater chance of developing IHD, and particularly MI. The relevance of this 'variability-gene' concept has been discussed in the field of IHD since its suggestion by Berg in 1987 (Berg, 1987; Humphries, 1994).

This can be examined by a 'genotype–stress' study, which has already been shown by several groups to be an excellent way of magnifying modest genotype effects on traits. We examined the effects of acute intensive exercise on plasma fibrinogen levels and the relationship of these responses to the fibrinogen G-455A genotype (Montgomery *et al.*, 1996). A group of 156 male British Army recruits were studied at the start of their 10 week basic training (which emphasizes physical fitness) and between 0.5 and 5 days after a major 2 day strenuous military exercise. Compared to baseline values, fibrinogen concentrations were significantly lower (12%, $P = 0.04$) at day 5 after the military exercise, consistent with a beneficial effect of training. However, within 12 h of the completion of the 2 day military exercise, fibrinogen levels were 14.5% higher than those at 5 days, suggesting that the acute response had already begun. Levels were significantly higher on days 1–3 after the military exercise (also supporting an 'acute phase' response to strenuous exercise) and were maximal on days 1 and 2 (27% and 37%, $P < 0.001$ respectively). As shown in *Figure 4*, the fibrinogen raising effect associated with the A allele was seen at baseline and after 5 days, but in both cases the effect was small and nonsignificant in this small sample. However, the degree of rise at 2–3 days after the military exercise was strongly related to the presence of the A allele. Although the group was small, the differences were statistically significant ($P = 0.01$), and the size of the effect is sufficient to be of biological significance. In particular, the men homozygous for the A allele (representing 4% of the population based on the observed allele frequencies) had fibrinogen levels which by day 2 had risen by more than 100% compared with their 'untrained' levels, and which were amongst the highest levels in the whole sample. These findings support an

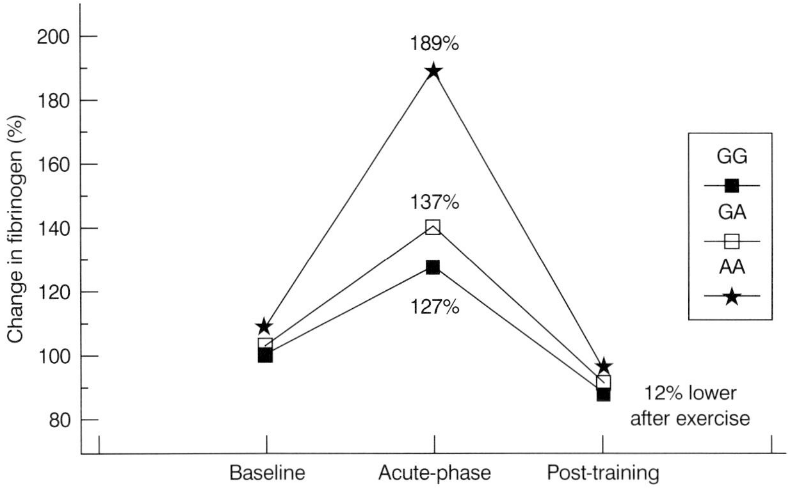

Figure 4. The mean levels of plasma fibrinogen concentration at pretraining, 2–3 days and 5 days after an intensive 2 day military exercise (data from Montgomery *et al.*, 1996)

association of the A allele with greater fibrinogen level 'responsiveness' after a cytokine inducing event. In support of this gene–environment interaction concept, there is good evidence from several studies that the size of the A raising effect is also modulated by factors such as sex (Humphries *et al.*, 1995; Tygbaerg Hansen *et al.*, 1997), age (Thomas *et al.*, 1996), and diabetes (Snowden *et al.*, 1992), as well as other inducers of 'inflammatory stress' such as smoking and the presence of IHD itself (Behague *et al.*, 1996; Green *et al.*, 1993).

9. When has a candidate gene won the election?

Before a mutation can be accepted as a proven IHD risk factor, it will be necessary to agree on the criteria that must be met. For this purpose the criteria proposed for causality of an infectious agent can be adapted.

Reproducibility. The reproducibility of the effect in different laboratories and in different settings is one piece of evidence, but the problem of gene–environmental interaction makes this less clear cut. As a minimum, it would be reasonable to propose that a statistically significant finding should be reproduced in three independent samples (preferably from different laboratories). If a particular mutation is associated with the same effect in different ethnic groups (even though its frequency may be quite different) this adds further weight to causality, as has been shown for the fibrinogen G-455A in Caucasians and Japanese (Iso *et al.*, 1995).

Plausibility. Biological plausibility of the effect of the mutation is important, and thus a mutation in the promoter of a gene that is having a demonstrated effect *in vitro* on transcription should affect the levels of a plasma risk factor, but not the function of the protein. Mutations that delete amino acids or that cause a premature stop codon are highly likely to effect function, as are nonconservative missense mutations and those changing important cysteine disulfide bridges. Conservative amino acid changes and wobble position changes are less likely to be functional, but some have been reported to alter function dramatically (Chen *et al.*, 1998). Sequence changes in introns may affect splicing efficiency or cause exon skipping (Fujimaru *et al.*, 1998), but many appear to be without effect. For every mutation it would be ideal to confirm in effect *in vitro* expression studies, but for some the effects on function may be relatively mild (e.g. ± 10% of activity), which would be difficult to prove using *in vitro* assays, but could still be of clinical importance in predicting atherosclerosis development over the lifetime of an individual. Improving the sensitivity and biological relevance of such *in vitro* assays is one of the aspects of the research into the genetic determinants of IHD risk that still requires considerable development.

Graded response. A graded response of the mutation would be supportive evidence, such that carriers of a mutation would be expected to have levels (or activity) of a protein intermediate between noncarriers and homozygotes for the mutation. This is often the case for promoter mutations or mutations affecting protein stability, for example the fibrinogen promoter G-445A (Humphries *et al.*, 1995). However, mutations in enzymes usually result in a 'recessive' pattern of inheritance when the product of the enzyme is the trait that is measured. A good example of this is one of the key enzymes of homocysteine metabolism methylene tetrahydrofolate reductase (MTHFR) where a missense mutation (Ala122Val) causes a thermolabile variant of the enzyme which is associated with hyperhomocysteinemia, and risk of IHD (Frosst *et al.*, 1995) and venous thrombosis (Margaglione *et al.*, 1998). Homozygotes for the rare allele thermolabile variant (approximately 10% of the population) have significantly elevated levels of plasma homocysteine compared to noncarriers and carriers of the variant (Gudnason *et al.*, 1997). Presumably this is because in carriers the slight reduction of MTHFR enzyme levels is sufficient to maintain homocysteine plasma homeostasis in the environment of adequate dietary folate levels, while in variant homozygotes, where MTHFR levels are considerably lower, the level of the enzyme is insufficient to cope even in the setting of an adequate folate intake and homocysteine levels are high.

Reversibility. Reversibility of effect is obviously not technically possible for a gene mutation in a patient, but where the strong effect of a mutation on an intermediate trait is the result of gene–environment interaction, removal of the high risk environment should have a large effect on the trait. As an example of this, as shown in *Figure 3*, the triglyceride raising effect associated with being a carrier of the LPL-Asn9 or -Ser291 mutations, is exacerbated by obesity. This predicts that weight loss in carriers should be followed by a greater fall in triglyceride levels than following a similar weight loss in noncarriers, and experiments to confirm this prediction could be an extremely powerful argument in support of causality.

Expression. Expression *in vitro* and *in vivo* would be the 'gold standard' proof of causality. None of the criteria above rule out the possibility that the variation is not itself causal but is rather a marker for another mutation elsewhere in the gene. Formally, this can only be proved by making a gene construct with the 'wild type' and 'mutant' sequence and expressing it *in vivo*. If the phenotype of interest is MI in a human subject, clearly this cannot be done for ethical and logistic reasons. Such experiments can be carried out in knockout and transgenic animals, such as rabbits or mice, but the animal models do not result in MI, and so at best can give information about effects of these mutations on intermediate traits. Even so, the findings from animals need to be extrapolated with care to humans because of the many differences in the metabolic pathways and particularly in the 'back-up' redundancy system in the different species.

Taking all of these different aspects into account, it should be possible to agree on the necessary criteria for acceptability, and this might most usefully be done using the 'Consensus conference' approach that has been used for other risk factors such as cholesterol, fibrinogen and homocysteine.

10. Conclusions

Over the last 15 years the analytical strategy of using candidate gene polymorphisms to carry out frequency and association studies in unrelated subjects has proved extremely fruitful. In part this is because IHD has so many 'intermediate' phenotypes whose metabolic makeup (and thus the genes involved in their control) are so well understood. Although some associations have not been repeatable in subsequent work many have proved to be robust, and, for example, the fibrinogen raising effect associated with the A-455 allele and the triglyceride raising effect associated with the LPL-Asn9 variant, have now been demonstrated beyond any reasonable doubt, although the precise molecular mechanism of both of these effects remains to be elucidated. However, the magnitude of the effects is modulated (both up and down) by many environmental factors (and probably by genetic factors yet to be discovered), and this means that in some subjects who have a particular (genotype) life style or environment these mutations are making a significant contribution to risk, while in others they may be making essentially no contribution. An approach which has already proved to be most useful in the field of IHD has been to 'stress' individuals by exercise (Montgomery *et al.*, 1996), a fatty meal (Gerdes *et al.*, 1997) or a high fat diet, and note that some individuals have a modest response and others a very large response. These gene–environment interactions are likely also to be the key to understanding the 'causes' determining most other multifactorial disorders. As well as being an explanation as to why some studies fail to reproduce the IHD risk association reported by others, this also opens up the field for fascinating molecular experiments to determine the mechanisms of these interactions and to develop genotype specific life style or pharmocological approaches to reduce levels of IHD intermediate phenotypes and thus lower risk of IHD.

Acknowledgments

The work in the author's laboratory was supported by grants from the British Heart Foundation (RG95007).

References

Austin, M.A., King, M.C., Vranizan, K.M., Newman, B. and Krauss, R.M. (1988) Inheritance of low-density lipoprotein subclass patterns: results of complex segregation analysis. *Am. J. Hum. Genet.* **43**: 838–846.

Austin, M.A., Wijsman, E., Guo, S.W., Krauss, R.M., Brunzell, J.D., Deeb, S. (1991) Lack of evidence for linkage between low-density lipoprotein subclass phenotypes and the apolipoprotein B locus in familial combined hyperlipidemia. *Genet. Epidemiol.* **8**: 287–297.

Austin, M.A., Newman, B., Selby, J.V., Edwards, K., Mayer, E.J. and Krauss, R.M. (1993) Genetics of LDL subclass phenotypes in women twins. Concordance, heritability, and commingling analysis. *Arterioscler. Thromb.* **13**: 687–695.

Austin, M.A., Talmud, P.J., Luong, L.A., Haddad, L., Day, I.N., Newman, B., Edwards, K.L., Krauss, R.M. and Humphries, S.E. (1998) Candidate-gene studies of the atherogenic lipoprotein phenotype: a sib-pair linkage analysis of DZ women twins. *Am. J. Hum. Genet.* **62**: 406–419.

Behague, I., Poirier, O., Nicaud, V., Evans, A., Arveiler, D., Luc, G., Cambou, J-P., Scarabin, P.Y., Bara, L., Green, F. and Cambien, F. (1996) β-Fibrinogen gene polymorphisms are associated with plasma fibrinogen and coronary artery disease in patients with myocardial infarction. *Circulation* **93**: 440–448.

Berg, K. (1987) Genetics of coronary heart disease and its risk factors. *CIBA Foundation Symposium* **130**: 14–133.

Berg, K. and Kierulf, P. (1989) DNA polymorphisms at fibrinogen loci and plasma fibrinogen concentration. *Clin Genet* **36**: 229–235.

Berg, K., Malinow, M.R., Kierulf, P. and Upson, B. (1992) Population variation and genetics of plasma homocyst(e)ine level. *Clin Genet* **41**: 315–321.

Bouchard, C. (1997) Genetic determinants of regional fat distribution. *Hum. Reprod.* **12(suppl 1)**: 1–5.

Chen, W., Kubota, S., Ujike, H., Ishihara, T. and Seyama, Y. (1998) A novel Arg362Ser mutation in the sterol 27-hydroxylase gene (CYP27): its effects on pre-mRNA splicing and enzyme activity. *Biochemistry* **37**: 15050–15056.

Dalmon, J., Laurent, M. and Courtois G. (1993) The human β fibrinogen promoter contains a hepatocyte nuclear factor 1-dependent interleukin-6 responsive element. *Mol. Cell. Biol.* **13**: 1183–1193.

Danesh, J., Collins, R. and Peto, R. (1997) Chronic infections and coronary heart disease: is there a link? *Lancet* **350**: 430–436.

Davignon, J., Gregg, R.E. and Sing, C.F. (1988) Apolipoprotein E polymorphism and atherosclerosis. *Arteriosclerosis* **8**: 1–21.

Day, I.N.M., Whittall, R.A., O'Dell, S.D., Haddad, L., Bolla, M.K., Gudnason, V. and Humphries, S.E. (1997) Spectrum of LDL receptor gene mutations in heterozygous familial hypercholesterolaemia. *Hum. Mut.* **10**: 116–127.

Fisher, R.M., Mailly, F., Peacock, R.E., *et al.* (1995) Interaction of the lipoprotein lipase asparagine 291→ serine mutation with body mass index determines elevated plasma triacylglycerol concentrations: a study in hyperlipidaemic subjects, myocardial infarction survivors and healthy adults. *J. Lipid Res.* **36**: 2104–2112.

Fisher, R.M., Humphries, S.E. and Talmud, P.J. (1997) Common variation in the lipoprotein lipase gene: effects on plasma lipids and risk of atherosclerosis. *Atherosclerosis* **13**: 145–159.

Friedlander, Y. and Kark, J.D. (1987) Complex segregation analysis of plasma and lipoprotein variables in a Jerusalem sample of nuclear families. *Hum. Hered.* **37**: 7–19.

Freidlander, Y., Elkana, Y., Sinnreich, R. and Kark, J.D. (1995) Genetic and environmental sources of fibrinogen variability in Israeli families. The Kibbutzim Family Study. *Am. J. Hum. Genet.* **56**: 1194–1206.

Frosst, P., Blom, H.J., Milos, R., *et al.* (1995) A candidate genetic risk factor for vascular disease: a common mutation in methylenetetrahydrofolate reductase. [letter] *Nature Genet.* **10**: 111–113.

Fujimaru, M., Tanaka, A., Choeh, K., Wakamatsu, N., Sakuraba, H. and Isshiki, G. (1998) Two mutations remote from an exon/intron junction in the beta-hexosaminidase beta-subunit gene affect 3′-splice site selection and cause Sandhoff disease. *Hum Genet* **103**: 462–469.

Fumeron, F., Betoulle, D., Luc, G., Behague, I., Ricard, S., Poirier, O., Jemaa, R., Evans, A., Arveiler, D. and Marques Vidal, P. (1995) Alcohol intake modulates the effect of a polymorphism of the cholesteryl ester transfer protein gene on plasma high density lipoprotein and the risk of myocardial infarction. *J. Clin. Invest.* **96**: 1664–1671.

Gerdes, C., Fisher, R.M., Nicaud, V., Boer, J., Humphries, S.E., Talmud, P.J. and Faergeman, O. *on behalf of the EARS Group* (1997) Lipoprotein lipase variants D9N and N291S are associated with increased plasma triglyceride and lower high-density lipoprotein cholesterol concentrations: studies in the fasting and post-prandial states; the European Atherosclerosis Research Studies. *Circulation* **96**: 733–740.

Goldstein, J.L. and Brown, M.S. (1989) Familial hypercholesterolemia. In: *The Metabolic Basis of Inherited Disease*, 6th Edn (eds. C.R. Scriver, A.L. Beaudet, W.S. Sly, and D. Valle). McGraw-Hill Book Co, New York, pp. 1215–1250.

Green, F., Hamsten, A., Blomback, M. and Humphries, S E. (1993) The role of β-fibrinogen genotype in determining plasma fibrinogen levels in young survivors of myocardial infarction and healthy controls from Sweden. *Thromb. Haemost.* **70**: 915–920.

Green, F., Kelleher, C., Wilkes, H., Temple, A., Meade, T. and Humphries S. (1991) A common genetic polymorphism associated with lower coagulation factor VII levels in healthy individuals. *Arterioscler. Thromb.* **11**: 540–546.

Greenberg, D.A. (1993) Linkage analysis of 'necessary' disease loci versus 'susceptibility' loci. *Am. J. Hum. Genet.* **52**: 135–143.

Gu, C., Todorov, A. and Rao, D.C. (1996) Combining extremely concordant sibpairs with extremely discordant sibpairs provides a cost effective way to linkage analysis of quantitative trait loci. *Genet. Epidemiol.* **13**: 513–533.

Gudnason, V., Thormar, K. and Humphries, S.E. (1997) Interaction of the cholesteryl ester transfer protein I405V polymorphism with alcohol consumption in smoking and non-smoking healthy men, and the effect on plasma HDL cholesterol and apoAI concentration. *Clin. Genet.* **51**: 15–21.

Haddad, L., Day, I.N.M., Hunt, S., Williams, R.R., Humphries, S.E. and Hopkins, P.N. (1999) Evidence for a third genetic locus causing familial hypercholesterolemia: a non-LDLR, non-apoB kindred. *J. Lipid Res.* **40**: 1113–1122.

Hamsten, A., Iselius, L., Dahlen, G. and de Faire, U. (1986) Genetic and cultural inheritance of serum lipids, low and high density lipoprotein cholesterol and serum apolipoproteins A-I, A-II and B. *Atherosclerosis* **60**: 199–208.

Hamsten, A., Iselius, L., de Faire, U. and Blomback, M. (1987) Genetic and cultural inheritance of plasma fibrinogen concentration. *Lancet* **ii**: 998–990.

Hasstedt, S.J., Albers, J.J., Cheung, M.C., Jorde, L.B., Wilson, D.E., Edwards, C.Q., Cannon,W.N., Ash, K.O. and Williams, R.R. (1984) The inheritance of high density lipoprotein cholesterol and apolipoproteins A-I and A-II. *Atherosclerosis* **51**: 21–29.

Hixson, J.E., and the Pathobiological Determinants of Atherosclerosis in Youth (PDAY) Research Group (1991). Apolipoprotein E polymorphisms affect atherosclerosis in young males. *Arteriosler. Thromb.* **11**: 1237–1244.

Hobbs, H.H., Brown, M.S. and Goldstein, J.L. (1992) Molecular genetics of the LDL receptor gene in familial hypercholesterolaemia. *Hum Mut* **1**: 445–466.

Hokanson, J.E. and Austin, M.A. (1996) Plasma triglyceride level is a risk factor for cardiovascular disease independent of high-density lipoprotein cholesterol level: a meta-analysis of population-based prospective studies. *J Cardiovasc Risk* **3**: 213–219.

Hong, Y., Pedersen, N.L., Egberg, N. and de-Faire, U. (1997) Moderate genetic influences on plasma levels of plasminogen activator inhibitor-1 and evidence of genetic and environmental influences shared by plasminogen activator inhibitor-1, triglycerides, and body mass index. *Arterioscler. Thromb. Vasc. Biol.* **17**: 2776–2782.

Humphries, S.E. (1994) Coronary artery disease and the variability gene concept: the effect of smoking on plasma levels of high density lipoprotein and fibrinogen. In: *Genotype to Phenotype* (eds S.E. Humphries and S. Malcolm). BIOS Scientific Publishers, Oxford, pp.123–141.

Humphries, S.E., Lane, A., Green, F.R., Cooper, J. and Miller, G.J. (1994) Factor VII coagulant activity and antigen levels in healthy men are determined by interaction between factor VII genotype and plasma triglyceride concentration. *Arteriosler. Thromb.* **14**: 193–198

Humphries, S.E., Ye, S., Talmud, P., Bara, L., Wilhelmsen, L. and Tiret, L. (1995) European Atherosclerosis Research Study: Genotype at the fibrinogen locus (G_{-455}-A β-gene) is associated with differences in plasma fibrinogen levels in young men and women from different regions in

Europe: Evidence for gender–genotype–environment interaction. *Arterioscler. Thromb. Vasc. Biol.* **15**: 96–104.

Humphries, S.E., Thomas, A., Montgomery, H.E., Green, F., Winder, A. and Miller, G. (1997) Gene-environment interaction in the determination of plasma levels of fibrinogen. *Fibrinolysis and Proteolysis* **11(suppl 1)**: 3–7

Humphries, S.E., Nicaud, V., Margalef, J., Tiret, L. and Talmud, P.J. ***for the EARS Group*** (1998) Lipoprotein lipase gene variation is associated with a paternal history of premature coronary artery disease and fasting and postprandial plasma triglycerides. *Arterioscler. Thromb. Vasc. Biol.* **18**: 526–534.

Humphries, S.E., Peacock, R., Gudnason, V. (1999) The genetic determinants of hyperlipidaemia. In: *Genetic Basis of Atherosclerosis* (eds. J. Betteridge, R. Illingworth and J. Shepherd). Arnold, London, pp. 127–162.

Iacoviello, L., Di Castelnuovo, A., De Knijff, P., D'Orazio, A., Amore C., Arboretti, R., Kluft, C., Benedetta Donati, M. (1998) Polymorphisms in the coagulation factor VII gene and the risk of myocardial infarction. *N. Engl. J. Med.* **338**: 79–85.

Iso, H., Folsom, A., Winklemann, J.C., Koike, K., Harada, S., Greenberg, B., Sato, S., Shimamoto, T., Lida, M. and Komachi, Y. (1995) Polymorphisms of the beta fibrinogen gene and plasma fibrinogen concentration in Caucasian and Japanese population samples. *Thromb. Haemost.* **73**: 106–111.

Karpe, F., Lundahl, B., Ehrenborg E., Eriksson, P. and Hamsten, A. (1998) A common functional polymorphism in the promoter region of the microsomal triglyceride transfer protein gene influences plasma LDL levels. *Arterioscler. Thromb. Vasc. Biol.* **18**: 756–761.

Khoury, M.J. and Wagener, D.K. (1995) Epidemiological evaluation of the use of genetics to improve the predictive value of disease risk factors. *Am. J. Hum. Genet.* **56**: 835–844.

Kruglyak, L. and Lander, E.S. (1995) Complete multipoint sib-pair analysis of qualitative and quantitative traits. *Am. J. Hum. Genet.* **57**: 439–454.

Kuivenhoven, J.A., Jukema, J.W., Zwinderman, A.H., de Knijff, P., McPherson, R., Bruschke, A.V., Lie, K.I. and Kastelein, J.J. (1998) The role of a common variant of the cholesteryl ester transfer protein gene in the progression of coronary atherosclerosis. The Regression Growth Evaluation Statin Study Group. *N. Engl. J. Med.* **338**: 86–93.

Lamon-Fava, S., Jimenez, D., Christian, J.C., Fabsitz, R.R., Reed, T., Carmelli, D., Castelli, W.P., Ordovas, J.M., Wilson, P.W. and Schaefer, E.J. (1991). The NHLBI twin study: heritability of apolipoprotein A-I, B, and low density lipoprotein subclasses and concordance for lipoprotein (a). *Atherosclerosis* **91**: 97–106.

Lane, A., Humphries, S.E. and Green, F.R. (1993) Effect on transcription of two common genetic polymorphism adjacent to the promoter region of the B-fibrinogen gene. *Thromb. Haemost.* **69**: 1539.

Leppert, M., Breslow, J.L., Wu, L., Hasstedt, S., O'Connell, P., Lathrop, M., Williams, R.R., White, R. and Lalouel, J-M. (1988) Inference of molecular defect of apolipoprotein B in hypobetalipoproteinemia by linkage analysis in a large kindred. *J. Clin. Invest.* **82**: 847–851.

Liu, W., Qian, C. and Francke, U. (1997) Silent mutation induces exon skipping of fibrillin-1 gene in Marfan syndrome. *Nature Genet.* **16**: 328–329.

Luc, G., Bard, J.M., Arveiler, D., Evans, A., Cambou, J.P., Bingham, A., Amouyel, P., Schaffer, P., Ruidavets, J.B., Cambien, F. (1994) Impact of apolipoprotein E polymorphism on lipoproteins and risk of myocardial infarction. The ECTIM Study. *Arterioscler Thromb* **14**: 1412–1419.

Mailly, F., Turgrul, Y., Reymer, P.W.A., *et al.* (1995) A common variant in the gene for lipoprotein lipase (Asp9→Asn): functional implications and prevalence in normal and hypertriglyceridemic subjects. *Arterioscler. Thromb. Vasc. Biol.* **15**: 468–478.

Mansfield, M.W., Stickland, M.H. and Grant, P.J. (1995) Environmental and genetic factors in relation to elevated circulating levels of plasminogen activator inhibitor-1 in Caucasian patients with non-insulin-dependent diabetes mellitus. *Thromb. Haemost.* **74**: 842–847.

Marenberg, M.E., Risch, N., Berkman, L.F., Floderus, B. and de Faire, U. (1994) Genetic susceptibility to death from coronary heart disease in a study of twins. *N. Engl. J. Med.* **330**: 1041–1046.

Margaglione, M., D'Andrea, G., d'Addedda, M., Giuliani, N., Cappucci, G., Iannaccone, L., Vecchione, G., Grandone, E., Brancaccio, V. and Di Minno, G. (1998) MTHFR and risk: The methylenetetrahydrofolate reductase TT677 genotype is associated with venous thrombosis independently of the coexistence of the FV Leiden and the prothrombin A20210 mutation. *Thromb. Haemost.* **79**: 907–911.

Mayeux, R., Saunders, A.M., Shea, S., Mirra, S., Evans, D., Roses, A.D., Hyman, B.T., Crain, B., Tang, M.X. and Phelps, C.H. (1998) Utility of the apolipoprotein E genotype in the diagnosis of Alzheimer's disease. Alzheimer's Disease Centers Consortium on Apolipoprotein E and Alzheimer's Disease. *N. Engl. J. Med.* **338**: 506–511.

Meade, T.W., Mellows, S., Brozovic, M., Miller, G.H., Chakrabarti, R.R., Haines, A.P., Stirling,Y., Imeson, J.D. and Thompson, S.G. (1986) Haemostatic function and ischaemic heart disease: principal results of the Northwick Park Heart Study. *Lancet* **ii**: 533–537.

Meade, T.W., Imeson, J. and Stirling, Y. (1987) Effects of changes in smoking and other characteristics on clotting factors and risk of ischaemic heart disease. *Lancet* **ii**: 986–988.

Mennen, L.I., de Maat, M.P., Schouten, E.G., Kluft, C., de Jong, P.T., Hofman, A. and Grobbee, D.E. (1997) Coagulation factor VII, serum-triglycerides and the R/Q353 polymorphism: differences between older men and women. *Thromb Haemost* **78**: 984–986.

Moll, P.P., Michels, V.V., Weidman, W.H. and Kottke, B.A. (1989). Genetic determination of plasma apolipoprotein and AI in a population-based sample. *Am. J. Hum. Genet.* **44**: 124–139.

Moll, P.P. (1993) Biometrical searches for new genes with effects on quantitative risk factors for atherosclerosis: problems and prospects. *Curr. Opin. Lipidol.* **4**: 144–151.

Montgomery, H.E., Clarkson, P., Nwose, O.M., *et al.* (1996) The acute rise in serum fibrinogen concentration with exercise is influenced by the G-453-A polymorphism of the beta-fibrinogen gene. *Arterioscler. Thromb. Vasc. Biol.* **16**: 386–391.

Montgomery, H., Clarkson, P., Dollery, C.M., *et al.* (1997) Association of angiotensin-converting enzyme gene I/D polymorphism with change in left ventricular mass in response to physical training. *Circulation* **96**: 741–747.

Myant, N.B. (1993) Familial defective apolipoprotein B-100: a review, including some comparisons with familial hypercholesterolaemia. *Atherosclerosis* **104**: 1 18.

Neil, A., Humphries, S.E., Betteridge, D.J.B. and Thompson, G. *on behalf of the Simon Broome Register Group*. (1999) Mortality in treated heterozygous familial hypercholesterolaemia: implications for clinical management. *Atherosclerosis* **142**: 105–112.

Pairitz, G., Davignon, J., Mailloux, H. and Sing, C.F. (1988) Sources of interindividual variation in the quantitative levels of apolipoprotein B in pedigrees ascertained through a lipid clinic. *Am. J. Hum. Genet.* **43**: 311–321.

Pajukantam, P., Nuotio, I., Terwilliger, J.D., *et al.* (1998) Linkage of familial combined hyperlipidaemia to chromosome 1q21–q23. *Nature Genet.* **18**: 369–373.

Panahloo, A., Mohamed-Ali, V., Lane, A., Green, F., Humphries, S.E. and Yudkin, J.S. (1995) Determinants of plasminogen activator inhibitor 1 activity in treated NIDMM and its relation to a polymorphism in the plasminogen activator inhibitor 1 gene. *Diabetes* **44**: 37–42.

Patterson, D. and Slack, J. (1972) Lipid abnormalities in male and female survivors of myocardial infarction and their first degree relatives. *Lancet* **i**: 1393–1399.

Penrose, L.S. (1953) The detection of autosomal linkage in data which consist of brothers and sisters of specified parentage. *Ann. Eugenics* **6**: 133–138.

Price, D.T. and Ridker, P.M. (1997) Factor V Leiden mutation and the risks for thromboembolic disease: a clinical perspective. *Ann. Intern. Med.* **127**: 895–903.

Reed, T., Tracey, R.P. and Fabsitz, R.R. (1994) Minimal genetic influences on plasma fibrinogen level in adult males in the NHLBI twin study. *Clin. Genet.* **45**: 71–77.

Rees, A., Stocks, J., Sharpe, C.R., Vella, M., Shoulders, C.C., Katz, J., Jowett, N.I. and Baralle Galton, D.J. (1985) Deoxyribonucleic acid polymorphism in the apolipoprotein A-1-C-III gene cluster. *J. Clin. Invest.* **76**: 1090–1095.

Ridker, P.M., Cushman, M., Stampfer, M.J., Tracey, R.P. and Hennekens, C.H. (1997) Inflammation, aspirin and the risk of cardiovascular disease in apparently healthy men. *N. Engl. J. Med.* **336**: 973–979.

Risch, N. and Merikangas, K. (1996) The future of genetic studies of complex human diseases. *Science* **273**: 1516–1517.

Ross, R. Atherosclerosis – an inflammatory disease. (1999) *N. Engl. J. Med.* **340**: 115–126.

Rotter, J.I., Bu, X., Cantor, R.M., Warden, C.H., Brown, J., Gray, R.J., Blanche, P.J., Krauss, R.M. and Lusis, A.J. (1996) Multilocus genetic determinants of LDL particle size in coronary artery disease families. *Am. J. Hum. Genet.* **58**: 585–594.

Roy, S.N., Mukhopadhyay, G. and Redman, C.M. (1990) Regulation of fibrinogen assembly. Transfection of HepG2 cells with Bβ cDNA specifically enhances synthesis of the three component chains of fibrinogen. *J. Biol. Chem.* **265**: 6389–6393.

Samani, N.J., Thompson, J.R., O'Toole, L., Channer, K. and Woods, K.L. (1996) A meta-analysis of the association of the deletion allele of the angiotensin-converting enzyme gene with myocardial infarction. *Circulation* **94**: 708–712.

Shepherd, J., Cobbe, S.M., Ford, I., Isles, C.G., Lorimer, A.R., MacFarlane, P.W., McKillop, J. and Packard, C.J. ***for the West of Scotland Coronary Prevention Study Group.*** (1995) Prevention of coronary heart disease with pravastatin in men with hypercholesterolaemia. *N. Engl. J. Med.* **333**: 1301–1307.

Sigurdsson, G. Jr., Gudnason, V., Sigurdsson, G. and Humphries, S.E. (1992) Interaction between a polymorphism of the apo A-I promoter region and smoking determines plasma levels of HDL and apo A-I. *Atherosler. Thromb.* **12**: 1017–1022.

Snowden, C., Houlston, R., Laker, M.F., Kesteven, P., Alberti, K.G.M.M. and Humphries, S.E. (1992) Plasma fibrinogen levels and fibrinogen genotype in non-insulin dependent diabetics. *Disease Markers* **10**: 1–9.

Talmud, P., Bujac, S.R., Hall, S., Humphries, S.E. and Miller, G.J. (2000) Substitution of asparagine for aspartic acid at residue 9 (D9N) of lipoprotein lipase markedly augments risk of IHD in male smokers. *Atherosclerosis* **149**: 75–81.

Thomas, A., Kelleher, C., Green, F., Meade, T.W. and Humphries, S.E. (1991) Variation in the promoter region of the β-fibrinogen gene is associated with plasma fibrinogen levels in smokers and non-smokers. *Thromb. Haemost.* **65**: 487–490.

Thomas, A.E., Kelleher, C.C., Green, F. and Humphries, S.E. (1996) Association of genetic variation at the β-fibrinogen gene locus and plasma fibrinogen levels: interaction between allele frequency of the G/A^{-455} polymorphism, age and smoking. *Clin. Genet.* **50**: 184–190.

Thompson, G.R., Maher, V.M.G., Matthews, S., *et al.* (1995) Familial hypercholesterolaemia regression study: a randomised trial of low density-lipoprotein apheresis. *Lancet* **345**: 811–816.

Tiret, L., Bonnardeaux A., Poirier, O., Ricard, S., Marques Vidal, P., Evans, A., Arveiler, D., Luc, G., Kee, F. and Ducimetiere, P. (1994) Synergistic effects of angiotensin-converting enzyme and angiotensin-II type 1 receptor gene polymorphisms on risk of myocardial infarction. *Lancet* **344**: 910–913.

Tybjaerg Hansen, A. and Humphries, S. (1992) Familial defective apolipoprotein B-100: a single mutation that causes hypercholesterolaemia and premature coronary artery disease. *Atherosclerosis* **96**: 91–107.

Tybjaerg Hansen, A., Agerholm Larsen, B., Humphries, S., Abildgaard, S., Schnohr, P. and Nordestgaard, B.G. (1997) A common mutation (G-455-A) in the beta-fibrinogen is an independent predictor of plasma fibrinogen, a risk factor for ischemic heart disease. *J. Clin. Invest.* **99**: 3034–3039.

Urbanek, M., Legro, R.S., Driscoll, D.A., Azziz, R., Ehrmann, D.A., Norman, R.J., Strauss, J.F., Spielman, R.S. and Dunaif, A. (1999) Thirty-seven candidate genes for polycystic ovary syndrome: Strongest evidence for linkage is with follistatin. *Proc. Natl Acad. Sci. USA Genetics* **96**: 8573–8578.

Utermann, G., Haibach, C., Trommsdorff, M., Kochl, S., Lingenhel, A., Abe, A. and Kraft, H.G. (1995) Genetic architecture of the atherogenic lipoprotein(a). *Ann. N Y Acad. Sci.* **748**: 301–311.

Varret, M., Rabès, J.P., Saint Jore, B., *et al.* (1999) A third major locus for autosomal dominant hypercholesterolemia maps to 1p34.1–p32. *Am. J. Hum. Genet.* **64**: 1378–1387.

Weiss, E.J., Bray, P.F., Tayback, M., Schulman, S.P., Kickler, T.S., Becker, L.C., Weiss, J.L., Gerstenblith, G. and Goldschmidt Clermont, P.J. (1996) A polymorphism of a platelet glycoprotein receptor as an inherited risk factor for coronary thrombosis. *N. Engl. J. Med.* **334**: 1090–1094.

Whitfield, J.B. and Martin, N.G. (1985) Genetic and environmental influences on the size and number of cells in the blood. *Genet Epidemiol* **2**: 133–144.

Wilson, A.F., Elston, R.C., Tran, L.D. and Siervogel, R.M. (1991) Use of the robust sib-pair method to screen for single-locus, multiple-locus, and pleiotropic effects: application to traits related to hypertension. *Am. J. Hum. Genet.* **48**: 862–872.

Xu, S. and Gessler, D.D. (1998) Multipoint genetic mapping of quantitative trait loci using a variable number of sibs per family. *Genet. Res.* **71**: 73–83.

Zhang, H., Henderson, H., Gagne, S.E., Clee, S.M., Miao, L., Liu, G. and Hayden, M.R. (1996) Common sequence variants of lipoprotein lipase: standardized studies of in vitro expression and catalytic function. *Biochim. Biophys. Acta.* **1302**: 159–166.

16

Overview

Pak Sham and Tim Bishop

1. Introduction

The diseases discussed in this book share several key characteristics. They are common in the general population and have major public health implications. They are complex in the sense of not being entirely attributable to a single genetic defect or environmental factor, but they all show evidence consistent with a role for genetic inheritance. These characteristics are those that we applied in deciding the content of this book; our choices are not all-encompassing,but the contents do reflect the ongoing application of modern genetic techniques to understanding the etiology of these diseases. In the book, there are indeed examples of single gene causation, pedigrees in which disease susceptibility segregates as a simple Mendelian trait (for example, susceptibility to early onset familial Alzheimer's disease or breast and/or ovarian cancer), but such families are invariably 'rare'. The conclusions from epidemiological and family studies reviewed in the various chapters indicate that the vast majority of cases with common diseases are likely to have a combination of several risk-increasing genetic variants and environmental factors. Strictly speaking, the term 'multifactorial' should refer only to this latter set of cases.

A number of clinical and genetic epidemiological features recur consistently among the multifactorial diseases reviewed in this book. First, there are often diagnostic uncertainties in that the boundaries between caseness and normality, and between differential diagnoses, are fuzzy. An exception is cancer, for which accurate diagnoses are possible through pathological examination. Goldgar (Chapter 4) attributes the relatively rapid progress in cancer genetics partly to reliable diagnosis. For other diseases, notably psychiatric disorders (Chapters 9, 10, and 14), diagnostic reliability has been improved by the introduction of operational diagnostic criteria. These criteria, based primarily on clinical symptoms, are to some extent arbitrary and not necessarily optimal for genetic studies. In principle, twin and family studies may help to delineate a promising phenotype via heritability and segregation analyses. This more expensive but less arbitrary approach to phenotype definition is an obvious strategy, given the apparent difficulties in mapping genes for many syndrome-based common diseases. An example of this approach is reviewed in Chapter 14, which summarizes the strong

Analysis of Multifactorial Disease, edited by T. Bishop and P. Sham.

evidence for a common genetic predisposition for major depression and generalized anxiety disorder.

In reality, the ability to dissect the genetic contributions to etiology must be a function of the precision with which we can identify caseness and the complexities of genetic involvement. For the cancer studies (Chapter 4) and Alzheimer's disease, the fact that in particular families monogenic inheritance suffices to explain the inheritance is sufficient to overcome the complexities of phenocopies and variation in age of onset.

Multifactorial diseases are usually only moderately familial. In single-gene disorders, the risk of disease in a sibling of an affected individual is typically elevated many thousand-fold above the risk in the general population. In contrast, the sibling relative risk for multifactorial disorders is typically only around 10 (schizophrenia and bipolar disorder), and sometimes as low as 2 or 3 (rheumatoid arthritis). Segregation analyses usually fail to find convincing evidence for the involvement of a major locus (for example, see Chapter 7 on type II diabetes and Chapter 11 on atopy and asthma). Twin and adoption studies find heritability estimates that range from 30% (type II diabetes) to 80% (bipolar disorder), typically with large confidence intervals. Taken together, these findings suggest the involvement of multiple genetic variants, each having only a minor or modest effect on risk.

Multifactorial diseases also tend to be associated with multiple epidemiological risk factors. Such factors include cigarette smoking, dietary deficiencies, excess exposure to toxins and radiation, lack of physical exercise, childhood trauma, chronic stress and recent life events. Although these risk factors may seem to be 'environmental', the variability in exposure to these factors among individuals in the population is likely to have both genetic and environmental components. The possible influence of genetic constitution on environmental exposure, sometimes referred to as 'the nature of nurture', has been recognized particularly in behavioral phenotypes (Plomin, 1994). Environmental risk factors may therefore represent possible pathways through which the effect of a genetic variant on risk may be mediated.

Genetic constitution may have an influence not only on the probability of exposure to environmental risk factors, but also on the effects of the exposure. Interesting examples of gene-environment interactions are given in Chapter 15 in relation to ischemic heart disease. As more disease-related genetic variants are discovered, more gene-environment interactions of this type are anticipated.

Some multifactorial disorders are closely related to quantitative phenotypes (for example, ischemic heart disease with plasma lipid levels, asthma with serum IgE levels, and depression with neuroticism score). A closely related quantitative phenotype is potentially more informative for genetic analysis primarily because the contrast between individuals at the two extremes of a continuous distribution is expected to be greater than that between affected and unaffected individuals. The advantages of using quantitative phenotypes for genetic analysis are discussed in Chapters 2 and 3.

2. Current status of genetic mapping in multifactorial diseases

One fact that emerges consistently from the chapters is that, when single-gene dominant forms of disease are present, these tend to be easy to recognize and to map by classical linkage analysis; such insights are the first true successes of the Human Genome Project. This is the case, for example, in several forms of cancer, early onset familial Alzheimer's disease, and certain syndromal forms of epilepsy and obesity. These successes are to a large extent attributable to the availability of large, multiplex families. As Rice *et al.* (Chapter 3) point out, a fully informative meiosis can potentially contribute 0.3 to the lod score, so that a family with 10 fully informative meioses is in itself sufficient to establish linkage. The discovery of a disease-causing gene in some families has led to the identification of the disease-causing genes in other families either through identifying a process (as for the discovery of presenilin 2 by a search for homology with presenilin 1, which was initially mapped by linkage analysis (Chapter 5)) or by indicating which families should be excluded in a subsequent linkage search, thereby reducing genetic heterogeneity (as for BRCA2 after the mapping of BRCA1 (Chapter 4)).

Contrary to early expectation, the identification of genes for rare Mendelian forms has seldom provided any useful clue to the nature of the common genetic variants that influence risk in the population. It appears that a mutation in these genes either causes major dysfunction or has no effect. For instance, for dominantly inherited cancer predisposition, susceptibility appears to be due to failure of basic biological processes such as cell cycle regulation or DNA repair (see Chapter 4); such processes do not seem to show intermediate levels of impairment. An exception to this may be the I1307K variant of APC. Truncating mutations in APC lead to familial adenomatous polyposis, a syndrome present in perhaps one in 8000 live births, which is associated with almost certain bowel cancer if the bowel is not surgically excised. However, this variant may be more common in some populations (e.g. the Ashkenazim), but be associated with more modest risk of cancer (Laken *et al.*, 1997).

In contrast to monogenic forms that segregate in a minority of families, an entirely positional approach to gene mapping using linkage analysis has seldom conclusively mapped the risk-increasing variants of truly multifactorial diseases. A comment that recurs in many chapters is that no initial linkage finding replicates in all subsequent studies. When the results from all the genome scans on a disorder are summarized, the typical picture that emerges is a multiplicity of weak linkage hotspots with little consistency between studies. These summary tables are often confusing and frustrating, as they seldom allow confident statements about the presence or absence of linkage.

Hence, there is an impression of discontinuity between rare mutations that are disease-causing, and common variants that only have modest influence on risk. One factor that may have contributed to this impression is the very rapid fall in the power of linkage analysis with decreasing effect size – there may only be a small range between an effect that is readily detectable, and one that requires samples at least two orders of magnitude greater. Both theoretical considerations and empirical findings from genome scans suggest that, in addition to using improved designs and larger samples for linkage studies, we should lower our expectation for what linkage can deliver. For most multifactorial diseases, linkage is unlikely to provide conclusive mapping, but it may help to identify 'hotspots' for further intensive search. Thus, an initial linkage finding that replicates in half of subsequent studies would usually be considered a strong 'hotspot'.

A consequence of the limited power of linkage for genes of small effect is the increasing use of association analysis for the confirmation and fine mapping of linkage hotspots and to examine candidate genes. An understanding of the pathophysiology of the disease confers an enormous advantage for association studies. There has been more success in identifying risk-increasing genetic variants for diseases such as ischemic heart disease, epilepsy and obesity, for which detailed pathophysiological models are available, than for more poorly understood disorders such as rheumatoid arthritis, schizophrenia and bipolar disorder. Pathophysiological knowledge may suggest candidate loci of various levels of credibility. Chapter 6 summarizes the impressive evidence for the involvement of ion channels in cell membranes in epilepsy, while Chapter 13 reviews the role in obesity of signalling proteins involved in hunger and satiety mechanisms. Pathophysiological knowledge may also reveal intermediate phenotypes (also called endophenotypes) that may partially mediate the effect of a genetic variant on disease risk. The advantages of using endophenotypes in genetic analysis are illustrated in ischemic heart disease in Chapter 15.

The theoretical prediction that association has potentially far greater power than linkage for detecting a gene of minor effect has been confirmed in Type I diabetes, where IDDM2 was consistently demonstrated by association but not linkage (Chapter 8). Similarly, despite the weak linkage finding for the ApoE locus in Alzheimer's disease, the subsequent association analyses provided strong and unequivocal evidence. The primary problem for association analysis is that, unlike linkage, a systematic genome scan will require an enormous number of markers, with estimates ranging from 50 000 (Chapter 4) to 500 000 (Chapter 3). For this reason, most current association analyses are focused on candidate loci. It is probably a matter of time, however, before the prospect of genome-wide association screens becomes reality. Already, a consortium is in place to create a resource of 300 000 SNPs in the next 2 years (see Chapters 1 and 6).

3. Issues in study design

It is clear from the reviews of diseases that researchers are well aware of the design issues discussed in Chapter 3. In general, it seems fair to say that the design issues involved in linkage studies are now well worked out. The larger the family, the more informative it is for linkage (and, overall, the more likely that a genetic susceptibility is involved), but the optimal degree of affection density depends on the underlying genetic model, such as the number of loci involved, allele frequencies, and the degree of dominance. Increasing the number of markers increases linkage information, but beyond 400 microsatellites or 1000 SNPs the gains are typically small. Allele frequencies must be carefully specified and can be estimated internally from the data if necessary. Typing unaffected individuals does not usually give much extra power, but may help to protect against false positives due to misspecification of allele frequencies or population stratification. Sequential testing strategies, such as grid-tightening and sample-splitting, may reduce genotyping costs while maintaining power.

The application of linkage analyses to the diseases reviewed in the various chapters seems to follow a similar pattern. First, large multiplex families are ascertained and screened for linkage. When these families become difficult to find, efforts are diverted to the collection of smaller pedigrees consisting of affected sib pairs (preferably), with or without additional affected relatives. This is clearly an entirely reasonable, pragmatic strategy. However, the failure to find large multiplex pedigrees suggests that any genetic effect is likely to be minor or modest, so that large numbers of small pedigrees are likely to be required for linkage to be detectable. Recently, sample sizes have increased to the level of 1000 affected sib pairs, sufficient for detecting a locus accounting for a sibling relative risk of 1.5. It may not be possible to increase sample size much further simply because of the limited number of affected sib pairs in any population. To this end, we welcome the increasing recognition of the value of well characterized collections and of the organization of consortia. The need for multiple groups to participate in both sample collection and analysis is well recognized scientifically and we would like to encourage funding bodies to recognize their responsibilities for the development (and cooperative utilization) of such resouces.

In contrast to linkage studies, there appears to be more uncertainty regarding the design of association studies. The uncertainty concerns the choice of populations, the use of related versus unrelated controls, the type of marker (whether microsatellite or SNP), and marker density. This uncertainty stems from our limited knowledge of the histories of human populations. Many genetic epidemiologists, including Newton Morton whose Foreword is included in this book, started their careers as population geneticists, but then turned their attention to the issue of disease with the promise of the new genetic information. In fact, the information gained to date from studying disease indicates the deficit of information available to us about our own heritage. Issues such as the value of focusing on isolated populations for genetic association studies (see section 5 of Chapter 3) highlight our limited understanding of the ancestral history of these populations. The hope is that the greater use of molecular genetic information by population geneticists will eventually lead to a much greater understanding of the structure and history of human populations, informing the design of association studies.

Despite current uncertainties, it is probably safe to recommend the use of relatively homogeneous or isolated populations, where these are available. This might be combined, for example, with a search for inbred individuals for homozygosity mapping. Other populations that might be particularly valuable are ones that have experienced significant recent admixture from two ancestral populations. Population admixture is a mechanism for generating linkage disequilibrium, so that there may be a greater degree of linkage disequilibrium between marker and disease alleles within these populations. This may be relevant to the success in some recent studies that concentrate on specific ethnic groups (for example, studies on obesity in Mexican-Americans).

The potential problem of spurious association arising from hidden population stratification has led to a large literature on various study designs and analytic methods that use related controls, such as parents or siblings. However, most of the association studies reviewed in the various chapters are of the standard case-control design. The biggest reason for this is probably the ease of collection of cases and unrelated controls, as compared to cases with parents or siblings, for adult-onset diseases. In an effort to balance type 1 and type 2 error rates, many researchers would opt for greater power at the expense of some false positives, at least in the initial stages of a study (see Section 5 of Chapter 10 for a discussion of multistage strategy for association analysis). Furthermore, there are safeguards for case-control studies, using traditional epidemiological methods of matching or adjustment for confounders. For genetic data there is the additional luxury of internal control provided by other genetic markers; methods for robust association analysis based on this background genetic information are now being developed (see Section 7 of Chapter 2).

The marker density required to reliably detect linkage disequilibrium is another controversial issue. Both theoretical and empirical estimates vary widely between studies, from a few thousand base pairs to several hundred thousand base pairs. For this reason, most current association studies tend to be restricted to the SNPs and microsatellites in small genomic regions defined either by previous linkage findings or the presence of candidate loci. However, there is a degree of optimism (e.g. Chapter 6), based in large part on the development of new technology such as high-throughout genotyping and DNA pooling, for genome-wide association analyses to become feasible in future.

Another important design issue in genetic studies of multifactorial disorders is how to take account of the role of the environment. Although environmental risk factors are clearly involved in the etiology of multifactorial disorders, most studies reviewed in this volume have not incorporated environmental measures. For some disorders, there is a lack of well-established environmental factors of large effect that are also reasonably easy to measure. Nevertheless, the increasing popularity of association analysis means that both genetic and environmental factors can be examined in the same study. As in factorial designs, this allows not only the examination with the potential interactions, but also enhances the power of detecting main effects.

4. Issues in statistical analysis

Both parametric (model-based) and nonparametric (model-free) methods of linkage analysis have undergone considerable refinement in recent years (see Chapters 2 and 3). Indeed, as allele-sharing methods are becoming multipoint and likelihood-based, and parametric models are becoming more arbitrary, the distinction between parametric and nonparametric methods is becoming blurred. Furthermore, the powers of the latest of these analytical methods have been shown to be quite close to each other, and to the optimum that can be achieved assuming full knowledge of the genetic model. Of course, the instinct of many investigators is not to trust a single method entirely, and so it is not unusual to subject the same data to both parametric and nonparametric methods of analysis.

There have also been enormous developments in the computational efficiency of linkage analysis. Multipoint analysis involving a dozen or more markers has become feasible through the use of the Lander-Green algorithm. This algorithm is, however, inherently inefficient for large pedigrees.

There has been a great deal of recent work on the analytical methods for association studies using related controls. The HRR, TDT and related tests have become standard analytical techniques, as can be confirmed by scanning through the chapters in this volume. The HRR and the TDT have the desirable feature of simplicity and comprehensibility (see Foreword by Newton Morton). As more complex designs involving larger families and multiple markers are increasingly being used, there is a need for the development of more sophisticated methods of analysis. Potentially fruitful areas of statistical development are numerous and include the analysis of multiple, tightly-linked loci for haplotype associations, the use of background markers to adjust for hidden population stratification, and the analysis of pooled DNA data. Although important developments are being made in these methods, they have not yet reached wide-spread use.

A relatively recent development is the adaptation of variance component models to incorporate genetic marker information for linkage and association analysis. This is a particularly attractive approach for quantitative traits that are approximately normally distributed. There is some evidence for the increasing popularity of this method (e.g. Chapter 13).

5. Future prospects

Like most of the contributors to this volume, we are naturally optimistic about the future of genetic research into multifactorial disorders. Current developments in genomics will provide very powerful tools for genetic mapping and ultimately the study of gene function. To take advantage of these technological developments, it is important to collect large samples of clinical and control subjects, according to sound design principles. However, we must warn against over-impatience, as experience has shown that progress is often slower than anticipated. There are many reasons why the dramatic success of mapping genes with high penetrance would not be carried over immediately to multifactorial disease and so the limited success to date should not be interpreted, as yet, as failure of the approaches or as an indication of biological complexities beyond resolution.

Acknowledgments

We would like to take this opportunity to thank all of the authors for the contributions and for their willingness to participate in this venture. We know the many hours and revisions that have gone into producing these chapters, and their good humor and professionalism have been a pleasure to encounter.

References

Laken, S.J., Petersen, G.M., Gruber, S.B. ***et al.*** (1997) Familial colorectal cancer in Ashkenazim due to a hypermutable tract in APC. *Nature Genet.* **17**: 7983.

Plomin, R. (1994) *Genetics and Experience: The developmental interplay between nature and nurture*. Sage Publications, Newbury Park, CA.

Index